# ADVANCES IN
# NUCLEAR PHYSICS

VOLUME 4

# Contributors to this Volume

**B. J. Allen**
*Physics Division*
*Oak Ridge National Laboratory*
*Oak Ridge, Tennessee*

**Elizabeth Urey Baranger**
*Laboratory of Nuclear Science*
*and Physics Department*
*Massachusetts Institute of Technology*
*Cambridge, Massachusetts*

**C. A. Barnes**
*Kellogg Laboratory*
*California Institute of Technology*
*Pasadena, California*

**Wiesław Czyż**
*Institute of Nuclear Physics*
*Cracow, Poland*

**J. H. Gibbons**
*Physics Division*
*Oak Ridge National Laboratory*
*Oak Ridge, Tennessee*

**E. C. Halbert**
*Electronuclear Division*
*Oak Ridge National Laboratory*
*Oak Ridge, Tennessee*

**Daphne F. Jackson**
*Department of Physics*
*University of Surrey*
*Guildford, England*

**R. L. Macklin**
*Physics Division*
*Oak Ridge National Laboratory*
*Oak Ridge, Tennessee*

**J. B. McGrory**
*Physics Division*
*Oak Ridge National Laboratory*
*Oak Ridge, Tennessee*

**S. P. Pandya**
*Electronuclear Division*
*Oak Ridge National Laboratory*
*Oak Ridge, Tennessee*

**B. H. Wildenthal**
*Electronuclear Division*
*Oak Ridge National Laboratory*
*Oak Ridge, Tennessee*

# ADVANCES IN NUCLEAR PHYSICS

Edited by
## Michel Baranger

*Department of Physics*
*Massachusetts Institute of Technology*
*Cambridge, Massachusetts*

## Erich Vogt

*Department of Physics*
*University of British Columbia*
*Vancouver, B.C., Canada*

## VOLUME 4

℗ SPRINGER SCIENCE+BUSINESS MEDIA, LLC 1971

Library of Congress Catalog Card Number 67-29001

ISBN 978-1-4615-8230-4       ISBN 978-1-4615-8228-1 (eBook)
DOI 10.1007/978-1-4615-8228-1

© 1971 Springer Science+Business Media New York
Originally published by Plenum Press, New York in 1971
Softcover reprint of the hardcover 1st edition 1971

# ARTICLES PUBLISHED IN EARLIER VOLUMES

## Volume 1

## Volume 2

## Volume 3

# ARTICLES PLANNED FOR FUTURE VOLUMES

Two-Nucleon Transfer Reactions and the Pairing Model
*R. Broglia, O. Hansen, and K. Riedel*

Nuclear Fission
*A. Michaudon*

Structure of Light Nuclei and Clustering Nature
*A. Arima, H. Horiuchi, K. Kubodera, and N. Takigawa*

Variational Techniques in the Nuclear Three-Body Problem
*L. Delves*

Nuclear Matter Calculations
*D. Sprung*

Heavy-Ion In-Beam Spectroscopy
*R. Diamond and F. Stephens*

Nuclear Gamma-Ray Spectroscopy with Ge(Li) Detectors.
*G. T. Ewan, A. E. Litherland, and T. K. Alexander*

# PREFACE TO VOLUME 4

The six articles of the present volume certainly represent the varied approach and the wide range of subject matter which have been the aim of this series. On the one hand there are three articles describing the detailed application of the independent particle model to nuclei, an idea which has been the central interest and the crowning success of nuclear physics over the past two decades. On the other hand there are three articles whose aim is to make accessible to nuclear physicists topics, such as high-energy scattering or astrophysical reactions, whose position in the field of nuclear physics has probably been too peripheral in recent years.

The nuclear shell model has proved a rich gold mine which is still not nearly exhausted. One aspect of this story is the use of new techniques to uncover further rich veins. In the early years of the shell model the spectroscopic information obtained from direct reactions was largely ignored. A decade ago French and Macfarlane and others began a systematic attack on this form of spectroscopy. In the first chapter of the present volume Daphne Jackson describes the great deal of information which has been obtained recently from knockout reactions.

The two final chapters focus attention on a narrow range of topics. They are intended to show in all detail how far the shell model can go in describing the properties of a group of nuclei. Some of the material here is new and could have been published as research papers rather in a review. But the approach is didactic. In the editors' view this close a look at the successes and shortcomings of the shell model is essential to an understanding of where nuclear physics stands at this time.

The chapter by Czyż was planned to present a physical description of high-energy scattering to a wide audience of nuclear physicists. This subject has been pioneered by Glauber and is of great current interest, even though it has not been easily accessible. The present article is not formal and not encyclopedic. In a rapidly changing field it is intuitive, at time almost personal, but the editors feel it conveys much of the central physics of the subject in an exciting way.

The two articles on astrophysical reactions are aimed at making the recent advances in this subject available to nuclear physicists in their own language. This has been for many years now an important application of nuclear physics—one might call it "theological engineering" for the light it is beginning to throw on the early history of the universe. There is perhaps no aspect of nuclear physics which does not have important macroscopic consequences in the stars. The editors hope that in addition to providing enjoyment and information the present articles will help to channel ideas from nuclear physics into further developments in astrophysics.

The varied fare of the series has been possible through the continued cooperation of many authors and of our publisher. We have not always succeeded in quantizing the flow of articles so that there is no delay to any author. Nor have we yet been able in our selection of authors and topics to convey nuclear physics adequately as an experimental science. In spite of these faults, we have had remarkable success in persuading most of the authors whom we approached, not only to write articles, but to write them in a restricted time interval. This interest of the authors confirms the editors' views about the need for this series. We shall continue to be grateful to our colleagues who suggest possibilities for future articles.

M. Baranger

E. W. Vogt

January 4, 1971

# PREFACE TO VOLUME 1

The aim of *Advances in Nuclear Physics* is to provide review papers which chart the field of nuclear physics with some regularity and completeness. We define the field of nuclear physics as that which deals with the structure and behavior of atomic nuclei. Although many good books and reviews on nuclear physics are available, none attempts to provide a coverage which is at the same time continuing and reasonably complete. Many people have felt the need for a new series to fill this gap and this is the ambition of *Advances in Nuclear Physics*. The articles will be aimed at a wide audience, from research students to active research workers. The selection of topics and their treatment will be varied but the basic viewpoint will be pedagogical.

In the past two decades the field of nuclear physics has achieved its own identity, occupying a central position between elementary particle physics on one side and atomic and solid state physics on the other. Nuclear physics is remarkable both by its unity, which it derives from its concise boundaries, and by its amazing diversity, which stems from the multiplicity of experimental approaches and from the complexity of the nucleon–nucleon force. Physicists specializing in one aspect of this strongly unified, yet very complex, field find it imperative to stay well-informed of the other aspects. This provides a strong motivation for a comprehensive series of reviews. Additional motivation arises from outside the community of nuclear physicists, through the inevitable occurrence of the nucleus as an accessory or as a tool in other fields of physics, and through its importance for terrestrial and stellar energy sources.

We hope to provide a varied selection of reviews in nuclear physics with a varied approach. The topics chosen will range over the field, the emphasis being on physics rather than on theoretical or experimental techniques. Some effort will be made to include regularly topics of great current interest which need to be made accessible by adequate reviews. Other reviews will attempt to bring older topics into clearer focus. The aim will be to attract the interest of both the active research worker and the research student. Authors will be asked to direct their article toward the maximum number of readers by separating clearly the technical material from the more basic aspects of the subject and by adopting a pedagogical point of view rather than giving a simple recital of recent results.

Initially, the *Advances* are scheduled to appear about once a year with approximately six articles per volume. To ensure rapid publication of the papers, we shall use the "stream" technique, successfully employed for series in other fields. A considerable number of planned future articles constitute the source of the stream. The flow of articles from the source takes place primarily to suit the convenience of the authors, rather than to include any particular subset of articles in a given volume. Any attempt at a systematic classification of the reviews would result in considerable publication delays. Instead, each volume is published as soon as an appropriate number of articles have been completed; but some effort is made to achieve simultaneity, so that the spread in completion dates of the articles in a given volume is much less than the interval between volumes.

A list of articles planned for future volumes is given on page v. The prospective articles together with those in this first volume still fall far short of our long-range aims for coverage of the field of nuclear physics. In particular, we definitely intend to present more articles on experimental topics. We shall eagerly receive and discuss outside suggestions of topics for additional papers, and especially suggestions of suitable authors to write them.

The editors owe a great deal to the authors of the present volume for their cooperation in its rapid completion, and to many colleagues who have already given advice about the series. In embarking on this venture, we have had the support of Plenum Press, a relatively new publisher in the field of physics, and of its vice-president, Alan Liss, who has an almost unmatched background in physics publications.

M. BARANGER
E. VOGT

October 15, 1967

# CONTENTS

Chapter 2

HIGH-ENERGY SCATTERING FROM NUCLEI
Wieslaw Czyż

Chapter 3

NUCLEOSYNTHESIS BY CHARGED-PARTICLE
REACTIONS
C. A. Barnes

## Chapter 4

# NUCLEOSYNTHESIS AND NEUTRON-CAPTURE CROSS SECTIONS

## B. J. Allen, J. H. Gibbons, and R. L. Macklin

## Chapter 5

# NUCLEAR STRUCTURE STUDIES IN THE $Z = 50$ REGION
## Elizabeth Urey Baranger

## Chapter 6

# AN $s$-$d$-SHELL-MODEL STUDY FOR $A = 18$–$22$

## E. C. Halbert, J. B. McGrory, B. H. Wildenthal, and S. P. Pandya

## Chapter 1

# THE INVESTIGATION OF HOLE STATES IN NUCLEI BY MEANS OF KNOCKOUT AND RELATED REACTIONS

## Daphne F. Jackson

*Department of Physics*
*University of Surrey*
*Guildford, England*

## 1. INTRODUCTION

There is a wide variety of reactions which can give rise to the excitation of hole states; for example, pickup and knockout caused by many different projectiles, photodisintegration, kaon capture, and pion absorption. Some of these reactions occur through different mechanisms, some occur through the same mechanism but involve very different regions of momentum transfer or have different degrees and areas of spatial localization, but all give information about nuclear structure in essentially the same way. When a single nucleon or group of nucleons is removed from a nucleus, by whatever means, information is obtained about its wave function in the nucleus and the hole state created by its removal. The hole state created by the removal of these nucleons is not, in general, an eigenstate of the residual nucleus, but the hole strength is distributed over several states of the residual nucleus. The amount of hole strength each particular final state contains is measured in terms of the spectroscopic factors and comparison with the experimental spectroscopic factors gives information about the parentage of the target nucleus and the relevance, or otherwise, of various nuclear models (MF 60, Mac 64).

As an example of the range of reactions under discussion and the information which can be obtained from them, we consider the removal

1

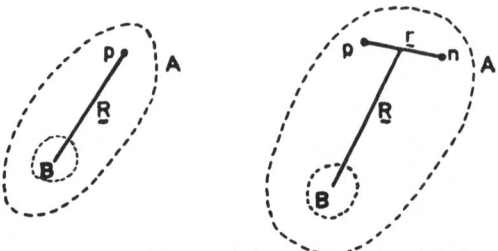

**Fig. 1. Diagram showing the relevant coordinates for the process** $A \to p + B$ **and for the process** $A \to (p + n) + B$.

of a single proton from a target $A$ leaving the residual nucleus $B$.[†] This can occur through a knockout reaction such as $(p, 2p)$ or $(e, ep)$ and through a pickup reaction such as $(d, {}^3He)$. It can also occur through the absorption of a negative kaon leading to the emission of a sigma hyperon and a pion, and through the absorption of a photon leading to the $(\gamma, p)$ reaction. In all cases, a study of the energy spectrum of the outgoing particles shows which of the excited states of the residual nucleus are excited and with what strength, and may also lead to the identification of states not already observed in other reactions. Comparison of the relative strength of excitation of different states in the same reaction, or the same state in different reactions, indicates the effect of various selection rules. Measurement of the $Q$-value for the transitions yields information about the proton binding energy in different single-particle states in the target nucleus. Finally, a study of the angular distribution corresponding to the excitation of each state yields a determination of the experimental spectroscopic factor and the angular momentum transfer, and also yields information on the single-particle wave function of the proton in the target nucleus. The simultaneous removal of a proton and neutron from nucleus $A$ can proceed through an even greater variety of reactions such as $(p, {}^3He)$, $(p, pd)$, $(\gamma, d)$, $(\gamma, np)$, $(\pi^+, 2p)$, etc., but examination of the energy spectra and angular distributions yields essentially the same spectroscopic information concerning the residual nucleus. In this case, however, information is obtained about the relative motion of the two nucleons in the target nucleus and also about the motion of their center of mass, as shown in Fig. 1.

[†] We use the letter $A$ throughout to indicate a target nucleus containing $A$ nucleons, and the letter $B$ to indicate the residual nucleus which may contain $A - 1$, $A - 2$, etc. nucleons depending on the particular reaction.

The simplest picture of these reactions is that of a one-step mechanism involving a direct interaction between the projectile and the nucleon or nucleons which are removed from the target nucleus. This picture is obscured to some extent by the presence of the other nucleons in the nucleus. The distortion effect due to the interaction of the projectile with the target nucleus and of the outgoing particles with the residual nucleus through the relevant optical potentials can be taken into account using the standard techniques of the distorted wave theory of nuclear reactions. There may also be effects due to exchange within the target nucleus and from core excitation, which proceeds through a two-step mechanism. In addition, those processes in which the final system contains two or more light reaction products as well as the residual nucleus may be affected by the final state interaction between the reaction products.

Aspects of single-nucleon and multi-nucleon transfer reactions are reviewed elsewhere in this series (GT 69, ET 69). In this article the emphasis is therefore on knockout reactions although many other processes, including transfer reactions, are mentioned in order to illustrate the relationship between the various reactions which give information on hole states.

## 2. FORMALISM FOR KNOCKOUT AND PICKUP REACTIONS

### 2.1. The Matrix Element and Overlap Integral

The distorted-wave matrix element for a one-step direct reaction has essentially three components:

(i) The distorted wave functions $\chi^{\pm}$ describe the motion of the incoming particle and the outgoing reaction products. For medium-energy nuclear projectiles, these wave functions are obtained from an exact or approximate solution of the Schrödinger equation with the relevant optical potentials, which must be known for a wide range of energies and projectiles. But at energies above about 200 MeV a modified Schrödinger equation or Klein–Gordon equation should be used. If the projectiles are electrons or muons, the Dirac equation must be used at all energies and the distorting potential is just the Coulombic potential of the target or residual nucleus.

(ii) The overlap integral represents the overlap of the nuclear wave function $\Phi_A{}^i$ for the initial state of the target nucleus and the wavefunction $\Phi_B{}^f$ for the particular final state of the residual nucleus. The formal properties of this integral can be deduced from a general formalism such as that given in this section, but the detailed construction of the overlap integral for a given reaction requires the assumption of a suitable nuclear model.

(iii) The interaction between the projectile and the active nucleons in the target is represented by a sum of two-body interactions. The two-body interaction is represented in distorted wave Born approximation (DWBA) by a potential and in distorted wave impulse approximation (DWIA) by the two-body $t$-matrix.

The simplest expression for the matrix element is obtained in zero-range DWBA and has the form

$$T_{if} \propto \sum_i \langle \chi^-(\mathbf{r})\phi(\mathbf{r}_{B+1}, \ldots, \mathbf{r}_A)\Phi_B^f(\mathbf{r}_1, \ldots, \mathbf{r}_B) \mid V_0\, \delta(\mathbf{r} - \mathbf{r}_i) \mid$$
$$\times\; \chi^+(\mathbf{r})\Phi_A^i(\mathbf{r}_1, \ldots, \mathbf{r}_A) \rangle$$
$$\propto V_0 \langle \chi^-\, \phi\; \Phi_B^f \mid \chi^+\, \Phi_A^i \rangle$$

where $\phi$ represents the internal wave function of the group of nucleons removed from the nucleus $A$. If a finite-range correction is made in DWBA using the local energy approximation (LEA) of Buttle and Goldfarb (BG 64) an additional factor, whose effect is discussed in Section 2.5, is introduced. If it is assumed that the two-body $t$-matrix is a function of momentum transfer $q$ only, the same form is appropriate for the matrix element in impulse approximation except that $V_0$ is replaced by the $t$-matrix $t(q)$. Finite-range corrections to DWIA are discussed in Sections 2.5 and 3.2. In the case of knockout reactions which have a three-body final state the symbol $\chi^{-*}$ represents more than one outgoing particle so that approximations are necessary to put the total wave function for the final state in product form, and these are discussed in Section 3. Some corrections due to exchange effects are discussed in Section 4.4.

We have not attempted to give a formal derivation of the distorted wave matrix element since we are interested in the nuclear structure information which is mainly, if not wholly, contained in the overlap integral (Ber 65, PS 65). However, it will be necessary from time to time to consider the implications of the underlying assumptions of the reaction theory, and to discuss the validity of some of the approximations used. In general, studies of inelastic scattering and reactions indicate that at low and medium projectile energies the modification of the two-body interaction due to the presence of other nucleons in the nucleus and the blocking of intermediate virtual states by the exclusion principle is such that the two-body interactiou is best described phenomenologically by some effective interaction $V$, i.e., through DWBA. At sufficiently high projectile energies, the struck nucleon or nucleons may be regarded as essentially free so that the two-body interaction may be summed to all orders in $V$ and replaced by the free two-body $t$-matrix. For elastic and inelastic scattering this procedure appears to be successful at incident energies of 150 MeV and above, but it cannot be assumed *a priori* that the same is true for reactions since the actual two-body scattering may be far off the energy-shell for free two-body scattering. In some sections of this article, plane waves will be used instead of distorted waves: this is done solely in order to derive simple formulas which can easily be interpreted, and these formulas are never adequate for accurate comparison with experimental data, however high the incident energy.

We now construct the overlap integral for the process $A \rightarrow x + B$ in which a group of nucleons represented by $x$ are removed from the target $A$ in state $i$ to leave the residual nucleus $B$ in state $f$. We represent the internal coordinates of $B$ by $\xi$, the internal coordinates of $x$ by $\mathbf{r}$, and the relative coordinates between the two centers of mass by $\mathbf{R}$. For simplicity, the spin coordinates are omitted. The overlap integral then has the form

$$\Psi_{fi}(\mathbf{r}, \mathbf{R}) = \int \Phi_B{}^f(\xi) \Phi_A{}^i(\xi, \mathbf{r}, \mathbf{R}) \, d\xi \tag{1}$$

It is sometimes convenient to also define the reduced overlap integral

$$\Psi_{kfi}(\mathbf{R}) = \int \phi_x^{k*}(\mathbf{r}) \Psi_{fi}(\mathbf{r}, \mathbf{R}) \, d\mathbf{r} \tag{2}$$

where $\phi_x{}^k$ is the wave function for the system $x$ in state $k$. In order to examine the properties of the overlap integral we write the wave function for the initial state in the form

$$\Phi_A{}^i = \sum_{\alpha\beta\gamma} C^{\alpha\beta\gamma} \phi_x{}^\alpha(\mathbf{r}) \psi_{xB}^\beta(\mathbf{R}) \Phi_B{}^\gamma(\xi) \tag{3}$$

where $\alpha\beta\gamma$ represent all necessary quantum numbers and the constant $C$ includes the expansion coefficient and the coupling coefficients. The summation includes integration over continuum states and the functions $\phi_x$, $\Phi_B$ are fully antisymmetrized and normalized. Thus the expansion for $\Phi_A$ may be thought of as a two-cluster expansion or as a fractional parentage expansion in the shell model, as in the example given in Section 2.3. The Hamiltonian for the system $A$ can be written as

$$H_A = H_x + H_B + T_{xB} + V_{xB} + W \tag{4}$$

where $V_{xB}$ is an averaged potential describing the interaction of $x$ with $B$,

$W$ is the residual interaction, and $H_x$, $H_B$ describe the internal motion of $x$ and $B$ respectively, so that

$$H_A \Phi_A{}^i = E_A{}^i \Phi_A{}^i, \quad H_B \Phi_B{}^\gamma = E_B{}^\gamma \Phi_B{}^\gamma, \quad H_x \phi_x{}^\alpha = E_x{}^\alpha \phi_x{}^\alpha \qquad (5)$$

Hence, it is easy to show that when the residual interaction is neglected, the reduced overlap integral obeys the Schrödinger equation

$$(T_{xB} + V_{xB})\Psi_{kfi} = (E_A{}^i - E_B{}^f - E_x{}^k)\Psi_{kfi} \qquad (6)$$

and hence

$$\Psi_{kfi} \xrightarrow[R \to \infty]{} e^{-aR}/R \qquad (7)$$

where

$$a^2 = \frac{2\mu}{\hbar^2}\, |\, E_A{}^i - E_B{}^f - E_x{}^k\, | \qquad (8)$$

and $\mu$ is the reduced mass of the system $x + B$.

From this formulation it can be seen that the reduced overlap integral depends on the final state $f$ of the residual nucleus and on the state $k$ of the system $x$, so that a transition to a particular pair of states $f$ and $k$ picks out a particular term in the parentage expansion for $\Phi_A{}^i$. The asymptotic behavior of the reduced overlap integral is unambiguously determined by the separation energy and, if the residual interaction $W$ can be neglected, it can be generated in an effective one-body potential. During the past five years the theory of overlap integrals has progressed very considerably, and the present situation for various systems $x$ is described in Sections 2.2–2.4.

## 2.2. The Single-Nucleon Case

In the case of single-nucleon pickup or knockout reactions the system $x$ reduces to a single nucleon, so that the overlap integral becomes

$$\Psi_{fi}(\mathbf{R}) = \int \Phi_B{}^f(\boldsymbol{\xi})\Phi_A{}^i(\boldsymbol{\xi}, \mathbf{R})\, d\boldsymbol{\xi} \qquad (9)$$

where $\mathbf{R}$ represents the coordinate of the single nucleon in the nucleus $A$ relative to the center of mass of the $A - 1$ nucleons which form the residual nucleus $B$. The fractional parentage expansion for the ground-state wave function of the target nucleus in terms of the complete set of wave functions $\Phi_B{}^{J_p M_p}$ for the parent states $J_p$ in the residual nucleus is given by

$$\Phi_A{}^{J_A M_A}(\boldsymbol{\xi}, \mathbf{R}) = \sum_{jmJ_p} (J_p M_p jm \mid J_A M_A)\, \mathscr{Z}_{J_A J_p}(j)\Phi_B{}^{J_p M_p}(\boldsymbol{\xi})\phi^{jm}(\mathbf{R}) \qquad (10)$$

where $\phi^{jm}$ is a normalized single-particle wave function and $\mathscr{S}$ is the fractional-parentage coefficient. Substituting this expansion into Eq. (9) gives

$$\Psi_{f i}(\mathbf{R}) = \sum_{jm} (J_B M_B j m \mid J_A M_A) \mathscr{S}_{J_A J_B}(j) \phi^{jm}(\mathbf{R}) \tag{11}$$

which is the overlap integral for the transition to a definite parent state with $J_p = J_B$.

The differential cross section is obtained by squaring the matrix element and averaging and summing over initial and final spin states in the usual way. Using Eq. (11) for the overlap integral it is found that the cross section for a single-nucleon pickup or knockout reaction can always be written in the form

$$\frac{d^{\alpha+\beta}\sigma}{d^\alpha\Omega d^\beta E} = \sum_{lj} S(J_A J_B; lj)\sigma_{lj}(\theta) \tag{12}$$

where $\alpha$ and $\beta$ depend on the number of reaction products and the manner of detecting them, and $\theta$ is the scattering angle. The function $\sigma_{lj}(\theta)$ contains spherical harmonics and integrals over the radial parts of the distorted wave functions and overlap integral, so that for a given $lj$ it depends on the particular state which has been excited only rather weakly through the dependence of these radial functions on the excitation energy. The spectroscopic factor $S(J_A J_B; lj)$ is defined through the relation

$$S(J_A J_B; lj) = N_{lj} \mid \mathscr{S}_{J_A J_B}(lj) \mid^2 \tag{13}$$

where $N_{lj}$ is the number of active protons or neutrons in the relevant subshell of the target nucleus characterized by the quantum numbers $lj$. Thus the cross section has been factorized into a factor $\sigma_{lj}(\theta)$ which depends on the reaction mechanism and a factor $S(J_A J_B; lj)$, which carries the spectroscopic or nuclear structure information. In an alternative derivation of the cross section the $I$-spin formalism is used so that the spectroscopic factor and fractional-parentage coefficient are labeled by the additional quantum numbers $T_A$ and $T_B$, additional Clebsch–Gordan coefficients for the $I$-spin appear in the expression for the cross section, and $N_{lj}$ represents the number of active nucleons in the relevant subshell. The numerical values of the spectroscopic factor and the fractional-parentage coefficient are different in the two cases. In either case the spectroscopic factor is closely related to the reduced width $\theta^2$ and to the strength function (MF 60, McC 68).

The work of Berggren and of Pinkston and Satchler (Ber 65, PS 65) established that the behavior of $\phi^{jm}$ is exactly that of the reduced overlap integral defined by Eqs. (6)–(8). For a single nucleon, the internal energy $E_z{}^k$ is by definition zero and the asymptotic behavior is governed by the separation energy $S_{xA} = -(E_A{}^i - E_B{}^f)$, where the symbol $x$ now represents either a proton or a neutron. This justifies the use of the separation energy procedure (SEP) or well-depth procedure (WD) in which the radial part of $\phi^{jm}$ is generated in a Saxon–Woods potential whose parameters are chosen to give a nucleon binding energy equal to the separation energy. The number of nodes in the radial wave function is usually determined by the principal quantum number $n$ of the active nucleons in the subshell $lj$, although a more accurate treatment would imply a summation over this and higher values of $n$ and would thus take account of configuration mixing in the target nucleus.

It is customary to refer to $\phi^{jm}$ as the single-particle wave function of the bound nucleon, although it is not a shell-model wave function in the usual sense (Ber 65, BT 66). By this we mean that the description of the target nucleus $A$ in terms of the coordinates $\xi$ and $\mathbf{R}$ and the Hamiltonian given by Eq. (4) is in principle translation invariant, which is not the case when a shell-model Hamiltonian is used unless special procedures are adopted to project out the center of mass motion. A consequence of this is that the radial coordinate $\mathbf{R}$ is referred to the center of mass of the residual nucleus $B$ and not to the center of mass of the target nucleus $A$. Further, if the initial state $J_A$ has more than one parent state $J_p$ in the residual nucleus, the separation energy $-(E_A{}^i - E_B{}^f)$ for a transition to a definite final state $J_B$ is different from the Hartree–Fock energy $-(E_A{}^i - \sum_p \mathscr{S}^2_{J_A J_p} E_B{}^p)$ where the second term gives the centroid position of the parent states. The extent to which this difference in energy is significant depends on the strength of excitation of the various parent states and the magnitude of the energy differences between them in relation to the experimental energy resolution.

It is clear that although the SEP yields an overlap integral which has the correct asymptotic behavior it may not yield the correct interior behavior or the correct magnitude in the asymptotic region. Improvements to the single-particle model involve a more accurate treatment of the initial and final nuclear states through inclusion of the residual interactions. If we write the initial wave function in the form

$$\Phi_A{}^i = \sum_{\beta\gamma} \Psi^{\beta\gamma}_{xB}(\xi, \hat{\mathbf{R}}) R^\beta_{xB}(R)$$

where $\hat{\mathbf{R}}$ represents the angular coordinates of $\mathbf{R}$ and $\beta$, $\gamma$ are the quantum numbers defined through Eq. (3), the inclusion of the residual interaction $W$ defined in Eq. (4) yields an equation for the radial function of the form

$$(E_B^{\gamma} - E_A^{i} - T_{xB} - V_{xB})R_{xB}^{\beta}(R) = \int \Psi_{xB}^{\beta\gamma*}(\xi, \hat{R})W(\xi, R)\Phi_A^{i}(\xi, R)\, d\hat{R}\, d\xi \tag{14a}$$

$$= \sum_{\mu\nu,L} U_L^{\beta\gamma,\mu\nu}(R)R_{xB}^{\mu}(R) \tag{14b}$$

Some attempts have been made (Ros 67, ST 66) to solve the set of coupled equations (14b) for rather special cases. Prakash and Austern and Philpott, Pinkston, and Satchler (PA 69, PPS 69) have calculated the right-hand-side of Eq. (14a) using shell-model wave functions and a residual interaction between the valence nucleons, while Sugawara and Huby and Hutton (Sug 68, HH 66) used Hartree–Fock techniques to construct the overlap integral. Kawai and Yazaki (KY 67, IKY 69) have developed an alternate method starting from an expression for $R_{xB}(R)$ in the form

$$R_{xB}(R) = \frac{1}{E_A^{f} - E_B^{i} - T_{xB}} \langle \Phi_B^{f} | V_{xB} + W | \Phi_A^{i} \rangle \tag{15}$$

They also take $W$ to be a residual interaction between the valence nucleons and use shell-model wave functions for the target and residual nucleus. A review and comparison of these methods has been given by Philpott, Pinkston, and Satchler (PPS 68).

A comparison (Tow 69) of the sensitivity of the ${}^{40}$Ca$(p, d)$ reaction at various energies to the form of the overlap integral is shown in Fig. 2. The overlap integral is calculated by the SEP, the method of Prakash and Austern (MRP), and the effective binding energy procedure (EBEP). (The latter gives an incorrect asymptotic behavior and will not be discussed here.) In the calculation of the results shown in Fig. 2, as in many other calculations of this type, the $I$-spin formalism for the spectroscopic factor was used, so that the spectroscopic information is contained in the factor $F^2$ which is given by

$$F_{nlj}^2 = (T_B M_{TB} t m_t | T_A M_{TA})^2 S(J_A J_B; nlj) \tag{16}$$

It can be seen from the figure that there is much greater sensitivity in shape and magnitude at the lower energy than at the higher energy.

A parallel approach to the problem has involved the search for the best possible single-particle potentials by fitting a range of data such as nucleon separation energies, elastic electron scattering and nuclear reaction data (Elt 68a, Jac 68). Electron scattering gives information about the

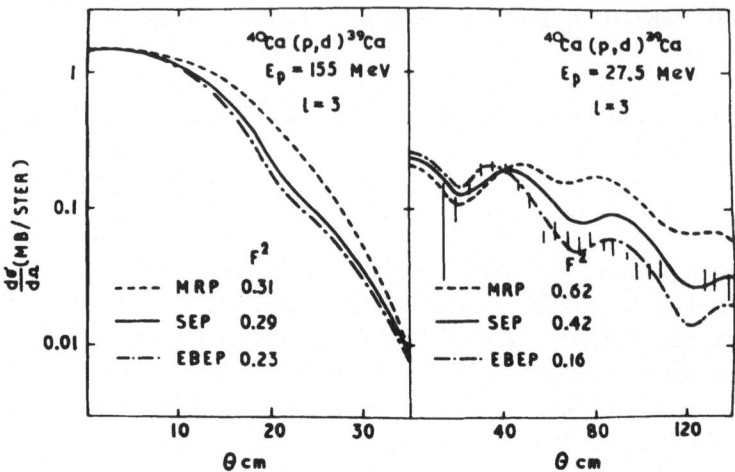

Fig. 2. A comparison of the sensitivity of the $^{40}$Ca($p$, $d$) reaction at different energies to the form of the overlap integral. The overlap integral is calculated by the matching radius procedure of Prakash and Austern (MRP), the separation energy procedure (SEP), and the effective binding energy procedure (EBEP). The quantity $F^2$ is given by $F_{lj}^2 = (T_A M_{T_A} t m_t \mid T_A M_{T_A})^2 S(J_A J_B : lj)$. [From I. S. Towner, *Nucl. Phys.* A126:97 (1969). Reproduced by permission of the author and the North-Holland Publishing Company.]

proton density distribution, which is the overlap of the ground-state wave function with itself, i.e.,

$$\varrho(r) = \sum_{i=1}^{Z} \int \Phi_A^{i*}(\xi) \Phi_A^{i}(\xi) \, \delta(\mathbf{r} - \xi_i) \, d\xi_i \qquad (17)$$

and so depends on all the parent states (ES 67, Ber 65). In the single-particle model the density (17) reduces to a product of single-particle wave functions

$$\varrho(\mathbf{r}) = \sum_{i=1}^{Z} \mid \phi_i^{nlj}(\mathbf{r}) \mid^2 \qquad (18)$$

These single-particle wave functions are generated in a set of state-dependent (i.e., energy-dependent) Saxon–Woods potentials and the parameters of the potentials are adjusted until there is agreement with the mean separation energies of the protons and the electron scattering data. Predictions for neutron single-particle states are obtained from the same potentials by correctly switching off the Coulomb interaction and assuming, for nuclei with a neutron excess, that the symmetry term has the same radial behavior as the central part of the potential. The use of these wave functions in the analysis of nuclear reactions will be discussed in Section 3. We note here, however, that although these wave functions represent a lowest order approximation to the overlap integrals, they have both the correct asymptotic behavior and something approaching the correct interior behavior since electron scattering is more sensitive to the interior of the nucleus than most nuclear reactions. This remark should be even more applicable to recent calculations with nonlocal potentials (Mel 69, EWB 69).

The attempt to improve the description of the overlap integrals by more accurate nuclear structure calculations and the attempt to improve the effective single-particle potentials are not alternatives, since the former still require a single-particle potential as can be seen from Eqs. (14) and (15). Provided that the separation energy is not markedly different from the Hartree–Fock energy, it may be expected that the more closely the single-particle potential reproduces the average behavior of the nucleons and approximates to the true Hartree–Fock potential, the more reliable are the refined nuclear structure calculations and the spectroscopic information obtained from them.

An examination of the formal properties of the single-particle potential and its relation to the Feynman diagrams which arise in nuclear structure calculations with realistic nucleon–nucleon forces has recently been given by Baranger (Bar 70) and Gross (Gro 69).

In the preceding discussion it has been assumed that the final state forms a bound system. However, in many reactions of interest this is not the case. For example, in a stripping reaction the stripped neutron may go into a continuum state or may go into a bound state in a particle unstable residual nucleus which decays sequentially. These processes have been studied by Huby and Mines and by Levin (HM 65, Lev 68). The formation of resonant states, long-lived compared with the time scale of direct reactions, has been discussed by Berggren (Ber 70) in the context of knockout reactions.

## 2.3. The Two-Nucleon Case

In the case of two-nucleon pickup and knockout reactions, such as $(p, {}^3\text{He})$ or $(p, pd)$, the system $x$ consists of a pair of nucleons and the energy $E_x{}^\alpha$ represents the relative energy of the pair of nucleons in state $\alpha$ inside the nucleus. The function $\phi_x(\mathbf{r})$ defined in Eq. (3) describes the relative motion of the pair in the nucleus and the function $\psi_{xB}(\mathbf{R})$ describes the motion of their center of mass.

The simplest process to consider is a knockout reaction leading to the emission of a bound deuteron, e.g., the $(p, pd)$ or $(e, ed)$ reaction. The energy $E_x{}^k$ is now the binding energy of the free deuteron and the asymptotic behavior of $\phi_x{}^k$ is determined by this energy. In the formalism of Section 2.1 the functions $\phi_x{}^\alpha$, $\phi_x{}^k$ are eigenstates of $H_x$ so that their overlap is given by

$$P(\alpha, k) = \int \phi_x^{k*}(\mathbf{r})\phi_x^{\alpha}(\mathbf{r})\, d\mathbf{r} \tag{19a}$$

$$= \delta_{\alpha k} \tag{19b}$$

and hence the reduced overlap integral defined in Eq. (2) is given by

$$\Psi_{kfi}(\mathbf{R}) = \sum_\beta C^{k\beta f}\psi_{xB}^\beta(\mathbf{R}) \tag{20}$$

Thus the reduced overlap integral describes the motion in the target nucleus of the center of mass of a pair of nucleons correlated as in the deuteron and has its asymptotic behavior determined by the separation energy for break up of the target nucleus into a free deuteron and a residual nucleus in state $f$. If the probability of finding such a pair of correlated nucleons is small, this will be reflected in the magnitude of the coefficients $C^{k\beta f}$.

In most calculations on deuteron knockout based on the shell model (BBR 65, Ber+ 65, and Sak 64a) the overlap integral has been constructed from shell-model wave functions using fractional parentage techniques and transforming the product of two single-particle wave functions of oscillator form into a product of functions of the relative coordinate and the center of mass coordinate using the Talmi coordinate transformation (Tal 52) and the Brody–Moshinsky transformation brackets (BM 61). This procedure is very frequently applied to the ground state of $^6$Li. The lowest shell-model configuration for this nucleus is $(1s)^4(1p)^2$ and in $LS$ coupling the ground state is almost a pure $^3S_1$ state. Thus a wave function of the form given by Eq. (3) can be constructed by taking $\Phi_B{}^\gamma$ to be the wave function of the $s$-shell core, so that $\gamma$ represents the quantum numbers of the core (in this case, the $\alpha$-particle), and by transforming the two $1p$ oscillator functions $\phi_{1p}(r_1/a)$ and $\phi_{1p}(r_2/a)$. This transformation gives

$$\phi_{1s}(r/\sqrt{2}\,a)\psi_{2s}(\sqrt{2}\,\bar{R}/a) - \phi_{2s}(r/\sqrt{2}\,a)\psi_{1s}(\sqrt{2}\,\bar{R}/a) \tag{21}$$

where $a$ is the oscillator length parameter, $\mathbf{r} = \mathbf{r}_1 - \mathbf{r}_2$, and $\bar{\mathbf{R}} = \frac{1}{2}(\mathbf{r}_1 + \mathbf{r}_2)$. Thus, in this example, the functions $\phi^\alpha\psi^\beta$ defined in Eq. (3) are both oscillator functions, the expansion over $\alpha$, $\beta$ contains two terms, and the quantum numbers $\alpha$, $\beta$ are those appropriate to $1s$ and $2s$ states. The coefficient $C^{\alpha\beta\gamma}$ is obtained as a product of the original coupling coefficients of the shell-model wave function and of the Brody–Moshinsky brackets of the transformation. Unfortunately, the new center of mass coordinate $\bar{\mathbf{R}}$ is not identical to $\mathbf{R}$ but is $\mathbf{R}_A + (A - 2)\mathbf{R}/A$ where $\mathbf{R}_A$ is the coordinate of the center of mass of the nucleus relative to the origin of the shell-model potential. This is a consequence of the fact that the description of the system given in terms of the Hamiltonian (4) is translationally invariant while that in terms of a shell-model Hamiltonian is not. The transformation of a product of more realistic single-particle wave functions has also been considered (MH 62), but in neither case will the transformation lead to a reduced overlap integral with the correct asymptotic behavior nor will the overlap with the free deuteron wave function be that required by Eq. (19b). However, there is no doubt that the overlap integral can be correctly treated within the framework of the shell model by using a sufficiently large configuration of single-particle states, and then expanding these states in terms of oscillator functions so that the center of mass motion can be separated out by standard techniques.

The cluster-model formalism (WK 59, NS 65, WM 66) has the advantage that it eliminates the problem of center-of-mass motion and can always provide a wave function for the target nucleus in the required system of relative coordinates. The usual procedure is to truncate the cluster expansion to a single term in which the clusters are in their respective ground states. For example, the cluster-model wave function for $^6$Li in the $\alpha + d$ representation is written in the form

$$\Phi = \mathscr{A}\{\phi_d(\mathbf{r})\Phi_\alpha(\xi)\psi_{d\alpha}(\mathbf{R})\chi(ST)\}$$

where $\mathscr{A}$ represents the antisymmetrization operator, $\chi$ is a charge-spin function, $\mathbf{r}$ and $\xi$ represent the internal coordinates of the deuteron cluster and $\alpha$-particle cluster respectively, and $\mathbf{R}$ represents the separation between the centers of mass of the two clusters. For the $^3S_1$ ground state of $^6$Li, the charge–spin function is coupled to $S = 1$, $T = 0$, and the relative motion described by $\psi_{d\alpha}$ is taken to be a $2s$ state. (Since in the shell model there are two nucleons in the $1p$ shell, the exclusion principle requires that in the cluster model the relative motion must be in a $2s$ and not a $1s$ state.) Similarly, the first excited state of $^6$Li could be described by the same cluster-model wave function but with the relative motion in a $1d$ state. If the functions $\phi_d$, $\Phi_\alpha$, $\psi_{d\alpha}$ are taken to be of Gaussian form with identical length parameters then, on antisymmetrization, this cluster-model wave function is identical to the shell-model wave function constructed from oscillator functions but with the center-of-mass motion removed. More complicated cluster-model wave functions with different length parameters for each function correspond to some degree of configuration mixing in the shell-model sense (Wil 62, TWP 62). The parameters of such cluster-model wave functions are determined by a variational calculation on the ground-state energy (TWP 61, Kop+ 61, STW 63); unfortunately, this variational method is not sensitive to the long-range behavior of the wave function.

A number of calculations on the $(p, pd)$ and $(e, ed)$ reactions have been carried out (Ruh+ 62, Ruh+ 63, Ber+ 65, JE 65, GOS 66, and Dwi 68) using a single term in the cluster-model expansion and wave functions of Gaussian form. The corresponding overlap integrals have incorrect asymptotic behavior and the calculations are in principle no more realistic than those of the simple shell model although in practice the relative motion of the clusters is often described by a Gaussian function of longer range than would arise in the oscillator shell model. The cluster-model approach can, of course, be made more realistic by retaining more terms in the cluster expansion (TP 60, OPW 65). The coefficients of the expansion are determined by a resonating-group method so that this approach still suffers from the insensitivity of a variational calculation to the important asymptotic region. The variational procedure may be avoided by using Young-diagram techniques (NS 65) or by solving coupled equations similar to the set defined by Eq. (14) (Jac 67a, Met 69).

The emission of unbound nucleon pairs occurs in such reactions as $(p, 3p)$, $(\gamma, np)$, $(\pi^+, 2p)$, etc. Since the two nucleons are not bound the function $\phi_x^k(\mathbf{r})$ represents a continuum state for the relative motion of the two nucleons; this is an eigenstate of $H_x$ and therefore includes the effect of the final state interaction. In practice, the overlap integral is almost invariably constructed using the oscillator shell model (Got 58, Gar+ 65, JK 64, KJ 63, Yu 66),[†] but the calculations and comparison with the data give clear indication of the inadequacy of this method. Attempts to improve the fit to the data have been made by multiplying the function $\phi_x^\alpha(\mathbf{r})$ by a correlation function $f(r)$, usually taken to be of the form (Dab 58)

$$
\begin{aligned}
f(r) &= 0, & r &\leq r_c \\
&= 1 - \exp(-\delta[(r/r_c)^2 - 1]), & r &> r_c
\end{aligned}
\tag{22}
$$

which evidently modifies the short-range behavior of $\phi_x^\alpha(\mathbf{r})$ but not the long-range behavior.

---

[†] For a complete list of calculations on the $(\pi, 2N)$ reaction see Koltun (Kol 69), Table XXII.

We now show some results to illustrate the magnitude of the errors which can be introduced when the overlap integral does not have the correct behavior. In order to do this we define the following Fourier transforms

$$G^\beta(Q) = \int e^{iQ \cdot R} \psi_{xB}^\beta(R) \, dR \tag{23}$$

$$F^{\alpha k}(q) = \int e^{i\frac{1}{2}q \cdot r} \phi_x^{k*}(r) \phi_x^\alpha(r) \, dr \tag{24}$$

$$a^\alpha(q) = \int e^{iq \cdot r} \phi_x^\alpha(r) \, dr \tag{25}$$

We consider the knockout of a deuteron from ${}^6$Li so that $\phi_x^k$ is the free deuteron wave function and

$$F_d(q) = \int e^{i\frac{1}{2}q \cdot r} |\phi_d^k(r)|^2 \, dr \tag{26}$$

is the deuteron form factor. The function $\psi_{xB}^\beta$ represents the relative motion of a deuteron and $\alpha$-particle in ${}^6$Li which we take to be a $2s$ state, as in the cluster model. Figure 3 shows results (Jac 67a) for $G^2(Q)$, normalized to unity at $Q = 0$, for various $2s$ wave functions:

(1) a square-well function with the correct separation energy of 1.47 MeV,

(2) a square-well function with a separation energy of 3.70 MeV, and

(3) an oscillator function $\psi_{2s}(\sqrt{2}R/a)$ with $a = 2.94$ fm.

The oscillator parameter used is an unusually large one, but if it were reduced the curve would fall less rapidly with $Q$. The experimental points are derived by dividing the experimental cross section for the ${}^6$Li$(p, pd)$ reaction (Ruh+ 62, Ruh+ 63) by the free cross section for proton–deuteron scattering and kinematic factors using Eq. (49) below. The validity of this procedure is discussed in Section 4.2.

Figure 4 shows the results (Jac 67a) for $a^\alpha(q)$, normalized to unity at $q = 0$, for various $1s$ wave functions for the relative motion of the two nucleons:

(1) an oscillator function $\phi_{1s}(r/\sqrt{2}a)$ with length parameter $a = 1.8$ fm,

(2) the product $\phi_{1s}(r/\sqrt{2}a)f(r)$ where $f(r)$ is the correlation function (22) with $r_c = 0.2$ fm and $\delta = 0.75$,

(3) a Hulthen function with binding energy 3.2 MeV, and

(4) a Hulthen function with the correct binding energy of 2.23 MeV for the deuteron.

The difference between the high-momentum components of these functions

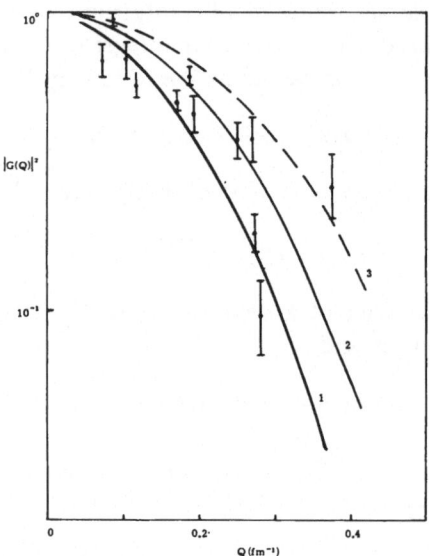

**Fig. 3.** The function $G^2(Q)$ normalized to unity at $Q = 0$ for various $2s$ wave functions for the relative motion of an $\alpha$-particle and a deuteron cluster in $^6$Li. Curve 1 corresponds to a square-well function with a separation energy of 1.47 MeV, curve 2 corresponds to a square-well function with a separation energy of 3.70 MeV, and curve 3 corresponds to an oscillator function with length parameter $a = 2.94$ fm. The experimental data is from the $^6$Li$(p, pd)$ reaction (Ruh+ 62, Ruh+ 63). [Reproduced by permission of the editor of *Il Nuovo Cimento*.]

is quite remarkable. As we have noted, the functions $\phi_d{}^\alpha$ and $\phi_d{}^k$ are in general not orthogonal functions because the overlap integral is constructed from model wave functions. This means that $F^{\alpha k}(q)/F_d(q)$ is not unity for deuteron knockout, but varies with $q$. Figure 5 shows the results (Jac 65a) obtained when the wave function for the deuteron cluster in $^6$Li is taken to be a Gaussian function.

For a two-nucleon pickup reaction, such as the $(p, t)$ reaction, the reduced overlap integral describes the motion in the nucleus of the center of mass of two neutrons correlated in the same way as two neutrons in the triton. The function $\phi_x{}^k(\mathbf{r})$ describes the motion of the two neutrons in the triton and $E_x{}^k$ is their relative energy. It is usual to take a shell-model approach to the overlap integral starting from the single-particle states $\phi^{n_1 l_1 j_1}(\mathbf{r}_1)$, $\phi^{n_2 l_2 j_2}(\mathbf{r}_2)$ for the two neutrons relative to the residual core. The simplest procedure (Gle 63, Gle 65) is then to take the functions $\phi^{n_1 l_1 j_1}$ and $\phi^{n_2 l_2 j_2}$ to be oscillator functions and transform them using the Talmi co-ordinate transformation and the Brody–Moshinsky transformation brackets in the manner we have already described for knockout reactions. This method can be improved by generating the single-particle functions in realistic potentials and then expanding these functions in terms of oscillator functions. Alternatively, the oscillator functions can be given Hankel function tails at a suitable matching radius. This raises the question, however, of what single-particle potentials and what separation energies should be

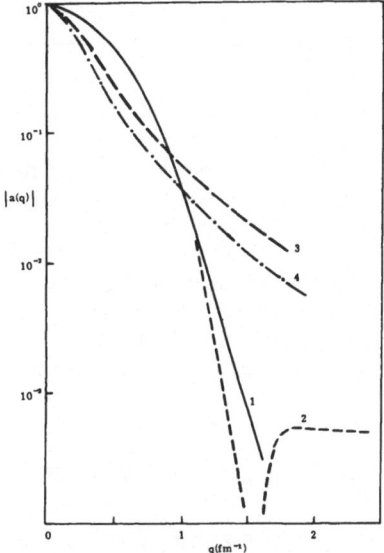

Fig. 4. The function $a(q)$, normalized to unity at $q = 0$, for various $1s$ wave functions for the relative motion of a pair of nucleons in the nucleus. Curve 1 corresponds to an oscillator function with length parameter $a = 1.8$ fm, curve 2 to the product of the same oscillator function and a correlation function with $r_c = 0.2$ fm and $\delta = 0.75$, curve 3 to a Hulthen function with a binding energy of 3.2 MeV, and curve 4 to a Hulthen function with a binding energy of 2.23 MeV. [Reproduced by permission of the editor of *Il Nuovo Cimento*.]

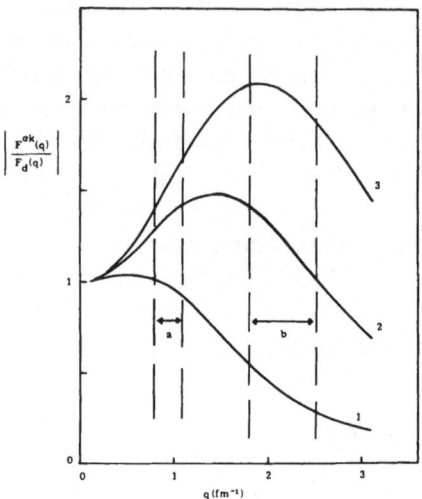

**Fig. 5.** The function $\left| F^{\alpha k}(q)/F_d(q) \right|^2$, normalized to unity at $q = 0$, for the deuteron cluster in $^6$Li. Curve 1 corresponds to a length parameter $\bar{a} = 0.23$ fm$^{-2}$ for the Gaussian wave function of the cluster, curve 2 to $\bar{a} = 0.46$ fm$^{-2}$, and curve 3 to $\bar{a} = 0.66$ fm$^{-2}$. The arrow (a) indicates the range of $q$ for the $(p, pd)$ reaction at 30 MeV and the arrow (b) indicates the range at 155 MeV. [Reproduced by permission of the editor of *Reviews of Modern Physics*.]

used. This point can best be examined if we write the overlap integral in the form

$$\Psi_{fi}(\mathbf{r}_1, \mathbf{r}_2) = \int \Phi_B{}^f(\boldsymbol{\xi}) \Phi_A{}^i(\mathbf{r}_1, \mathbf{r}_2, \boldsymbol{\xi}) \, d\boldsymbol{\xi} \qquad (27)$$

The equation for this overlap integral is (JG 69)

$$[\varepsilon - T_1 - T_2 - V(\mathbf{r}_1) - V(\mathbf{r}_2) - W(\mathbf{r}_1, \mathbf{r}_2)]\Psi_{fi}(\mathbf{r}_1, \mathbf{r}_2) = 0 \qquad (28)$$

where $\varepsilon = E_A{}^i - E_B{}^f$, $W(\mathbf{r}_1, \mathbf{r}_2)$ is the residual interaction between the two neutrons, and $V(\mathbf{r}_1)$ is the potential which binds each neutron to the core. The most common procedure (TH 69, RM 64a, BK 67, BH 68) is to omit the residual interaction but bind each neutron in a Saxon–Woods potential at a separation energy of $\frac{1}{2}\varepsilon$; this implies that the effect of the residual interaction on the energy has been included in the potentials $V(\mathbf{r}_1)$ and $V(\mathbf{r}_2)$, but configuration mixing can be taken into account by using a sum of products of these single-particle wave functions. Drisko and Rybicki (DR 66) take a single product of single-particle wave functions and bind each neutron at the separation energy $\varepsilon_n$ for a nucleus with one neutron outside the $A - 2$ core; this implies that $\langle W(\mathbf{r}_1, \mathbf{r}_2)\rangle = \varepsilon - 2\varepsilon_n$, which is the pairing energy of two neutrons. Jaffe and Gerace (JG 69) include the residual interaction and vary it to obtain the correct separation energy $\varepsilon$; this procedure also yields the expansion coefficients for the set of basis states.

The radial part of the overlap integral for the $^{42}$Ca$(p, t)^{40}$Ca reaction calculated by Jaffe and Gerace is shown in Fig. 6. Case I corresponds to their procedure of binding all the particles in one potential and including the residual interaction, while case II corresponds to the method of binding all the particles at $\frac{1}{2}\varepsilon$. As might be expected, there is a substantial difference in magnitude in the surface region, and this leads to a difference in magnitude for the predicted differential cross sections but a rather smaller effect on the shape of the cross sections.

The cross section for two-nucleon pickup in DWBA is inevitably more complicated than that for the single-nucleon case. In particular, it is not possible to factorize the cross section, as in Eq. (12), into the product of a

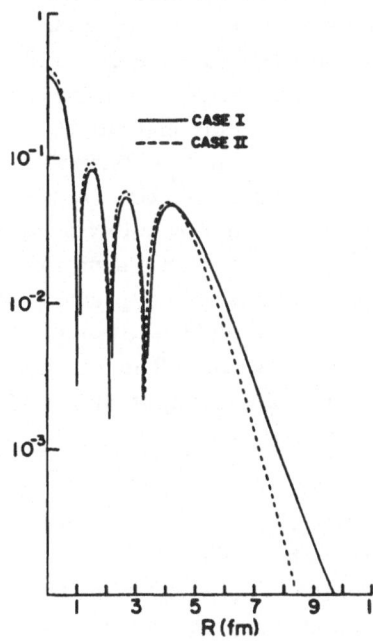

Fig. 6. The radial part of the overlap integral for the $^{42}$Ca$(p, t)$ reaction. Case I corresponds to the procedure of binding both neutrons in one potential and including the residual interaction, while case II corresponds to the method of binding both neutrons at half the separation energy. [From R. L. Jaffe and W. J. Gerace, *Nucl. Phys.* **A125**:1 (1969). Reproduced by permission of the authors and the North-Holland Publishing Company.]

term containing the nuclear structure information (the spectroscopic factor) and a term depending on the reaction mechanism. This leads to the possibility of obtaining additional nuclear structure information which is discussed in Section 4.

## 2.4. The Multi-Nucleon Case

The complexity of multi-nucleon pickup reactions such as $(p, \alpha)$, $(\alpha, {}^7\text{Li})$, etc., is such that the analyses are usually restricted to the removal of nucleons in the same shell and the overlap integrals are usually constructed from oscillator functions (Bay 64). Some progress has recently been made (Rot 69) in deriving spectroscopic factors for these reactions using an oscillator basis and the generalization of the Brody–Moshinsky transformation brackets due to Smirnov (Smi 61, Smi 62).

## 2.5. Distortion and Finite-Range Effects

For a realistic calculation of the cross section for any nuclear reaction it is necessary to take account of the distortion of the wave functions of the incident and scattered particles. One of the effects of this distortion is a reduction in the magnitude of the cross section owing to the absorption of particles from the elastically scattered beam. Another effect, which is particularly important in the context of the study of hole states, is the introduction of additional momentum components so that the momentum components actually "observed" in a knockout or transfer reaction are not exactly those predicted by the Fourier transform of the overlap integral. In order to illustrate this point we consider a knockout reaction $(a, ap)$ in which the knocked out proton is emitted at a rather low energy. The projectile $a$ need not be specified but we assume that its incident and outgoing energies are sufficiently high so that it may be described approximately by plane waves. The transition matrix element for this process is then given in impulse approximation by

$$T_{if} \propto F(\mathbf{Q} + \mathbf{k}_p) \int \chi_p^{-*}(\mathbf{k}_p, \mathbf{R}) \psi_p(\mathbf{R}) \, e^{i(\mathbf{Q} + \mathbf{k}_p) \cdot \mathbf{R}} \, d\mathbf{R} \qquad (29)$$

where $\mathbf{Q}$ is the recoil momentum of the residual nucleus, $\mathbf{k}_p$ is the momentum of the outgoing proton, $\mathbf{Q} + \mathbf{k}_p = \mathbf{k}_a - \mathbf{k}_a'$ is the momentum transfer, $F(\mathbf{Q} + \mathbf{k}_p)$ is the transform of the two-body interaction, $\psi_p$ is the overlap integral, and $\chi_p$ is the distorted wave function for the outgoing proton. If we expand the distorted wave in terms of plane waves the matrix element becomes[†]

$$T_{if} \propto F(\mathbf{Q} + \mathbf{k}_p) \int\!\!\int \psi_p(\mathbf{R})\, e^{i(\mathbf{Q}+\mathbf{k}_p-\mathbf{k})\cdot\mathbf{R}} a(\mathbf{k})\, d\mathbf{R}\, d\mathbf{k}$$

$$\propto F(\mathbf{Q} + \mathbf{k}_p) \int G(\mathbf{Q} + \mathbf{k}_p - \mathbf{k}) a(\mathbf{k})\, d\mathbf{k} \qquad (30)$$

where we have used Eq. (23). Thus we see that the distortion does indeed modify the momentum components of the overlap integral. When the distortion of all the particles involved in the reaction is taken into account, the effect becomes more complicated but it is often the case that the distortion of one particular particle contributes the most important effect. ·

The example given above is relevant to the (e, ep) experiment (Ama 67, Ama+ 64, 65a, 65b, 66a, 66b, 67) in which the electron energies are in the region of 500–600 MeV and the energy of the outgoing proton is around 100 MeV. A calculation for the $^{12}$C(e, ep) reaction using an approximate treatment of the proton distortion (Wat 68) shows that the additional momentum components introduced by distortion are indeed important, and casts doubt on earlier calculations (Cio 68) in which the distortion is represented by a constant reduction factor taken outside the integral over $\mathbf{R}$. However, accurate distorted wave calculations with an overlap integral having the correct asymptotic behavior (EG 69, 70) indicate that the use of a reduction factor is acceptable for the case of $1p$ knockout from $^{12}$C but is not satisfactory for knockout from more deeply bound states or from heavier nuclei.

The factorization of the matrix element (29) into the transform of the two-body interaction and the transform of the overlap integral is exact only when the plane wave impulse approximation (PWIA) is used, whereas in DWIA, additional approximations are required to obtain such a factorization (BH 59, BJ 63, Jac 66). Because of this factorization, the DWIA is sometimes regarded as a zero-range approximation although the dependence of the factor $F(\mathbf{Q} + \mathbf{k}_p)$ on momentum transfer implies that the finite-range behavior of the two-body interaction is taken into account in the same way as in a plane wave calculation. McCarthy (LM 64, LM 66) has formulated a finite-range distorted wave formalism for the (p, 2p) reaction, which is known as the distorted-wave $t$-matrix approximation (DWTA), and has compared DWIA and DWTA over a wide range of incident energies. A discussion of some finite-range effects for the (p, pd) and (p, pα) reactions has been given (Jac 66) but no calculations have been carried out for these reactions.

---

[†] The plane-wave approximation is recovered if $a(\mathbf{k})$ is replaced by $\delta(\mathbf{k} - \mathbf{k}_p)$.

An understanding of the interplay of finite-range and distortion effects has been obtained from exact finite-range DWBA calculations for pickup and stripping reactions (Aus+ 64, Dic+ 65) and from approximate calculations using the LEA (BG 64, Tow 67a, Jai 69a). In zero-range PWBA and DWBA the integral over the overlap integral and wave functions of the incident and scattered particles is multiplied by a constant which is the strength of the two-body interaction. In finite-range PWBA this constant is replaced by the Fourier transform of the two-body interaction which in general falls rapidly with increasing momentum transfer, i.e., with increasing scattering angle. However, the additional low momentum components introduced by distortion have an important effect at large angles where they prevent the cross section in DWBA from falling as rapidly as in PWBA. Thus the combined effect of distortion and finite-range correction in finite-range DWBA calculations for reactions at medium energies and low $Q$-values is to reduce the cross section fairly uniformly over the whole angular range. The high momentum components introduced by the distortion tend to be localized in the nuclear interior where the real part of the optical potential is largest in magnitude but the finite-range correction tends to reduce the contribution from high momentum transfers and hence to suppress that contribution to the nuclear matrix element from the interior region. This result gives some support for the use of cutoff radii in zero-range calculations to simulate the finite-range correction.

# 3. SINGLE-NUCLEON KNOCKOUT AND RELATED REACTIONS

## 3.1 Comparison of Knockout and Pickup Reactions

In order to make a simple comparison of the pickup and knockout reactions we examine the cross sections for the $(p, 2p)$ and $(p, d)$ reactions given by PWBA. For the $(p, 2p)$ reaction we take the momentum of the incident proton to be $\mathbf{p}_0$ and the momenta of the two outgoing protons to be $\mathbf{p}_1$ and $\mathbf{p}_2$ respectively. The matrix element can then be written as

$$T_{if} \propto \int V(\mathbf{s}) \, e^{i\mathbf{q} \cdot \mathbf{s}} \, d\mathbf{s} \int \psi_p^\beta(\mathbf{R}) \, e^{i\mathbf{Q} \cdot \mathbf{R}} \, d\mathbf{R} \qquad (31)$$

where $\mathbf{Q} = \mathbf{p}_0 - \mathbf{p}_1 - \mathbf{p}_2$ is the recoil momentum of the residual nucleus, $\mathbf{q} = \mathbf{p}_0 - \mathbf{p}_1$ is the momentum transfer, $\mathbf{s}$ is the proton–proton separation distance, $V(\mathbf{s})$ is the proton–proton interaction, and $\psi_p$ is the proton overlap integral. For the $(p, d)$ reaction we take the momentum of the incident proton to be $\mathbf{p}_0$ and that of the outgoing deuteron to be $\mathbf{p}_d$. The matrix element is given by

$$T_{if} \propto \int \phi_d^*(\mathbf{r}) V(\mathbf{r}) e^{i\mathbf{K} \cdot \mathbf{r}} d\mathbf{r} \int \psi_n(\mathbf{R}) e^{-i\mathbf{P} \cdot \mathbf{R}} d\mathbf{R} \tag{32}$$

and

$$\mathbf{K} = \mathbf{p}_0 - \tfrac{1}{2}\mathbf{p}_d, \qquad \mathbf{P} = \mathbf{p}_d - \mathbf{p}_0$$

$\mathbf{r}$ is the proton–neutron separation distance, $V(\mathbf{r})$ is the proton–neutron interaction, and $\psi_n$ is the neutron overlap integral. For both reactions, this simple formalism indicates that the nuclear structure information enters through the overlap integral. Using the single-particle model described in Section 2.2 we can express each overlap integral in terms of the appropriate coefficient of fractional parentage and a single-particle wave function for the proton or neutron. Thus the second integral in Eq. (31) is proportional to the transform of a proton single-particle wave function and expresses the probability of finding in the target nucleus a proton with momentum $-\mathbf{Q}$. Similarly, the second integral in Eq. (32) expresses the probability of finding in the target nucleus a neutron with momentum $\mathbf{P}$. We therefore call this integral the momentum distribution for the bound nucleon and denote it by $g_\beta^{PW}$, so that

$$g_\beta^{PW} = \int \psi^\beta(\mathbf{R}) e^{i\mathbf{Q} \cdot \mathbf{R}} d\mathbf{R} \tag{33}$$

If distorted waves are used to replace the factor $e^{i\mathbf{Q} \cdot \mathbf{R}}$ we denote the corresponding integral by $g_\beta$.

The comparison of Eqs. (31) and (32) reveals the basic similarity of the direct interaction picture of pickup and transfer reactions, and comparison of the energy spectra of the $^{12}C(p, d)$ and $^{12}C(p, 2p)$ reactions under the same conditions and good energy resolution (Pug+ 65) gave experimental confirmation of the similarity of the reaction mechanism. However, the energy–momentum conditions in the two reactions are quite different, and this has an important influence on the range of proton or neutron momentum components which are examined and on the effect of distortion and finite-range corrections on the angular distributions. For the $(p, d)$ reaction we have

$$P = (p_0^2 + p_d^2 - 2p_0 p_d \cos \phi)^{\frac{1}{2}}$$

$$p_d = \sqrt{2} p_0 (1 + Q_d/E_0)^{\frac{1}{2}}$$

where $Q_d = -S_{nA} + 2.23$ is the $Q$-value (in MeV) for the reaction, $S_{nA}$ is the separation energy of the neutron in the target nucleus, and $E_0$ is the incident proton energy. Thus if $Q_d < 0$, the region of small $P$ can be investigated using low- or medium-energy protons while the region of large $P$ can be studied with proton energies $E_0 \gg |Q_d|$, i.e., the same reaction at different energies is sensitive to different momentum components of the single-particle overlap integral. In the knockout reaction the presence of two fast particles in the final state allows much greater freedom in the range of the recoil momentum $Q$ which can be studied. In the symmetric coplanar $(p, 2p)$ experiment (Fig. 7) we have

$$|\mathbf{p}_1| = |\mathbf{p}_2|, \qquad \theta_1 = -\theta_2 = \theta, \qquad \varepsilon = 0 \tag{34}$$

so that

$$Q = p_0 - 2p_1 \cos \theta \tag{35}$$

$$p_1 = p_2 = \tfrac{1}{2}\sqrt{2}\,p_0(1 + Q_p/E_0)^{\tfrac{1}{2}} \tag{36}$$

where $Q_p = -S_{pA}$ is the $Q$-value for the reaction, and $S_{pA}$ is the proton separation energy. Using these formulas we can compare, for example, the $^{12}C(p, 2p)$ and $^{12}C(p, d)$ reactions at 155 MeV: in the former, the range of $Q$ covered is $-150 < Q < 150$ MeV/c, whereas the range of $P$ in the latter reaction is $140 < P < 530$ MeV/c. It follows that at this energy the states rich in low momentum components, i.e., the $1s$ states, will show up much more strongly in the $(p, 2p)$ reaction than in the $(p, d)$ reaction. This prediction is confirmed by the experimental results. It also follows that the $(p, 2p)$ reaction will be most sensitive to the surface behavior of the overlap integral and the reaction will show surface localization, while the $(p, d)$ reaction at the same energy will be sensitive to the interior behavior of the

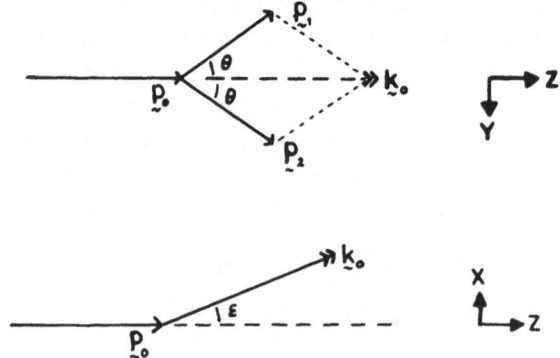

Fig. 7. Definition of momenta and angles for the $(p, 2p)$ reaction.

overlap integral as well. This surface localization is well-established from DWIA calculations for the $(p, 2p)$ reaction (Jac 67b, Jai 69a) and comparison with DWBA calculations for the $(p, d)$ reaction (Tow 67a, Tow 69) shows that the latter have a greater contribution from the interior region. In the $(p, 2p)$ calculations, the additional low momentum components introduced by the distortion have an important effect on the angular distribution, whereas in the $(p, d)$ calculations it is the additional high momentum components, introduced particularly by the distortion of the deuteron, which have the most significant effects.

## 3.2. Special Features of Knockout Reactions

The symmetric coplanar case of the $(p, 2p)$ reaction has a rather special symmetry, since a rotation of $k\pi$, where $k$ is any integer, about the direction of the incident beam transforms the system into one which is physically indistinguishable, provided that polarizations are not measured. If we neglect spin and characterise the proton overlap integral by the quantum numbers $nlm$, the rotational symmetry about a $z$-axis taken along the direction of the incident beam gives the condition that (Mar 58, JB 65)

$$g_{nlm} = (-1)^m g_{nlm}$$

i.e.,

$$g_{nlm} \equiv 0, \quad \text{if } m \text{ is odd} \tag{37}$$

In the plane wave approximation it follows from Eq. (33) that $g_{nlm}$ is non-zero only when $m = 0$, but in an accurate distorted wave calculation there are contributions from the other even $m$ components. It is interesting to note that this difference is a consequence of the introduction of additional *transverse* momentum components by the distortion (Jac 67c, EJ 67). If the outgoing protons are not symmetric in energy and angle, or if the incident and outgoing protons are not coplanar, this symmetry is lost and the recoil momentum

$$\mathbf{Q} = \mathbf{p}_0 - (\mathbf{p}_1 + \mathbf{p}_2) = \mathbf{p}_0 - \mathbf{k}_0 \tag{38}$$

is no longer parallel or antiparallel to the direction $\mathbf{p}_0$ of the incident beam, as can be seen from Fig. 7. Thus, in this case also there are transverse components of the recoil momentum and even in a plane-wave approximation there can be contributions from all $m$ components. The symmetry is also lost if the two outgoing particles are not identical.

Examination of Eq. (33) for the momentum distribution shows that the cross section for the $(p, 2p)$ reaction has a maximum at $Q = 0$ if the overlap integral has $s$-state components whereas the cross section has a minimum if the overlap integral contains no $s$-state components. In the plane wave approximation this minimum is predicted to be zero. The low (longitudinal) momentum components introduced by the distortion serve to fill in the minimum, although the effect of distortion is usually insufficient to explain the amount of filling in observed in most counter experiments with symmetric coplanar geometry unless optical potentials with unusually large real parts are used (McC 65, LM 64). It has also been noted that the effect on distorted wave calculations of the finite experimental energy resolution (McC 65) and angular resolution (EJ 67) produces additional filling in, since both allow events which depart slightly from the symmetric coplanar geometry and have nonzero transverse recoil momentum $Q_x$ or $Q_y$ when the longitudinal momentum $Q_z$ is zero. More recent experiments using a propane bubble chamber (YH 65, YH 67) and using spark chambers (Jam+ 69a) indicate that for $^{12}C$, at least, the filling in is almost completely eliminated when the momenta of the protons are precisely determined. Studies of the $(p, 2p)$ reaction with a coplanar geometry in which the outgoing protons are symmetric in angle but not in energy (GWS 67) show that most of the effect of unequal energy sharing appears in the kinematic factors but the filling in of the minimum increases as the ratio of the energies departs from unity.

The angle $\theta_m$ at which the recoil momentum is zero in a symmetric coplanar experiment is obtained from Eqs. (35) and (36) as

$$\cos \theta_m = \tfrac{1}{2}[E_0/(E_0 - S_{pA})]^{\frac{1}{2}}$$

For zero separation energy $S_{pA}$ the angle $\theta_m$ must be 45°, but for a bound proton with $S_{pA} > 0$ the angle $\theta_m$ is less than 45°. The angle $\theta_m$ is shifted slightly by distortion and relativistic corrections to the energies.

The first integral in Eq. (31) can be rewritten in the form

$$T_{pp} = \int e^{-i\frac{1}{2}(\mathbf{p}_1 - \mathbf{p}_2) \cdot \mathbf{s}} V(\mathbf{s}) e^{i\frac{1}{2}(\mathbf{p}_0 + \mathbf{Q}) \cdot \mathbf{s}} d\mathbf{s} \tag{39}$$

where $\tfrac{1}{2}(\mathbf{p}_0 + \mathbf{Q})$, $\tfrac{1}{2}(\mathbf{p}_1 - \mathbf{p}_2)$ are the momenta of the scattered proton before and after collision in the proton–proton center of mass system. Using impulse approximation $|T_{pp}|^2$ can be formally expressed in terms of the free proton–proton cross section, (BJ 63, MHR 58) but because of the energy required to release the bound proton the matrix element $T_{pp}$ is off the energy-shell for free $p$–$p$ scattering. The scattering angle in the $p$–$p$

system is given by

$$\cos \bar{\theta} = (\mathbf{p}_0 + \mathbf{Q}) \cdot (\mathbf{p}_1 - \mathbf{p}_2)/|\,\mathbf{p}_0 + \mathbf{Q}\,|\,|\,\mathbf{p}_1 - \mathbf{p}_2\,| \qquad (40)$$

and it can be seen from Fig. 7 that in the symmetric coplanar experiment $(\mathbf{p}_1 - \mathbf{p}_2)$ is perpendicular to $(\mathbf{p}_0 + \mathbf{Q})$; hence $\bar{\theta} = 90°$ and the proton–proton scattering occurs in singlet-even states. For proton energies below 200 MeV in the lab system the free $p$–$p$ cross section increases with decreasing relative energy so that the effect of the momentum $-\mathbf{Q}$ of the proton in the nucleus is to depress the cross section for a relative momentum $|\,\mathbf{p}_0 + \mathbf{Q}\,| > p_0$ compared to that for $|\,\mathbf{p}_0 + \mathbf{Q}\,| < p_0$. The effect of some of the approximations and uncertainties in this procedure have been examined by Jain (Jai 69a) who has shown that DWIA calculations at 400 MeV and above should be quite reliable but below 200 MeV there are rather large uncertainties due to off-energy-shell effects, particularly at small angles. A more accurate treatment of $T_{pp}$ can be obtained by using a pseudopotential or effective local $t$-matrix which has the correct on-energy-shell behavior. If it can be assumed that this pseudopotential gives also reasonably correct off-energy-shell behavior, this procedure allows an accurate treatment of the finite range of the interaction which introduces a contribution from triplet-odd states of the $p$–$p$ system. This DWTA method has been developed and studied extensively by McCarthy and coworkers (LM 64, LM 66, DM 68, McC 68) who have shown that there are departures from DWIA at energies below about 100 MeV.

An exact description of the three-body final state of a knockout reaction leads to the appearance in the Hamiltonian of a term which couples the motion of the particles. This coupling may appear in the kinetic-energy or the potential-energy terms depending on the choice of coordinates for the system, and it is only when this term is neglected that the wave function for the final state can be written in product form. This problem has been discussed in detail elsewhere (JB 65, Jac 68). We note here only that there are two approximations commonly in use:

(i) The kinetic-energy approximation (BJ 63, LM 64) yields a final-state wave function in the form of a product of two distorted wave functions, one for each outgoing particle, and the wave functions of the residual nucleus. This approximation is valid in the static limit of an infinitely heavy residual nucleus but not in the plane wave limit, and provides the most general and flexible description.

(ii) The di-proton approximation (Jac 67b) yields a product of a distorted wave function for the center of mass of two outgoing particles and the wave function of the residual nucleus. This approximation is valid in the plane wave limit and leads to much simpler calculations but is applicable only to experiments which are symmetric though not necessarily coplanar. A comparison of these approximations at 460 MeV has been given by Jain (Jai 69a).

In both approximations for the $(p, 2p)$ reaction the energy of the outgoing protons or di-proton in the $A + 1$ center-of-mass system varies with scattering angle. Since the local optical potential for protons is energy dependent this means that in any calculation of the angular distribution the optical potentials for the final state should be varied as the scattering angle is varied. The effect of this energy dependence is to increase the distortion

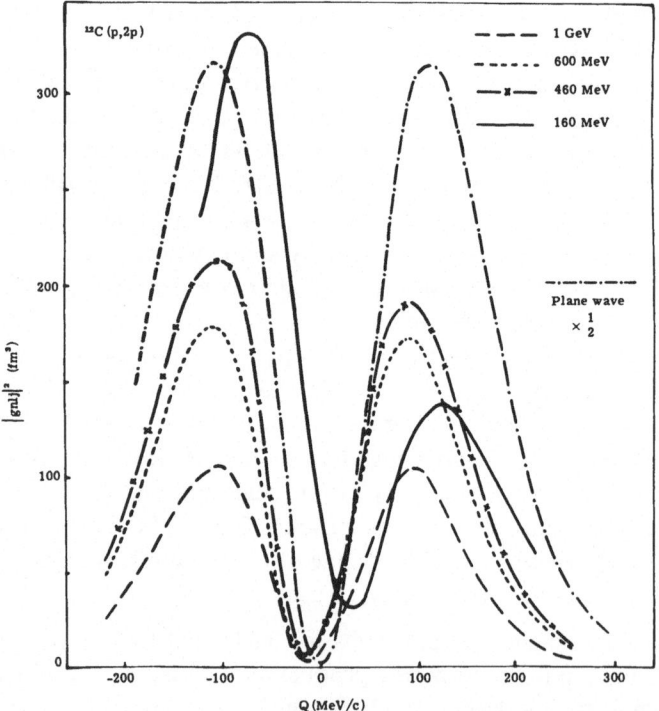

**Fig. 8. The distorted momentum distribution** $\mid g_{nlj} \mid^2$ **for a** $1p$ **proton in** $^{12}$C **calculated at incident proton energies of 160 MeV, 460 MeV, 600 MeV, and 1 GeV. The plane-wave calculation is also shown but is reduced by a factor of $\frac{1}{2}$. [From D. F. Jackson and B. K. Jain,** *Phys. Letters* **27B**:147 (1968). Reproduced by permission of North-Holland Publishing Company.]

and hence reduce the cross section for protons scattered through angles $>45°$, compared with those scattered through angles $<45°$. This effect is illustrated in Fig. 8 which shows the distorted momentum distribution for a $1p$ proton in $^{12}C$ at various energies as a function of recoil momentum $Q$ (JJ 68). It can be seen that in the plane wave case the minimum occurs exactly at $Q = 0$ and the maxima on either side are of equal magnitude. In contrast, the curve calculated at 160 MeV is seriously affected by the distortion; the maxima and minimum are displaced, the two maxima are of different magnitudes, and the minimum is filled in. These distortion effects become less important at higher energies where the main effect is a reduction in the overall magnitude. The ratios of the left-hand maxima for the distorted wave calculation to the plane wave value are, for increasing energy, 0.531, 0.34, 0.28, and 0.17.

Attempts are now being made to reformulate the theory of knockout reactions using Fadeev's treatment of the three-body problem (DG 66, McC 68, KK 70, Meb 69). In the static limit, this approach leads to a lowest-order term in the form of the DWTA or DWIA together with correction terms which represent excitation of the core. A formulation of PWIA derived from the Fadeev equations has been used to study off-energy-shell effects in the $(p, 2p)$ reaction (RSL 70). The results indicate that these effects are unimportant above 200 MeV but become significant for decreasing incident energy and increasing separation energy. The interplay between distortion and off-shell effects has not yet been studied by this method.

## 3.3 Spectroscopic Studies

In the earliest $(p, 2p)$ experiments on $1p$-shell nuclei (MHR 58, Gar+ 62) the summed energy spectra showed two broad peaks for $A \leq 12$ and three peaks for $A > 12$, which were interpreted as due to knockout from $s_{\frac{1}{2}}$, $p_{\frac{3}{2}}$, and $p_{\frac{1}{2}}$ states. As the energy resolution of $(p, 2p)$ experiments improved, these peaks showed further structure and it was at first thought that this was due to a high degree of configuration mixing which would introduce additional single-particle states into the ground state of the target nucleus. It became clear, however, that these extra peaks are due to the splitting of the single hole strength among the possible excited states of the residual nucleus and that the position of the peaks can be associated with the energies of the excited states (BB 62, Die 62, Jac 65c). Detailed spectroscopic studies of some energy spectra for both the $(p, 2p)$ and $(e, ep)$ reactions have since been carried out (BBR 65, ZS 66, Cio 68, BKM 69) for light nuclei.

Spectroscopic studies of the angular distributions observed in the $(p, 2p)$ reaction have only recently been attempted and almost all analyses prior to 1967 were carried out using sum rules. In this approach it is assumed that a particular peak in the energy spectrum and the corresponding angular distribution include contributions from all allowed final states of the residual nucleus and that the measured separation energy is equal to the mean separation energy. The radial part of the overlap integral is evaluated at the mean separation energy so that the sum over the cross sections for transitions to different final states with spin $J_B$ is given by

$$\sum_{J_B} \frac{d^3\sigma}{d\Omega_1 d\Omega_2 dE} (J_A J_B; lj) = \sum_{J_B} S(J_A J_B; lj)\sigma_{lj}(\theta) \tag{41a}$$

$$= \mathscr{S}(lj)\sigma_{lj}(\theta) \tag{41b}$$

In Section 2.2 the expression for the cross section was derived without including isospin, and in this case the sum rule $\mathscr{S}$ is given by

$$\mathscr{S}(lj) = N_{lj}(\text{protons}) \tag{42}$$

where $N_{lj}(\text{protons})$ represents the number of *protons* in the $lj$ subshell. In fact, most spectroscopic calculations for the $(p, 2p)$ reaction have used the isospin formalism which implies the explicit inclusion of the isospin quantum numbers and coupling coefficients. If the sum over both the final spin $J_B = J_A \pm \frac{1}{2}$ and the final isospin $T_B = T_A \pm \frac{1}{2}$ is taken, the sum rule $\mathscr{S}$ is now equal to $N_{lj}$, the number of active *nucleons* in the $lj$ subshell. However, since $m_t = \frac{1}{2}$ for the proton only those final states with $T_B > M_{TA} + \frac{1}{2}$ can be formed in the $(p, 2p)$ reaction. The expression for the summed cross section is now

$$\sum_{J_B} \frac{d^3\sigma}{d\Omega_1 d\Omega_2 dE} (J_A T_A J_B T_B; lj) = \sum_{J_B} S(J_A T_A J_B T_B; lj)$$

$$\times (T_B M_{TB} \tfrac{1}{2} \tfrac{1}{2} | T_A M_{TA})^2 \sigma_{lj}(\theta) \tag{43}$$

and the relevant sum rules for $1p$-shell nuclei are (Tow 67b) for $T_A = 0$, $T_B = \frac{1}{2}$:

$$\mathscr{S}(p_{\frac{1}{2}}) = N_{lj} \tag{44a}$$

$$\mathscr{S}(s_{\frac{1}{2}}) = \sum_{J_B} (2J_B + 1)(2T_B + 1)/(2J_A + 1)(2T_A + 1) = 4 \tag{44b}$$

and for $T_A = \frac{1}{2}$, $M_{TA} = \frac{1}{2}$, $T_B = 1$, and $M_{TB} = 1$

$$\mathscr{S}(p_{\frac{1}{2}}) = \tfrac{3}{4}(N_{lj} - 1) \tag{44c}$$

$$\mathscr{S}(s_{\frac{1}{2}}) = \sum_{J_B} (2J_B + 1)(2T_B + 1)/(2J_A + 1)(2T_A + 1) = 3 \tag{44d}$$

In the more recent experiments on the $(p, 2p)$ reaction (TSR 63, Tyr+ 66, Ard+ 67, GWS 67) some individual final states have been resolved and the analysis by Berggren (Tyr+ 66) in terms of occupation numbers revealed the need for a spectroscopic analysis. Such analyses have been carried out for $1p$-shell nuclei by Jain and Jackson (JJ 67) using spectroscopic factors calculated from the fractional-parentage coefficients of Cohen and Kurath (CK 65) and single-particle wave functions for the bound proton derived from fits to electron scattering so that there are, in principle, no adjustable parameters in the calculation. Nuclei in the same shell have been studied by Kolybasov and Smorodinskaya (KS 67a, KS 69) who fit the data to obtain values for the reduced widths and a nuclear radius parameter for a Butler-type momentum distribution. In many cases, the data still correspond to more than one excited state so that it is necessary to add the cross sections for these states, and the agreement with the data is only moderate.

In Table I we give the ratio for some $1p$-shell nuclei of the sum of the experimental spectroscopic factors to all observed states to the sum rule given by Eqs. (44a)–(44d). The experimental spectroscopic factors are in most cases in satisfactory agreement with those predicted from the work of Cohen and Kurath although there are rather large uncertainties in some cases (JJ 67). These results suggest (Jac 68) that the observed states do not exhaust the sum rule and that there must be additional $1s$ and $1p$ hole strength which has not so far been detected. This means that estimates of the mean separation energies may be in doubt and the assumption that the position of the peak in a poor resolution experiment yields a reliable value for the mean separation energy (ES 67, Elt 68a) can not at present be verified. This has serious implications for Hartree–Fock calculations and for estimates of rearrangement and total energy (BS 65, 67, Köh 66, Elt 67).

The $(p, 2p)$ reaction leading to low-lying states of $2s1d$-shell nuclei has also been studied (Jai 68a, KS 67b). The qualitative features of the data are quite well reproduced, and for $^{24}$Mg and $^{23}$Na the experimental spectroscopic factors are in reasonable agreement with those predicted from the $SU_3$ and rotational models. The agreement between the spectroscopic factors deduced from the $(p, 2p)$, $(d, {}^3\mathrm{He})$, and $({}^3\mathrm{He}, \alpha)$ reactions on $^{28}$Si is far from satisfactory (Jai 68a) and the same applies to the $(p, 2p)$ and $(d, {}^3\mathrm{He})$ reactions on $^{24}$Mg (Wag 69, Ard+ 69). These discrepancies are mainly due to the complexity of the target nuclei and the difficulty of separating uncertainties in the description of the reaction mechanism and of the nuclear wave functions. It seems likely that much better energy resolution will be required in the $(p, 2p)$ reaction before this problem is resolved.

It is unfortunately true that the energy resolution of the existing data

**TABLE I. Ratio of Observed Spectroscopic Factors in the $(p, 2p)$ Reaction to the Theoretical Sum Rule**

| Target nucleus | $J_A^\pi$ | $T_A$ | Residual nucleus | $J_B^\pi$ | $T_B$ | Subshell | $\mathscr{S}_{exp}/\mathscr{S}_{theo}$ |
|---|---|---|---|---|---|---|---|
| $^6$Li | $1^+$ | 0 | $^5$He | $\frac{3}{2}^+$ $\frac{1}{2}^+$ | $\frac{1}{2}$ $\frac{1}{2}$ | $s_{\frac{1}{2}}$ | 0.70 |
| | | | | $\frac{3}{2}^-$ $\frac{1}{2}^-$ | $\frac{1}{2}$ $\frac{1}{2}$ | $p_{\frac{3}{2}}$ | ~1.0 |
| $^7$Li | $\frac{3}{2}^-$ | $\frac{1}{2}$ | $^6$He | $2^-$ $1^-$ | 1 1 | $s_{\frac{1}{2}}$ | 0.47 |
| | | | | $0^+$ $2^+$ | 1 1 | $p_{\frac{3}{2}}$ | — |
| $^9$Be | $\frac{3}{2}^-$ | $\frac{1}{2}$ | $^8$Li | $2^-$ $1^-$ | 1 1 | $s_{\frac{1}{2}}$ | 0.27–0.52 |
| | | | | $2^+$ $1^+$ $3^+$ | 1 1 1 | $p_{\frac{3}{2}}$ | 0.88 |
| $^{11}$B | $\frac{3}{2}^-$ | $\frac{1}{2}$ | $^{10}$Be | $2^-$ $1^-$ | 1 1 | $s_{\frac{1}{2}}$ | — |
| | | | | $0^+$ $2^+$ | 1 1 | $p_{\frac{3}{2}}$ | 0.73–0.90 |
| $^{12}$C | $0^+$ | 0 | $^{11}$B | $\frac{1}{2}^+$ | $\frac{1}{2}$ | $s_{\frac{1}{2}}$ | 0.66 |
| | | | | $\frac{3}{2}^-$ $\frac{1}{2}^-$ | $\frac{1}{2}$ $\frac{1}{2}$ | $p_{\frac{3}{2}}$ | 0.90 |

in energy regions above 150 MeV is not adequate for a detailed spectroscopic study. However, at 50 MeV much better energy resolution has been obtained and data on $^{12}$C and $^{89}$Y are available (Pug+ 67, Ric+ 69). In $^{12}$C, the strong excitation of the $\frac{3}{2}^-$ ground state of $^{11}$B is clearly seen, but the energy spectrum also reveals the excitation of states at 2.14 MeV ($\frac{1}{2}^-$), 4.46 MeV ($\frac{5}{2}^-$), 5.14 MeV ($\frac{3}{2}^-$), and 6.76 MeV ($\frac{7}{2}^-$). Of these states, the ones with spins of $\frac{1}{2}^-$ and $\frac{3}{2}^-$ can be predicted with about the right strength from intermediate coupling calculations (AK 64, Mac 64, CK 65), but those with spins of $\frac{5}{2}^-$ and $\frac{7}{2}^-$ are forbidden within the framework of the lowest shell-model configuration and a one-step reaction mechanism. Calculations which include $2p$–$2h$ configurations in the ground state of $^{12}$C (GP 63) cannot give sufficient strength for these $j$-forbidden states[†] and it seems probable that their presence is due to core excitation proceeding through a two-step mechanism.

At incident energies around 50 MeV the rather simple quasi-elastic features of the $(p, 2p)$ reaction which we have deduced from Eq. (31) are suppressed and the shape of the angular distribution is largely determined by the distortion and the behavior of the proton–proton interaction (Pug+ 67, DM 68). This means that uncertainties in the description of the reaction mechanism will obscure interpretation of the structure of the nuclear states, and *vice versa*. In the case of the $^{89}$Y$(p, 2p)$ reaction, however, the single-particle description should be valid and the spectroscopic factor for the transition to the ground state of $^{88}$Sr should be near to unity. This reaction is therefore being studied in detail by McCarthy (MT 69) at 50 MeV in order to refine the treatment of the reaction mechanism.

A study of the localization of the $(p, 2p)$ reaction (Jac 69) has shown that this occurs in the region just outside the halfway radius of the nuclear matter distribution. Similar calculations for the $(\alpha, \alpha p)$ reaction show that this reaction would be localized much further out in the vicinity of the nuclear strong-absorption radius and that the localization would be much sharper. In the latter case there is little sensitivity to any feature of the wave function of the bound proton except its magnitude in the extreme surface region. These characteristics are due to the strong absorption of the incident and scattered $\alpha$-particle and suggest that the $(\alpha, \alpha p)$ reaction will not be influenced by uncertainties in the behavior of the overlap integral or the distorted waves in the interior region; instead, this reaction should give very reliable information about spectroscopic factors in a manner comparable with stripping below the Coulomb barrier. Because the knockout experiments involve angular correlation measurements they should, in principle, reveal information about the population of nuclear substates. This information could be obtained either by comparing the symmetric coplanar $(p, 2p)$ reaction with the nonsymmetric or noncoplanar case (Jac 67c) or by comparing the symmetric coplanar $(p, 2p)$ and $(\alpha, \alpha p)$ reaction. For example if we consider the knockout of a $p_{\frac{3}{2}}$ proton from a $0^+$ ground state it follows from Eq. (37) that the only substates which will be populated in the symmetric coplanar $(p, 2p)$ reaction are those with $M_f = \pm\frac{1}{2}$ but as soon as the symmetry is removed the substates with $M_f = \pm\frac{3}{2}$ will also be populated. Experiments of this type have been carried out by Hourany (Hou 70). The

---

[†] The same conclusion has been reached in a discussion (Tow 67a) of the excitation of $j$-forbidden states in $^{11}$B through the $^{12}$C$(p, d)$ reaction.

shapes of the observed angular distributions are in agreement with theoretical predictions but there are some discrepancies with the magnitudes. There is as yet no clear evidence for any $j$-dependence of the $(p, 2p)$ cross section, although there is a suggestion of such an effect in the 460 MeV data on $^{16}$O (Tyr+ 65). Since, as shown in Section 3.2, the outgoing protons in a symmetric experiment are in a relative singlet state, any $j$-dependence is most likely to be observed in a nonsymmetric experiment or in the $(\alpha, \alpha p)$ reaction.

## 3.4. Proton States

A considerable amount of information on proton separation energies is now available from measurements on the $(p, 2p)$, $(e, ep)$, and $(d, {}^3\text{He})$ reactions, and analyses of the data for these reactions and for elastic electron scattering in terms of a single-particle model yields information about the proton single-particle wave functions. For most nuclei in the $1p$ shell, and some in the $2s1d$ shell, the same wave functions give a good description of the elastic electron scattering and the $(p, 2p)$ reaction in the energy range 160–460 MeV (JS 63, Elt 68a, Jac 68). It is not surprising that there are difficulties with nuclei such as $^{28}$Si and $^{24}$Mg for which a single-particle model is of doubtful relevance, but it is a little disconcerting that the wave function of the $1d_{\frac{5}{2}}$ proton state in $^{40}$Ca predicted from the analysis of electron scattering (ES 67) does not yield agreement with the angular distribution for the $(p, 2p)$ reaction[†] (Jai 68b). The single-particle model does, however, give a good description of the proton states in $^{208}$Pb, for which the single-particle potentials derived by fitting the separation energies (Ros 68, BG 69) yield proton distributions in agreement with elastic electron scattering data (BG 69, Elt 68b) and proton wave functions in agreement with the data for the $(d, {}^3\text{He})$ reaction at 50 MeV (Par+ 69).

Information on the separation energies of protons in the more loosely bound states comes mainly from the $(p, 2p)$ reaction in the energy region of 150–200 MeV (Tyr+ 65, Rio 65, Ard+ 67, Ruh+ 67) and the $(d, {}^3\text{He})$ reaction at 52 MeV (Mai+ 69, Wag 69, Kas+ 69) and 34 MeV (HNB 67, NH 68, WN 68a,b). The results are generally in good agreement for protons in the outer shells of $1p$- and $2s1d$-shell nuclei, and for protons in the $2s1d$-shell in $2p1f$-shell nuclei. There is a rather marked discrepancy between the results obtained for the $1p$ nucleons in $2s1d$-shell nuclei from the $(p, 2p)$

---

[†] This conclusion is based on the assumption, possibly doubtful, that the data (Tyr+ 66) do not contain any contribution from the $2s_{\frac{1}{2}}$ knockout.

reaction and the $(d, {}^{3}\text{He})$ reaction, which appears to arise from the splitting of the $1p$ hole strength over a wide range of excitation energy so that the $1p$ and $2s1d$ hole states overlap to such an extent that they cannot be resolved in the $(p, 2p)$ experiment (Ard+ 69, Wag 69).

Information on the separation energies of tightly bound protons comes from the $(p, 2p)$ reaction at 460 MeV (Tyr+ 65) and 387 MeV (Jam+ 69b) and from the $(e, ep)$ reaction (Ama+ 64, 65a,b, 66a,b, 67). The results are in good agreement for the $1s$ and $1p$ states in light nuclei up to and including the $1p$ state in ${}^{40}\text{Ca}$, but for the $1s$ state in ${}^{40}\text{Ca}$ and the $1s$ and $1p$ states in heavier nuclei there is some divergence. The $(e, ep)$ data appear to increase linearly with $A$ while the $(p, 2p)$ data appear to saturate at a value of $\sim$55 MeV for the $1s$ states and $\sim$43 MeV for the $1p$ states. In view of the uncertainties on these data[†] it is probably unwise to press either judgment or interpretation too far at this stage but we may note that the $(p, 2p)$ results are associated with angular correlation data which allow identification of the proton states whereas the $(e, ep)$ measurements are not as yet accompanied by angular distributions. Preliminary results from measurements of energy spectra and angular distributions for the $(p, 2p)$ reaction at 600 MeV carried out by the Caen–CERN–Uppsala collaboration confirm the results obtained at 387 MeV for light and medium-mass nuclei. A very large number of events has been obtained and the error on the location of the peak position is of the order of 1 MeV.

The saturation of the separation energy for $1s$ and $1p$ states, as observed in the $(p, 2p)$ experiment, is in accord with the trend observed for the $1d$ and $2s$ states (Ruh+ 67) and removes the need to invoke rearrangement (Wag 69) or some other process to explain the discrepancy between the behavior of the $2s$ and $1d$ states and the nonsaturating behavior of the $1p$ states as observed in the $(e, ep)$ experiment. The actual magnitudes of the proton separation energies in heavier nuclei are, of course, of considerable relevance to Hartree–Fock and similar calculations. It appears that single-particle energies in agreement with the most recent $(p, 2p)$ data can be obtained within the framework of Hartree–Fock theory by allowing depletion of the normally occupied states, and that the average depletion is consistent with the values calculated from realistic nucleon–nucleon interactions (Bec 70).

---

[†] It is important to note that the experimental uncertainties on these data are the uncertainties in the location of the center of gravity for each hole state, and that these are of the order of 4 to 14 MeV. In addition, the observed widths of the states are greater than the experimental resolution.

## 3.5. Neutron States

Neutron states are studied through the $(p, d)$ and $(d, t)$ reactions and these reactions yield information mainly about the neutrons in outer shells. There is at present no detailed and reliable information about the wave functions and separation energies of neutrons in inner shells, as there is for protons, although there are a number of predictions based on calculations with single-particle potentials (Ros 68, BG 69).

In light nuclei with $N = Z$ it is reasonable to use the same single-particle potentials for protons and neutrons and adjust the depth is each case to give the correct asymptotic behavior. This approach has been used by Towner (Tow 67a, 69) to fit data on the $(p, d)$ reaction at 155 MeV (Bac+ 69a) and has proved fairly successful. The data at 155 MeV are rather insensitive to the procedure adopted to construct the overlap integral, as can be seen from Fig. 2, and the data at 100 MeV (Lee+ 67) show only slightly more structure than is apparent at 155 MeV. This suggests that the energy region of 100–150 MeV is not a particularly good region in which to study the neutron overlap integral, although this lack of sensitivity to the form of the overlap integral should make the extraction of the spectroscopic factors more reliable. In this energy region the main uncertainty at present arises from lack of knowledge of the deuteron optical potential.

The $(d, {}^3\text{He})$ and $(d, t)$ reactions offer an attractive method of comparing proton and neutron overlap integrals. Figure 9 shows the cross sections measured at a deuteron energy of 28 MeV (Gai+ 68) and plotted against momentum transfer. The data are reasonably well reproduced by a DWBA calculation and the ratios of the cross sections are roughly proportional to the ratios of the squares of the proton and neutron single-particle wave functions at some radius in the asymptotic region. A study of the $(d, t)$ and $(d, {}^3\text{He})$ reactions on ${}^{208}\text{Pb}$ at 50 MeV (Par+ 69) has shown that these reactions are localized in the extreme surface region and this explains why the ratio of the cross sections is proportional to the squares of the asymptotic tails of the wave functions. It was found the neutron wave functions predicted from Rost's single-particle potential (Ros 68) for ${}^{208}\text{Pb}$ do not give agreement with the $(d, t)$ reaction.

The observation that the single-particle model for ${}^{208}\text{Pb}$ apparently gives a good description of a variety of data involving protons but fails in processes involving neutrons (Bat 70) is very surprising. However, our knowledge of neutron wave functions and neutron density distributions in nuclei is very scanty compared with that for protons and information on the separation energies of low-lying neutron states is quite negligible. A detailed

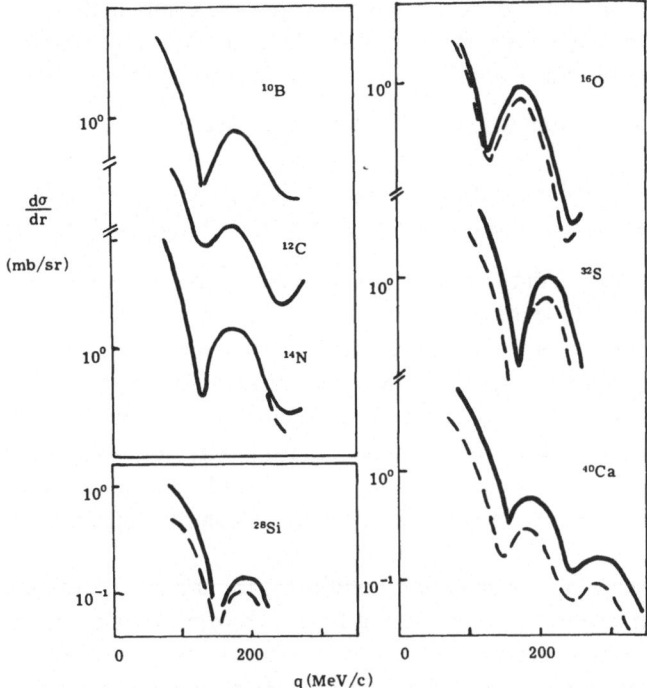

Fig. 9. A comparison of the $(d, {}^{3}\text{He})$ reaction (full line) and $(d, t)$ reaction (dashed line) at 28 MeV. The cross sections for the $(d, t)$ reaction have been multiplied by the ratio of the momenta of the outgoing $^{3}\text{He}$ and triton. [From M. Gaillard et al., Nucl. Phys. A119:161 (1968). Reproduced by permission of the authors and the North-Holland Publishing Company.]

comparison of $(d, t)$ and $(d, {}^{3}\text{He})$ reactions is feasible and would be very valuable. The best source of information about inner shells is undoubtedly the single-nucleon knockout reaction and coincidence measurements on the $(p, pn)$ reaction are now urgently required, even if these measurements are feasible only with rather poor energy resolution.

## 4. CLUSTER KNOCKOUT AND RELATED REACTIONS

### 4.1. Comparison of Knockout and Absorption Processes

In this section we consider those direct reactions which lead to the ejection of a group of two or more nucleons from the nucleus. We exclude from the discussion processes such as evaporation and spallation, and also knockout reactions such as $(p, 3p)$ in which there are more than two fast particles in the final state. The remaining processes are of three types:

(i) Quasi-elastic knockout reactions, such as $(p, pd)$, $(p, p\alpha)$, and $(\alpha, 2\alpha)$.

(ii) Photodisintegration processes, such as $(\gamma, np)$ and $(\gamma, d)$.

(iii) Pion absorption processes, such as $(\pi^+, 2p)$ and $(\pi^-, 2n)$.

As we have seen from the general discussion given in Section 2, the study of these reactions leads to information about the distribution of the hole strength among the states of the residual nucleus, the parentage of the ground state of the target nucleus, and the wave function of the group of nucleons in the target nucleus. It is convenient to refer to the group of nucleons as a "cluster" in the target nucleus; this is not intended to imply any preference for the cluster model over the shell model, although it is certainly to be hoped that these reactions will give results which distinguish between these models.

The energy and momentum conditions for the knockout reactions are very different from those pertaining to the absorption of pions or photons. For example, if a pion is absorbed by two nucleons in such a way that the two nucleons emerge almost back-to-back and the recoil momentum of the residual nucleus is small, the conservation laws for energy and momentum require that the two nucleons have a relative momentum in the nucleus of at least 750 MeV/c. Thus the absorption process focuses attention on the high-momentum components or short-range behavior of the part of the nuclear wave function or, more accurately, the part of the overlap integral which describes the relative motion of the two nucleons, and this is the source of the expectation that such processes could yield information about short-range correlations in nuclei. In the knockout reactions, the presence of two outgoing particles in addition to the residual nucleus allows much greater flexibility in the choice of the momentum range to be studied, although the emphasis has generally been on the lower region of recoil momentum. In the symmetric coplanar $(p, pd)$ reaction at 155 MeV the recoil momentum is in the range $-80 < Q < 80$ MeV/c; the corresponding momentum transfer is in the range 380–490 MeV/c and hence it follows from Eq. (24) that the reaction is sensitive to these momentum components of the relative wave function of the two nucleons. At 30 MeV the momentum transfer in the $(p, pd)$ reaction is in the range 175–215 MeV/c.

The matrix elements for the knockout and absorption processes are formally rather similar and it is illuminating to classify the methods used in their analysis according to philosophy rather than type of reaction. There are essentially two methods and these are described in Sections 4.2 and 4.3.

## 4.2. The Quasi-free or Peripheral Model

The essential feature of this model is the assumption that the internal wave function of the group of nucleons to be ejected resembles very closely that of the corresponding free particle. Thus in the quasi-deuteron model of the $(\gamma, np)$ reaction due to Gottfried (Got 58) it is assumed that the pair of nucleons in the target nucleus resemble very closely a free deuteron, at least at small separation distances. The cross section for the process can then be expressed in terms of the cross section $(d\sigma/d\Omega)_{\gamma d}$ for the photodisintegration of the deuteron and, for plane waves, is given by

$$d^2\sigma \propto F(\mathbf{P}) \left(\frac{d\sigma}{d\Omega}\right)_{\gamma d} d\varepsilon_n \, d\Omega_n \qquad (45)$$

where $\mathbf{P}$ is the momentum of the center of mass of the nucleon pair in the nucleus before ejection, i.e.,

$$\mathbf{P} = \mathbf{p}_1 + \mathbf{p}_2 - \mathbf{k}_\omega \qquad (46)$$

where $\mathbf{k}_\omega$ is the photon momentum and $\mathbf{p}_1$, $\mathbf{p}_2$ are the momenta of the outgoing nucleons. In the analysis of the photodisintegration process it is customary to use closure to sum over all final states and in this case the function $F(\mathbf{P})$ is essentially the square modulus of the Fourier transform of the wave function of the center of mass of the nucleon pair in the nucleus. The equivalent approach to pion absorption was introduced by Brueckner, Serber, and Watson (BSW 51) and developed in later papers (Eck 63, Eri 64, Div 65). It is assumed that the absorption process involves two nucleons which are in a relative $s$-state and that the interaction is short range. The behavior of the relative wave function for these two nucleons is absorbed into a set of phenomenological amplitudes which are determined from the analysis of pion production in nucleon–nucleon collisions which is the exact inverse of the capture process on two free nucleons. The results obtained for various nuclei are reviewed by Koltun (Kol 69).

The arguments for the validity of this quasi-free approach to absorption processes are based on the important role of high relative momentum, or internucleon distances of $<0.5$ fm. It is, therefore, essential to ensure that there are sufficient high-momentum components in the nuclear wave function, and the assumption that the short-range correlations between a pair of nucleons in the nucleus are the same as those between two free nucleons in the same spin and $I$-spin state seems a very reasonable approximation.[†] When these high momenta are less important, for example in photodisintegration at low energies, the quasi-deuteron model seems to break down (Fuj 62).

The cross section for a knockout reaction can be derived using the same basic assumption. We consider the $(p, pd)$ reaction leading to a definite final state and take the momenta of the incoming proton, outgoing proton, and outgoing deuteron to be $\mathbf{p_0}, \mathbf{p_1}, \mathbf{p_2}$ respectively, so that the recoil momentum of the residual nucleus is

$$\mathbf{Q} = \mathbf{p_0} - \mathbf{p_1} - \mathbf{p_2} \tag{47}$$

(i.e., $-\mathbf{Q}$ is the momentum of the center of mass of the pair of nucleons or deuteron cluster before collision), and the momentum transfer is

$$\mathbf{q} = \mathbf{p_0} - \mathbf{p_1} \tag{48}$$

The cross section for the $(p, pd)$ reaction is then given by (JE 65, Jac 65a)

$$\frac{d^3\sigma}{d\Omega_1 d\Omega_2 dE} = K \, | \, G(Q) \, |^2 \left( \frac{d\sigma}{d\Omega} \right)_{pd} \tag{49}$$

where $K$ is a kinematic factor and $(d\sigma/d\Omega)_{pd}$ is the cross section for free proton–deuteron scattering evaluated at a scattering angle in the proton–deuteron center-of-mass system of

$$\bar{\theta} = (\mathbf{p_0} + \tfrac{1}{2}\mathbf{Q}) \cdot (\mathbf{p_1} - \tfrac{1}{2}\mathbf{p_2}) / | \, \mathbf{p_0} + \tfrac{1}{2}\mathbf{Q} \, | \, | \, \mathbf{p_1} - \tfrac{1}{2}\mathbf{p_2} \, | \tag{50}$$

As in the case of the $(p, 2p)$ reaction, the relative momentum before collision is not the same as the relative momentum after collision. Hence the collision in the nucleus is off the energy shell for the free collision and there is again an ambiguity in the choice of the relative energy at which $(d\sigma/d\Omega)_{pd}$ is to be evaluated.

The structure of the function $G(Q)$ is most easily understood if we examine the plane wave expression

$$G(Q) = \iint e^{i\mathbf{Q} \cdot \mathbf{R}} \phi_a^{k*}(\mathbf{r}) \Psi_{fi}(\mathbf{r}, \mathbf{R}) \, d\mathbf{r} \, d\mathbf{R} \tag{51}$$

---

[†] It should be stressed that the assumption of the quasi-free model is not that there are no correlations but rather that the two-nucleon correlations in the nucleus are the same as in the free system.

where $\phi_d{}^k$ is the wave function of the free deuteron and $\Psi_{f_i}$ is the overlap integral defined in Eq. (1). Now using Eq. (3), we have

$$G(Q) = \sum_{\alpha\beta} C^{\alpha\beta f} \int \phi_d^{k*}(\mathbf{r})\phi_d{}^\alpha(\mathbf{r}) \, d\mathbf{r} \int e^{i\mathbf{Q}\cdot\mathbf{R}} \psi_{dB}^\beta(\mathbf{R}) \, d\mathbf{R} \qquad (52a)$$

$$= \sum_{\alpha\beta} C^{\alpha\beta f} P(\alpha, k) G^\beta(Q) \qquad (52b)$$

where $P(\alpha, k)$ is defined by Eq. (19a) and $G(Q)$ by Eq. (23) and, if complete overlap of the wave functions of the free deuteron and deuteron cluster is assumed so that $P(\alpha, k)$ is given by Eq. (19b), Eq. (52) reduces to

$$G(Q) = \sum_\beta C^{k\beta f} G^\beta(Q) \qquad (53)$$

Hence in the quasi-free model the dependence of the cross section on the recoil momentum $Q$ gives information about the momentum components of the reduced overlap integral defined in Eq. (20), i.e., on the motion of the center of mass of the deuteron cluster in the target nucleus. The factorization obtained in Eq. (52a) is valid in PWIA as long as exchange between the residual nucleus and the deuteron cluster is neglected, but to obtain the factorization in DWIA it is necessary to assume that the two-body interaction is of sufficiently short range that the effect of distortion does not change significantly over this range. With this assumption, distortion may be included by replacing the factor $e^{i\mathbf{Q}\cdot\mathbf{R}}$ by products of distorted waves. Only the kinetic-energy approximation is useful for the description of the $(p, pd)$ and $(p, p\alpha)$ reactions but the di-proton model could be applied to the $(\alpha, 2\alpha)$ reaction.

Because of the different momenta involved, the arguments for using the quasi-free model for knockout reactions must be rather different from those used to justify it for absorption processes. It certainly provides the simplest intuitive description of knockout reactions, and represents an extreme form of impulse approximation which might be expected to be valid in cases of small separation energy and a high degree of clustering. A variety of experiments to test the validity of the quasi-free model have been carried out. The most common procedure is to use Eq. (49) to derive the experimental distorted momentum distribution $|G(Q)|^2$ from the measured cross section and to show that this experimental distribution has the behavior expected from some general and reasonable assumptions about the reduced overlap integral. This procedure has produced acceptable results for the $^6\text{Li}(p, pd)$, $^6\text{Li}(p, p\alpha)$, and $^9\text{Be}(p, p\alpha)$ reactions at 155 MeV (Ruh+

63), the $^9Be(p, p\alpha)$ reaction at 57 MeV (Roo+ 68), the $^9Be(p, p\alpha)$ and $^{16}O(p, p\alpha)$ reactions at 160 MeV (Kan 68), the $^6Li(\alpha, 2\alpha)$, $^7Li(\alpha, 2\alpha)$, and $^9Be(\alpha, 2\alpha)$ reactions at 55 MeV (Piz+ 69), and the $^6Li(\alpha, \alpha d)$ reaction at 24 MeV (Bäh+ 69). The results obtained in this way for the $^6Li(p, pd)$ reaction are shown in Fig. 3. It can be seen from this figure that the general trend of the experimental results and of the theoretical curves are in agreement but the detailed fit is not good, and this immediately raises the question of whether the discrepancy is due to an inadequacy of the overlap integral or of the quasi-free model. A more satisfactory test of the quasi-free model has been devised by the Maryland group (Roo+ 69) who have examined the $^6Li(p, p\alpha)$ reaction at an incident energy of 61.5 MeV and

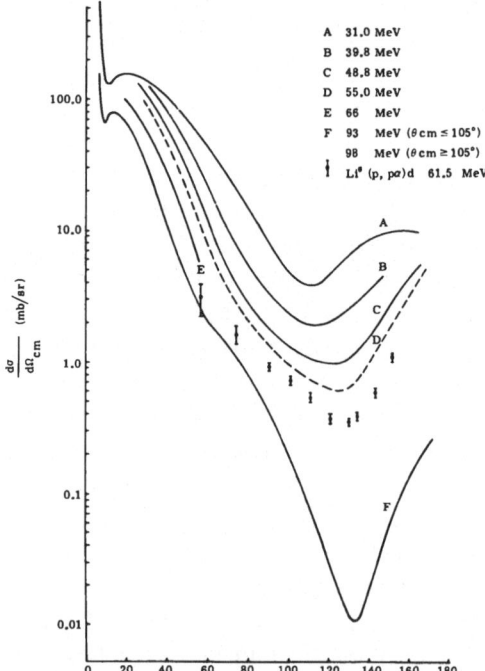

Fig. 10. The cross section at $Q = 0$ for the $^6Li(p, p\alpha)$ reaction at 61.5 MeV divided by the kinematic factor plotted against scattering angle in the proton–alpha center-of-mass system, compared with the cross section for free proton–alpha scattering at various energies. [From P. G. Roos, H. Kim, M. Jain, and H. M. Holmgren, *Phys. Rev. Letters* **22**:241 (1969). Reproduced by permission of the authors and the editor of *Physical Review Letters.*]

TABLE II. Separation Energies for Breakup in Light Nuclei

| Breakup process | Separation energy (MeV) |
| --- | --- |
| $^{16}O \rightarrow {}^{12}C + \alpha$ | 7.16 |
| $^{12}C \rightarrow {}^{8}Be + \alpha$ | 7.37 |
| $^{9}Be \rightarrow {}^{5}He + \alpha$ | 2.53 |
| $^{7}Li \rightarrow {}^{5}He + d$ | 9.68 |
| $^{7}Li \rightarrow t + \alpha$ | 2.47 |
| $^{6}Li \rightarrow d + \alpha$ | 1.47 |
| $^{6}Li \rightarrow {}^{3}He + t$ | 15.79 |

various scattering angles chosen so that the recoil momentum $Q$ was always zero, which in this case corresponds to a maximum in $G(Q)$. They plot the cross section for the knockout reaction divided by the kinematic factor, which from Eq. (49) should be equal to a constant $|G(0)|^2$ times the free cross section for $p$–$\alpha$ scattering if the quasi-free model is valid. It can be seen from Fig. 10 that their results agree very closely with the behavior of the free cross section. A somewhat more detailed examination of the $_9Li(\alpha, 2\alpha)$ reaction has been carried out by the same group (Pug+ 69) with equally encouraging results, although it must be noted that the breakup of $^6Li$ into an $\alpha$-particle and a deuteron is a particularly advantageous case for the quasi-free model owing to the very low separation energy. (Some separation energies for the breakup of light nuclei are given in Table II.)

The derivation of the cross section in the factorized form given by Eq. (49) is equivalent to the representation of the knockout reaction $A(p, pd)$ by the Feynman diagram shown in Fig. 11, in which the left-hand vertex represents the dissociation of the target nucleus $A \rightarrow B + d$ and the right-hand vertex represents the scattering of the incident proton from the deuteron. This is the pole graph for the reaction and the corrections to it are the triangle graphs which represent, in the language of distorted wave theory, contributions from distortion, core excitation, and rescattering.[†] The validity of the simple pole graph for a given reaction may be investigated by applying the Trieman–Yang test (TY 62), which in the case of the knockout or

---

[†] In the calculations reported on the $(p, 2p)$ reaction (KS 69) the contributions from the triangle graph are represented by constants so that it is not yet possible to make a meaningful comparison with conventional distorted-wave calculations. A comparison of the contributions to the $(\alpha, 2\alpha)$ reaction from pole and triangle graphs has recently been made (ML 70) using dispersion theory.

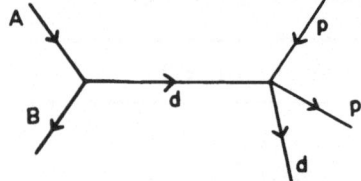

Fig. 11. Pole graph for the reaction $A(p, pd)B$.

similar reaction requires that the cross section is invariant under rotation of the plane containing the outgoing particles about the direction of the exchanged particle in a coordinate system in which the projectile is at rest (SKA 65, Sha 66). The Trieman–Yang test is a necessary condition for the validity of the pole or peripheral mechanism, and has been successfully applied to the $^6Li(\pi^+, 2p)$ reaction at 76 MeV (Cha+ 67) and to the $^6Li(p, pd)$ reaction at 19 MeV (LPB 69).

Even if the quasi-free model or pole approximation does not appear to be valid for a given reaction it does not necessarily follow that it is impossible to extract information on the spectroscopic factor or the momentum distribution, but such information can only be obtained if a more accurate description of the reaction mechanism is available.

## 4.3. The Microscopic Model

In this model it is assumed that the interaction between the projectile and the cluster is sensitive to the internal structure of the cluster which is different from the free particle, and that the object of the analysis is to study this difference. The interaction is expressed as a sum of two-body interactions between the projectile and the individual nucleons in the cluster, and the matrix element then depends on both the wave function of the center of mass of the cluster and its internal wave function. The majority of calculations on photodisintegration at medium energies and on pion absorption leading to the emission of two nucleons have been carried out by this method (KJ 63, JK 64, Koh 65, Sak 65, Kop 67, EE 70, CS 70, Web 70). These calculations use an overlap integral constructed from oscillator functions and modified by the correlation function defined in Eq. (22) to introduce more high momentum compónents, but since the final state interaction between the two outgoing nucleons and pion rescattering (KJ 63, KR 66, Che 68) also introduce additional high-momentum components, the interpretation of the data in terms of nuclear structure is not straightforward.

In the microscopic description of the knockout reaction the interaction is just the sum of the two-nucleon interactions. The cross section for the $(p, pd)$ reaction in impulse approximation becomes

$$\frac{d^3\sigma}{d\Omega_1 d\Omega_2 dE} = K\,|\,G(Q, q)\,|^2\,|\,t(q)\,|^2 \qquad (54)$$

where $K$ is a kinematic factor, $Q$ is the recoil momentum, and $q$ is the momentum transfer, as before. The function $|\,t(q)\,|^2$ represents the two-nucleon $t$-matrix averaged over the spin and isospin states of the deuteron, and is connected to the cross section for free $p$–$d$ scattering by the relation

$$\left(\frac{d\sigma}{d\Omega}\right)_{pd} \propto |\,F_d(q)t(q)\,|^2 \qquad (55)$$

where $F_d$ is the deuteron form factor defined by Eq. (26). Thus Eq. (54) can be rewritten in the form

$$\frac{d^3\sigma}{d\Omega_1 d\Omega_2 dE} = K'\left|\frac{G(Q, q)}{F_d(q)}\right|^2 \left(\frac{d\sigma}{d\Omega}\right)_{pd} \qquad (56)$$

In the plane wave approximation the function $G(Q, q)$ is given by

$$G(Q, q) = \iint e^{i\mathbf{Q}\cdot\mathbf{R}}\,e^{\frac{1}{2}i\mathbf{q}\cdot\mathbf{r}}\phi_d^{k*}(\mathbf{r})\Psi_{fi}(\mathbf{r}, \mathbf{R})\,d\mathbf{r}\,d\mathbf{R}$$

$$= \sum_{\alpha\beta} C^{\alpha\beta f}G^\beta(Q)F_{if}^{\alpha k}(q) \qquad (57)$$

where we have used Eqs. (1), (3), (23), and (24). As we have already noted, the wave functions $\phi_d^k$ for the free deuteron and $\phi_d^\alpha$ for the deuteron cluster used in most calculations are not the same, and the ratio $F_{if}^{\alpha k}(q)/F_d(q)$ must therefore not be neglected. The results shown in Fig. 5 suggest that this term does not have a drastic effect on the shape of the cross section but it can have a large effect on the magnitude.

The arguments for the use of this model to describe knockout reactions are essentially the same as those for the use of a microscopic description of nuclear reactions in general. In the special case when there is a strong overlap between one of the $\phi^\alpha$ and the wave function of the knocked out particle this model reduces to the quasi-free model. Thus the microscopic model involves an assumption about the reaction mechanism but the quasi-free model involves an additional assumption about the target nucleus. The cross section (56) for the microscopic model does not show a factorization into the contribution from the two vertices in the pole graph, although the plane wave approximation (57) for $G(Q, q)$ allows factorization of the amplitude. In an exact treatment of distortion and finite-range effects it is no longer possible to factor $G(Q, q)$, but there are plausible approximations (Jac 66) which allow factorization and throw these effects into the functions $G^\beta(Q)$. Almost all the calculations in the microscopic model so far reported have used plane waves (Sak 64b, JE 65, Dwi 68, JSB 69).

## 4.4. The Effect of Exchange

Exchange contributions to the cross section can arise as a result of exchange within the target (target exchange) or of exchange between the projectile and the target nucleons (projectile exchange). The contributions from projectile exchange to the $^6$Li$(p, pd)$ have been shown to be quite negligible for incident energies of 100 MeV and above (Dwi 68), and the contributions to the $^6$Li$(p, p\alpha)$ reaction at 61.5 MeV are also negligible (Nag 69). Both of these calculations use PWBA. However, in impulse approximation the use of the free two-body cross section taken from experiment automatically takes account of projectile exchange. In the rest of this discussion we therefore consider only target exchange.

An investigation of the effect of target exchange in knockout reactions from $^6$Li and $^7$Li has been made by Dwight (Dwi 68) using the microscopic model and the PWBA. The lithium isotopes have been most thoroughly studied experimentally (Ruh+ 63, HCP 66, Bac+ 68, Roo+ 69, and Bac+ 69c) and some relevant information is given in Table III. The overlap integrals were constructed from cluster-model wave functions as described in Section 2.3 using a single $\alpha + d$ configuration for $^6$Li and an $\alpha + t$ configuration for $^7$Li. Each cluster function was taken to be Gaussian and the two-nucleon potential was also taken to be of Gaussian form. The procedure adopted was to fix the size of the $\alpha$-cluster to be equal to that of the free $\alpha$-particle or to be equal to the $s$-shell size predicted by fits to the elastic electron scattering data with shell-model wave functions. The other parameters were varied in such a way that the rms radius of the nucleus remained constant at the measured value, so that an increase in clustering implies an increase in the separation of the clusters and a simultaneous decrease in the size of the triton or deuteron cluster. The effect of including target exchange (i.e., exchange between the $\alpha$-particle and deuteron cluster) for the $^6$Li$(p, pd)$ reaction is to decrease the cross section at 50 MeV but to increase it at 100 MeV and above. The effect decreases with increased clustering. The magnitude of the cross section increases with increased clustering, but the effect on the shape of the cross section is small. For the $^7$Li$(p, pd)$ reaction the inclusion of the exchange term does affect the shape

TABLE III. Maximum Cross Sections in $\mu$b$\cdot$sr$^{-2}$ MeV$^{-1}$ for Various Knockout Reactions on $^6$Li and $^7$Li

| Reaction | Maximum cross section at 155 MeV | Maximum cross section at 50–60 MeV | Behavior at $Q = 0$ |
|---|---|---|---|
| $^6$Li$(p, pd)$ | 750 | 1500 | maximum |
| $^6$Li$(p, p^3$He$)$ | 37 | — | maximum |
| $^6$Li$(p, p\alpha)$ | 620 | ~500 | maximum |
| $^7$Li$(p, pd)$ | ~90 | 40 | maximum |
| $^7$Li$(p, pt)^a$ | ~35 | 40 | minimum |

[a] In this case the maximum occurs at $Q = \pm 50$ MeV/c.

of the cross section at the lower energies with the half-width of the angular distribution depending quite strongly on the amount of clustering. It may be noted that this calculation shows that the $\alpha + t$ model for $^7$Li does lead to a maximum in the $(p, pd)$ cross section at $Q = 0$ in agreement with the experimental data; this corrects an error in earlier work (JE 65) on this reaction.

The most startling result obtained in these calculations occurs for the $^7$Li$(p, pt)$ reaction. At an incident energy of 50 MeV the predicted cross section has a minimum at $Q = 0$ and maxima at $Q = \pm 60$ MeV/c but at higher energies the effect of the exchange terms is to convert the minimum at $Q = 0$ into a maximum. Now the experimental data for this reaction at 50 MeV (HCP 66) show a minimum at $Q = 0$ and maxima at $Q = \pm 50$ MeV/c, and the very recent data at 155 MeV show almost identical behavior (Rad 69). A study of the energy dependence of this reaction between 50 and 150 MeV would be reassuring but the present evidence suggests that the calculations of Dwight overestimate the effect of target exchange. An explanation of this may possibly be found in the work of Jain, Sarma, and Banerjee who have studied the $^6$Li$(p, pd)$ reaction using the microscopic model (JSB 69) and the quasi-free model (JSB 70). Their results show that when the radial part of the reduced overlap integral is modified to have the correct asymptotic behavior the magnitudes of the exchange terms are very much reduced. The physical explanation of this is presumably that in these reactions with very low separation energies the clusters are fairly well separated in the target nucleus and the effect of exchange between them is therefore rather small.

## 4.5. Spectroscopic Studies

Most of the early analyses of data on the knockout reaction were concerned with the determination of the angular momentum of the cluster relative to the residual nucleus and with attempts to reproduce the observed momentum distribution with rather simple cluster-model or shell-model expressions for the overlap integral. The spectroscopic factors and sum rules for knockout of various clusters from light nuclei have now been evaluated in the framework of the oscillator shell model (NS 65, Ber+ 65) and comparison with the data on knockout of $\alpha$-particles from $^{12}$C (JP 63, IHG 63) using the quasi-free model shows that this approach can predict a sufficiently large value for the effective number of clusters[†] in light nuclei provided that antisymmetrization is included. A similar theoretical approach (BBR 65) has provided predictions for the relative reduced widths for knockout of various clusters from light nuclei and of the excitation spectrum for these reactions. Few of these predictions have yet been tested.

The experiments on the $^{9}$Be$(p, p\alpha)$ reaction at 57 MeV (Roo+ 68) and at 160 MeV (Kan 68) yield very similar results for the momentum distribution, and the effective number of $\alpha$-particles in $^{9}$Be deduced using the quasi-free model are also in good agreement. The experimental data on the $^{12}$C$(p, p\alpha)$ reaction (JP 63, Kan 68, Eps 69, GK 70) are not in agreement but there is strong evidence for important contributions from sequential decays which obscure the knockout events. Studies of the $(\alpha, 2\alpha)$ and $(p, p\alpha)$ reactions on $^{6}$Li and $^{7}$Li at 60 MeV (Jai 69b, Jai+ 70) indicate that the quasi-free model works well for the $(p, p\alpha)$ reaction but that there are considerable uncertainties in the $(\alpha, 2\alpha)$ reaction owing to off-energy-shell effects. The work on the $^{6}$Li$(\alpha, 2\alpha)$ reaction has been extended by Watson (Wat 69) with particular reference to off-shell effects and the cluster model for $^{6}$Li. The $(\alpha, 2\alpha)$ reaction on light nuclei has also been studied in the low-energy region of 25–28 MeV and analyses (HS 65, Kud 65, BM 68) show that the shape and magnitude of the cross section is strongly affected by the final state interaction of the two outgoing $\alpha$-particles, the asymptotic behavior of the overlap integral, and most of all by the off-energy-shell effects. The $^{12}$C$(\alpha, 2\alpha)$ reaction has been studied at 90 MeV (Jac+ 70) and the excitation energy spectrum of the residual $^{8}$Be nucleus can be reproduced using the quasi-free model and cluster model wave functions for $^{12}$C. The $^{28}$Si$(\alpha, 2\alpha)$ reaction has been studied at 104 MeV (PEV 69). The population of the ground state and excited states of $^{24}$Mg and the values of the

---

[†] The term "effective number of clusters" is used in several papers (NS 65, Ber+ 65, and BBR 65) to represent the number given by the theoretical sum rule.

effective number of clusters for each transition are interpreted in terms of the shell model while the angular distributions are fitted using the quasi-free model and a cluster model for the momentum distribution.

There is as yet very little experimental information on the knockout of clusters from heavy nuclei although it is clear from the work of Igo, Hansen, and Gooding (IHG 63) that such experiments are feasible. A study of the knockout reaction has been made (Jac 65b) in the quasi-free model and some predictions for the probability for emission of $\alpha$-particles have been given (GLT 62, MSN 64).

Many of the studies of the $(\pi, NN)$ reaction have been concerned with high-momentum components of the relative motion of the two nucleons in the nucleus. However, spectroscopic studies of the momentum distribution of the center of mass of the two nucleons (Pel 68) and of the energy spectrum (Kop 66, Fav+ 67, GZ 69) show that the shell model gives a reasonably good description of the excitation of low-lying two-hole states in light nuclei.

## 4.6. Comparison with Pickup Reactions

In order to make a comparison with the description of pickup reactions we consider the $(p, t)$ reaction and take the coordinates $\mathbf{r}$, $\mathbf{R}$, $\boldsymbol{\rho}$ to specify the separation between the two neutrons, between their center of mass and the residual nucleus, and between the proton and the center of mass of the two neutrons. In the zero-range DWBA a $\delta$-function $\delta(\boldsymbol{\rho})$ is introduced and this corresponds to the use of a zero-range interaction between the incident proton and the center of mass of the two picked up neutrons. With a suitable choice for the wave function of the free triton the integration over the $\mathbf{r}$-dependent parts of the triton wave function and the overlap integral can be carried out to leave an integration over the coordinate $\mathbf{R}$ involving the distorted waves and the part of the overlap integral which describes the motion of the center of mass of the two neutrons. Thus this treatment is equivalent to a distorted wave treatment of a cluster knockout reaction in the quasi-free model. In the point-triton approximation (RM 64) the whole triton wave function is taken to be of zero range through the use of a product of $\delta$-functions, $\delta(\mathbf{r})\,\delta(\boldsymbol{\rho})$. The difference in the two approximations appears through the difference in the magnitude of each pair of integrals over $\mathbf{r}$. In either case, the overlap integral is required as a function of $\mathbf{r}$ and $\mathbf{R}$ and is constructed from the single-particle states $(n_1 l_1 j_1)(n_2 l_2 j_2)$ for the two neutrons in the manner described in Section 2.3.

The quantum numbers of the transferred pair of neutrons are $L$, $S$, $J$, and $T$, where $\mathbf{L} = \mathbf{l}_1 + \mathbf{l}_2$, etc. It is usually assumed that the two neutrons are in a relative $s$-state, and the possible values of $LSJT$ are further restricted by the requirement that the wave function of the transferred pair must be antisymmetric. This leads to the allowed values of $S$ and $T$ given in Table IV. The cross section for the two-nucleon pickup reaction can be written in the form (Bay 64, TH 69)

$$\frac{d\sigma}{d\Omega} \propto \sum_{LSJTA} f(S, T) \left| \sum_{l_1 l_2} S^{\frac{1}{2}}(l_1 l_2; LSJT) B_L^A(\theta) \right|^2 \qquad (58)$$

where the coefficient $f(S, T)$ depends on the particular reaction, $S(l_1 l_2; LSJT)$ is the spectroscopic factor, and $B_L^A$ depends on the reaction mechanism, the incident energy, and the $Q$-value. In contrast to the cross sections for single-nucleon pickup or knockout reactions and for cluster knockout in the quasi-free model there is no longer a factorization into a term containing the nuclear structure information and a term depending on the reaction mechanism. This greatly increases the complexity of the analysis and makes it impossible to determine the individual spectroscopic factors directly from experiment but it also implies that two-nucleon stripping is much more sensitive to the various components of the wave function than is single-nucleon stripping and can, in principle, be used to test the relative phases of these components. Two-nucleon fractional parentage coefficients for the $1p$ shell have been tabulated by Cohen and Kurath (CK 70).

From the selection rules given in Table IV it can be seen that the $(p, t)$ and $(p, {}^3He)$ reactions allow excitation of states with $T_B = |M_{TA}| + 1$, and this property has been exploited to excite analogue states. For example, $T = \frac{3}{2}$ states in ${}^7Li$, ${}^7B$, ${}^{11}C$, and ${}^{11}B$ have been observed using the $(p, t)$ and $(p, {}^3He)$ reactions on ${}^9Be$ and ${}^{13}C$ (DCP 66, Cos+ 68) and $T = 2$ states have been observed in $2s1d$-shell nuclei using the $(p, t)$ reaction (CPG 64). It can also be seen that the $(p, {}^3He)$ reaction proceeds through both the $S = 0$ and $S = 1$ transfer whereas the $(p, t)$ reaction proceeds

TABLE IV. Quantum Numbers of the Transferred Pair in a Two-Nucleon Transfer Reaction

| Reaction | $S$ | $T$ |
|---|---|---|
| $(p, t)$ | 0 | 1 |
| $(p, {}^3He)$, $(n, t)$ | 0 | 1 |
|  | 1 | 0 |
| $(n, {}^3He)$ | 0 | 1 |
| $(d, \alpha)$ | 1 | 0 |

through $S = 0$ only. This leads to the possibility of $S$-forbidden transitions in the $(p, t)$ reaction (CDP 66). The properties make the two-nucleon pickup reaction a valuable spectroscopic tool and make it possible to deduce some nuclear structure information almost directly from the selection rules observed experimentally.

Detailed interpretation of the data for two-nucleon pickup on medium and heavy nuclei has proceeded in terms of the shell model and the pairing model (Gle 63, Bay 64, Boh 68, Nat 68, BHR 70).

## 5. OTHER REACTIONS

There are a number of other reactions which can give, or may give, information similar to that obtained from knockout and pickup reactions. For completeness we list these here with brief comments about the present state of theory and experiment.

### 5.1. Pion Reactions

Pion absorption on a single nucleon can occur only if the nucleon has momentum in the nucleus in excess of $\sim$500 MeV/c. Thus this process might be used to study the high-momentum components of single-nucleon states. Measurements on the $^{12}C(\pi^+, p)$ reaction have been made (WBG 68) and have been interpreted in terms of absorption on a single neutron. The observed high momentum components are substantially in excess of those predicted using oscillator functions and plane waves (LE 66) but preliminary distorted wave calculations (JE 69) indicate that a large proportion of these components are introduced by the distortion of the incoming pion. The $(p, \pi^+)$ reaction also involves large momentum transfers, and studies of this reaction on $^{12}C$, $^{13}C$, and $^{14}N$ (Dom+ 70) indicate that the experimental data reveal high-momentum components in excess of these predicted by using plane wave and oscillator functions.

Single-nucleon knockout reactions $(\pi, \pi N)$ have been studied at various pion energies (RM 64, Chi+ 68, 69) but there is at present considerable controversy over the extent to which this process proceeds by quasi-elastic knockout (Dal 68, Bre+ 69, KS 69b, Kol 69).

### 5.2. Kaon Capture

Kaon capture in heavy nuclei has been studied extensively in order to obtain information on the neutron distribution in heavy nuclei. In addition, there is a small amount of information on $K^-$ capture in light nuclei leading to the emission of a $\Sigma\pi$ pair (Bru+ 63, Dav+ 67a, Dav+ 67b) and assuming that the $K^-$ is captured at rest, measurement of the momenta of the pion and sigma hyperon yield a direct measurement of the (distorted) momentum distribution of the capturing nucleon. The results for this momentum distribution in $^{12}$C (Bru+ 63) can be reproduced provided that the nucleon wave function has the correct asymptotic behavior and the distortion of the outgoing pion and sigma hyperon are taken into account (Ada 63).

## 5.3. Coulomb Disintegration

The disintegration of light nuclei in the Coulomb field of a heavy nucleus has been studied, and the cross section for direct breakup into two fragments depends on the reduced overlap integral (HW 65, ABH 69). Calculations have been made for $^6$Li (HW 65), and for $^7$Li and $^7$Be (HW 69) using cluster-model wave functions of Gaussian form, and have given reasonable agreement with the data.

## 5.4. Inelastic Scattering to the Continuum

Measurements of the spectra for inelastic proton scattering (WR 66) and inelastic electron scattering (BIB 64) at high excitation energies show a broad peak with a maximum at approximately the energy to be expected for free nucleon–nucleon scattering at that angle. This peak is known as the quasi-elastic peak and is interpreted using the impulse approximation so that the width of the peak is determined by the spread of the nucleon momentum components in the nucleus. Simple DWIA calculations based on the single-particle model (Kro 68, de F 69) yield reasonable agreement with the data, and the single-particle momentum components are in general consistent with those deduced from the studies of the $(p, 2p)$ and $(e, ep)$ experiments.

# 6. SUMMARY AND CONCLUSIONS

The early data on single-nucleon knockout did not merit a detailed spectroscopic analysis. However, comparison with the $(d, {}^3\text{He})$ data demonstrates that such a study is necessary if consistent conclusions are to be drawn from knockout and pickup reactions. The theoretical formalism is established and some preliminary spectroscopic studies have been made on the best available data, but experiments with better energy resolution are really required. The existing data have, however, given valuable information on the single-proton overlap integral and momentum components which is roughly consistent with that obtained from other reactions and electron scattering, and have given extremely valuable information on proton separation energies. In contrast, the present knowledge on neutron states is still inadequate and knowledge of neutron separation energies for the lowest single-particle states is almost negligible. Coincidence measurements on the $(p, pn)$ reaction are urgently needed.

Spectroscopic studies of cluster knockout reactions are not yet meaningful. Data with better energy resolution at energies above 100 MeV are required so that the uncertainties due to off-energy-shell effects which beset the interpretation of data in the 20–60 MeV region can be avoided. More tests of the quasi-free or peripheral model are required, especially for reactions with larger separation energies than the popular case of $^6\text{Li} \rightarrow \alpha + d$, and more studies of the energy dependence of these reactions are required to indicate the importance, or otherwise, of exchange contributions.

The formal theory of the overlap integral for single-nucleon reactions is well established and methods of improving on the single-particle model are being developed. The formal theory of the overlap integral for two-nucleon and cluster reactions is now being studied and there is some indication that the description of knockout reactions may prove more straightforward than the description of transfer reactions. The formalism of the cluster model offers the most natural framework for the description of cluster knockout reactions but the cluster-model wave functions so far used have mostly been too crude to give reliable results. More serious is the lack of tabulated fractional-parentage coefficients and spectroscopic factors for two-nucleon and cluster reactions and the variety of notations and phase conventions in those tabulations which are available. Despite these present deficiencies there is no doubt that knockout and related reactions can reveal important features of nuclear structure and can provide a stringent test of our ability to represent them.

## ACKNOWLEDGMENTS

This article is based on a talk given at the Symposium on Pick Up and Quasi-free Knock Out of Particles and Clusters at the University of Maryland in April 1969. I am indebted to the organizers of that symposium for the invitation which stimulated this work. I am also indebted to the many authors who have made information available to me in the form of preprints, theses, and private correspondence.

# REFERENCES

ABH 69    A. Aurdal, J. Bang, and J. M. Hansteen, *Nucl. Phys.* **135**:632 (1969).

Ada 63    R. K. Adair, *Phys. Letters* **6**:86 (1963).

AK 64     D. Amit and A. Katz, *Nucl. Phys.* **58**:388 (1964).

Ama 67    U. Amaldi, in *Proceedings of the International School Of Physics "Enrico Fermi,"* Course No. 36, Academic Press, New York (1967), p. 284.

Ama+ 64   U. Amaldi, G. Campos Venuti, G. Cortellessa, G. Fronterotta, A. Reale, P. Salvadori, and P. Hillman, *Phys. Rev. Letters* **13**:431 (1964).

Ama+ 65a  U. Amaldi, G. Campos Venuti, G. Cortellessa, G. Fronterotta, A. Reale, and P. Salvadori, *Proc. Acad. Linei* **38**:499 (1965).

Ama+ 65b  U. Amaldi, G. Campos Venuti, G. Cortellessa, G. Fronterotta, A. Reale, and P. Salvadori, *Proc. Acad. Linei* **39**:470 (1965).

Ama+ 66a  U. Amaldi, G. Campos Venuti, G. Cortellessa, E. de Sanctis, S. Frullani, R. Lombard, and P. Salvadori, *Phys. Letters* **22**:593 (1966).

Ama+ 66b  U. Amaldi, G. Campos Venuti, G. Cortellessa, E. de Sanctis, S. Frullani, R. Lombard, and P. Salvadori, *Proc. Acad. Linei* **41**:494 (1966).

Ama+ 67   U. Amaldi, G. Campos Venuti, G. Cortellessa, E. de Sanctis, S. Frullani, R. Lombard, and P. Salvadori, *Phys. Letters* **25B**:24 (1967).

Ard+ 67   M. Arditi, H. Doubre, M. Riou, D. Royer, and C. Ruhla, *Nucl. Phys.* **A103**:319 (1967).

Ard+ 69   M. Arditi, L. Bimbot, M. Doubre, N. Frascaria, J. P. Garron, M. Riou, and D. Royer, preprint.

Aus+ 64   N. Austern, R. Drisko, E. Halbert, and G. R. Satchler, *Phys. Rev.* **133**:B3 (1964).

Bac+ 69a  D. Bachelier, M. Bernas, I. Brissaud, C. Detraz, and P. Radvanyi, *Nucl. Phys.* **A126**:60 (1969).

Bac+ 69b  D. Bachelier, M. Bernas, C. Detraz, P. Radvanyi, and M. Roy, *Phys. Letters* **26B**:283 (1968).

Bähr+ 69a K. Bähr, T. Becker, O. M. Bilaniuk, and R. Jahr, *Phys. Rev.* **178**:170 (1969).

Bar 70    M. Baranger, *Nucl. Phys.* **A149**:225 (1970).

Bay 64    B. F. Bayman, in *Proceedings of the International Conference on Nuclear Spectroscopy with Direct Reactions*, (F. E. Throw, ed.), Argonne National Laboratory Report ANL 6878 (1964), p. 335.

BB 62     V. V. Balashov and A. N. Boyarkina, *Nucl. Phys.* **38**:629 (1962).

BBR 65    V. V. Balashev, A. N. Boyarkina, and I. Rotter, *Nucl. Phys.* **59**:417 (1965).

Ber 65    T. Berggren, *Nucl. Phys.* **72**:337 (1965).

Ber 70    T. Berggren, preprint.

Ber+ 65   P. Beregi, N. S. Zelenskaya, V. G. Neudatchin, and Yu F. Smirnov, *Nucl. Phys.* **66**:513 (1965).

BG 64     P. J. A. Buttle, and L. J. B. Goldfarb, *Proc. Phys. Soc.* **83**:701 (1964).

BG 69     C. J. Batty and G. W. Greenlees, *Nucl. Phys.* **A133**:673 (1969).

BH 59     J. S. Blair and E. M. Henley, *Phys. Rev.* **112**:2029 (1959).

BH 68     B. F. Bayman and N. M. Hintz, *Phys. Rev.* **172**:1113 (1968).

BIB 64    G. R. Bishop, D. B. Isabelle, and C. Bétourné, *Nucl. Phys.* **54**:97 (1964).

BJ 63     T. Berggren and G. Jacob, *Nucl. Phys.* **47**:481 (1963).

BK 67     B. F. Bayman and A. Kallio, *Phys. Rev.* **156**:1121 (1967).

BKM 69    V. V. Balashev, N. M. Kabachnik, and V. I. Markov, *Nucl. Phys.* **A129**: 369 (1969).

BM 61     T. Brody and M. Moshinsky, *Tables of Transformation Brackets for Nuclear Shell-Model Calculations*, Universidad Nacional Autonoma de Mexico (1961).

BM 68     V. V. Balashov and D. W. Meboniya, *Nucl. Phys.* **A107**:369 (1968).

Boh 68    A. Bohr, in *Proceedings of the Dubna Conference on Nuclear Structure*, International Atomic Energy Agency, Vienna (1968), p. 179.

Bre+ 69   T. Bressani, C. Charpak, J. Favier, L. Massonet, W. E. Meyerhof, and C. Zupancic, *Nucl. Phys.* **B9**:429 (1969).

Bru+ 63   Bruxelles, Oxford, and University College (London) Collaboration, in *Proceedings of the Sienna International Conference on High-Energy Physics* (1963), p. 199.

BS 65     D. M. Brink and N. Sherman, *Phys. Rev. Letters* **14**:393 (1965).

BS 67     D. M. Brink and N. Sherman, *Nucl. Phys.* **A94**:385 (1967).

BSW 51    K. A. Brueckner, R. Serber, and K. M. Watson, *Phys. Rev.* **84**:258 (1951).

BT 66     T. Berggren and H. Tyren, *Annual Reviews of Nuclear Science* **16**:153 (1966).

CDP 66    J. Cerny, C. Detraz, and R. H. Pehl, *Phys. Rev.* **152**:950 (1966).

Cha+ 67   G. Charpak, J. Favier, L. Massonet, and C. Zupancic, in *Proceedings of the Gatlinburg International Nuclear Physics Conference* (R. L. Becker, ed.) Academic Press, New York (1967), p. 465.

Che 68    Il. T. Cheon, *Nucl. Phys.* **A121**:679 (1968).

Chi+ 68   D. J. Chivers, J. J. Domingo, E. M. Rimmer, R. C. Witcomb, B. W. Allardyce, and N. W. Tanner, *Phys. Letters* **26B**:573 (1968).

Chi+ 69   D. J. Chivers, E. M. Rimmer, B. W. Allardyce, R. C. Witcomb, J. J. Domingo, and N. W. Tanner, *Nucl. Phys.* **A126**:129 (1969).

Cio 68    C. Ciofi degli Atti, *Nucl. Phys.* **A106**:215 (1968).

CK 65     S. Cohen and D. Kurath, *Nucl. Phys.* **73**:1 (1965).

Cos+ 68   S. W. Cosper, R. L. McGrath, J. Cerny, C. C. Maples, G. W. Goth, and D. G. Fleming, *Phys. Rev.* **176**:1113 (1968).

CPG 64    J. Cerny, R. H. Pehl, and G. T. Garvey, *Phys. Letters* **12**:234 (1964).

Dab 58    J. Dabrowski, *Proc. Phys. Soc.* **71**:658 (1958).

Dal 68    O. D. Dalkarov, *Phys. Letters* **26B**:610 (1968).

Dav 67a   D. F. Davis, L. Chinn, J. J. Lord, and R. J. Piserchio, *Nuovo Cimento* **47**:573 (1967).

Dav 67b   D. H. Davis, S. P. Lovell, M. Csejthey-Barth, J. Sacton, and G. Schorocheff, *Nucl. Phys.* **B1**:434 (1967).

de F 69   T. de Forest, *Nucl. Phys.* **A132**:305 (1969).

Dic+ 65   J. Dickens, R. Drisko, E. Halbert, and G. R. Satchler, *Phys. Letters* **15**: 337 (1965).

Die 62    K. Dietrich, *Phys. Letters* **2**:139 (1962).

Div 65    P. P. Divakaran, *Phys. Rev.* **139**:B387 (1965).

DG 66     L. R. Dodd and K. R. Greider, *Phys. Rev.* **146**:675 (1966).

DM 68     P. A. Deutchman and I. E. McCarthy, *Nucl. Phys.* **A112**:399 (1968).

DR 66     R. M. Drisko, and F. Rybicki, *Phys. Rev. Letters* **16**:275 (1966).

Dwi 68    J. R. Dwight, Ph. D. Thesis, University of London (1968), unpublished.

Eck 63    S. G. Eckstein, *Phys. Rev.* **129**:413 (1963).

EG 69     C. D. Epp and T. A. Griffy, *Bull. Am. Phys. Soc.* **14**:572 (1969).

EG 70        C. D. Epp and T. A. Griffy, *Phys. Rev.* **C1**:1633 (1970).
EJ 67        L. R. B. Elton and D. F. Jackson, *Phys. Rev.* **155**:1070 (1967).
Elt 67       L. R. B. Elton, *Phys. Letters* **25B**:60 (1967).
Elt 68a      L. R. B. Elton, in *Proceedings of the International Conference on Electromagnetic Sizes of Nuclei* (D. J. Brown, M. K. Sundaresan, and R. D. Barton, eds.) Carleton University, Ottawa (1968), p. 267.
Elt 68b      L. R. B. Elton, *Phys. Letters* **26B**:689 (1968).
Eps+ 69      M. Epstein, M. D. Holgren, M. Jain, H. G. Pugh, P. G. Roos, N. S. Wall, C. D. Goodman, and C. A. Ludemann, *Phys. Rev.* **178**:1698 (1969).
Eri 64       M. Ericson, *Compt. Rend.* **258**:1471 (1964).
ES 67        L. R. B. Elton and A. Swift, *Nucl. Phys.* **A94**:52 (1967).
ET 69        B. Elbek, and P. O. Tjøm, in *Advances in Nuclear Physics* (M. Baranger, and E. Vogt, eds.) Plenum Press, New York (1969), Vol. 3, p. 259.
EWB 69       L. R. B. Elton, S. J. Webb, and R. C. Barrett, in *Proceedings of the Third International Conference on High-Energy Physics and Nuclear Structure* (S. Davons, ed.) Plenum Press, New York (1969), p. 67.
Fuj 62       S. Fujii, *Nuovo Cimento* **25**:995 (1962).
Gai+ 68      M. Gaillard, R. Bouché, L. Feuvrais, P. Gaillard, A. Guichard, M. Gusakow, J. L. Leonhardt, and J. R. Pizzi, *Nucl. Phys.* **A119**:161 (1968).
Gar+ 62      J. P. Garron, J. C. Jacmart, M. Riou, J. C. Teillac, and K. Strauch, *Nucl. Phys.* **37**:126 (1962).
Gar+ 65      J. Garvey, B. H. Patrick, J. G. Rutherglen, and I. L. Smith, *Nucl. Phys.* **70**:241 (1965).
Gle 63       N. K. Glendenning, in *Annual Reviews of Nuclear Science* **13**:191 (1963).
Gle 65       N. K. Glendenning, *Phys. Rev.* **137**:B102 (1965).
GLT 62       H. Gauvin, M. Lefort, and X. Tarrago, *Nucl. Phys.* **39**:447 (1962).
GOS 66       T. A. Griffy, R. J. Oakes, and H. M. Schwarz, *Nucl. Phys.* **86**:313 (1966).
Got 58       K. Gottfried, *Nucl. Phys.* **5**:557 (1958).
GP 63        A. Goswami and M. K. Pal, *Nucl. Phys.* **44**:294 (1963).
Gro 69       D. H. E. Gross, *Phys. Letters* **30B**:16 (1969).
GT 69        C. Glashausser and J. Thirion, in *Advances in Nuclear Physics* (M. Baranger, and E. Vogt, eds.) Plenum Press, New York (1969), Vol. 2, p. 79.
GWS 67       B. Gottschalk, K. H. Wang, and K. Strauch, *Nucl. Phys.* **A90**:83 (1967).
HCP 66       D. L. Hendrie, M. Chabre, and H. G. Pugh, UCRL Report 16580 (1966) unpublished.
HH 66        R. Huby and J. L. Hutton, *Phys. Letters* **19**:660 (1966).
HM 65        R. Huby and J. R. Mines, *Rev. Mod. Phys.* **37**:406 (1965).
HNB 67       J. C. Hiebert, E. Newman, and R. H. Bassel, *Phys. Rev.* **154**:898 (1967).
HS 65        J. Huira and I. Shimodaya, *Prog. Tho. Phys.* **34**:861 (1965).
HW 65        J. M. Hansteen and H. Wittern *Phys. Rev.* **137**:B524 (1965).
HW 69        J. M. Hansteen and H. Wittern, *Z. Phys.* (1969), to be published.
IGH 63       G. Igo, T. J. Gooding, and L. Hansen, *Phys. Rev.* **131**:337 (1963).
IKY 69       M. Igarishi, M. Kawai, and K. Yazaki, *Progr. Theor. Phys.* (*Japan*) **42**:254 (1969).
Jac 65a      D. F. Jackson, *Rev. Mod. Phys.* **37**:393 (1965).
Jac 65b      D. F. Jackson, *Nuovo Cimento* **40**:109 (1965).
Jac 65c      D. F. Jackson, *Nuovo Cimento* **41**:86 (1965).
Jac 66       D. F. Jackson, *Proc. Phys. Soc.* **88**:101 (1966).

Jac 67a      D. F. Jackson, *Nuovo Cimento* **51B**:49 (1967).

Jac 67b      D. F. Jackson, *Nucl. Phys.* **A90**:209 (1967).

Jac 67c      D. F. Jackson, *Phys. Rev.* **155**:1065 (1967).

Jac 68      D. F. Jackson, *Advances in Physics* **17**:481 (1968).

Jac 69      D. F. Jackson, *Nucl. Phys.* **A123**:273 (1969).

Jai 68a      B. K. Jain, *Nucl. Phys.* **A116**:256 (1968).

Jai 68b      B. K. Jain, Ph. D. Thesis, University of Surrey (1968), unpublished.

Jai 69a      B. K. Jain, *Nucl. Phys.* **A129**:145 (1969).

Jai 69b      M. Jain, Ph. D. Thesis, University of Maryland (1969), unpublished.

Jam+ 69a      A. N. James, P. T. Andrews, P. Butler, N. Cohen, and B. G. Lowe, *Nucl. Phys.* **A133**:89 (1969).

Jam+ 69b      A. N. James, P. T. Andrews, P. Kirkby, and B. G. Lowe, *Nucl. Phys.* **A138**: 145 (1969).

JB 65      D. F. Jackson and T. Berggren, *Nucl. Phys.* **62**:353 (1965).

JE 65      D. F. Jackson and L. R. B. Elton, *Proc. Phys. Soc.* **85**:659 (1965).

JE 69      W. B. Jones, and J. M. Eisenberg, *Nucl. Phys.* **87**:331 (1966).

JG 69      R. L. Jaffe and W. J. Gerace, *Nucl. Phys.* **A125**:1 (1969).

JJ 67      B. K. Jain and D. F. Jackson, *Nucl. Phys.* **A99**:113 (1967).

JJ 68      D. F. Jackson and B. K. Jain, *Phys. Letters* **27B**:147 (1968).

JK 64      R. I. Jibuti and T. I. Kopaleishvili, *Nucl. Phys.* **59**:337 (1964).

JP 63      A. N. James and H. G. Pugh, *Nucl. Phys.* **42**:441 (1963).

JS 63      A. Johansson and Y. Sakamoto, *Nucl. Phys.* **42**:625 (1963).

JSB 69      A. K. Jain, N. Sarma, and B. Banerjee, *Nuovo Cimento* **62**:219 (1969).

JSB 70      A. K. Jain, N. Sarma, and B. Banerjee, *Nucl. Phys.* **A142**:330 (1970).

Kan 68      S. Kannenberg, Ph. D. Thesis, Northeastern University (1968), unpublished.

KJ 63      T. I. Kopaleishvili and R. I. Jibuti, *Nucl. Phys.* **36**:56 (1963).

KK 70      P. A. Kazaks and R. D. Koshel, *Phys. Rev.* **C1**:1906 (1970).

Koh 65      T. Kohmura, *Prog. Theo. Phys.* **34**:234 (1965).

Köh 66      H. S. Köhler, *Nucl. Phys.* **88**:529 (1966).

Kol 69      D. S. Koltun, in *Advances in Nuclear Physics* (M. Baranger and E. Vogt, eds.) Plenum Press, New York (1969), Vol. 3, p. 71.

Kop+ 61      T. I. Kopaleishvili, I. Sh. Vashakidze, V. I. Mamasakhlisov, and G. A. Chilashvili, *Nucl. Phys.* **23**:430 (1961).

KR 66      D. S. Koltun, and A. Reitan, *Phys. Rev.* **141**:1413 (1966).

Kro 68      F. R. Kroll, M. Sc. Thesis, University of Maryland (1968), unpublished.

KS 67a      V. M. Kolybasov, and N. Ya. Smorodinskaya, *Yad. Fiz.* **5**:777 (1967).

KS 67b      Y. Kudo, and S. Suekane, *Prog. Theo. Phys.* **38**:520 (1967).

KS 69a      V. M. Kolybasov and N. Ya Smorodinskaya, *Nucl. Phys.* **A136**:165 (1969).

KS 69b      V. M. Kolybasov and N. Ya Smorodinskaya, *Phys. Letters* **30B**:11 (1969).

Kud 65      Y. Kudo, *Prog. Theo. Phys.* **34**:942 (1965).

KY 67      M. Kawai and K. Yazaki, *Prog. Theo. Phys.* **38**:850 (1967).

LE 66      J. Le Tourneaux and J. M. Eisenberg, *Nucl. Phys.* **87**:331 (1966).

Lee+ 67      J. K. P. Lee, S. K. Mark, P. M. Portner, and R. B. Moore, *Nucl. Phys.* **A106**:357 (1967).

Lev 68      F. S. Levin, *Nucl. Phys.* **A115**:449 (1968); *Annals of Phys.* **46**:41 (1968).

LM 64      K. L. Lim and I. E. McCarthy, *Phys. Rev.* **133**:B1006 (1964).

LM 66      K. L. Lim and I. E. McCarthy, *Nucl. Phys.* **88**:433 (1966).

LPB 69      R. L. Leibert, K. H. Purser, and R. L. Burman, in *Proceedings of the International Conference on the Properties of Nuclear States*, (M. Harvey *et al.*, eds.) University of Montreal Press (1969), p. 758.

Mac 64      M. H. Macfarlane, in *Proceedings of the International Conference on Nuclear Spectroscopy with Direct Reactions* (F. E. Throw, ed.), Argonne National Laboratory Report ANL 6878 (1964) 249.

Mai+ 69     G. Mairle, G. Th. Kaschl, H. Link, H. Mackh, U. Schmidt-Rohr, G. J. Wagner, and P. Turek, *Nucl. Phys.* **A134**:180 (1969).

Mar 58      Th. A. J. Maris, *Nucl. Phys.* **9**:577 (1958).

McC 65      I. E. McCarthy, *Rev. Mod. Phys.* **37**:388 (1965).

McC 68      I. E. McCarthy, *Introduction to Nuclear Theory*, Wiley, New York (1968), Chap. 14.

Meb 69      J. V. Meboniya, *Phys. Letters* **30B**:153 (1969).

Mel 69      H. Meldner, *Phys. Rev.* **178**:1815 (1969).

Met 69      N. P. Mett, private communication.

MF 60       M. H. Macfarlane and J. B. French, *Rev. Mod. Phys.* **32**:567 (1960).

MH 62       V. A. Madsen and E. M. Henley, *Nucl. Phys.* **33**:1 (1962).

MHR 58      Th. A. J. Maris, P. Hillman, and H. Tyren, *Nucl. Phys.* **7**:1 (1958).

MSN 64      L. Mailing, Yu F. Smirnov, and V. G. Neudatchin, *Phys. Letters* **11**:49 (1964).

MT 69       I. E. McCarthy and A. W. Thomas, *Nucl. Phys.* **A135**:463 (1969).

Nag 69      K. Nagetani, *Phys. Letters* **29B**:468 (1969).

Nat 68      O. Nathan, in *Proceedings of the Dubna Conference on Nuclear Structure*, International Atomic Energy Agency, Vienna (1968), p. 191.

NH 68       E. Newman and J. C. Hiebert, *Nucl. Phys.* **A110**:366 (1968).

NS 65       V. G. Neudatchin and Yu F. Smirnov, *Atomic Energy Review* **3**:157 (1965).

OPW 65      S. Okai, S. H. Park, and K. Wildermuth, *Z. Phys.* **184**:451 (1965).

PS 65       W. T. Pinkston and G. R. Satchler, *Nucl. Phys.* **72**:641 (1965).

PS 69       A. Prakash and N. S. Austern, *Annals of Physics* **51**:418 (1969).

Par+ 69     W. C. Parkinson, D. L. Hendrie, H. H. Duhm, J. Mahoney, J. Saudinos, and G. R. Satchler, *Phys. Rev.* **178**:1976 (1969).

PPS 68      R. J. Philpott, W. T. Pinkston, and G. R. Satchler, *Nucl. Phys.* **A119**:241 (1968).

Pug+ 65     H. G. Pugh, D. L. Hendrie, M. Chabre, and E. Boschitz, *Phys. Rev. Letters* **14**:434 (1965).

Pug+ 67     H. G. Pugh, D. L. Hendrie, M. Chabre, E. Boshitz, and I. E. McCarthy, *Phys. Rev.* **155**:1054 (1967).

Pug+ 69     H. G. Pugh, J. W. Watson, D. A. Goldberg, P. G. Roos, D. I. Bonbright, and R. A. J. Riddle, *Phys. Rev. Letters* **22**:408 (1969).

Ric+ 69     K. Richie, R. M. Eisberg, M. Makino, and C. Waddell, *Nucl. Phys.* **A131**:501 (1969).

Rio 65      M. Riou, *Rev. Mod. Phys.* **37**:375 (1965).

RM 64a      J. R. Rook and D. Mitra, *Nucl. Phys.* **51**:96 (1964).

RM 64b      P. L. Reeder and S. S. Markowitz, *Phys. Rev.* **133**:B639 (1964).

Roo+ 68     P. G. Roos, H. G. Pugh, M. Jain, H. D. Holmgren, M. Epstein, and C. A. Ludemann, *Phys. Rev.* **176**:1246 (1968).

Roo+ 69     P. G. Roos, H. Kim, M. Jain, and H. D. Holmgren, *Phys. Rev. Letters* **22**:242 (1969).

Ros 67     E. Rost, *Phys. Rev.* **154**:994 (1967).
Ros 68     E. Rost, *Phys. Letters* **26B**:184 (1968).
Rot 69     I. Rotter, *Nucl. Phys.* **A135**:378 (1969).
Ruh+ 62    C. Ruhla, M. Riou, J. P. Garron, J. C. Jacmart, and L. Massonet, *Phys. Letters* **2**:44 (1962).
Ruh+ 63    C. Ruhla, M. Riou, M. Gusakow, J. C. Jacmart, M. Liu, and L. Valentin, *Phys. Letters* **6**:282 (1963).
Ruh+ 67    C. Ruhla, M. Arditi, H. Doubre, J. C. Jacmart, M. Liu, R. A. Ricci, M. Riou, and J. C. Roynett, *Nucl. Phys.* **A95**:526 (1967).
Sak 64a    Y. Sakamoto, *Phys. Rev.* **134**:B1211 (1964).
Sha 66     I. S. Shapiro, in *Proceedings of the International School of Physics "Enrico Fermi"*, Academic Press, New York (1966), p. 210.
SKA 65     I. S. Shapiro, V. M. Kolybasov, and G. R. Augst, *Nucl. Phys.* **61**:353 (1965).
Smi 61     Yu. F. Smirnov, *Nucl. Phys.* **27**:177 (1961).
Smi 62     Yu. F. Smirnov, *Nucl. Phys.* **39**:346 (1962).
ST 66      R. Stock and T. Tamura, *Phys. Letters* **22**:304 (1966).
STW 63     E. W. Schmid, Y. C. Tang, and K. Wildermuth, *Phys. Letters* **7**:263 (1963).
Sug 68     K. Sugawara, *Nucl. Phys.* **A110**:305 (1968).
Tal 52     I. Talmi, *Helveta Phys. Acta* **25**:185 (1952).
TH 69      I. S. Towner and J. C. Hardy, *Advances in Physics* **18**: 401 (1969).
Tow 67a    I. S. Towner, *Nucl. Phys.* **A93**:145 (1967).
Tow 67b    I. S. Towner, private communication.
Tow 69     I. S. Towner, *Nucl. Phys.* **A126**:97 (1969).
TP 60      T. A. Tombrello and G. C. Phillips, *Nucl. Phys.* **19**:555 (1960).
TSR 63     G. Tibell, O. Sundberg, and P. V. Renberg, *Arkiv. Fysik* **25**: 433 (1963).
TWP 61     Y. C. Tang, K. Wildermuth, and L. D. Pearlstein, *Phys. Rev.* **123**:548 (1961).
TWP 62     Y. C. Tang, K. Wildermuth, and L. D. Pearlstein, *Nucl. Phys.* **32**:504 (1962).
TY 62      S. B. Trieman and C. N. Yang, *Phys. Rev. Letters* **8**:140 (1962).
Tyr+ 66    H. Tyren, S. Kullander, O. Sundberg, R. Ramachandrian, P. Isacsson, and T. Berggren, *Nucl. Phys.* **79**:321 (1966).
Wag 69     G. J. Wagner, Princeton University Report PUC-937-354 (1969).
Wat 68     A. Watt, *Phys. Letters* **27B**:190 (1968).
WBG 68     T. R. Witten, M. Blecher, and K. Gotow, *Phys. Rev.* **174**:1166 (1968).
WN 68      B. H. Wildenthal and E. Newman, *Phys. Rev.* **167**:1027 (1968).
WM 66      K. Wildermuth and W. McClure, *Cluster Representation of Nuclei*, Springer-Verlag, Berlin (1966).
Wil 62     K. Wildermuth, *Nucl. Phys.* **31**:478 (1962).
WK 59      K. Wildermuth and Th. Kannellopoulos, CERN Report 59-23 (1959).
WR 66      N. S. Wall and P. G. Roos, *Phys. Rev.* **150**:811 (1966).
YH 65      T. Yuasa and E. Hourany, *Phys. Letters* **18**:146 (1965).
YH 67      T. Yuasa and E. Hourany, *Nucl. Phys.* **A103**:577 (1967).
Yu 66      D. U. L. Yu, *Annals of Physics* **38**:392 (1966).
ZS 66      N. S. Zelenskaya and Yu. F. Smirnov, *Bull. Acad. Sci. (USSR) (Phys. Ser.)* **30**:291 (1966).

# SUPPLEMENTARY REFERENCES

Bac+ 69c    D. Bachelier, M. K. Brussel, P. Radvanyi, M. Roy, M. Sowinski, M. Bernas,
            and I. Brissaud, *Proceedings of the Third International Conference on High
            Energy Physics and Nuclear Structure* (S. Devons, ed.), Plenum Press, New
            York (1969), p. 318.

Bat 70      C. J. Batty, *Phys. Letters* **31B**:496 (1970).

Bec 70      R. L. Becker, *Phys. Letters* **24**:400 (1970); *Phys. Letters* **32B**:263 (1970).

CK 70       S. Cohen and D. Kurath, *Nucl. Phys.* **A141**:145 (1970).

CS 70       D. L. Cheshire and S. E. Sobottka, *Nucl. Phys.* **A146**:129 (1970).

Dom+ 70     J. J. Domingo, B. W. Allardyce, C. H. Q. Ingram, S. Rohlin, N. W. Tanner,
            J. Rohlin, E. M. Rimmer, G. Jones, and J. P. Girardeau-Montaut, *Phys.
            Letters* **32B**:309 (1970).

EE 70       W. Elsaesser and J. M. Eisenberg, *Nucl. Phys.* **A144**:441 (1970).

Fav+ 67     J. Favier, T. Bressani, G. Charpak, L. Massonet, W. E. Moyerhof, and
            C. Zupancic, *Phys. Letters* **25B**:409 (1967).

GK 70       B. Gottschalk and S. Kannenberg, *Phys. Rev.* **C2**:24 (1970).

GZ 69       N. F. Golovanova and N. S. Zelenskaya, *Soviet Journal of Nucl. Phys.*
            **8**:158 (1969).

Hou 70      E. Hourany, Thèse, Université de Paris (1970), unpublished.

Jac+ 70     C. Jacquot, Y. Sakamoto, M. Jung, C. Baixeras-Aiguabella, L. Girardin,
            and H. Braun, *Nucl. Phys.* **A148**:325 (1970).

Jai+ 70     M. Jain, P. G. Roos, H. G. Pugh, and H. D. Holmgren, *Nucl. Phys.* **A153**:
            49 (1970).

Kas+ 69     G. Th. Kaschl, G. Mairle, U. Schmidt-Rohr, G. J. Wagner, and P. Turek,
            *Nucl. Phys.* **A136**:286 (1969).

Kop+ 66     T. I. Kepaleishvili, I. Z. Machabeli, G. Sh. Goksadze, and N. B. Krupen-
            nikova, *Phys. Letters* **22**:181 (1966).

Kop 67      T. I. Kopaleishvili, *Nucl. Phys.* **B1**:335 (1967).

ML 70       B. Mithra and R. Laverrière, *Nucl. Phys.* **A155**:535 (1970).

Pel 68      F. Pellegrini, *Nuovo Cimento* **54B**:335 (1968).

Pev 69      R. D. Plieninger, W. Eichelberger, and E. Velten, *Nucl. Phys.* **A137**:20 (1969).

Piz+ 69     J. R. Pizzi, M. Gaillard, P. Gaillard, A. Guichard, M. Gusakow, G. Re-
            boulet, and C. Ruhla, *Nucl. Phys.* **A136**:496 (1969).

Sak 64b     Y. Sakamoto, *Nuovo Cimento* **33**:566 (1964).

Sak 65      Y. Sakamoto, *Nuovo Cimento* **37**:774 (1965).

RSL 70      E. F. Redish, G. J. Stephenson, and G. M. Lerner, *Phys. Rev.* (1970), to be
            published.

Wat 69      J. W. Watson, Ph.D. Thesis, University of Maryland (1969), unpublished.

Web 70      H. J. Weber, *Annals of Phys.* **57**:322 (1970).

*Chapter 2*

# HIGH-ENERGY SCATTERING FROM NUCLEI

## Wieslaw Czyż

*Institute of Nuclear Physics*
*Cracow, Poland*

## 1. INTRODUCTION

This article is mostly about high-energy *nuclear* scattering. It also tries to point out many common features of high-energy scattering processes which sometimes lie very far from nuclear physics, e.g., similarities between Delbrük scattering and hadron–deuteron scattering (see Section 4). This approach results in a very strong emphasis on generality of the problem and cuts down any careful analysis of more specific problems. This is so, because the author's intention was to write a reasonably compact article, not a book. It is a high price to pay; the author is fully aware of that.

The analysis presented here is based on the Glauber model. It was presented many years ago by Glauber in his lectures given at University of Colorado (Gla 59, see also Gla 55) and is based on an earlier work of Molière (Mol 47) and parallels some other authors' work in a similar direction [notably Schiff (Sch 56), see also (Wu 57, Sax 57, Sch 68)]. In this article we refer to this model as the Glauber model or diffractive multiple-scattering model, although the author has made a rather personal choice of the extensions and extrapolations of Glauber's model and of the examples used to illustrate it.

The construction of the article is as follows. First the diffractive multiple-scattering model (Glauber model) is developed in a simple form, and then the existing experimental evidence for this model is discussed. Then follow extensions, applications to fields of physics other than nuclear, prospects of future applications, and a very modest discussion of the applicability limits of the model. In organizing the material this way we wanted to emphasize that even the simple version of the model takes us very far and is very successful. After establishing this fact one can afford to be choosy and sophisticated: *"primum edere deinde philosophari."*

## 2. MODELS OF HIGH-ENERGY SCATTERING. SCATTERING OF OBJECTS WITH "PREEXISTING" SUBUNITS

When a fast particle scatters from a nucleus it scatters from a collection of bound nucleons. Similarly, when a nucleus scatters from a nucleus one may think of this process as being composed of many nucleon–nucleon scattering processes. This is a very far reaching assumption which leads to many very definite predictions for high-energy nuclear processes.

Let us suppose that we can, indeed, compute the nuclear amplitudes from the "elementary" amplitudes of the components of the colliding objects (that is to say, from nucleon–nucleon amplitudes in the case of nucleus–nucleus or nucleon–nucleus scattering and the incident hadron–nucleon amplitudes in the case of hadron–nucleus scattering). Then the only degrees of freedom which will enter into the calculations will be the coordinates and quantum numbers of the nucleons which compose the colliding nuclei.

All the other degrees of freedom (the most important among them are, presumably, mesonic degrees of freedom), which play a very important role in the high-energy hadronic scattering, are taken into account in the nucleon–nucleon or hadron–nucleon elastic scattering amplitudes. At very high energies the production processes influence in an essential way these elementary elastic amplitudes and perhaps even determine them completely. Hence, in such an approach, all productions off the individual nucleons in the nucleus are taken into account. The coherent productions off the target nucleus, however, are not taken into account and are not fed back into the elastic hadron–nucleus amplitude if we are to construct it from the elementary hadron–nucleon elastic amplitudes.

In order to see better which production processes are, in a consistent way, taken into account in our model we look at the unitarity relation in the elastic scattering amplitude constructed from the individual amplitudes (e.g., the nucleon–nucleon amplitudes in the case of nucleus–nucleus collisions). The differential cross section, $d\sigma/d\Omega$, for elastic scattering of any

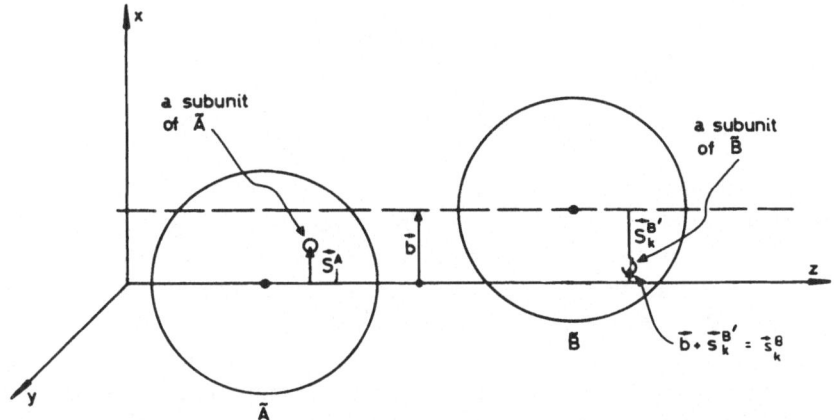

**Fig. 1. Definitions of the coordinates used in the formulas for the amplitudes of two colliding composite objects.** If we call $\gamma_{jk}$ the profile function of the collision of the subunits $j$ and $k$, from the geometry depicted above we see that the argument of $\gamma_{jk}$ is $\gamma_{jk}(\mathbf{b} - \mathbf{s}_j{}^A + \mathbf{s}_k{}^B)$.

composite system $\tilde{A}$ by a composite system $\tilde{B}$ can be expressed in terms of the elastic scattering amplitude $M(\mathbf{k}', \mathbf{k})$, or $M(\mathbf{k}, \boldsymbol{\Delta})$,

$$\frac{d\sigma}{d\Omega} = |M|^2$$

where $\mathbf{k}$ is the relative momentum of $\tilde{A}$ and $\tilde{B}$ before and $\mathbf{k}'$ their relative momenta after the collision. $\boldsymbol{\Delta}(= \mathbf{k} - \mathbf{k}')$ is the recoil momentum, or momentum transfer, of the collision.

A general form of the scattering amplitude of our model can be given in terms of the two-dimensional picture of the collision shown in Fig. 1. This picture views the collision of the two systems in terms of their impact parameter $\mathbf{b}$, which is a two-dimensional vector. Although we shall come back later to the problem of the validity of this picture, let us say here that it is a very good approximation for small-angle high-energy scattering, where the longitudinal momentum transfer is virtually zero. A very readable account of the application of this model to simple potential scattering can be found in (Gla 59), which also describes this model in terms of other approximations for potential scattering and applies the model to a wide range of problems. Here we are interested in the following general form of $\tilde{M}(\mathbf{k}, \boldsymbol{\Delta})$ given in terms of the impact parameter $\mathbf{b}$, the momentum transfer $\boldsymbol{\Delta}$, and a "profile" function:

$$\tilde{M}(\mathbf{k}, \boldsymbol{\Delta}) = \frac{ik}{2\pi} \int d^2b \, e^{i\boldsymbol{\Delta} \cdot \mathbf{b}} \Gamma(b) \qquad (2.1)$$

The profile of the colliding objects is taken to be*

$$\Gamma(b) = \langle e^{i\mathbf{p}_f^A \cdot \mathbf{r}_A} \Phi_A(s^{A'}) \, e^{i\mathbf{p}_f^B \cdot \mathbf{r}_B} \Phi_B(s^{B'}) \, | $$
$$\times \, \hat{\Gamma}(b; s^A, s^B) \, | \, e^{i\mathbf{p}_i^A \cdot \mathbf{r}_A} \Phi(s^{A'}) \, e^{i\mathbf{p}_i^B \cdot \mathbf{r}_B} \Phi_B(s^{B'}) \rangle \qquad (2.2)$$

The operator $\hat{\Gamma}$ is constructed here from the elastic scattering amplitudes of pairs of subunits (pairs of nucleons in the case of colliding nuclei) of the object $\tilde{A}$ (e.g., nucleus $\tilde{A}$) and the object $\tilde{B}$ (e.g., nucleus $\tilde{B}$). Its explicit form in the diffractive multiple scattering model is given below. Here we shall need only to have a profile operator $\hat{\Gamma}$ which depends on the positions in space of the subunits of the scattering objects. If the subunits' positions were fixed in space, the complete elastic scattering amplitude would be given by (2.1) with $\hat{\Gamma}(b; s^A, s^B)$ instead of $\Gamma(b)$. Since they more inside of the colliding objects, an average has to be taken as shown in (2.2). All the integrations over the longitudinal degrees of freedom (z-coordinate) of the wave functions are assumed to be performed in (2.2) ($\hat{\Gamma}$ does not depend on them!). An explicit example is found in Appendix B.

Now let us explain further the symbols used in Eq. (2.2); $\mathbf{p}_i^A$, $\mathbf{p}_i^B$, $\mathbf{p}_f^A$, and $\mathbf{p}_f^B$ are the initial and final momenta, and $\Phi_A(s^{A'})$ and $\Phi_B(s^{B'})$ are the internal ground-state wave functions of the colliding objects. They depend on "$A - 1$" and "$B - 1$" ($A$ and $B$ being the number of subunits in $\tilde{A}$ and $\tilde{B}$, respectively) *internal* coordinate vectors, $s_j^{A'}$ and $s_k^{B'}$. These internal coordinate vectors are two-dimensional and are in the plane spanned by the impact parameter vector $\mathbf{b}$, i.e., the $x$, $y$ plane of Fig. 1. The $z$-axis of Fig. 1 is perpendicular to this plane. In the scattering process the $z$-axis lies along the direction of the incident beam, or more suitably, along the average, $\mathbf{k} + \mathbf{k}'$, of the incident and emerging beam. In the two-dimensional space each coordinate is referred to the center of mass (c.m.) vectors, $\mathbf{r}_A$ and $\mathbf{r}_B$ of the colliding objects. Thus we have

$$s_j^{A'} = s_j^A - \mathbf{r}_A, \qquad s_k^{B'} = s_k^B - \mathbf{r}_B \qquad (2.3)$$

where

$$\mathbf{r}_A = \frac{1}{A} \sum_j s_j^A, \qquad \mathbf{r}_B = \frac{1}{B} \sum_k s_k^B \qquad (2.4)$$

---

* From (2.1) and (2.2), employing the general property $\hat{\Gamma}(b; s_j^A, s_k^B) = \hat{\Gamma}(b - s_j^A + s_k^B)$ [compare (2.16)] and definitions (2.3) and (2.4), one obtains that $\tilde{M}$ contains the momentum-conserving Dirac δ-functions. Henceforth, we shall skip them and use only the remaining part of (2.1), which we call $M$, and which contains only the internal coordinates $s^{A'}$, $s^{B'}$ and wave functions $\Phi_A$ and $\Phi_B$ [see also the Appendix of (Gla 67a)], and whose absolute square gives the elastic scattering cross section.

Having written down the very general scattering amplitude (2.1) which forms our model, we now consider the "optical theorem" which imposes particle conservation on the scattering amplitude. The optical theorem relates the total cross section, $\sigma_T$, to the forward scattering amplitude $M(\mathbf{k}, \Delta = 0)$ (here we skip the momentum-conserving $\delta$-function, which means replacing $\tilde{M}$ by $M$.)

$$\text{Im } M(\mathbf{k}, \Delta = 0) = \frac{k\sigma_T}{4\pi} \tag{2.5}$$

From (2.1), (2.2), and (2.5) we get the following unitarity relation:

$$\sigma_T = 2 \int \text{Re}\langle \hat{\Gamma} \rangle \, d^2b$$
$$\equiv \int d^2b |\langle \hat{\Gamma} \rangle|^2 + \int d^2b [\langle \hat{\Gamma}^\dagger \hat{\Gamma} \rangle - |\langle \hat{\Gamma} \rangle|^2] + \int d^2b \{2\text{Re}\langle \hat{\Gamma} \rangle - \langle \hat{\Gamma}^\dagger \hat{\Gamma} \rangle\} \tag{2.6}$$

where

$$\langle \hat{\Gamma} \rangle = \langle \Phi_A \Phi_B | \hat{\Gamma} | \Phi_A \Phi_B \rangle \tag{2.7}$$

The three terms in (2.6) have a straightforward interpretation (Czy 69d):

(i)   The first term, $\int d^2b |\langle \hat{\Gamma} \rangle|^2$, is the total elastic cross section.

(ii)   The second term

$$\int d^2b [\langle \hat{\Gamma}^\dagger \hat{\Gamma} \rangle - |\langle \hat{\Gamma} \rangle|^2] \equiv \int d^2b \sum_{n \neq 0} |\langle n | \hat{\Gamma} | 0 \rangle|^2$$

where $|0\rangle = |\Phi_A \Phi_B\rangle$ and $|n\rangle$ are all possible excited states of the colliding objects (this set of states includes the ones in which only one object is excited). This term therefore represents the contribution to the total cross section coming from all possible excitations of the two colliding objects. Note however that $\hat{\Gamma}$, which depends on the positions of the subunits, generates only those excited states which are simply all the possible configurations of the same set of subunits different from the ground-state configuration. For instance, in the case of nuclear scattering, the states $|n\rangle$ are all the possible nuclear excited states in which no particle production occurs (e.g., mesons). If one adds the ground state to them, one obtains a complete set of states: $\sum_n |n\rangle\langle n| = 1$.

(iii)   The third term can be interpreted as a total reaction cross section

$$\int d^2b \{2 \text{Re}\langle \hat{\Gamma} \rangle - \langle \hat{\Gamma}^\dagger \hat{\Gamma} \rangle\} = \int d^2b \sigma_r(b) \tag{2.8}$$

for producing various particles off the individual subunits. This follows from the observation that the following relation exists between the profile $\Gamma(b)$ and the so-called reflection coefficient $\eta(b)$ (in the impact-parameter representation):

$$\Gamma(b) = \langle \hat{\Gamma} \rangle = 1 - \eta(b) = 1 - \langle \hat{\eta} \rangle \tag{2.9}$$

where $\hat{\eta}(b; s^{A'}, s^{B'}) = 1 - \hat{\Gamma}(b; s^{A'}, s^{B'})$. One obtains this relation by performing a transition from the partial wave expansion of the scattering amplitude

$$M(k, \theta) = \frac{i}{2k} \sum_{l=0}^{\infty} (2l + 1)(1 - \langle \hat{\eta}_l \rangle) P_l(\cos \theta),$$

to the impact-parameter representation (2.1).

As the partial reaction cross section is

$$\sigma_{l,r} = \pi \lambda^2 (2l + 1)(1 - |\langle \hat{\eta}_l \rangle|^2) \tag{2.10}$$

we get from (2.9) and (2.10) the relation (2.8). Note that we first take the ground-state average of the operator $\hat{\eta}$ and then compute $\sigma_r(b)$.

The total reaction cross section (2.8) takes care of a very large class of production processes, but from its construction in our model we can see that certain production processes are *not* included in (2.8). In order to see this let us observe that $\hat{\Gamma}(b, s^{A'}, s^{B'})$ is constructed from the individual profiles of the *elastic* "subunit from $\tilde{A}$ — subunit from $\tilde{B}$" scattering amplitudes (see below). That means that if we fit the individual theoretical amplitudes to the experimental data (to, e.g., the elastic nucleon–nucleon cross section, in the case of nucleus–nucleus or nucleon–nucleus scattering), we do take into account the "shadow" of *all* inelastic processes in the elastic scattering amplitudes of individual subunits. This is, presumably, the most important contribution of the inelastic processes to the elastic scattering amplitude of $\tilde{A}$ from $\tilde{B}$ at a few GeV. At higher energies (and small momentum transfers) however, the processes producing particles on *one* subunit and reabsorption on the *others* (so that the net result is elastic scattering) are *not* taken into account as long as the profile operator $\hat{\Gamma}(b; s^{A'}, s^{B'})$ is constructed, as is done below, from the individual *elastic* scattering amplitudes. Thus, for example, the so-called "inelastic shadow effects" (Pum 68) need a more sophisticated model of $\hat{\Gamma}(b; s^{A'}, s^{B'})$ than the one discussed below.

Note that if one introduces the Van Hove overlap function (Van 64) $\omega(b)$, defined as follows:

$$\sigma_{\text{inel}} = 2\pi \int_0^{\infty} db \, b \, \omega(b) \tag{2.11}$$

one gets from (2.6) the following unitarity relation:

$$2 \operatorname{Re}\Gamma(b) = | \Gamma(b) |^2 + \omega(b) \qquad (2.12)$$

$\omega(b)$ is composed of two physically different contributions,

$$\omega(b) = \omega_1(b) + \omega_2(b) \qquad (2.13)$$

$$\omega_1(b) = \langle \hat{\Gamma}^+(b, s^{A'}, s^{B'})\hat{\Gamma}(b, s^{A'}, s^{B'})\rangle - |\langle\hat{\Gamma}(b, s^{A'}, s^{B'})\rangle|^2 \qquad (2.14)$$

$$\omega_2(b) = 2 \operatorname{Re}\langle\hat{\Gamma}(b, s^{A'}, s^{B'})\rangle - \langle \hat{\Gamma}^+(b, s^{A'}, s^{B'})\hat{\Gamma}(b, s^{A'}, s^{B'})\rangle \qquad (2.15)$$

The overlap function $\omega_1(b)$ is produced by the coherent excitations of the target (rearrangement of subunits). Thus in the case of nuclear collisions $\omega_1(b)$ comes from all the nuclear excitations without production of any new particles. On the other hand, $\omega_2(b)$ corresponds to all production processes on the subunits of the target (in the sense discussed above).

Thus, we come to the following picture of high-energy nuclear scattering. There are two groups of degrees of freedom, whose roles are well separated in the limit of very high energy scattering. The first group consists of the degrees of freedom which describe the nucleons, their positions, etc. The second group consists of the degrees of freedom connected with all the possible production processes from nucleons (irrespective of their positions, which are defined by the first group of degrees of freedom). This separation into two groups of degrees of freedom implies the existence of two physically different contributions to the overlap function $\omega_1(b)$ and $\omega_2(b)$, defined by (2.14) and (2.15).

In all the formulas discussed above, the only degrees of freedom of the subunits of the colliding objects were their position coordinates. In principle, one can introduce the other degrees of freedom, such as spin and isospin quantum numbers, into the profiles, but we shall stick to the simplest model, which does not deal explicitly with these degrees of freedom. This does not mean that we shall neglect the spin effects altogether. We shall discuss, for example, the scattering from deuterium (see Section 3), where the spin (equal to 1), combined with a deformation ($S + D$ ground state), produces an important effect. In the general discussion of the high-energy-scattering model, however, we shall limit ourselves to the simplest possible formulation in order to emphasize the most essential features of the physics behind these processes.

One constructs the profile operator (Gla 59, Fra 68a, Czy 69a,f, Har 69b, Kof 69) on the basis of the additivity of phase shifts of all possible "subunit $\tilde{A}$ − subunit $\tilde{B}$" scattering amplitudes, or by computing the probabilities of all possible multiple collisions assuming that the individual profiles are certain probability distributions (Czy 69a). In either case we get identical formulas. In particular, for the scattering-amplitude operator expressed through the profile operator $\Gamma$ we get (the coordinates $s_j^A$ and $s_k^B$ are defined in Fig. 1, cf. Appendix A):

$$\hat{M}(k, \Delta; s^A, s^B) = \frac{ik}{2\pi} \int d^2b e^{i\Delta \cdot \mathbf{b}} \left\{ 1 - \prod_j^A \prod_k^B [1 - \gamma_{jk}(\mathbf{b} - s_j^A + s_k^B)] \right\}$$

$$= \frac{ik}{2\pi} \int d^2b e^{i\Delta \cdot \mathbf{b}} \left\{ 1 - \prod_j^A \prod_k^B \left[ 1 - \frac{1}{2\pi ik} \int d^2\delta_{jk} e^{-i\delta_{jk} \cdot (\mathbf{b} - s_j^A + s_k^B)} f_{jk}(\delta_{jk}) \right] \right\}$$

$$= \hat{M}(k, \Delta; s^{A'}, s^{B'}) e^{i\Delta \cdot \mathbf{r}_A - i\Delta \cdot \mathbf{r}_B} \tag{2.16}$$

where $f_{jk}(\delta)$ is the elastic scattering amplitude of the $j$th subunit of $\tilde{A}$ against the $k$th subunit of $\tilde{B}$ and $\gamma_{jk}$ is the corresponding profile. We can see from (2.16) that if the c.m. coordinate dependence of the $\tilde{A}$ and $\tilde{B}$ wave functions are factored as in (2.2) we get the momentum-conserving Dirac $\delta$-function in front of the transition matrix element. In Appendix A the factorization of the c.m. coordinates in $\hat{M}$, shown in (2.16), is proved.

In the case of light nuclear targets the factorization [as shown in (2.2)] of the $\mathbf{r}_A$ and $\mathbf{r}_B$ variables in the nuclear wave functions (and consequently in $\hat{M}$) is essential (see, e.g., Czy 68a). If, however, the colliding objects have many subunits (many nucleons, as in the case of the collision of nuclei), giving $A, B \gg 1$, the factorization of the centers of mass of $\tilde{A}$ and $\tilde{B}$ is unimportant; in many cases this considerably simplifies the calculations [for more quantitative discussion cf. (Czy 69a)].

We may also remind the reader at this place that formula (2.16) follows from the assumption that the total "$\tilde{A} - \tilde{B}$" interaction (potential) is a sum of all individual interactions (potentials)

$$V = \sum_{j,k} v_{jk}(\mathbf{r}_j - \mathbf{r}_k) \tag{2.17}$$

Then, the total phase shift is, in the high energy limit (Gla 59), ($\hbar = c = 1$),

$$\chi(b; s^{A'}, s^{B'}) = -\frac{1}{v} \int_{-\infty}^{+\infty} dz \, V(b, z; s^{A'}, s^{B'}) \tag{2.18}$$

where $z$, because of the infinite limits of integration, one may take as the $z$-component of the distance between the centers of mass of $\tilde{A}$ and $\tilde{B}$, $\mathbf{b}$ is the projection of this distance on the plane perpendicular to the direction of the collision (hence $\mathbf{b} = \mathbf{r}_A - \mathbf{r}_B$), and $v$ is the velocity. From

(2.17) and (2.18) it follows that $\chi$ is a sum of the individual phase shifts $(\chi = \sum_{j,k} \chi_{jk})$, which through

$$\hat{M} = \frac{ik}{2\pi} \int d^2b \, e^{i\mathbf{\Delta}\cdot\mathbf{b}}(1 - e^{i\chi(b)}) \tag{2.19}$$

leads to (2.16).

From the discussion following (2.7) we may conclude that the following three cross sections deserve a more complete discussion. First, the elastic cross section (here, as before, we denote $|\Phi_A\Phi_B\rangle = |0\rangle$ and skip the explicit dependence of the operators on the positions of subunits)

$$\frac{d\sigma_{\text{el}}}{d\Omega} = \left| \frac{ik}{2\pi} \int d^2b \, e^{i\mathbf{\Delta}\cdot\mathbf{b}}\langle 0 | \hat{\Gamma}(b) | 0\rangle \right|^2 \tag{2.20}$$

Second, the so-called "scattering cross section"

$$\frac{d\sigma_{\text{sc}}}{d\Omega} = \left(\frac{k}{2\pi}\right)^2 \int d^2b \, d^2b' \, e^{i\mathbf{\Delta}\cdot(\mathbf{b}-\mathbf{b'})}\langle 0 | \hat{\Gamma}^+(b)\hat{\Gamma}(b') | 0\rangle \tag{2.21}$$

which, as may be seen from (iii), includes the elastic cross section plus all (purely nuclear, in the case of nuclear scattering, rearrangement of subunits, in general) inelastic contributions. Hence, in the case of nuclear scattering, it is essentially a poor energy resolution $(\Delta E \sim 100 \text{ MeV})$ cross section (one of the easiest to measure for very high energy projectiles). Third, the inelastic cross section

$$\frac{d\sigma_{\text{inel}}}{d\Omega} = \frac{d\sigma_{\text{sc}}}{d\Omega} - \frac{d\sigma_{\text{el}}}{d\Omega} \tag{2.22}$$

$$\frac{d\sigma_{\text{inel}}}{d\Omega} = \left(\frac{k}{2\pi}\right)^2 \int d^2b \, d^2b' \, e^{i\mathbf{\Delta}\cdot(\mathbf{b}-\mathbf{b'})}$$
$$\times [\langle 0 | \hat{\Gamma}^+(b)\hat{\Gamma}(b') | 0\rangle - \langle 0 | \hat{\Gamma}^+(b) | 0\rangle\langle 0 | \Gamma(b') | 0\rangle] \tag{2.23}$$

which, upon integration over $\Delta$, becomes the second term of (2.6). It thus represents the contribution of the $\omega_1(b)$ part of the overlap function [compare (2.14)] to the total cross section.

It is perhaps worth mentioning that the $\omega_2(b)$ part of the overlap function (2.15) has never been measured or theoretically analyzed in the case of high-energy nuclear scattering.

Let us notice that the formulas (2.16), (2.20), (2.21), and (2.23) are in a form which can be directly applied to analyze hadronic scattering if one accepts that the hadrons are made up of quarks. If so, we would have a very close analogy with scattering of light nuclei. The absolute values of the cross sections would be very drastically influenced by the fact that only small number of subunits are involved [that is to say, the constraints introduced by the factorization of the c.m. motion are very important (Czy 69a)] and, as in nuclear case, the meson production from quarks plus the excitations which consist of rearrangement of subunits (it seems very appropriate to call them diffractive excitations) would be the main contributors to the Van Hove overlap function. Thus one would also expect the copious production of mesons from quarks to determine the diffractive peak, very much like in the case of high-energy nuclear scattering.

A word about the c.m. motion correction in (2.21) and (2.23) is in order here. From the properties of the $\hat{M}$ operator of (2.16) we see that we *do not need* to factor the c.m. motion in $d\sigma_{sc}/d\Omega$ (2.21); the exponents of the last expression in (2.16) cancel out. In $d\sigma_{inel}/d\Omega$ of (2.23) however, one has to keep the c.m. factors in the second term (subtraction of the elastic scattering contribution).

Let us now discuss several special cases of Eqs. (2.20), (2.21), and (2.23), which are of considerable interest in high-energy nuclear and hadronic scattering.

Let us first of all discuss hadron–nucleus elastic scattering. In this case we should put, for example, $B = 1$ in our formulas and drop $\Phi_B$ from the wave function (2.2). The recent results (Czy 69b, Gla 69) seem to indicate that these cross sections, at not too large momentum transfers and for farily large nuclei, can be quite reliably computed from the knowledge of the single-nucleon density distributions and a simple functional representation of the incident-hadron–target-nucleon (the subunit of target nucleus) amplitude. For instance, at small momentum transfers this amplitude can be accurately given by the following profile:

$$\gamma_{n,p}(b) = \frac{(1 - i\alpha_{n,p})\sigma_{n,p}}{4\pi a_{n,p}} e^{-b^2/2a_{n,p}} \tag{2.24}$$

where the indices $n, p$ refer to the target neutron or proton respectively, $\alpha$ is the ratio of the real to the imaginary part of the amplitude, and $a$ is the "slope" of the incident-hadron–target-nucleon elastic cross section. Equation (2.24) gives, through the relation (2.16), the elastic scattering amplitude which perfectly fits the data at small momentum transfers [cf. the collection of experimental data given in (Bel 68)]. If one next considers larger momentum transfers, where such cross sections exhibit some structure [cf. $\pi$–$p$, $p$–$p$, $p$–$\bar{p}$, etc., large momentum transfer data in (Bel 68, All 68, 69b)], one has to use much more complex inputs for $\gamma_{n,p}$ or $f_{n,p}$. But such a degree of sophistication had to be used, until now, only in the poor energy resolution proton–deuteron scattering cross sections, at large momentum transfers (All 69a).

We shall now discuss, in a little more detail, two special cases which are important, instructive, and which may provide us with considerable physical insight into high-energy scattering of composite objects which might be useful in the extensions and generalizations of the model. These two special cases are: scattering of hadrons from deuterium, and scattering of hadrons from large nuclei.

First hadron–deuteron scattering. This problem has a fairly long history, which goes back to Glauber's papers (Gla 55, Gla 60), and a flood of more recent papers (see references of Section 3). Since we do know the structure of the deuteron quite well it is important to check, on all available experimental materials, our understanding of the high-energy scattering. Such evidence is collected in Section 3. It does give very strong support to the model. In spite of the fact that there are only two nucleons in the deuteron, it is a very complicated target because of its ground-state spin 1 and the deformation due to the presence of the $D$ state in the deuteron ground state. So, a complete calculation is quite complicated, especially if the incident particle has a spin (e.g., when it is a proton). We shall not go into all the details of the problem. One may find them in (Fra 66a, Fra 66b, Gla 67b, Fra 69, Har 68, Kuj 68, Fra 68b, Mic 69, Alb 69). We shall stress only a few important points.

First of all we can easily factor the c.m. coordinates, introduce the relative $n$–$p$ coordinate s and, from (2.16), write the deuteron profile operator

$$\hat{\Gamma}(\mathbf{b}, \mathbf{s}) = \gamma_n\left(\mathbf{b} - \frac{\mathbf{s}}{2}\right) + \gamma_p\left(\mathbf{b} + \frac{\mathbf{s}}{2}\right) - \gamma_n\left(\mathbf{b} - \frac{\mathbf{s}}{2}\right)\gamma_p\left(\mathbf{b} + \frac{\mathbf{s}}{2}\right) \quad (2.25)$$

Although there is no spin dependence in this profile, one can use it to compute from (2.20) and (2.21) the $d\sigma_{el}/d\Omega$ and $d\sigma_{sc}/d\Omega$ cross sections, taking into account the most important effects of the deuteron ground-state spin ($= 1$) and deformation. So, if we are not interested in the fine effects of the hadron–deuteron scattering, a very careful treatment of the spin dependencies of the incident-hadron–target-nucleons amplitudes seems to be unnecessary, as the following arguments show. Firstly, the $p$–$d$ data compared with the $\pi$–$d$ data show them to look very much alike (see Section 3). Hence the spin $\frac{1}{2}$ of the proton does not seem to play an important role. Secondly, if one takes the profiles from experiment (more precisely, from the elastic cross sections) one does already take into account some average of the contributions of the spin amplitudes. Thirdly, if one limits oneself to the very small angle scattering of the incident pion beam, one does not need to consider any spin-dependent amplitude because it goes to zero as $\sin \theta$ ($\theta =$ scattering angle).

One obtains, for example, the elastic cross section by sandwiching $\hat{M}$, gotten from $\hat{\Gamma}$ according to (2.16) and (2.20), between the deuteron ground states, which one can write as follows:

$$\psi_m(r) = (4\pi)^{-1/2}r^{-1}[u(r) + 8^{-1/2}S_{12}w(r)]\chi_{1,m} \tag{2.26}$$

where $u$ and $w$ are the radial $S$ and $D$ functions, with

$$S_{12} = r^{-2}[3(\boldsymbol{\sigma}_1\cdot\mathbf{r})(\boldsymbol{\sigma}_2\cdot\mathbf{r}) - \boldsymbol{\sigma}_1\cdot\boldsymbol{\sigma}_2] \tag{2.27}$$

where $\sigma_1$ and $\sigma_2$ are the Pauli spin operators, and $r$ is the neutron–proton relative coordinate. $\chi_{1,m}$ is the spin function for spin $= 1$ with the magnetic quantum number $m$. The elastic cross section is then

$$\frac{d\sigma_{\text{el}}}{d\Omega} = \frac{1}{3} \sum_{m,m'} |\langle m \mid \hat{M} \mid m'\rangle|^2 \tag{2.28}$$

Note that (2.28) includes all possible $m \neq m'$ transitions, which correspond to the deuteron spin-flip scattering and behave entirely differently from the $m = m'$ transitions. Franco and Glauber (Fra 69) analyze these two physically different groups of contributions in detail.

The $\Delta m = 0$ scattering is, to a very good approximation, the one which has been considered in all the earlier papers which neglected the $D$ state and the deuteron spin. This (non-spin-flip) scattering dominates and determines the most prominent features of the cross section: a steep part at small momentum transfers, due predominantly to single scattering, and a flatter part at large momentum transfers, due predominantly to double scattering (see Fig. 4–7). The $\Delta m = 0$ part of the cross section (2.28) also exhibits a well-defined minimum in the break region, where the single- and double-scattering contributions are of the same order of magnitude. This minimum is not observed in experiment because it is filled by the $\Delta m \neq 0$ part of the cross section (2.28).

Since the elastic cross section is dominated in this break region by $\Delta m \neq 0$ (spin-flip) amplitudes, one may expect some strong polarization (or alignment of the deuteron spin) phenomena to occur. There is a possibility, therefore, to produce and analyze strongly aligned beams of high-energy deuterons by elastic scattering from protons. A detailed discussion of the properties of all these amplitudes in elastic handron–deuteron scattering can be found in (Har 68, 69, Fra 69, Alb 69, Mic 69). Any future development of polarized deuteron targets should allow a substantially more detailed check of the theory.

Not all measured cross sections in hadron–deuteron scattering are, however, sensitive to the existence of the $D$ state and spin $= 1$ of the deuteron ground state. Neither the total cross section nor the scattering cross section $d\sigma_{sc}/d\Omega$ of Eq. (2.21) are influenced in any important way by the deformation of the ground state.

In such a case one can use just a spinless deuteron ground-state wave function $\varphi(\mathbf{r})$ ($\mathbf{r}$ being the relative coordinate), or the corresponding from factor

$$S(\boldsymbol{\Delta}) = \int d^3r \, |\, \varphi(\mathbf{r})\,|^2 e^{i\boldsymbol{\Delta} \cdot \mathbf{r}} \qquad (2.29)$$

and get from (2.25), (2.20), and the optical theorem (2.5) the following expression for the hadron–deuteron total cross section:

$$\sigma_T^{(d)} = \sigma_T^{(n)} + \sigma_T^{(p)} + \frac{2}{k^2} \int d^2\Delta \, S(\boldsymbol{\Delta}) \, \mathrm{Re}[f_n(\boldsymbol{\Delta})f_p(-\boldsymbol{\Delta})] \qquad (2.30)$$

where the inverse of relation (2.1) between the profiles and amplitudes was employed to explicitly introduce amplitudes into expression (2.30):

$$\gamma(b) = \frac{1}{2\pi i k} \int d^2\delta \, e^{-i\boldsymbol{\delta} \cdot \mathbf{b}} f(\delta) \qquad (2.31)$$

Although expression (2.30) contains many approximations it provides us with a very good first approximation, since the only model-dependent term (the last term; double-scattering contribution) is very small (see Section 3). In the future, when very accurate measurements of $\sigma_T^{(d)}$ will be available, one will have to consider all the intricacies of the double-scattering corrections including spins, deformations, exchange effects (Gla 67), and perhaps some other, more fundamental, modifications.

A somewhat similar situation occurs in the case of $d\sigma_{sc}/d\Omega$. Employing (2.21), (2.25), (2.29), and (2.31) one gets

$$
\begin{aligned}
\frac{d\sigma_{sc}}{d\Omega} = {}& |\, f_n(\boldsymbol{\Delta})\,|^2 + |\, f_p(\boldsymbol{\Delta})\,|^2 + 2S(\boldsymbol{\Delta}) \, \mathrm{Re}[f_n(\boldsymbol{\Delta})f_p{}^*(\boldsymbol{\Delta})] \\
& - \frac{1}{\pi k} \, \mathrm{Im} f_n{}^*(\boldsymbol{\Delta}) \int d^2\delta \, S\!\left(\boldsymbol{\delta} - \tfrac{1}{2}\boldsymbol{\Delta}\right) f_n\!\left(\tfrac{1}{2}\boldsymbol{\Delta} + \boldsymbol{\delta}\right) f_p\!\left(\tfrac{1}{2}\boldsymbol{\Delta} - \boldsymbol{\delta}\right) \\
& - \frac{1}{\pi k} \, \mathrm{Im} f_p{}^*(\boldsymbol{\Delta}) \int d^2\delta \, S\!\left(\boldsymbol{\delta} + \tfrac{1}{2}\boldsymbol{\Delta}\right) f_n\!\left(\tfrac{1}{2}\boldsymbol{\Delta} + \boldsymbol{\delta}\right) f_p\!\left(\tfrac{1}{2}\boldsymbol{\Delta} - \boldsymbol{\delta}\right) \\
& + \frac{1}{(2\pi k)^2} \int d^3r \, |\, \varphi(\mathbf{r})\,|^2 \left|\, \int d^2\delta \, e^{i\boldsymbol{\delta} \cdot \mathbf{s}} f_n\!\left(\tfrac{1}{2}\boldsymbol{\Delta} + \boldsymbol{\delta}\right) f_p\!\left(\tfrac{1}{2}\boldsymbol{\Delta} - \boldsymbol{\delta}\right)\right|^2
\end{aligned}
$$

$$ (2.32) $$

Here, again, the single-scattering contributions [the first line of (2.32)] very strongly dominate the cross section. This is so because at small $\Delta$ $d\sigma_{sc}/d\Omega \approx d\sigma_{el}/d\Omega$, and we know that the single scattering determines $d\sigma_{el}/d\Omega$ at small $\Delta$. At large $\Delta$, again the first two terms dominate because they do not decrease with $\Delta$ as sharply as $S(\Delta)$ (nucleons are smaller than the deuteron).

The dominance of the first two terms also implies that the diffractive structure is washed away regardless whether there is a deformation of the ground state and the ground-state spin is 1. The results of such calculations (compared with experimental data) are shown in Fig. 8 below.

The light nuclei targets should be treated similarly to the deuteron, that is to say their ground-state wave functions have to be used and the full multiple scattering expression of the amplitude operator (2.16). Obviously every target nucleus has its specific properties and must be treated separately. In Section 3 a representative selection of data on hadron–light-nuclei scattering cross sections is presented. Here we shall discuss large spherical spin-zero nuclei where some asymptotic expressions have a very good chance of being accurate. To calculate (2.20) or (2.21) with a complicated ground-state wave function and a complex functional form of $f_{jk}$'s is usually a very involved numerical task, often impossible to perform even with the fast present-day computers. One can, however, make several drastic simplifications and yet keep a high degree of accuracy of the calculations for certain cross sections.

If we assume that the absolute value squared of the ground-state wave function of the nucleus factors into a product of the single-particle densities ($\psi_0^*\psi_0 \cong \prod_j^A \varrho_j(r_j)$, this approximation is discussed in more detail in Section 3) we get from (2.16), (2.20), and (2.24) the following important formula for the phase shift $\chi_s(b)$ generated by strong interactions:

$$i\chi_s(b) = -\tfrac{1}{2}N(1 - i\alpha_n)\sigma_n\tilde{\varrho}_n(b) - \tfrac{1}{2}Z(1 - i\alpha_p)\sigma_p\tilde{\varrho}_p(b) \qquad (2.33)$$

The indices $n$ and $p$ refer, again, to the neutrons and protons of the target nucleus, $N$ and $Z$ are the number of neutrons and protons, respectively, and $\tilde{\varrho}_{n,p}$ are given by the following formula:

$$\tilde{\varrho}_{n,p}(b) = \frac{2}{[(1 - i\alpha_{n,p})\sigma_{n,p}]} \int d^2s\,\gamma_{n,p}(s) \int_{-\infty}^{+\infty} dz\varrho_{n,p}(\mathbf{b} - \mathbf{s}, z) \qquad (2.34)$$

with $\gamma_{n,p}$ given by (2.24) and $\varrho_{n,p}$ being the neutron (proton) single-particle densities. The formula (2.33) was obtained in the limit of $A$ (the number of nucleons in the target nucleus) approaching $\infty$ (optical limit). In this limit [for more details of this and similar limits see (Czy 69a)] the products of (2.16) can be expressed by the exponential functions and $\chi$ of (2.19) identified. The error one makes by replacing the multiple scattering amplitude $(M^{\mathrm{ms}})$ by its optical limit $(M^{\mathrm{opt}})$ was estimated in (Czy 69a) and is as follows:

$$M^{\mathrm{ms}} - M^{\mathrm{opt}} \approx \mathcal{O}\left(\frac{[\ln(\sigma A^{1/3} r_0)]^{1/4}}{A}\right) \tag{2.35}$$

where $r_0 \cong 1.1$ fm and $\sigma$ is an average of the $\sigma_n$ and $\sigma_p$ total cross sections. For nuclei like $^{208}$Pb this is less than 1% difference.

Let us notice that through the relation (2.18) the relation (2.33) *uniquely* determines an optical potential. Since $\varrho_{n,p}(\mathbf{r})$ are much broader than $\gamma_{n,p}(b)$ we can use only the forward hadron–nucleus scattering amplitude and get for the optical potential the well-known approximate expression discussed, for example, in (Gla 59) [see also (Fol 69)]:

$$-\frac{1}{\hbar v} V_{\mathrm{opt}}(\mathbf{r}) \cong \frac{2\pi A}{k} f(0)\varrho(\mathbf{r}) \tag{2.36}$$

A very complete and interesting discussion of the construction of the optical potentials from the hadron–nucleon interactions for high-energy nuclear scattering and its relations with the method presented here is found in (Fol 69).

If one calculates cross sections for light nuclear targets one has to use the multiple scattering formula. The most important difference between the multiple scattering and the optical limit formulas comes in this case, however, not so much from the failure of the $A = \infty$ approximation as from the lack of factorization of the c.m. motion. For example, in the cases of hadron–deuteron, deuteron–deuteron, $^4$He–$^4$He, and hadron–$^4$He scattering, there are substantial differences between the optical limit and the multiple scattering formulas. This problem was considered in more detail in (Czy 69a).

Section 3 gives many examples of elastic scattering cross sections calculated from multiple scattering amplitudes and shows their sensitivity (or insensitivity) to the details of the ground-state wave functions.

Let us discuss now the corrections which the Coulomb interaction introduces into the formulas discussed above. We shall assume that, in the high-energy limit, the total phase shift is the sum of the Coulomb phase shift $\chi_C$ (the phase shift if the strong interactions did not exist) and the strong interaction phase shift $\chi_S$ (the phase shift if the Coulomb interactions did not exist). This is an assumption. Notice that if one could represent the

interaction between the incident hadron and the target nucleus as a sum of
the strong interaction and the Coulomb interaction potentials, the additivity
of the corresponding phase shifts would follow from the high-energy limit
expressions for the phase shift:

$$\chi(b) = -\frac{1}{v} \int_{-\infty}^{+\infty} dz \left[ V_S\left(\sqrt{b^2 + z^2}\right) + V_C\left(\sqrt{b^2 + z^2}\right) \right] = \chi_S(b) + \chi_C(b) \tag{2.37}$$

One can thus take the Coulomb phase shift of the two extended composite
charged particles to be

$$\chi_C(b) = -\frac{e_A e_B}{v} \int_{-\infty}^{+\infty} dz \int d^3r' d^3r'' \frac{\varrho_A(\mathbf{r'})\varrho_B(\mathbf{r''})}{|\mathbf{r} + \mathbf{r'} - \mathbf{r''}|} \tag{2.38}$$

where $e_A$ and $e_B$ are the charges of the two particles and $\varrho_A$ and $\varrho_B$ are their
density distributions. This phase shift should be added to the strong interaction phase shift calculated in whatever way is appropriate (e.g., taken
from (2.33) for heavy nuclear targets, or extracted from the profile $\Gamma$ calculated for the multiple scattering case). There are, of course, many points
concerning reliability of such evaluation of $\chi_C$ which should be discussed.
For instance, one can also calculate $\chi_C$, similarly to $\chi_S$, from the individual
Coulomb amplitudes. One can show (Les 69c) that the results of such calculations are approximately the same. For a more detailed results, one has
still to wait.

The Coulomb phase shift profoundly changes the structure of the
elastic scattering amplitude in many cases. One can see this as follows:
$\chi_C$ is purely real, but $\chi_S$ is not. Let us split $\chi_S$ into its imaginary and
real parts

$$\chi_S(b) = i\xi(b) + \bar{\xi}(b) \tag{2.39}$$

From (2.19) and (2.39) we get the real and imaginary parts of the amplitude $M$:

$$\text{Im } M = k \int_0^\infty db\, b J_0(\Delta b)\{1 - e^{-\xi(b)} \cos[\chi_C(b) + \bar{\xi}(b)]\} \tag{2.40}$$

$$\text{Re } M = k \int_0^\infty db\, b J_0(\Delta b)\, e^{-\xi(b)} \sin[\chi_C(b) + \bar{\xi}(b)] \tag{2.41}$$

For $\Delta \neq 0$ (nonforward scattering) the first term in the integrand of (2.40)
can be dropped [because it is proportional to $\delta(\Delta)$] and the only difference
between Im $M$ and Re $M$ is the overall sign and replacement of $\sin[\chi_C(b)$
$+ \bar{\xi}(b)]$ by $\cos[\chi_C(c) + \bar{\xi}(b)]$ or *vice versa*. The $\xi(b)$ and $\bar{\xi}(b)$ decrease

with increasing $b$ and become zero for $b \gg R$ ($R$, nuclear radius). Instead $|\chi_C(b)|$ increases with $b$ [e.g., for point charges and large $b$ we have $\chi_C \approx (e_A e_B / v) \ln(kb)$]. So it may sometimes happen that $\chi_C + \tilde{\xi}$ depends very weakly on $b$ and is approximately equal $\frac{1}{2}\pi n$ ($n = 1, 2, \ldots$) or $\pi n$. In the first case the Re $M$ contribution dominates the cross section, in the second case the Im $M$ contribution dominates. So it may happen that if we do not take the Coulomb interaction into account and $\tilde{\xi}$ is small, which is usually the case, $\cos \tilde{\xi} \gg \sin \tilde{\xi}$ and Im $M$ dominates, but if we add $\chi_C$ and $\cos(\chi_C + \tilde{\xi}) \ll \sin(\chi_C + \tilde{\xi})$ the roles of the Im $M$ and Re $M$ switch. This is precisely the case in $p$–$^{208}$Pb scattering which is discussed in Section 3 [for more detailed discussion of this point see (Czy 69c, Czy 69b)].

Since $\chi_C$ changes sign if the charge of one of the colliding particles changes its sign, we can see that the balance between Im $M$ and Re $M$ may be influenced by such a change ($\chi_C + \tilde{\xi}$ goes to $-\chi_C + \tilde{\xi}$, which may change the relative importance of $\cos[\chi_C(b) + \tilde{\xi}(b)]$ and $\sin[\chi_C(b) + \tilde{\xi}(b)]$). This balance, in turn, determines the depths of the diffractive minima. Thus the influence of the Coulomb interaction should be readily seen at and around diffractive minima (Czy 69b, Czy 69c). That this is indeed the case is shown in Section 3. One may also mention here that even in the case of scattering of very light nuclei the Coulomb interaction introduces large corrections. For example, in the case of elastic $^4$He–$^4$He scattering these corrections amount to $\sim$100% at the first diffractive minimum (Czy 69e), but outside of the minima they are negligible.

Let us now consider, in the case of medium and heavy nuclei, the nuclear inelastic cross section (no production of pions, or any other particles) which is a part of $d\sigma_{sc}/d\Omega$ [compare (2.23)]:

$$\frac{d\sigma_{sc}}{d\Omega} = \frac{d\sigma_{el}}{d\Omega} + \frac{d\sigma_{inel}}{d\Omega} \tag{2.42}$$

Above we have been discussing $d\sigma_{el}/d\Omega$ the elastic scattering cross section. The separation shown in (2.42) is very convenient because some factors important in $d\sigma_{el}/d\Omega$ are unimportant in $d\sigma_{inel}/d\Omega$ and *vice versa*. Let us mention here a few important features of $d\sigma_{inel}/d\Omega$ and then discuss them in more detail. In contrast to elastic scattering, (*i*) the Coulomb inteactions virtually do not play any role except, perhaps, at very small momentum transfers, (*ii*) at small momentum transfers the nucleon–nucleon correlations (hence the Pauli exclusion principle, in particular) play an important role and, percentagewise, amount to corrections similar to those in single-scattering processes (e.g., quasi-elastic electron scattering), (*iii*) the parameters $\alpha$ [see (2.24)] play a negligible role, and (*iv*) the parameters $a$ [the slopes of the incident-hadron–target-nucleon elastic cross sections, see (2.24)] play a much more important role in $d\sigma_{\text{inel}}/d\Omega$ than in $d\sigma_{\text{el}}/d\Omega$.

Let us first discuss the role of Coulomb interactions in $d\sigma_{\text{inel}}/d\Omega$. As is shown in [(Czy 69b), cf. Appendix A] almost all Coulomb corrections cancel out and, to a first approximation, we have (for medium $\Delta$)

$$\frac{d\sigma_{\text{inel}}}{d\Omega} = \mathscr{A}\left[\frac{N}{A}\,|f_n(\Delta)|^2 + \frac{Z}{A}\,|f_p(\Delta)|^2\right] \qquad (2.43)$$

where $f_n$ and $f_p$ are the elastic scattering incident-hadron–nucleon (neutron or proton) amplitudes, $A = N + Z$ ($N$, number of neutrons, $Z$, number of protons), and $\mathscr{A}$, the "effective number" of nucleons (Gla 67a) in the target nucleus which undergo quasi-elastic collisions with the incident hadrons, is given as follows (Gla 67a):

$$\mathscr{A} = A \int d^2b\varrho(b)\, e^{-\sigma\varrho(b)} \qquad (2.44)$$

where $\sigma$ is an average of $\sigma_n$ and $\sigma_p$.

One gets the same formula (2.43) as a first approximation to $d\sigma_{\text{inel}}/d\Omega$ if one neglects the Coulomb interactions (Gla 67a). The dominance of quasi-elastic scattering in inelastic cross sections was already shown in the example of $d\sigma_{\text{sc}}/d\Omega$ calculated for the deuteron target [cf. (2.32)]. However, in the case of heavier nuclei, the screening of nucleons by each other is quite appreciable and the factor $\mathscr{A}$ takes care of this phenomenon. The dominance of quasi-elastic scattering makes points (*i*) through (*iv*) plausible. First, the Coulomb interaction corrections come only from the hadron–proton scattering amplitudes and for heavy nuclei and incident protons of $\sim$20 GeV we get from (2.43) a correction on the order of 10% for $\Delta^2 \approx 0.01$ (GeV/c)$^2$, and on the order of 1 to 2% for $\Delta^2 \approx 0.04$ (GeV/c)$^2$ (Czy 69b) (we compare the results with $f_p$, which includes Coulomb phase shift, with the results for $f_p = f_n$). Since (2.43) is a *screened single* scattering, one may suspect that the role the various correlations play in $d\sigma_{\text{inel}}/d\Omega$ is similar to the quasi-elastic electron–nucleon scattering (which is essentially a *single-scattering* process). Although it is still impossible to give very reliable numbers, from the example of $^{16}$O quoted in Section 3 and the discussion given in (Gol 68, Fol 69, Czy 69b) it follows that such similarity does indeed exist. Points (*iii*) and (*iv*) again follow naturally from (2.43) because $\alpha$ is not (whereas $a$ is) important in a single-scattering process.

So, the first approximation formula (2.43) does indeed give us a qualitative explanation of several properties of $d\sigma_{\text{inel}}/d\Omega$ but, in order to make any serious comparison with experiment, one has to do better than that. One of the possible improvements is to explicitly include the multiple scattering contributions to all orders. In some special cases, one can even do it analytically. For instance, if one takes the ground-state density as a product of Gaussian densities

$$\psi_0{}^*\psi_0 = \prod_{j=1}^{A} \varrho(r_j) = \mathcal{N} \prod_{j=1}^{A} \exp(-r_j{}^2/R^2) \qquad (2.45)$$

one can give an analytic expression for $d\sigma_{\text{inel}}/d\Omega$ in (2.23), provided a Gaussian profile is used for the hadron–nucleon profile (2.24). This was pointed out in (Czy 67c) and also, more recently, in (Tre 69a). If one uses, however, some more sophisticated forms of $\psi_0$ one has to evaluate (2.23) numerically. Many results for $d\sigma_{\text{sc}}/d\Omega$ and $d\sigma_{\text{inel}}/d\Omega$ shown in Section 3 [those taken from (Czy 69b)] were obtained after numerical computations of the integrals of (2.23) ($\psi_0{}^*\psi_0$ was taken there in form of products of single-particle densities, as shown in the first half of (2.45); for example, in the case of heavy nuclei the Fermi density distribution was used). From comparison [cfr. (Czy 69b)] of these "exact" computations with the formula (2.43) one sees that large deviations from (2.43) should be expected at small $\Delta$'s.

The other obvious improvement to be made is to include nucleon correlations in $\psi_0{}^*\psi_0$. Several attempts in this direction have been made (Gol 68, Boc 69, Mon 69, Fol 69, Czy 69b). There are, however, considerable difficulties to be solved. The most straightforward, and probably the most effective line of attack, is to expand the ground-state density $\psi_0{}^*\psi_0$ (Fol 69)

$$
\begin{aligned}
| \, \psi_0(\mathbf{r}_1, \ldots, \mathbf{r}_A) \, |^2 &= \varrho^{(A)}(\mathbf{r}_1, \mathbf{r}_2, \ldots, \mathbf{r}_A) \\
&= \varrho(\mathbf{r}_1)\varrho(\mathbf{r}_2)\ldots\varrho(\mathbf{r}_2) + \sum_{\substack{\text{all pairs of} \\ \text{contractions}}} [\varrho(\mathbf{r}_1)\varrho(\mathbf{r}_2)\ldots\varrho(\mathbf{r}_A)]
\end{aligned}
$$

$$+ \text{ higher correlation terms} \qquad (2.46)$$

where

$$\varrho(\mathbf{r}) = \int d^3\mathbf{r}_2 \ldots d^3\mathbf{r}_A \varrho^{(A)}(\mathbf{r}, \mathbf{r}_2, \ldots, \mathbf{r}_A) \qquad (2.47)$$

and a contraction of a pair is defined by

$$\Delta(\mathbf{x}, \mathbf{y}) = \varrho^{(2)}(\mathbf{x}, \mathbf{y}) - \varrho(\mathbf{x})\varrho(\mathbf{y})$$

where

$$\varrho^{(2)}(\mathbf{x}, \mathbf{y}) = \int d^3r_3 \ldots d^3r_A \varrho^A(\mathbf{x}, \mathbf{y}, \mathbf{r}_3, \ldots, \mathbf{r}_A) \qquad (2.48)$$

One can extend the expression (2.46) to include the three-body correlations and so on. If we keep only two terms as shown in (2.46) we reproduce exactly the expectation values of the one-body single-scattering
and two-body double-scattering operators. This follows from the identities

$$\int d^3r_1 \ldots d^3r_A \varrho^A(\mathbf{r}_1 \ldots \mathbf{r}_A) \sum_{j=1}^{A} \mathscr{O}(\mathbf{r}_j) = A \int d^3r \varrho(\mathbf{r}) \mathscr{O}(\mathbf{r}) \qquad (2.49)$$

and

$$\int d^3r_1 \ldots d^3r_A \varrho^A(\mathbf{r}_1 \ldots \mathbf{r}_A) \sum_{i<j=1}^{A} \mathscr{O}(\mathbf{r}_i, \mathbf{r}_j)$$

$$= \frac{A(A-1)}{2} \int d^3x \, d^3y \, \mathscr{O}(\mathbf{x}, \mathbf{y})[\varrho(\mathbf{x})\varrho(\mathbf{y}) + \varDelta(\mathbf{x}, \mathbf{y})]$$

$$= \frac{A(A-1)}{2} \int d^3x \, d^3y \, \mathscr{O}(\mathbf{x}, \mathbf{y})\varrho^{(2)}(\mathbf{x}, \mathbf{y}) \qquad (2.50)$$

If we do not include in $\varrho^A$ of (2.46) the triple and higher many-body correlations, the expectation values of the triple and higher multiplicity scattering
contributions are only approximately calculated. One may hope, however,
that the most important effects of the correlations are included in (2.46).
The papers (Gol 68, Fol 69, Mon 69, Boc 69), and to some extend (Czy 69b)
accept this point of view. In any case, it would seem that the most important
two-body correlations are due to the Pauli exclusion principle and that the
dynamical short-range correlations play a rather insignificant role (Mon 69,
see also Czy 69b). Hence the situation found many years ago in the inelastic
electron scattering from nuclei (McV 62) seems to repeat itself in the case
of high-energy hadron–nucleus scattering. So, the effect of the Pauli exclusion principle correlations, computed for the case of $^{16}$O [Section 3, Figs. 14
and 15, see (Czy 69b)] by taking a completely antisymmetric ground-state
wave function and including the multiplicity of scattering to all orders, does
presumably give a reasonably accurate idea about the role of correlations
in $d\sigma_{el}/d\Omega$, $d\sigma_{sc}/d\Omega$, and $d\sigma_{inel}/d\Omega$ cross sections. Unfortunately, it is difficult
to repeat the same kind of computations for heavy nuclei.

There is a very interesting possibility of extending the high-energy
multiple scattering model to describe the scattering of hadrons (e.g., pion–
nucleon, nucleon–nucleon) in the limit of very high energy. This was
suggested in (Cho 67, Cho 68, Dur 68) and then extended and supplemented by many additional assumptions by very many authors. We shall
describe here the original (and the simplest) version, which also follows
from formula (2.16). First of all, from our experience in applying the

multiple scattering model to nuclear scattering we know that, in the case of elastic scattering, the single-particle densities determine the cross section to a good accuracy (compare Section 3). From (2.16) we get then the following expression for the elastic scattering amplitude:

$$
M = \frac{ik}{2\pi} \int d^2b\, e^{i\Delta \cdot b} \Big\{ 1 - \int \prod_{l=1}^{A} d^2s_l^{A} \varrho_{A,l}(s_l^{A}) \prod_{m=1}^{B} d^2s_m^{B} \varrho_{B,m}(s_m^{B}) \\
\times \prod_{j=1}^{A} \prod_{k=1}^{B} [1 - \gamma_{jk}(\mathbf{b} - \mathbf{s}_j^{A} + \mathbf{s}_k^{B})] \Big\} \tag{2.51}
$$

In (Czy 69a) it was shown that if both $A$ and $B$ become arbitrarily large and all the single-particle densities in $\tilde{A}$ and $\tilde{B}$ separately are assumed the same, (2.51) becomes the so-called optical limit amplitude

$$
M^{\mathrm{opt}} = \frac{ik}{2\pi} \int d^2b\, e^{i\Delta \cdot b} \{ 1 - e^{-\varkappa \int d^2s d^2s' \varrho_A(s) \gamma(b - s + s') \varrho_B(s')} \} \tag{2.52}
$$

where $\varkappa$ is, in general, a complex constant. In the case of pure absorption $\varkappa$ is real ["the subunit of $\tilde{A}$ − subunit of $\tilde{B}$" scattering amplitude is purely imaginary, thus $\alpha = 0$ in (2.24)]. If we parametrize the "subunit of $\tilde{A}$ − subunit of $\tilde{B}$" profile as in (2.24) we have

$$
\varkappa = \tfrac{1}{2}(1 - i\alpha)AB\sigma \tag{2.53}
$$

If $\gamma$ is a very sharply peaked function, compared to $\varrho_A$ and $\varrho_B$, we get

$$
M^{\mathrm{opt}} \approx \frac{ik}{2\pi} \int d^2b\, e^{i\Delta \cdot b} \{ 1 - e^{-\varkappa \int d^2s \varrho_A(s) \varrho_B(b - s)} \} \tag{2.54}
$$

The formula (2.52) in the case of hadron–nucleus scattering becomes

$$
M^{\mathrm{opt}} \approx \frac{ik}{2\pi} \int d^2b\, e^{i\Delta \cdot b} \{ 1 - e^{-\varkappa \int d^2s \varrho_A(s) \gamma(b - s)} \} \tag{2.55}
$$

This formula has been tested in the case of nuclear collisions [(Czy 68a, Czy 69a, Les 69b), see also Section 3] and found in good agreement with experiment [the expression (2.33) for the hadron–nucleus phase shift is just the exponent of (2.55)].

The formulas (2.52), (2.54), and (2.55) depend on very few characteristics of the internal structure of colliding particles and they describe collisions of systems with very many internal degrees of freedom. It is very tempting, therefore, to use them to calculate the nucleon–nucleon or meson–nucleon elastic cross sections. Due to strong interactions one may think that very many internal degrees of freedom take part in these scattering processes. The formulas (2.52) and (2.54) lead to exactly the same results as obtained in (Cho 68, Dur 68) for the high-energy $p$–$p$ scattering. If we accept that the profiles of the subunits are much narrower than $\varrho_A(s)$ and $\varrho_B(s)$, and use (2.54), we can express the amplitude through only one

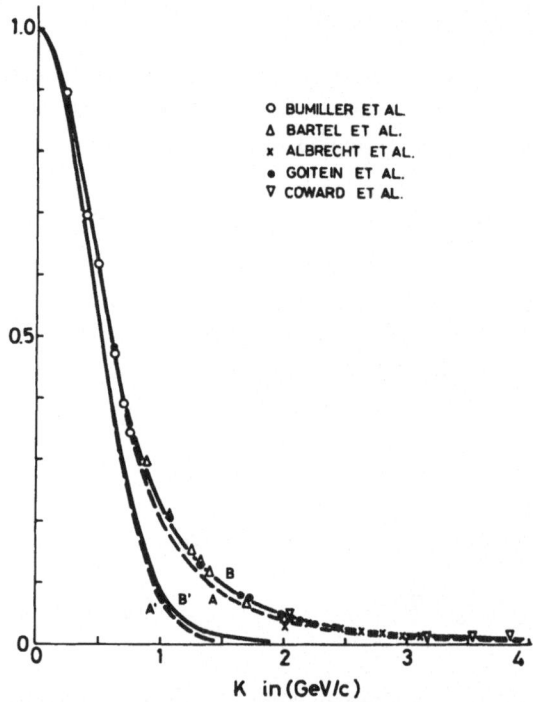

Fig. 2. The proton charge form factor $F_1$ as computed from Eq. (2.54) solved with the help of (2.56) for the proton form factor $F_1(K)$. ($K$ is here the momentum transfer.) The curves (A) and (B) were obtained (Cho 68) with the following fits to the $p$–$p$ elastic cross section:

(A) $(d\sigma/dt)_{pp} = 79.04 \exp(-10.3t) mb\ (\text{GeV}/c)^{-2}$

(B) $(d\sigma/dt)_{pp} = 79.04[\exp(-5.15t) + 0.015$
$\exp(-2t)]^2 mb\ (\text{GeV}/c)^{-2}$

(A') and (B') are computed from (2.54) with only the lowest nonvanishing terms of the integrand kept and the $p$–$p$ cross sections as given above.

complex constant $\varkappa$ (which is not completely free because the measured total cross section imposes a constraint on it). The convolution of the two densities can be replaced by the form factors which are known from the experiment

$$\int d^2s\varrho_A(\mathbf{b} - \mathbf{s})\varrho_B(\mathbf{s}) = (2\pi)^{-2} \int d^2qe^{-i\mathbf{q}\cdot\mathbf{b}}F_A(\mathbf{q})F_B(\mathbf{q}) \qquad (2.56)$$

if we accept that the hadronic matter density is approximately the same as the charge density. It was shown (Cho 68, Dur 68) that from (2.54) and (2.56), with realistic proton form factors, one gets a diffractive structure which may be identified with the one seen in experiment (All 68). One must remember, however, that only qualitative features of such results are significant. The model is essentially applicable to very high energies and the existing data ($\sim$20 GeV) may not yet lie in this asymptotic region.

One can also solve (2.54) and (2.56) for $F_A(\mathbf{q})F_B(\mathbf{q})$ and from for example, the $p$–$p$ high-energy cross section, compute $F_p{}^2(\mathbf{q})$. One gets in this way a very impressive fit to the existing measurements of $F_p{}^2(\mathbf{q})$ [(Cho 68), see Fig. 2].

In such a treatment of high-energy hadron–hadron collisions, one treats them as if they were composed of subunits just as the nuclei are composed of nucleons. In other words we accept that these subunits preexist. One may also apply the methods of the multiple collision model to the systems which do not have any preexisting subunits (e.g., photons), but which can fluctuate. For more details see Section 4.

## 3. EXISTING EVIDENCE SUPPORTING OUR UNDERSTANDING OF THE HIGH-ENERGY SCATTERING

We shall collect the existing evidence for the model of high-energy nuclear collisions presented in the previous section listing all the target nuclei which have been investigated so far and briefly describing and discussing the results. We shall limit ourselves to the experiments performed at energies not lower than 1 GeV incident energy. We shall make very few exceptions from this rule, deviating only in cases when the results are, for some reason or other, very closely connected with the above 1 GeV experiments.

In any case, we shall not consider the phenomena connected with the resonant scattering which, in the case of $\pi^\pm$ scattering, is a dominant feature at energies below 1 GeV.

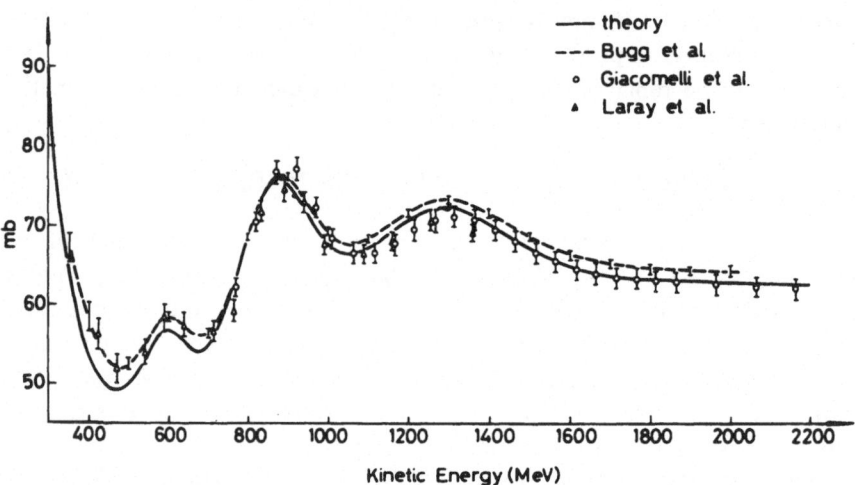

Fig. 3. Theoretical and experimental values for the total $\pi$–$d$ cross section as a function of the kinetic energy of the pion. The dashed line is the mean value of $\pi^+$–$d$ and $\pi^-$–$d$ as measured by Bugg *et al.*, the solid line was computed in (Fäl 68) with the "Fermi motion" taken into account.

## 3.1. Scattering from Deuterons*

This is the "oldest" target and the experimental data are comparatively abundant. First, the total cross sections. As is shown by formula (2.30) the total cross section in our high-energy scattering model is composed of two main contributions: the single scattering and the double scattering. The single scattering contributes, in the case of few GeV pions, about 96% of the total cross section, and the double scattering about 4% (Fäl 68). The single-scattering term is independent of any specific multiple scattering model. Thus the test of the model is rather uncertain because it depends on a very small effect and the absolute normalization of the experimental data is also uncertain to about 1.7% for $\pi^\pm$–$d$ total cross section measurements at a few GeV (Fäl 68, Car 68, Gia 68, Ler 63). Figure 3 shows comparison between the high-energy scattering model and the experimental data. There are also some older calculations, compared with experiment, of the total cross sections for antiprotons (Fra 66a) and protons (Bug 66). The overall agreement with experiment is good, but one should remember that in view of the smallness of the effect and experimental uncertainties this is, so far,

---

* The cross sections shown in the figures are plotted against anlges ($\theta$), momentum transfers ($\Delta$, $q$), or the squares of the momentum transfers ($\Delta^2 = q^2 = -t$).

not a very critical test of the theoretical model. One may worry whether the incident-hadron–neutron cross sections employed in the relation (2.30) are well enough known. A partial answer to this question is contained in (Eng 68) where a very good agreement between $n$–$p$, $n$–$d$ and $p$–$p$, $p$–$d$ total cross sections has been found [for different results cf. (Kre 68)].

The total cross sections for nuclei heavier than the deuteron become more suitable for testing the high-energy scattering model because in these cases the effect of screening is much more pronounced than in the deuteron case (see Table 1).

A much richer source of information comes from the differential cross sections. We present $\pi$–$d$ (Figs. 4, 5, and 6) and $p$–$d$ (Fig. 7) elastic cross sections at various energies, and a poor energy resolution $p$–$d$ cross section

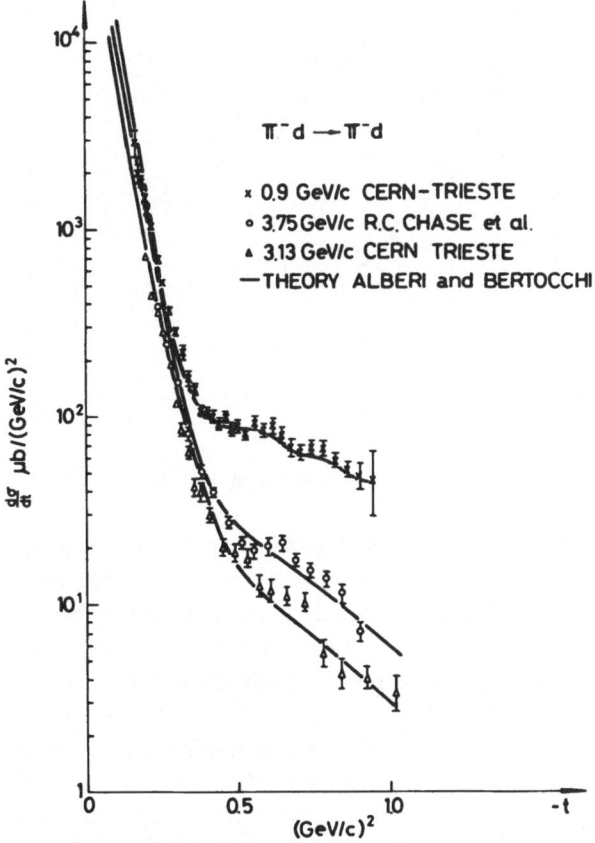

Fig. 4. Comparison of the measured $\pi^-$–$d$ elastic cross sections (Bra 69) with theory (Alb 69).

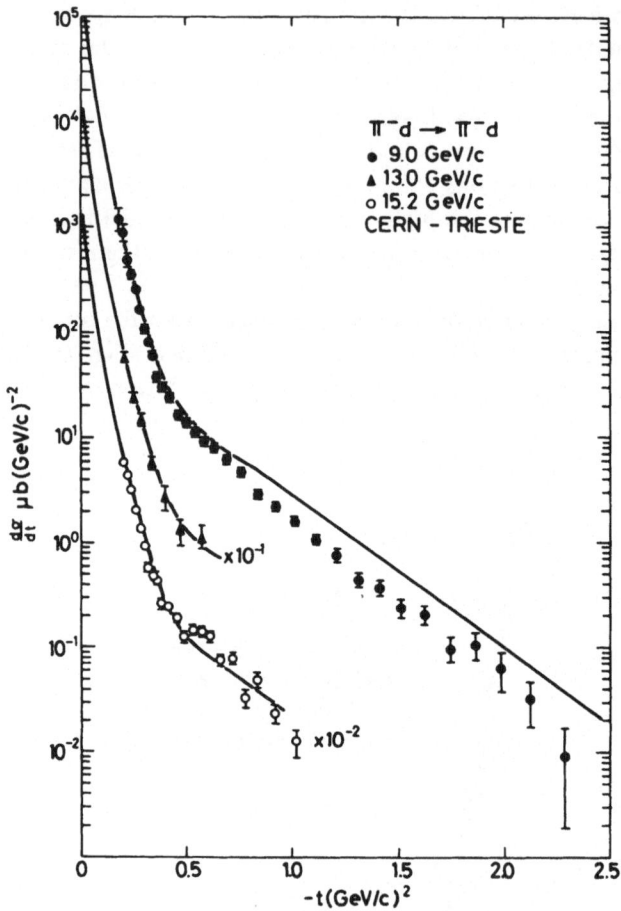

**Fig. 5. Continuation of Fig. 4.**

at 19.2 GeV/c Fig. (8). Before going into any details of these results let us remark that deuteron target is especially important because we know its structure quite well and there are only two nucleons inside. So, if our model of high-energy scattering is correct we ought to be able to explain all scattering phenomena. It is true that the ground-state spin (= 1) and deformation (*D* state) complicate analysis considerably but the uncertainties can, in principle, be estimated and the deuteron has been and, probably, will be one of the most important testing grounds of the model presented in Section 2.

The results presented in Figs. 4, 5, 6, 7, 8 very strongly suggest that our model of Section 2 agrees closely with the experimental results. They also exhibit (compare, e.g., Fig. 7) the effect of $D$ state in the deuteron ground state. Inclusion of the $D$ state *and* spin 1 of the ground state in elastic scattering gives, in addition to the profile $\langle m \mid \Gamma \mid m \rangle$ [dominated by the $S$ state, $m$ being the magnetic quantum number of the ground-state angular momentum and $\Gamma$ the deuteron profile operator (2.25)], another deuteron spin-flip profile $\langle m \mid \Gamma \mid m' \rangle$ for $m \neq m'$, which behaves quite differently than the first one and whose contribution should be added *incoherently* to the first one (Har 68). This contribution, which comes from the deuteron deformation, fills the dip which *always* exist if we take a nondeformed deuteron ground state (only the $S$ state, neglecting the $D$ state). It is worth noticing at this point that even introducing all of the complicated spin effects does not wash away the minimum as long as we have a spherically symmetric ground state ($S$ state only) (Fra 68b, Kuj 68). One needs de-

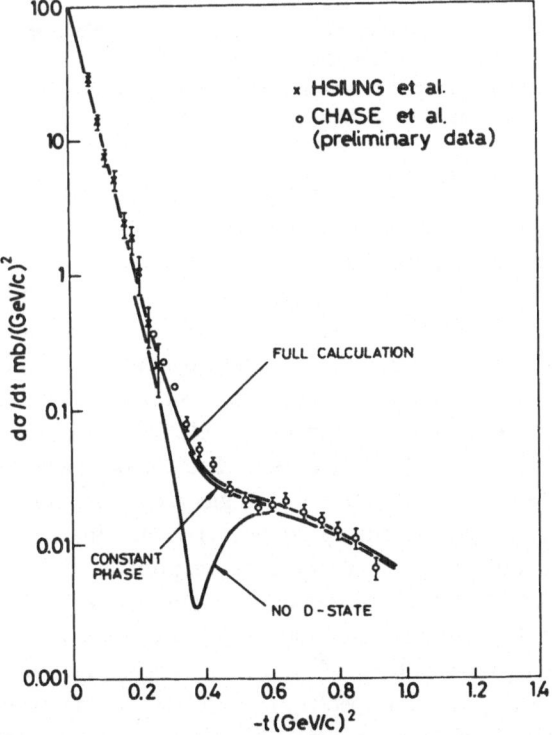

Fig. 6. Comparison of the measured $\pi$–$d$ elastic cross sections at 3.65 and 3.75 GeV/c incident momentum (Cha 69, Hsi 68) with theory (Mic 69). Theoretical curve without $D$-state is also shown.

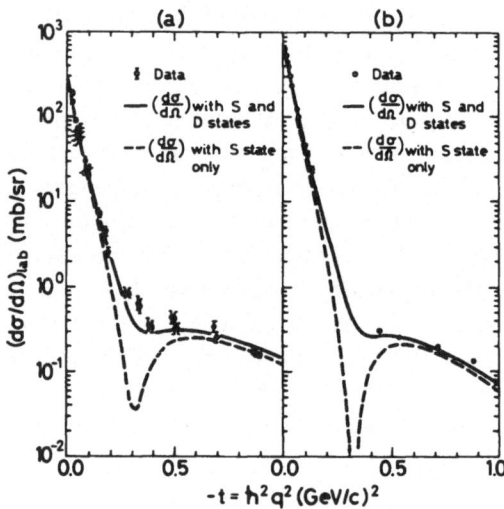

Fig. 7. Comparison of the measured $p$–$d$ elastic cross sections. (a) 1 GeV (Ben 67) and (b) 2 GeV (Col 67, Zol 66, and Kir 66), with theory (Fra 69). Theoretical curves without $D$-state are also shown.

formation ($D$ state) *and* spin 1 of the ground state to fill the minimum and to obtain a good agreement with experiment. This is supported by the experimental results and calculations shown in Figs. 4, 5, 6, 7.

Figures 4 through 8 confirm our understanding of the high-energy scattering of hadrons from deuterons. First of all, the fundamental concept of the existence of single- and double-scattering contributions in the hadron–deuteron scattering amplitude is very strongly supported by the experimental results shown in Figs. 4–8. In the purely elastic scattering cross sections (Fig. 4–7) two different slopes are seen. The steeper part (at smaller momentum transfers) is dominated by single-scattering contributions. The flatter part (at larger momentum transfers) is dominated by the double-scattering contributions. The recent data on poor energy resolution cross section of 19.2 GeV/c protons on deuterium [Fig. 8, (All 69a)] further support our multiple scattering model. This experiment covered the momentum transfer range $0.06 \leq \varDelta^2 \leq 1.6$ (GeV/c)$^2$ and did approximately separate, in the momentum spectrum of the scattered proton, single and double scattering. Since the poor resolution scattering cross section ($d\sigma_{\text{sc}}/d\Omega$) is given by the sum rule (2.32), which is rather insensitive to the details of the ground-state wave function, one can reliably calculate $d\sigma_{\text{sc}}/d\Omega$; the results shown in Fig. 8 nicely agree with all the predictions of the model.

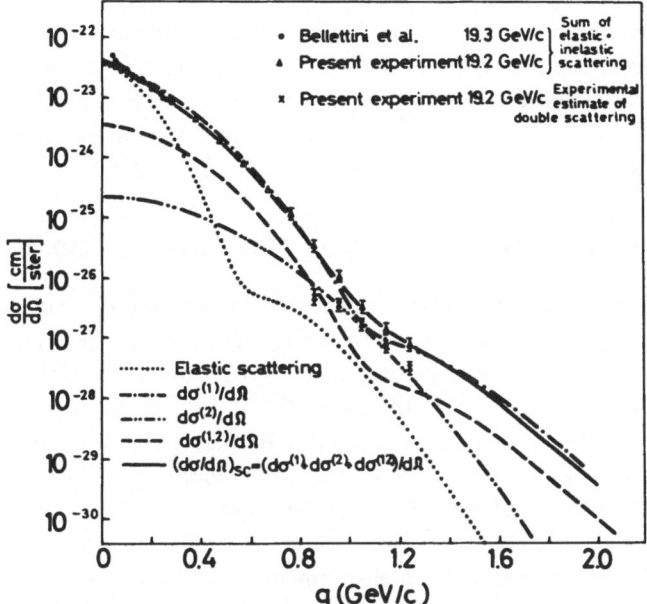

**Fig. 8. Poor energy resolution** $p$–$d$ **cross section measured and computed by the CERN group (All 69a).** The double scattering $(d\sigma^{(2)}/d\Omega)$ contribution to the cross section was experimentally separated from the single scattering contribution $(d\sigma^{(1)}/d\Omega)$ and the interference term $(d\sigma^{(1,2)}/d\Omega)$.

It is also worth noting that the structure seen in the proton–proton elastic scattering cross section (All 68, 69b) is also seen in $d\sigma_{sc}/d\Omega$ of $p$–$d$ scattering at 19.2 GeV (All 69a). Measurements of $d\sigma_{sc}/d\Omega$ without separating single and double scattering, however, for 5.53 GeV/c pions on deuterons (Cha 69), do not exhibit as good agreement with the computed cross section as in the case of protons. However, one must keep in mind that the $\pi$–nucleon amplitudes in (Cha 69) were taken without any structure (a Gaussian dependence on the momentum transfer squared was assumed), which is a very drastic oversimplification.

There also exist experimental results on proton–deuteron scattering at 0.582 GeV (Bos 68, Bos 69) which look very similar to the 1 GeV $p$–$d$ scattering (Ben 67), and also measurements of proton polarization at 0.544 GeV, which were compared with calculations (Bos 69, Rem 68, and 69). Also deuteron–deuteron scattering at six laboratory momenta, from 0.68 to 2.12 GeV/c, was measured (Gos 69).

Let us finally mention that in 1 GeV $p$–$d$ backward scattering (Ben 67), hence in the region where the multiple-scattering model presented in Section 2 is presumably inapplicable, there exists a well-established peak explained (Ker 69) by the presence (with 1/2 to 1% probability) of the $N^*(1688)$ resonance in the deuteron ground state. This interpretation has been further strengthened recently by the analysis of the large angle $p–d$ scattering at 580 MeV (Vin 70).

## 3.2. Scattering from ³He, ⁴He, ⁶Li, ⁷Li, and ⁹Be Targets

As it was already explicitly illustrated in the case of deuteron, there are very essential differences between the purely elastic $d\sigma_{el}/d\Omega$ good energy resolution and the poor energy resolution scattering cross sections without meson production $d\sigma_{sc}/d\Omega$. In the case of ³He and ⁴He nuclei, only purely elastic cross sections were measured (Bos 68, 69, and Pal 67) whereas ⁶Li and ⁹Be targets were used only in the poor energy resolution scattering experiments [to the best of our knowledge (Bel 66)]. It is always desirable to have consistent explanation for both cross sections $d\sigma_{el}/d\Omega$ and $d\sigma_{sc}/d\Omega$ plus all possible polarization measurements for each incident beam and each target nucleus, but we still do not have such complete series of experiments.

Elastic scattering of 0.582 GeV protons from ³He (Bos 69) exhibits different slopes of $d\sigma_{el}/d(\Delta^2)$ but no diffractive minima. From our experience with deuterium in the previous section one may expect this to happen. Presumably, ³He is not spherical. This, combined with the fact that its ground-state spin is 1/2, will force us (even if we neglect spin interactions of the incident proton and target nucleons) to add *incoherently* the non-target spin-flip and target spin-flip contributions and this may have the effect of destroying the diffractive minima.

The first measurements of $d\sigma_{el}/d(\Delta^2)$ for 1 GeV protons scattered from ⁴He were reported at the Gatlinburg Conference on Nuclear Structure in 1966 (Pal 66) and were promptly interpreted in terms of the diffractive multiple-scattering model (Glauber model, Czy 67a, 67b, Bas 67). It was also quickly realized (Czy 67a,b,d, Wil 67, Bas 68) that if one uses the best available data for the density of ⁴He obtained from the elastic electron scattering (Fro 67) the agreement with experiment improves and becomes very impressive. Figures 9 and 10 illustrate this last point. It was also pointed out (Czy 68a, 69a) that the most important physical reason for the original failure (Pal 66, 67) of the optical model to agree, even approximately, with the data was the lack of interrelations between the coordinates of the four nucleons forming the ⁴He nucleus (the constraint imposed

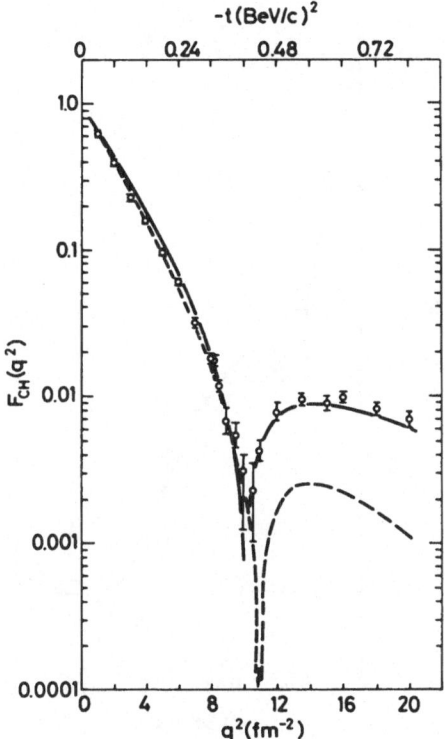

**Fig. 9. Elastic electron–⁴He form factor.** The experimental data is from (Fro 67). The solid curve using the double-Gaussian density [(Bas 68), Eq. (4.11)]. The dashed curve was obtained after correlations were introduced into the ground state density [(Bas 68), Eq. (4.10)].

on the nucleons by the correct treatment of the c.m. motion; lack of such a constraint is inherent in the optical model, which treats the target as composed of spatially extended stuff interacting with the incoming particle).

The more nucleons there are in the target nucleus the less important this constraint is and for ¹⁶O (see below) it is already negligible. The spin effects, for $p$–⁴He case, were estimated (Fra 68b, Kuj 68) and found important, especially at larger momentum transfers. Unfortunately, in these last calculations, the simplest possible (Gaussian) ground-state wave functions for ⁴He were used. This was due to a considerable increase of complexity of computations when one uses the input nucleon–nucleon scattering ampli-

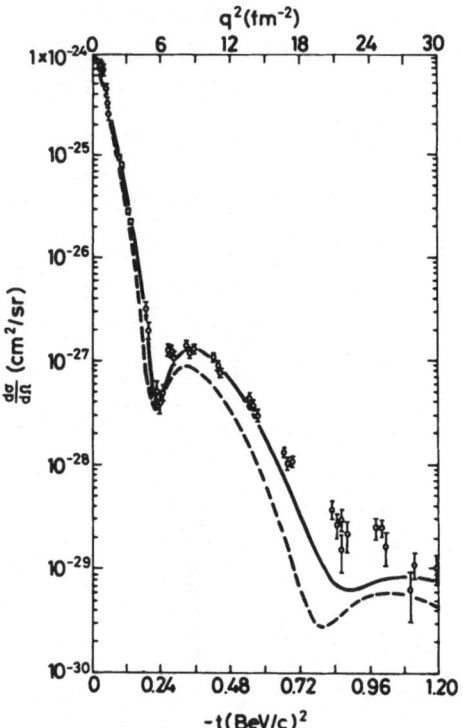

**Fig. 10. Elastic proton–⁴He cross section at 1 GeV.** The experimental data taken from (Pal 67). The solid curve evaluated with the same double-Gaussian density as in Fig. 9 [(Bas 68), Eq. (4.11)]. The dashed curve was calculated with correlations introduced [(Bas 68), Eq. (4.10)].

tudes in their full and glorious complexity. So, for $p$–⁴He scattering, we have computations with good densities but without spin effects, and computations with spin effects but with incorrect densities. Some recent computations by Ford and Pentz of $p$–⁴He cross section and polarization at 0.587 and 0.544 GeV respectively, were also reported (Bos 69).

A few general remarks are in order here. Firstly, the computations which do not contain the spin variables explicitly do take *some* of the spin effects into account. In such calculations [e.g., (Bas 67, Czy 67a)] one takes nucleon–nucleon amplitudes from the nucleon–nucleon cross sections; thus there exist some contributions of the spin-dependent amplitudes in these calculations, giving in some sense an average effect of the spins. So, the proton–nucleus amplitude also does contain some average spin contributions. It may turn out that, in the limit of very high energy, where the spin-flip part of the proton–$^4$He amplitude is negligible (it goes to zero with scattering angle because it contains as a factor $\sin\theta$), such an approach may to be quite adequate. To support this hypothesis let us point out that in the case of $\pi$–$^4$He scattering, when there exists only one amplitude, the calculations which take into account the spin effects of the individual $\pi$–nucleon amplitudes (For 68) give virtually the same results as the simplest calculations which neglect the spins. In these cases (low-energy scattering) where $\sin\theta$ is not negligible and the spin-flip part of the proton–nucleus amplitude has to be taken into account (one may expect large polarization effects there) one probably has to worry about spins. One would like to have, however, some simplified version of treatment of spins because con-

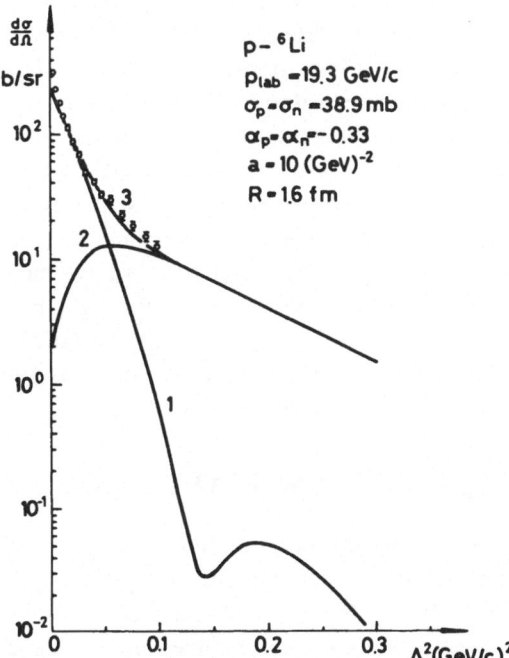

Fig. 11. Comparison of the theory (Czy 69b) with the data for proton–$^6$Li poor resolution experimental data at 19.3 GeV/c (Bel 66). (1) Elastic scattering, (2) inelastic scattering [Eq. (2.23)], (3) scattering cross section [Eq. (2.21)]. The oscillator potential ground state density was used [(a) of Section 3] with $\beta$ coefficient for $^6$Li.

struction of the total nucleon–nucleus amplitude from the individual nucleon–nucleon amplitudes with all spin dependences included is a hair-raising method.

We are of the opinion that there are many cases (with only one total amplitude) in which one does not need to introduce spin explicitly into the calculations. In such cases, we have the very simple and attractive result that the only important input concerning the structure of the target is the single-particle density. If we accept that in elastic electron scattering from spin zero nuclei one measures the Fourier transform of this density (the charge form factor), we may conclude that the same form factor determines the electron and hadron–nucleus elastic cross sections. It is a very interesting question as to how general is the relation between the form factors and hadronic cross sections. We shall come back to this conjecture many times in this paper.

Now let us discuss briefly the poor energy resolution cross section $d\sigma_{sc}/d(\Delta^2)$ for light $^6$Li, $^7$Li, and $^9$Be nuclei (Bel 66) (there are no $d\sigma_{el}/d(\Delta^2)$ measured for these targets). First attempts to calculate $d\sigma_{sc}/d(\Delta^2)$ for light nuclei (Czy 67c) from the sum rule (2.21) showed a complete consistency with the calculations of the elastic scattering cross sections. They also showed comparative, insensitivity of $d\sigma_{sc}/d(\Delta^2)$ to the details of the ground-state wave function and the lack of diffractive structure (for light nuclei; for heavy nuclei the diffractive structure reappears, see below). Figure 11 shows $d\sigma_{sc}/d(\Delta^2)$ and $d\sigma_{el}/d(\Delta^2)$ for $^6$Li (Czy 69b). We can see the relative size of these two cross sections, very typical for light nuclei, and a very good agreement with experimental data (Bel 66). As far as $d\sigma_{sc}/d(\Delta^2)$ goes, there are inessential differences between $^6$Li on one hand and $^7$Li and $^9$Be on the other [for more details see (Czy 69b)].

## 3.3. Scattering from $^{12}$C and $^{16}$O Targets

The nucleus $^{12}$C is one of the very few targets for which $d\sigma_{el}/d(\Delta^2)$, the spectra of inelastically scattered protons, and the total cross section were measured (Pal 67, Fri 67, Igo 67, and Cor 68). The spectra of the outgoing protons were taken at different fixed angles, hence it was impossible to perform the sum over all excitations at precisely constant momentum transfer. Besides, the proton spectra, at large energy losses, were very strongly influenced by meson production, hence the contributions to the inelastic scattering coming from purely nuclear excitations and the meson production process could not be distinguished precisely. Consequently one could construct $d\sigma_{sc}/d(\Delta^2)$ only approximately (Czy 68b). Figure 12 [taken

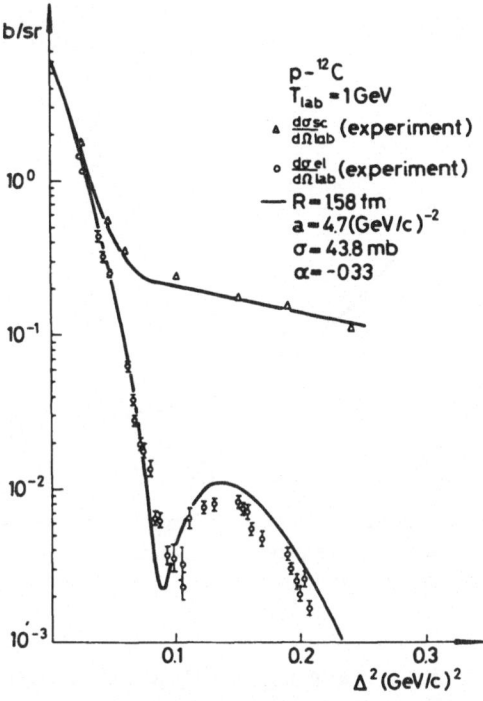

**Fig. 12.** Proton–¹²C good (points) and poor (trian-
gles) resolution scattering cross sections. Taken
from (Fri 67, Cor 68, and Czy 68b); theoretical
curves are taken from (Czy 69b). The oscillator
potential ground-state density was used [(a) of
Section 3] with $\beta$ coefficient for ¹²C.

from (Czy 69b)] shows calculated $d\sigma_{el}/d(\Delta^2)$ and $d\sigma_{sc}/d(\Delta^2)$ from the sum
rule against the available data at 1 GeV (Pal 67, Fri 67, and Cor 68). Figure
13 shows calculated $d\sigma_{sc}/d(\Delta^2)$ (Czy 69b) against the 20 GeV data (Bel 66).
From these two figures we can see that the agreement with all the existing
data is very good. The elastic cross section does not show an ideal fit but
one must remember that ¹²C is probably nonspherical, and that may change
the cross section the right way (Les 69a).

Now let us go over to the case of ¹⁶O target. This is an especially simple
target: spherical (as far as we can judge from the nuclear spectroscopic
data), spin zero, and doubly magic. However, only the proton elastic
scattering cross section and the total cross section have been measured
(Pal 67, Igo 67) at 1 GeV. Nevertheless, the numerical results one gets out

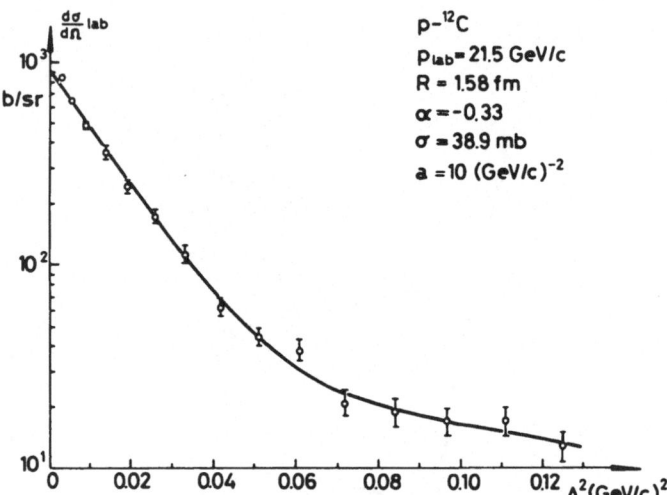

**Fig. 13. Poor energy resolution** $p$–$^{12}$C **scattering cross section at 21.5 GeV/c (Bel 66) compared with the calculations from (Czy 69b). The oscillator potential ground-state density was used.**

of our diffractive multiple scattering model should be treated pretty seriously because, due to the simplicity of the target nucleus, the numerical results have a good chance of proving (or disproving) the model. Let us start with Figs. 14 and 15 (Les 69b, Czy 69b) which show $d\sigma_{el}/d(\Delta^2)$ [with the experimental data (Pal 67)] and $d\sigma_{sc}/d(\Delta^2)$ for the following three cases of the ground state of $^{16}$O [for $d\sigma_{el}/d(\Delta^2)$ essentially the same results are contained in (Bas 68)]:

(a) The single-particle density model, which assumes:

$$\psi_0{}^*\psi_0 = \prod_{j=1}^{A} \varrho(r_j)$$

where

$$\varrho(r) = \varrho(0)\left(1 + \beta\,\frac{r^2}{R^2}\right) e^{-r^2/R^2}$$

(b) The product wave function $\psi_0 = \prod_{j=1}^{A} \varphi_j(r_j)$, where $\varphi_j(r_j)$ are the single-particle states of the oscillator-well shell model:

$$\varphi_0(\mathbf{r}) = 2\pi^{-1/4}R^{-3/2} \exp(-r^2/2R^2)\,Y_{00} \qquad \text{for } s \text{ shell}$$

$$\varphi_{1m}(\mathbf{r}) = 3^{-1/2}2^{3/2}\pi^{-1/4}R^{-5/2}r \exp(-r^2/2R^2)\,Y_{1m} \qquad \text{for } p \text{ shell}$$

(c) The antisymmetrized wave function:

$$\psi_0 = (A!)^{-1/2} \, \| \, \varphi_j(r_j) \, \|$$

where $\| \ldots \|$ is the determinant of the $\varphi_j$ functions defined in (b).

From Figs. 14 and 15 we can see that, first of all, if we take the best ground-state wave function (c) we get excellent agreement with experiment. Besides, in the case of $d\sigma_{el}/d(\varDelta^2)$ (Fig. 14) all three cases (a), (b), and (c) differ very little. Instead, in the case of $d\sigma_{sc}/d(\varDelta^2)$ (Fig. 15), although the cases (a) and (b) are very similar, the case (c) is definitely different both from (a) and (b). This means that, at rather small momentum transfer, correlations (in our case the Pauli-exclusion-principle correlations) do appreciably influence $d\sigma_{inel}/d(\varDelta^2) = d\sigma_{sc}/d(\varDelta^2) - d\sigma_{el}/d(\varDelta^2)$ [for a more detailed discussion see (Czy 69b, Czy 70)].

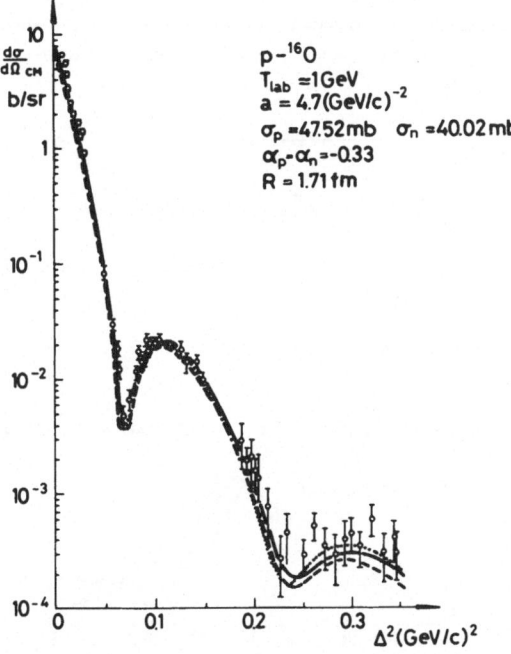

**Fig. 14.** Elastic $p$–$^{16}$O cross section at 1 GeV. The experimental points taken from (Pal 67). The three curves were computed (Les 69b) for the three models of the ground state (see the text): case (a), dashed line; case (b), dotted line; and case (c), solid line.

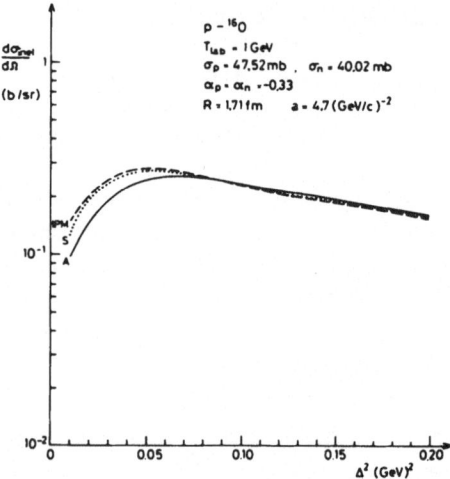

**Fig. 15. Inelastic $p$-$^{16}$O scattering at 1 GeV for the same cases as in Fig. 14.** IPM, independent particle model [see case (a) in the text]; S, product wave function [case (b) in the text]; A, completely antisymmetrized wave function [case (c) in the text]. All curves contain corrections due to the nucleus center-of-mass motion, as discussed in Section 2. In the case of $^{16}$O they are important for small momentum transfers.

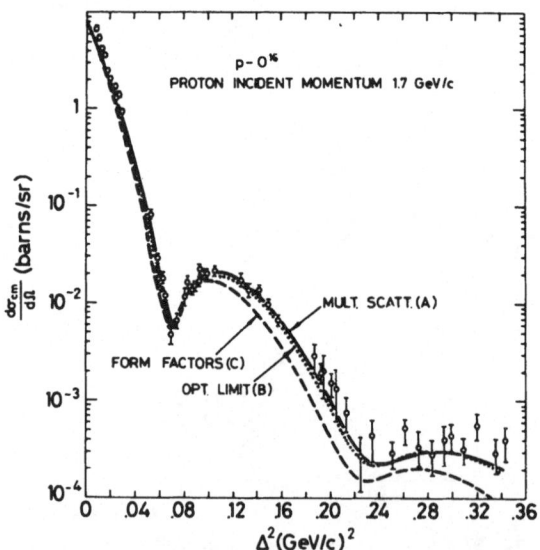

**Fig. 16. Elastic $p$–$^{16}$O scattering cross section at 1 GeV.** The experimental points taken from (Pal 67). The three theoretical curves come from (Czy 69a) and they are obtained from the usual multiple-scattering cross section: case (a) in the text is marked (A), formula (2.55) is marked (B), and formula (2.54) is marked (C).

Figure 16 shows comparison of the case (a) with the "optical limit" (Czy 69a) given by (2.55), the formula (2.54), which gives the elastic cross section in terms of the charge form factors of the colliding objects [in this case the proton and $^{16}O$ form factors are used, according to formule (2.54)], and the experimental points (Pal 67). We can again see a very good agreement between the multiple scattering formula and experiment, and also the fact that for $A = 16$ we are already very, very close to the limit of infinitely many subunits in the target (optical limit) and, also very close to the cross section (2.54) used in the case of nucleon–nucleon elastic scattering by Chou and Yang (Cho 68) and Durand and Lipes (Dur 68).

From our general considerations in Section 2 it follows that $^{16}O$ might

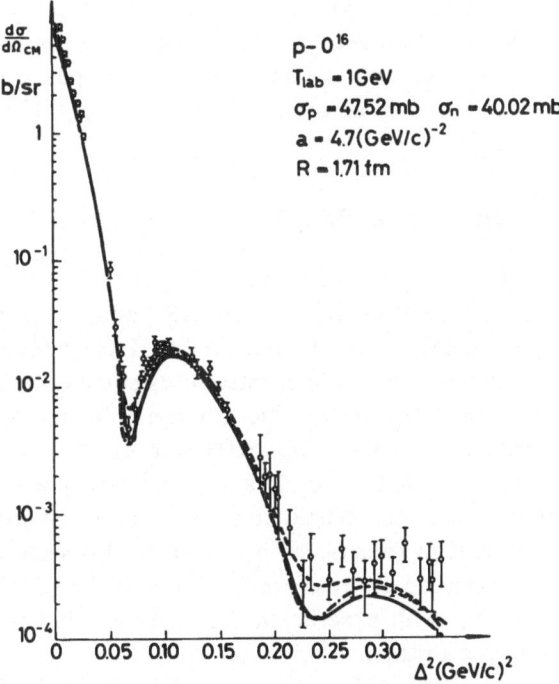

**Fig. 17. Influence of the Coulomb interaction on the elastic $p$–$^{16}O$ cross section at 1 GeV.** The experimental points taken from (Pal 67). All curves are calculated from the single-particle densities (Czy 69b) [see the case (a) in the text]. Solid line: Coulombic interactions included, $\alpha_p = -0.06$ and $\alpha_n = -0.4$, as given in (Bug 66); dashed line: Coulombic interactions included, $\alpha_p = \alpha_n = -0.33$; dotted–dashed line: no Coulombic interactions, $\alpha_p = \alpha_n = -0.33$.

turn out to be an excellent target for measurements of the real parts of the elastic incident-hadron–target-nucleon amplitudes. Since such measurements would rely very heavily on the computed values of $d\sigma_{el}/d(\Delta^2)$ at the diffractive minima, we show in Fig. 17 [taken from (Czy 69b)] the role of Coulomb interactions in the elastic $p$–$^{16}$O scattering at 1 GeV incident kinetic energy. The experimental points are also shown (Pal 67). We can see that any serious measurement of $\gamma_{n,p}(b)$ from $^{16}$O *must* include Coulomb interactions. The Coulomb interaction changes (pushes them up) the curves at the minima by about 100%. From the fact that the curve with $\langle\alpha\rangle_{av} = -0.33$ *without* Coulombic interaction is almost the same as the curve with $\langle\alpha\rangle_{av} = -0.23$ *with* Coulomb interaction, one sees that the Coulombic interaction acts as if a correction to the strong interaction $\alpha$ were present (amounting in this case to $-0.10$). The cross sections shown in Fig. 17 were calculated in the "optical limit". Reference (Les 69b) and Fig. 15 show that if we used the multiple scattering cross section with the properly antisymmetrized ground-state wave function, all the cross sections of Fig. 17 would be shifted upwards about 10%.

## 3.4. Scattering from $^{27}$Al, $^{64}$Cu, $^{208}$Pb, and $^{238}$U Targets

For these four target nuclei only the poor energy resolution cross sections [from the CERN experiment (Bel 66)] and the total cross sections (see the last part of this section) are available. Except for $^{27}$Al, the targets used were not pure isotopes, which introduces some uncertainties into the interpretation. In all these cases, the Coulomb interaction plays a very fundamental role and neither $d\sigma_{el}/d(\Delta^2)$ nor $d\sigma_{sc}/d(\Delta^2)$ can be reliably calculated without it. In Section 2, a method was presented which was used in computation of the elastic and scattering cross sections shown in Figs. 18 and 19. Out of four nuclei discussed in this section only two (experimental and theoretical) cross sections (on $^{64}$Cu and $^{208}$Pb) are shown here. The agreement with experiment for $^{27}$Al and $^{238}$U is also very good except near the first diffractive minimum of $^{238}$U.

One should perhaps make a comment here about the fits presented in Figs. 18 and 19 and the fits (to the same experimental data) discussed in the paper by Goldhaber and Joachain (Gol 68). The point made in their paper was that if one very carefully fits the calculated $d\sigma_{sc}/d(\Delta^2)$ to the experimental points at small momentum transfers (where their model is, according to the authors, very reliable) one is unable to fit the part of the cross section which exhibits maxima and minima using any "reasonable" (i.e., commonly accepted) choice of the neutron and proton densities.

One may make the following remarks. First of all, the model (Section 2) used in the fits shown in Figs. 18 and 19 was different from that used by Goldhaber and Joachain (Gol 68). There is, however, about a 10% difference [in the case of $^{208}$Pb target (Fig. 19)] between theory and experiment at very small momentum transfers (it is rather difficult to see it because of the logarithmic scale). But, one may have very serious doubts as to whether, within the framework of the existing models, the small momentum transfer region of $d\sigma_{sc}/d(\Delta^2)$ can be reliably calculated. The uncertainty comes

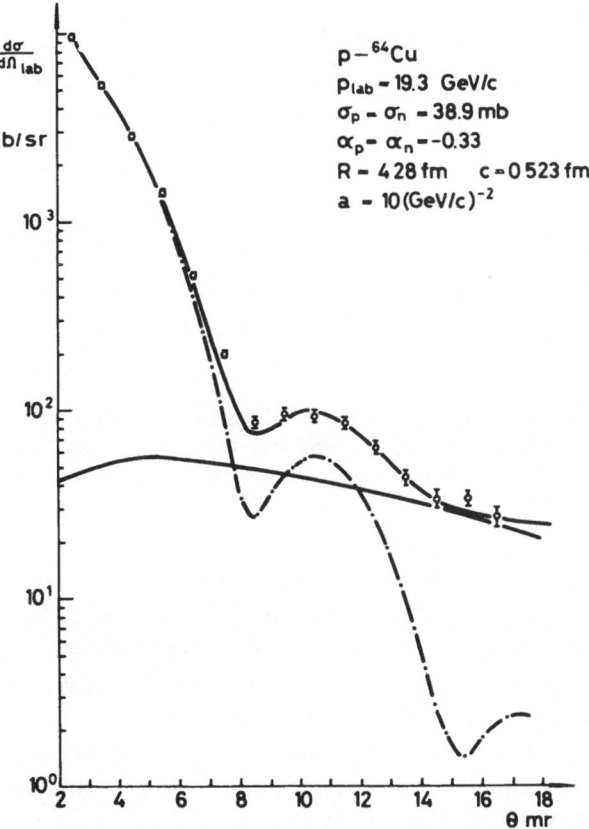

**Fig. 18. Comparison of the theoretical predictions with the experimental results for $^{64}$Cu.** [Poor energy resolution cross section at 19.3 GeV/c (Bel 66).] The solid line close to the experimental points: $D\sigma_{sc}/D\Omega$; the flat solid line: $d\sigma_{inel}/d\Omega$; and the dotted–dashed line: $d\sigma_{el}/d\Omega$. The Fermi distribution $\varrho(r) = \varrho_0(1 - \exp[(r - R)/c])$ was used as the ground-state density (Czy 69b).

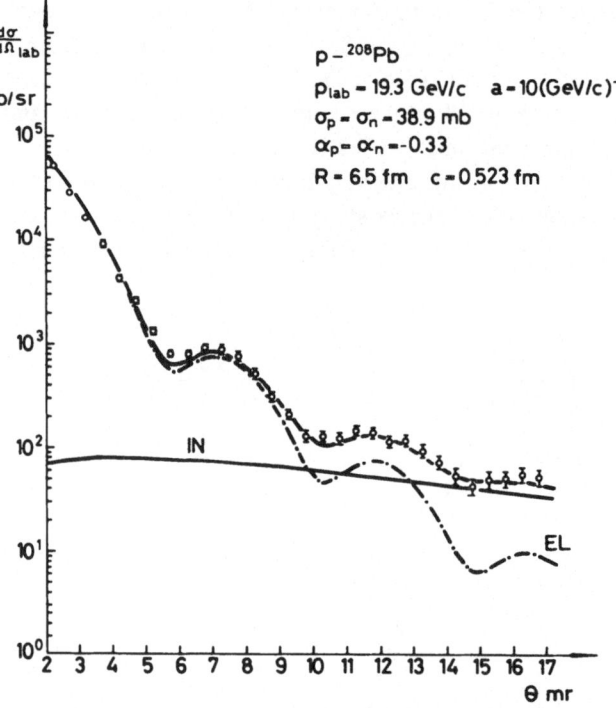

Fig. 19. The same as in Fig. 18 except the target is $^{207.3}$Pb (the calculations are for pure $^{208}$Pb). The Fermi distribution $\varrho(r) = \varrho_0(1 - \exp[(r - R)/c])$ was used as the ground-state density (Czy 69b).

from the very probable existence of the so-called coherent "inelastic shadow" effects (Pum 68, see also Czy 69b,c) which should already be important at small momentum transfers for $\sim$20 GeV incident hadrons. In any case it would seem that, at high incident energy, one should fit the calculated curve in the region of medium momentum transfer. At energies of few GeV however, where the coherent "inelastic shadow" effects are presumably nonexistent, the small momentum transfer region would indeed be the "reliable" region.

## 3.5. Total Cross Sections

Table 1 summarizes the existing data on the total cross sections (the deuteron target is not included because it was discussed separately at the beginning of this section) and results of calculations based on the diffractive multiple-scattering model described in Section 2. As one can see from Table 1, the agreement is good. There are essentially two groups of theoretically calculated total cross sections. One of them consists of the total cross sections which were computed together with $d\sigma_{el}/d(\Delta^2)$ and/or $ds_{sc}/d(\Delta^2)$. In other words, the same parameters were used to fit a very considerable amount of data beside the total cross section. These are cross sections for 20 GeV protons and 1 GeV protons. The other group consists of all the remaining results which were obtained just to fit the total cross section data ($d\sigma_{el}/d(\Delta^2)$ and $d\sigma_{sc}/d(\Delta^2)$ in these cases have not been measured to the best of the author's knowledge). One may generally say that there are no problems in fitting total cross sections. If one has a good fit to the, for example, elastic cross section, one gets, as a rule, a very good agreement with the measured total cross section using the same parameters and nuclear ground-state wave functions. So, there is, so far, a complete consistency with the results discussed in the preceding sections.

# 4. SCATTERING OF OBJECTS WITHOUT "PREEXISTING" SUBUNITS. SCATTERING FROM FLUCTUATIONS

In Section 2 a model of high-energy scattering of composite objects was discussed which assumed that $\tilde{A}$ and $\tilde{B}$ were composed of $A$ and $B$ subunits, respectively, and the total scattering amplitude was constructed from the individual "subunit of $\tilde{A}$ — subunit of $\tilde{B}$" scattering amplitudes. This model has been applied successfully to nuclear scattering process (the existing evidence is collected in Section 3) and extended to high-energy hadron–hadron scattering (Cho 68, Dur 68). It turned out to be a generalization of the Wu and Yang model of hadron–hadron scattering proposed a few years ago [(Wu 65), the Wu and Yang model is its first approximation]. In all the cases considered so far the subunits of scattering objects were assumed to exist before the collision took place ("preexisting" subunits). It turns out, however, that the same technique of constructing the total amplitude from the amplitudes of some subunits of the colliding objects may also be useful if these subunits show up only in fluctuations of the colliding objects.

One can see this in the example of scattering of photons from various targets. Since quantum electrodynamics describes photons and their interactions so well it would seem that the photon structure is as different from the structures of particles with preexisting subunits sitting inside them

**TABLE I. Total Cros**

| Target | 20 GeV protons | | 1 GeV protons | | 1.4 GeV neutrons | | 10 GeV/ neutrons |
|---|---|---|---|---|---|---|---|
| | Bel 66[e] | calculated[a] | Igo 67[e] | calculated | Coo 55[e] | calculated | Eng 68[e] |
| He | — | — | 152±8 | 145.6[b] | — | See | 141±6 |
| ⁶Li | 232±5 | 197 | 199±11 | — | — | (Abu 68), | — |
| ⁷Li | 250±5 | 228 | — | — | — | Fig. 1 | 237±7 |
| Be | 278±4 | 286 | — | — | 308±13 | | 271±6 |
| C | 335±5 | 342 | 370±9 | 358[b] | 378±10 | | 340±3 |
| O | — | — | 475±44 | 469.4[b] 467[c] | — | | — |
| Al | 687±10 | 657 | — | — | 703±18 | | 683±3 |
| Fe | — | — | — | — | — | | 1204±12 |
| Cu | 1360±20 | 1253 | — | — | 1388±39 | | 1364±14 |
| Sn | — | — | — | — | 2202±62 | | — |
| Pb | 3290±100 | 2965 | 3155±450 | — | 3209±55 | | 3146±50 |
| Bi | — | — | — | — | 3275±62 | | — |
| U | — | 3117 | — | — | 3640±91 | | — |

[a] L. Leśniak and H. Wołek, private communications. The same parameters and ground-stat
[b] (Les 69b). Compare also Fig. 1 of (Abu 68).
[c] (Bas 68).
[d] (Abu 69).
[e] Experimental.
[f] Absorption cross sections.

(atoms, nuclei) as it can be. But, in accordance with the standard concepts of the theories of quantized fields, the photon fluctuates: it becomes sometimes an electron–position pair, sometimes a pair of pions, sometimes a nucleon–antinucleon pair, sometimes a vector meson, etc.; Fig. 20 shows the corresponding graph. Let us limit ourselves to purely electromagnetic interactions. Then in the spirit of our model of very high energy scattering, it should be possible to express the photon scattering from an external Coulomb field (Delbrück scattering), for example, through the electron and

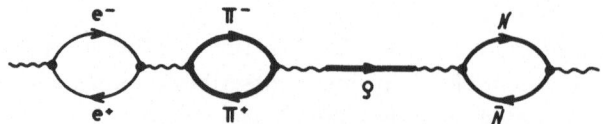

Fig. 20. A fluctuating photon.

**Sections in Milibarns**

| 27 GeV/c neutrons Jon 68[e] | 8.3 GeV/c neutrons Pan 62[e] | 20–65' GeV/c $K^-$ | | 20–65 'GeV/c $\pi^-$ | | 20–65' GeV/c $\bar{p}$ | |
|---|---|---|---|---|---|---|---|
| | | All 69c[e] | calculated[d] | All 69c[e] | calculated[d] | All 69c[e] | calculated[d] |
| — | — | — | — | — | — | — | — |
| — | — | 102 | 97.364 | 120 | 114.277 | 182 | 174.769 |
| — | — | — | — | — | — | — | — |
| 250±6 | — | 135 | 137.148 | 149 | 159.361 | 230±10 | 234.606 |
| 300±7 | 345±15 | 161 | 173.994 | 178 | 200.331 | 275±7 | 286.653 |
| — | — | — | — | — | — | — | — |
| 573±17 | 600±23 | 310 | 329.944 | 332 | 370.185 | 480±10 | 488.924 |
| 1023±15 | — | — | — | — | — | — | — |
| 1090±90 | 1217±48 | 590 | 631.864 | 630 | 689.020 | 880±50 | 842.794 |
| — | 1805±75 | 900±70 | 973.671 | 1080 | 1042.95 | 1280±150 | 1222.01 |
| 2630±120 | 2556±100 | 1350±50 | 1419.80 | 1530 | 1499.86 | 1650±100 | 1704.04 |
| — | — | — | — | — | — | — | — |
| 2770±150 | — | 1420 | 1555.40 | 1580 | 1638.06 | 1780 | 1848.73 |

densities were used as in (Czy 69b). Compare also Fig. 2 of (Abu 68).

position high-energy scattering amplitudes (from a Coulomb field) and the "wave function" of the electron–positron fluctuation. More than that through, essentially, the same two amplitudes and the same "wave function" of the fluctuation one *must be able* to express, if the model is right, the high-energy photon–photon scattering cross section (as a scattering of a two electron–positron pairs from each other) and the Compton cross section (as a scattering of an electron–positron pair from an electron). Such calculations should give the leading terms in the cross sections listed above. One could also include the fluctuations which contain, for example, *two* electron–positron pairs. They would, however, contribute only negligible corrections to the leading terms. Similarly, as in the cases of collisions of composite objects with preexisting subunits, the "ground-state wave functions" (or the fluctuation densities) and the subunit–subunit high-energy scattering amplitudes have to be taken from somewhere else (see also Appendix B). In the case of quantum electrodynamics they have been calculated [(Che 69a), and the references cited therein]. With this "input" taken from outside of the model we get expressions for all the above listed electro-dynamical processes in terms of these two universal quantities, which is a nontrivial result. These results are identical with those of Cheng and Wu (Che 69a,b) and Chang and Ma (Chan 69).

Indeed, let $\varrho(\mathbf{s}_1, \mathbf{s}_2)$ be the "spatial density" of the electron–positron fluctuation and its Fourier transform

$$\tilde{\varrho}(\boldsymbol{\delta}_1, \boldsymbol{\delta}_2) = \int d^2s_1 \, d^2s_2 \, e^{i\boldsymbol{\delta}_1 \cdot \mathbf{s}_1 + i\boldsymbol{\delta}_2 \cdot \mathbf{s}_2} \varrho(\mathbf{s}_1, \mathbf{s}_2) \tag{4.1}$$

The functions $\varrho$ and $\tilde{\varrho}$ depend also on the photon polarization before and after the fluctuation, but we shall not show this dependence explicitly because it is irrelevant in our considerations. Then in order to compute the Delbrück scattering or Compton scattering amplitude at high energy, one can use Eq. (2.16) assuming that, say, $\tilde{A}$ is composed of an electron-positron pair whose spatial distribution is given by $\varrho(\mathbf{s}_1, \mathbf{s}_2)$ and $\tilde{B}$ consists of a given Coulomb field or an electron (the algebra is analogous in these two cases). Introducing the following Fourier transforms of the profile functions:

$$\Sigma_{\pm}(\boldsymbol{\delta}) = \int d^2t \, e^{i\boldsymbol{\delta} \cdot \mathbf{t}}(1 - \gamma_{\pm}(t)) = \int d^2t \, e^{i\boldsymbol{\delta} \cdot \mathbf{t}} \, e^{i\chi_{\pm}(t)} \tag{4.2}$$

where "+" corresponds to positron and "−" to electron scattering (from the external Coulomb field or the target electron), and $\chi_{\pm}$ are the corresponding phase shifts, we get

$$M = \int d^2s_1{}^A \, d^2s_2{}^A \, \varrho(\mathbf{s}_1{}^A, \mathbf{s}_2{}^A)\hat{M}(k, \Delta; \mathbf{s}_1{}^A, \mathbf{s}_2{}^A)$$

$$= \frac{ik}{(2\pi)^3} \int d^2\delta_1 \, d^2\delta_2 \, \delta^{(2)}(\Delta - \boldsymbol{\delta}_1 - \boldsymbol{\delta}_2)\tilde{\varrho}(\boldsymbol{\delta}_1, \boldsymbol{\delta}_2)\{(2\pi)^4 \, \delta^{(2)}(\boldsymbol{\delta}_1) \, \delta^{(2)}(\boldsymbol{\delta}_2)$$

$$- \Sigma_{+}(\boldsymbol{\delta}_1) \, \Sigma_{-}(\boldsymbol{\delta}_2)\} \tag{4.3}$$

Similarly, for the photon–photon scattering amplitude we get from (2.16), assuming that both $\tilde{A}$ and $\tilde{B}$ are composed of an electron-positron pair,

$$M = \int d^2s_1{}^A \, d^2s_2{}^A \, d^2s_1{}^B \, d^2s_2{}^B \, \varrho(\mathbf{s}_1{}^A, \mathbf{s}_2{}^A)\varrho(\mathbf{s}_1{}^B, \mathbf{s}_2{}^B)$$

$$\times \hat{M}(k, \Delta; \mathbf{s}_1{}^A, \mathbf{s}_2{}^A, \mathbf{s}_1{}^B, \mathbf{s}_2{}^B)$$

$$= \frac{ik}{(2\pi)^7} \int d^2\delta_1 \, d^2\delta_2 \, d^2\delta_3 \, d^2\delta_4 \, \delta^{(2)}(\Delta - \boldsymbol{\delta}_1 - \boldsymbol{\delta}_2 - \boldsymbol{\delta}_3 - \boldsymbol{\delta}_4) \, \tilde{\varrho}(\boldsymbol{\delta}_1 + \boldsymbol{\delta}_2, \boldsymbol{\delta}_3 + \boldsymbol{\delta}_4)$$

$$\times \tilde{\varrho}(\boldsymbol{\delta}_1 + \boldsymbol{\delta}_3, \boldsymbol{\delta}_2 + \boldsymbol{\delta}_4)[(2\pi)^8 \, \delta^{(2)}(\boldsymbol{\delta}_1) \, \delta^{(2)}(\boldsymbol{\delta}_2) \, \delta^{(2)}(\boldsymbol{\delta}_3) \, \delta^{(2)}(\boldsymbol{\delta}_4)$$

$$- \Sigma_{+}(\boldsymbol{\delta}_1) \, \Sigma_{-}(\boldsymbol{\delta}_2) \, \Sigma_{-}(\boldsymbol{\delta}_3) \, \Sigma_{+}(\boldsymbol{\delta}_4)] \tag{4.4}$$

The formulas (4.3) and (4.4) were obtained in (Che 69a) and (Chan 69) from studies of the high-energy limits of quantum electrodynamical expressions for Delbrück, Compton, and $\gamma$–$\gamma$ amplitudes obtained by summing some infinite sets of diagrams.

As we already said, the high-energy scattering amplitudes of the subunits and the densities of the fluctuations have to be supplied from quantum electrodynamics or some other theory which tells us how to compute them (see also Appendix B). Actually, to guess the high-energy phase shifts $\chi_{\pm}$ is easy, because it turns out to be [(Che 69a), compare also relation (2.18)]:

$$\chi_{\pm}(b) = \pm Ze^2 \int_{-\infty}^{+\infty} dz(b^2 + z^2)^{-1/2} \tag{4.5}$$

where $Z = 1$, in the case of electron–electron (or positron–positron) scattering. The $\tilde{\varrho}(\delta_1, \delta_2)$ (or its equivalent) has been calculated in (Che 69a,b) and also in (Chan 69). One can express the photon impact factor $\mathscr{I}_\gamma$ given in (Che 69b) through $\tilde{\varrho}$ (Chan 69) as follows:

$$\mathscr{I}_\gamma(\delta_1, \delta_2) = \tfrac{1}{2}[\tilde{\varrho}(\delta_1 + \delta_2, 0) - \tilde{\varrho}(\delta_1, \delta_2)] \tag{4.6}$$

There has also been suggested a general rule of computing $\tilde{\varrho}(\delta_1, \delta_2)$ from a straightforward noncovariant perturbation theory (Che 69a). Such a rule makes it possible, in principle, to extend to all high-energy scattering processes the model of high-energy scattering of objects without preexisting subunits, which can fluctuate however. This approach was suggested by Cheng and Wu (Che 69a). In Appendix B we give an example of evaluation of the high-energy limit of Delbrück scattering amplitude, which uses a noncovariant perturbation theory to compute the initial and final photon "wave functions" and the Glauber-model Coulomb profiles of the electron–positron pair.

Before going any further this is perhaps the right moment to stress similarities and differences between the high-energy scattering of objects with preexisting subunits and objects whose subunits show up in their fluctuations. First of all, in both cases the total scattering amplitude is constructed from the scattering amplitudes of the subunits. That is why the Delbrück or Compton scattering amplitudes, for example, have the same structure as hadron–deuteron amplitude: one subunit scatters from an aggregate of two. But this "aggregate" is described quite differently in these two cases. Its structure is given by the ground-state wave function in the first case and by the density of a fluctuation (There are many different fluctuations possible), in the second. In the first case we have to deal with a genuine bound state, in the second with a virtual excited state.

Can a similar "fluctuation approach" be applied to describe very high energy hadron–hadron scattering? Although it is not possible to quote any concrete results as yet, it is perhaps worthwhile to make few remarks about the problem. It would seem that the recently proposed "parton model" in the version worked out in (Dre 69a,b; 70a,b), which has been used to explain the experimental data on the so-called "deep inelastic" electron–proton scattering, might provide us with a framework to compute scattering amplitudes of fluctuating hadrons. In the papers quoted above the authors introduce expansions of the states of colliding hadrons (in their case, colliding with electrons) in terms of various fluctuations, in a similar way as was discussed above (the constituents of these fluctuations are called partons). Hence, in the framework of the canonical quantum field theory of pseudo-scalar pions and nucleons with charge-symmetric $\gamma_5$ coupling, they have definite prescriptions for calculating analogues of the $\bar{\varrho}$ functions discussed above for the case of fluctuating photons. As we have already stressed, it is much too early to say anything about successful (or unsuccessful) application of such an approach to purely hadronic collisions. One may however suspect that, because of the large coupling constant one will have to include many (may be even infinitely many) different fluctuations in any hadronic scattering (note that this problem was absent in quantum electrodynamics due to the small coupling constant). The problem may then become so involved that any self-consistent determination of the hadron–hadron amplitudes from the subunit–subunit amplitudes will be outside of our reach. We may even end up with a model which will be effectively equivalent to the Chou and Yang (Cho 68) model. Notice also that, in any case, one would probably have all kinds of characteristic effects for multiple diffractive scattering (e.g., diffractive minima) present.

Let us go back to the fluctuating photon (Fig. 20). Above we discussed the electron–positron fluctuations which play a fundamental role in the photon scattering if only electromagnetic interactions are present. On the other hand the "strong" fluctuations (vector mesons, $\pi$ many-particle systems, pairs of baryons, etc.) may be more important if the target is a strongly interacting system, e.g., a nucleus. In fact, a nucleus is a very interesting target for photons because one may check whether the "strong" fluctuations in photon do exist. The effect of time dilatation plays a crucial role here. If the photon is very energetic, the fluctuations viewed from the rest system of the nucleus may last so long that if a fluctuation occurs while the photon goes through the nucleus the effective interaction is exclusively between the fluctuation subunits (vector mesons, many-hadron systems, etc.) and the nucleus. In other words, due to the time dilatation factor, $k/m$, for the very high energy photons the whole nucleus scatters a fluctuation, and the probability of having a very short fluctuation, which would begin and end deeply inside the target nucleus, becomes negligible. This implies that a certain fraction of the photon–nucleus interactions must be strong interaction processes. Thus the total absorption cross section at high energies should contain a component which varies with the number of nucleons in the nucleus $A$ as $2\pi R^2 = 2\pi r_0^2 A^{2/3}$ ($R$ being the radius of the nucleus). One should stress that the very fact of the existence of $A^{2/3}$ dependent term in photon–nucleus interaction does not prove any specific photon–hadron interaction model (e.g., the vector-meson-dominance model), it just supports the hypothesis of the existence of hadronic fluctuations in the photon.

There exists experimental evidence of such a phenomenon (Cal 69). From the photoabsorption cross section on hydrogen one may expect a photon mean free path in nuclear matter to be about 800 fm. Hence one would expect that the total photoabsorption cross section on a nucleus with $A$ nucleons should be proportional to $A$. Instead the $A$ dependence is about $A^{0.9}$, indicating that some shadowing process does take place [(Cal 69), see also Fig. 21]. This fact very strongly supports the picture of a fluctuating photon and the existence of hadronic fluctuations.

It is therefore of primary importance to analyze a specific models of photon–hadron interactions. In particular the vector-meson-dominance model deserves a special attention (see also Section 5). There have been many theoretical papers on this and related subjects and some representative ones are (Bell 64, Bell 68, Bro 69, Gri 69, Got 69, Nau 69, Sto 67).

We have discussed a model of a fluctuating photon. We have chosen this way of describing photon high-energy interactions because it fits nicely into the multiple-scattering description of high-energy scattering we have tried consistently to follow in this article. One may reach the same conclusions analyzing photon interactions in a multitude of ways, which very often differ only in the language they use but lead to the same results. For instance, one may also get the condition of the importance of hadronic fluctuations at high energy from an uncertainty principle argument. If the invariant mass of a hadronic fluctuation is $m$, the energy is not conserved at the fluctuation vertex by an amount

$$\Delta E \approx (m^2 + k^2)^{1/2} - k \approx \frac{m^2}{2k} \tag{4.7}$$

giving a time uncertainty $\Delta t \approx 2k/m^2$, which contains the time dilatation factor $(k/m)$ which, as discussed above, makes the fluctuation interact with the whole nucleus and produces the $A^{2/3}$ dependence of the photoabsorp-

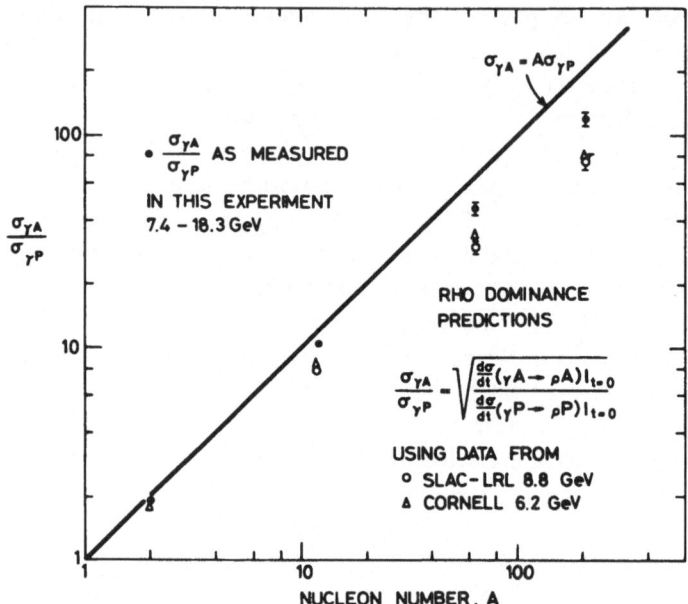

**Fig. 21.** The ratio of the total photoabsorption cross section of nucleus *A* to that of H. As observed (black points) (Cal 69), compared with result expected from a purely electromagnetic photon (line) or from ϱ-dominant photon using Cornell and SLAC–LRL data (open points).

tion cross section. This uncertainty in energy can also be translated into a spatial uncertainty of the fluctuation [see, e.g., (Bro 69)]

$$\Delta x \approx \frac{2k}{m^2} \qquad (4.8)$$

which leads to the same physical conclusions.

# 5. PROSPECTS OF WHAT NUCLEAR PHYSICS AND PARTICLE PHYSICS CAN LEARN FROM HIGH-ENERGY NUCLEAR SCATTERING

These two aspects of high-energy nuclear scattering are very intimately connected with each other: we have to know the hadron–nucleon amplitude in order to learn something new about the nuclear structure and, *vice versa*, we have to know the structure of the target nucleus in order to learn something new about the hadron–nucleon amplitude. One has, eventually, to obtain some "self-consistent" understanding of high-energy nuclear scattering based on knowledge of the hadron–nucleon amplitudes and the nuclear wave functions taken from sources other than high-energy scattering. The question is whether one can improve the present knowledge of these two sides of the problem. When we look at the experimental results and the theoretical interpretation collected in Section 3 we may conclude that, so far, we have managed to produce quite impressive evidence for the diffractive multiple scattering model described in Section 2. One must remember, however, that we get this impressive agreement virtually without any free parameters: we have taken them from various existing experiments, and hence we have not yet learned anything essentially new, either about the structure of nuclear targets or about the hadron–nucleon amplitudes. We merely checked that what we already know about them is enough to produce, within the model of scattering of composite objects, an explanation of the existing high-energy scattering data.

From this situation it follows that in order to learn new things, the accuracy of experimental data must be improved considerably and should be extended to other targets and projectiles.

Let us review a few obvious possibilities of attacking some of the poorly explored problems of nuclear structure. We shall assume in this part of our discussion that we know the hadron–nucleon amplitudes adequately. Since the elastic scattering cross section $d\sigma_{el}/d\Omega$ (and, to some extent, $d\sigma_{sc}/d\Omega$) are quite sensitive to the nuclear matter densities and we may accept that we know the proton densities from the electron elastic scattering; the neutron densities should, in principle, be measurable this way. So far (compare Section 3 and references quoted there), the assumption of the same neutron and proton densities was adequate to explain the existing experimental results. Notice that the difference of the neutron and proton matter radii obtained from an analysis of pionic atoms has been recently found insignificant (And 70). If the future measurements are to give more accurate data on proton and neutron matter densities it is essential to know extremely well the real part of the hadron–nucleon scattering amplitude. This is so because the diffractive structure of $d\sigma_{el}/d\Omega$ is very sensitive both to the nuclear density distributions and to the real part of the hadron–nucleon amplitude. Even if we know the hadron–nucleon amplitude very accurately the interpretation of the elastic scattering experiment is easy only if there is only *one* amplitude of the hadron–nucleus scattering to compute. This is the case when a spin-zero hadron (pion) scatters from a spherical spin-zero target nucleus (we shall call them "simple" targets).

There is also a very interesting and a very little explored area of high-energy hadron scattering from deformed nuclei. As it was shown in (Les 69a) the diffractive structure depends very definitely on the deformation of the target and looks different for prolate and oblate ellipsoidal shapes (with the same rms radius). The calculations for large deformations are comparatively easy to handle but for small and average deformations, where one has to project out the ground-state wave functions with correct total angular momentum, they become a major numerical problem. In the approximation of large deformations (the ground-state wave function has no definite angular momentum), one observes the following tendencies common to the cases of oblate and prolate deformation (Les 69a): as the deformation increases the first diffractive minimum shifts towards higher momentum transfers and the second diffractive maximum decreases. In the case of oblate deformations the shift of the first minimum depends very weakly on the magnitude of the deformation, whereas in the case of prolate deformations it depends very strongly.

The problem of investigating the short-range dynamical correlations between the nucleons in nuclei has also been for a long time one of little explored problems of nuclear physics. It would seem at first that one could extract some important information about short-range dynamical nucleon correlations both from $d\sigma_{el}/d\Omega$ and $d\sigma_{sc}/d\Omega$ (or $d\sigma_{inel}/d\Omega$) because the multiple scattering is so important. For instance the first diffractive minimum is due (if we use the language of the multiple-scattering model) to the interference of single- and double-scattering contributions. Hence the geometry of the screening should be very important in the high-energy scattering. In other words if, for example, a repulsive correlation prevents two nucleons from being close to each other, the contribution of the double scattering is cut down. The trouble, however, is that similar to the case of electron scattering (where, essentially, only single scattering is effective), the exclusion principle (statistical) correlations cover up the influence of dynamical correlations almost completely. As it was shown in many examples [see, e.g., (Bas 68, Les 69b, Czy 69b, Czy 70)] the most important effect of the exclusion principle is the change it introduces into the single-particle densities. For instance the fact that, in light nuclei after filling the $s$-shell, the $p$-shell is being filled with nucleons (because of the exclusion principle!) changes the single-particle densities enough to produce minima in electron form factors of these nuclei and also to introduce important factors in the high-energy elastic scattering cross sections of hadrons from nuclei (Bas 68, Les 69b). The explicit antisymmetrization changes $d\sigma_{el}/d\Omega$ very little (Bas 68, Les 69b), it does however change $d\sigma_{inel}/d\Omega$ quite appreciably at small momentum transfers but not more than 20 to 30% [compare Fig. 15]. The exclusion principle (statistical correlations) prevent two protons (or two neutrons) with the same spin to approach each other, hence they mask the role of the dynamical correlations to a large extent. Thus, crudely speaking, the situation is similar to the case of elastic and inelastic electron scattering from nuclei (McV 62). That does not mean, however, that we shall not learn anything new about nucleon dynamical correlations from the high-energy hadron–nucleus scattering. When used jointly with some other data, which are also influenced by the short-range dynamical correlations, such experiments will probably serve as an important source of information on this problem, which is very difficult to investigate both experimentally and theoretically. For instance, high-energy electron and pion scattering cross sections interpreted jointly may further our so far very scarce knowledge of such correlations. In order to do that we shall need more detailed and more precise sets of experiments, designed in a coherent way (covering, for instance, some specific momentum transfer regions or performing some well-defined sum rules which can be computed theoretically).

Note that in the case of very light nuclear targets one does not have statistical correlations to mask the dynamical ones. The ⁴He nucleus is especially interesting because its spin is zero, it is spherical, and its first excited state lies more than 20 MeV above the ground state allowing one to separate the elastic from inelastic scattering comparatively easily.

There is also a very attractive prospect of analyzing excited nuclear states. Some preliminary data on ¹²C are already available [(Fri 67); for theoretical analysis, see (Lee 68)]. For such spectroscopic research on a more extensive scale some very high energy resolution experimental setups are needed.

Let us now go over to the prospects of what particle physics can learn from high-energy nuclear scattering. First, we shall start with the question of whether we can learn something about the incident-hadron–target-nucleon amplitude from the hadron–nucleus cross sections. In order to learn something about the *elastic* hadron–nucleon amplitude one has to make sure that the total hadron–nucleus amplitude of the elastic cross section in question is, to a very good approximation, determined by the elastic hadron–nucleon amplitudes. In other words, processes through which, at a certain point inside the nucleus, the incident particle is converted into a different one or more particles, and then at some other point, still inside the nucleus, the product particles are converted back into the incident hadron (the whole process contributing to the elastic scattering amplitude) are of such importance that the extraction of *elastic* hadron–nucleon amplitudes becomes very uncertain, if not impossible. The so-called inelastic shadow effects belong to this category of processes and their contribution to the total cross sections has been estimated (Pum 68) to be important at very high energies ($\sim20\%$ reduction of the total $p$–Pb cross section at 30 GeV). These processes however disappear at lower energies [below 5 to 10 GeV according to (Pum 68)] and are limited to very small scattering angles. So, if we want to analyze a "clean" case of elastic hadron–nucleus scattering in terms of the elastic hadron–nucleon amplitudes *only*, we should be limiting ourselves to the momentum transfers $\Delta \gtrsim [(M + m)^2 - M^2]/2k$ and $\Delta R \lesssim 1$, where $\Delta$ is the three-momentum transfer required to coherently produce particles with mass $m$ at the incident particle momentum $k$ by the incident particles with mass $M$. The second condition is the coherence requirement. These two conditions imply that if we take for $m$ the mass of the $\varrho$ meson and $R \approx 6$ fm (which corresponds to a large nucleus), there should be very little of inelastic coherent shadow effect present at energies around 5 GeV.

Let us stress here that in Section 2 we did not include the inelastic shadow effect in our model and any analysis based on formulas of Section 2 contains this "original sin." We shall therefore limit ourselves to the cases where our model is presumably valid: the energies should either be not too high, or at high energies (e.g., CERN energies $\sim20$ GeV), the momentum transfers should not be too small. (Incidentally, this suggests that one should not perhaps attach too much importance to an ideal fit of the experimental points, at small momentum transfers, of the curves computed from our model.)

We shall proceed assuming that there exist cases where the inelastic effects inside the nucleus contribute very little to the elastic scattering amplitude, and we shall discuss such cases only.

There are many things one can, in principle, learn about hadron–nucleon amplitudes because of many very intricate interference effects which are sensitive to many details of these amplitudes and the corresponding cross sections. Not many things, however, can be reliably extracted from the hadron–nucleus amplitude at present because of experimental and theoretical uncertainties involved in any such detailed analysis.

One of the very attractive possibilities is to measure the ratio of the real to the imaginary part of the forward hadron–nucleon elastic scattering amplitude. This has been pointed out recently many times because the diffractive structure seen in the hadron–nucleus cross sections (see Section 3) is very sensitive to this ratio [denoted as $\alpha$ in (2.24)]. In fact $\alpha$ is the most important factor determining the depth of the diffractive minima for spherical spinless nuclear targets (Czy 69c). In order to have the theoretical interpretation as clean as possible one should design the experiments with spin-zero incident projectile (pions) and spin-zero spherical target nuclei. Then there is *only one* amplitude to compute and, for heavy nuclei, e.g., $^{208}$Pb, the expression for the strong interaction phase shift $\chi_s$ [compare (2.39)] is given by (2.33) (Czy 69b,c). If we accept the same $\sigma$'s and densities for protons and neutrons as in (2.33) we get

$$i\chi_s(b) = -\xi(b)(1 - i\alpha) \tag{5.1}$$

where

$$\alpha = A^{-1}(N\alpha_n + Z\alpha_p) \tag{5.2}$$

The existence of a Coulomb phase shift $\chi_c$ [compare (2.38), (2.40), (2.41)] helps extracting $\alpha$'s from the hadron–nucleus cross sections. This is especially so if one can switch the beam of the incident pions from $\pi^+$ to $\pi^-$ at the same energy. Assuming charge symmetry, $\alpha_n$ belongs to the $\pi^+-n$ or $\pi^--p$ pair, and $\alpha_p$ to the $\pi^+-p$ or $\pi^--n$ pair. If the elastic cross sections are measured for $\pi^+$ and then for $\pi^-$ beams and then subtracted from each other, the difference between them is *due entirely to the existence of* $\alpha$ [compare the expressions (2.41) and (2.42) for the scattering amplitude] and nothing else. This difference is appreciable at the diffractive minima (Czy 69b,c) especially for heavy nuclei. Since $d\sigma_{inel}/d\Omega$ virtually does not depend on the Coulomb interactions at medium momentum transfers (see Section 2), the same method may be applied to $d\sigma_{sc}/d\Omega$ (which is much easier to measure than $d\sigma_{el}/d\Omega$) and $\alpha$ from (5.2) may be gotten from the difference:

$$\frac{d\sigma_{sc}(\pi^+)}{d\Omega} - \frac{d\sigma_{sc}(\pi^-)}{d\Omega} \approx \frac{d\sigma_{el}(\pi^+)}{d\Omega} - \frac{d\sigma_{el}(\pi^-)}{d\Omega} \tag{5.3}$$

It is enough to measure (5.3) at two diffractive minima and then from (5.2) compute $\alpha_n$ and $\alpha_p$. Notice that one has to see clearly the diffractive structure in $d\sigma_{sc}/d\Omega$ in order to be able to measure the difference (5.3), hence heavy targets are preferable in applying this method. For instance, $^{208}$Pb is a very good target (spherical and spinless) and the diffractive structure in $d\sigma_{sc}/d\Omega$ is clearly seen (compare Fig. 19). One can, of course, extract $\alpha$ from the absolute value of $d\sigma_{el}/d\Omega$ at the minimum, but then a lot of various effects, which also fill the minima, have to be very accurately under control.

As was argued in Section 2 the $d\sigma_{sc}/d\Omega$ cross section for hadron–deuteron scattering is dominated by the single-scattering contribution, which is simply the sum of the hadron–proton and hadron–neutron scattering cross-sections. Furthermore the single- and double-scattering contributions can be experimentally separated (All 69a, Coc 62). From the experimental data shown in Fig. 8 (All 69a) one can see that even in $d\sigma_{sc}/d\Omega$ one can see the diffractive-like structure of the $p$–$p$ cross section. There is the hope therefore that, after subtracting (experimentally) the double-scattering contribution, one may be able to learn about the hadron–neutron elastic scattering amplitude. This very interesting question has recently been discussed (All 69a, Str 70) and the prospects look encouraging.

The next example of what new information we can learn from nuclear scattering about problems in particle physics is the determination of the total cross sections on nucleons of strongly interacting short-lived particles generated inside of nuclear targets. The idea is to analyze the production of such particles on nuclei as a two-step process: first, production of the short-lived particle, then, on its way out, the multiple scattering from the nucleons of the nucleus. The production cross sections computed this way strongly depend on the total cross section of individual nucleons, as it was in the case of hadron–nucleus scattering discussed above, and on the coupling constant for the production of our unstable particle from a nucleon. From experiments at several energies and on various nuclei one can in this way compute the total cross section and the coupling constant. Since it is impossible to have beams of short-lived particles, such an analysis offers a unique opportunity to measure the total cross sections of very short-lived strongly interacting particles.

Such an analysis had been originally given in (Dre 66) and (Ros 66) for the process of photoproduction of $\varrho$-mesons on nuclei and the total $\varrho$–nucleon cross section extracted. Subsequently there have been many papers on this and parallel subjects both theoretical and experimental [e.g., (Bel 68, Köl 68, Mar 68, Tre 69a,b, Bro 69, Got 69, Nau 69, Tin 68, McC 69 a,b, Bul 69) and references cited therein]. Note that these papers contain an analysis of a specific fluctuation $(\gamma$–$\sigma)$ of the photon. In fact some of the points discussed in Section 4 belong also to the theme of this section.

One could continue analyzing many other problems of particle physics [e.g., (Tre 69c)] whose understanding may be broaden and improved by putting them in the "nuclear shadow" (Bel 68), but we have to end somewhere.

# 6. CONCLUDING REMARKS

We have tried in this article, to extract the common features of the high-energy scattering in many areas of physics, some of them as remote from each other as, for example, quantum electrodynamics and nuclear physics. We have also tried to compare, wherever possible, the experimental data with predictions of the diffractive multiple scattering model. This led to a strong emphasis upon high-energy nuclear collisions because the main experimental material comes from there; also, the theoretical basis for the model is comparatively plausible. We hope that there is enough of such material in the article to justify its title.

Let us now discuss the very vague problems of the region of validity of the multiple scattering model. First of all let us limit ourselves to nuclear scattering because we have more experience with these problems. Strictly speaking, to the best of this author understanding, we simply do not know the validity regions. Yet a few aspects of this problem are perhaps worth describing. One can, for instance, formulate this problem precisely for potential scattering. Then we may then ask at what scattering angles do the formulas (2.18) and (2.19) give the amplitude which agrees (within given accuracy) with the exact amplitude computed from the Schrödinger equation (numerically, if necessary). Some discussions in this spirit have been published by many authors [see e.g., (Sch 65, Sax 57, Gla 59, Fes 66, Sch 68, Fes 70)]. More detailed estimates of the angular region of validity must depend on the detailed functional form of the potential $V$. But one can also give some very crude estimates of the angular region of validity. First of all let us assume that (Sch 56, Gla 59, Fes 66)

$$kR \gg 1, \qquad V_0/E \ll 1 \qquad (6.1)$$

where, as usual, $k$ is the wave vector of the incident particle, $R$, the range of forces exerted on this particle by the target, $V_0$, the depth (average) of the potential, and $E$, the energy of the incident particle. Under assumption (6.1) Schiff (Sch 56) got the following validity region for the scattering angle:

$$\theta \ll (kR)^{-1/2} \qquad (6.2)$$

Instead, in (Gla 59) the following two cases are distinguished [compare also (Fes 66)]:

$$\theta < (kR)^{-1/2} \qquad \text{if } (V_0R)/\hbar v < 1 \qquad (6.3)$$

and

$$\theta < (V_0/E)^{1/2} \qquad \text{if } (V_0R)/\hbar v > 1 \qquad (6.4)$$

where $v$ is the velocity of the projectile. Let us discuss the condition (6.2). If one identifies $R$ with the radii of the target nuclei and computes

$(kR)^{-1/2}$ for the cases discussed in Section 3, one comes to the conclusion that, for light nuclei, the diffractive minima of the cross sections are outside of the validity region of the model and, for heavier nuclei, the first diffractive minima are barely within this region (Czy 69g) (this is because the diffractive structure in $d\sigma_{el}/d\Omega$ moves, with increasing $R$, faster than $(kR)^{-1/2}$, toward small angles). It is, however, very likely that identifying $R$ with the radius of the nucleus we grossly underestimate the angular region of validity. We should, perhaps, use the nucleon–nucleon interaction radius, because the total potential is a sum of nucleon–nucleon interaction potentials. When we do that, the validity region increases considerably (Czy 69g). If we, on the other hand, want to distinguish between the two cases (6.3) and (6.4) we find that, in the case of nuclear scatterings (as $V_0 \approx 50$ MeV $\approx \frac{1}{4}$ fm$^{-1}$), for $R \approx 5$ fm and a fast incident particle ($v = 1$) $V_0 R/\hbar v \approx 1$, hence the situation essentially does not change.

If we use the picture of a multiple scattering in which one composes the total profile from individual nucleon profiles treating them as absorption probability distributions (Czy 69a), we can use the following criterion of validity: as long as the geometry of screening is not changed we can use the model. That gives the following limitation on the scattering angle: $\theta \ll$ arctan $(r_0/R)$, where $r_0$ is radius of the nucleon and $R$ that of the nucleus. In the case of a $^4$He target, for example, we get $\tan(r_0/R) \approx \frac{2}{3}$, and in the case of $^{208}$Pb we get $\tan(r_0/R) \approx \frac{1}{4}$. Hence the angular region of validity is in this case very large, and contains all the experimental data available (compare the data collected in Section 3). All these *qualitative* arguments strongly suggest that, if the target is composed of well-defined subunits, the high-energy scattering model is *probably* valid at angles larger than suggested by the standard (Sch 56) criterion $\theta \ll (kR)^{-1/2}$, where $R$ is the radius of the target.

There is also a very obvious question to ask: what is the relation of the diffractive multiple scattering model (Glauber model) to the well-known methods of treating multiple scattering when the interactions between pairs of particles are given by potentials (Gold 64)? Several papers dealt with this problem [see e.g., (Har 64, Abe 66, Pum 68a, Rem 68, and especially Har 69a)]. The conclusion is (Har 69a) that the Glauber model in potential theory with only two-body interactions is exact in the high-energy small-angle limit. That means that, in the language of standard potential multiple scattering theory (Gold 64), the off-shell contributions of the two-particle $t$ matrices to the double scattering are cancelled by the higher order terms in the multiple scattering series. For an explicit demonstration of such cancellations see (Har 69a).

The results quoted above show how careful one should be in calculating corrections to the Glauber model. But, as far as the regions of validity are concerned, we are still in the dark. First of all the results of the potential theories may not be very relevant to very high-energy hadron–nucleus collisions. The ground-state wave function of the nucleus may also contain some components with excited states of the nucleons (Ker 69, Vin 70), or some other strongly interacting particles. In the language of Section 5, such fluctuations are, presumably, rather improbable since the nuclei are comparatively loosely bound systems. There is also the so-called "inelastic shadow" effect we have already discussed a few times in the previous sections. It adds one more and a very serious uncertainty to our understanding (or lack thereof) of the regions of validity of the Glauber model.

The relativistic Feynman graph and field theoretic techniques have also been applied to analyze the high-energy hadron–nucleus scattering. We shall not go into these problems and we shall limit ourselves to quoting some representative papers on this subject [see (Abe 66, Gri 69, Gri 69a, Gri 69b, Wil 69)].

In concluding let us stress two points. First, it is a very difficult and a not well defined task to establish limits of the validity of the model presented in Section 2. But there are many indications that, at least in the region of energies and momentum transfers where the inelastic shadow effect is unimportant, that thus model works very well.*

The second point is this. Let us suppose that we know *very well* the ground-state wave function of a light composite object (a light nucleus or an atom) in its rest system. Let us suppose further that agreement is found between the measured and computed cross sections for this particle scattering from a known (infinitely heavy) target over a range of small to large momentum transfers. Let us assume also that we know very well the scatter-

---

* Let us notice that in the field theoretic models of high-energy multiple scattering discussed in Section 4 (Che 69a, Chan 69, and Dre 70) there are no off-mass-shell contributions to the amplitude: All the subunit particles are on the mass-shell and their scattering amplitudes are as if they were free [incidentally, this fact resembles the cancellation of the off-energy-shell contributions in potential scattering (Har 69a)]. This is so because in these models we do not have any intermediary structures (like nucleons in a nucleus) which may be identifield as subunits and whose internal degrees of freedom (e.g., mesonic degrees of freedom of nucleons in a nucleus) do not enter into computations of the scattering amplitudes of the colliding composite objects. In these field theoretic models the subunits are indivisible pointlike particles. In the limit of very-high-energy scattering, the corrections of the inelastic shadow effect have to be considered when the subunits have composite structure of their own, like nucleons in nuclei.

ing amplitudes of individual subunits so that we are able to perform a Glauber-like computations of the cross sections in question. But the cross sections in our model are calculated *with the same ground-state wave functions irrespective of the velocity of the recoil.* Hence, the agreement of the model and experiment at large momentum transfers would be very difficult to reconcile with the existence of, for instance, Lorentz contraction or other "deformations" of the wave function introduced by the high recoil velocity. Such deformations of the shape of the target might dramatically change the large-momentum-transfer behavior of the cross sections. It may turn out, therefore, that the high-energy scattering of light composite objects whose ground-state wave functions, in their rest frame, we do know pretty well may help us to test the Lorentz–Poincaré group at small distances (Chy 70).

## ACKNOWLEDGMENTS

The author is grateful to very many people (too many to list all the names) for discussions, support, and encouragement. His thanks are due to L. Leśniak and H. Wołek for their unfailing reliability during the years of common research and their contributions, essential to the understanding of many problems discussed in this article.

The author wants also to apologize to all people whose work has been overlooked, distorted, or not emphasized enough in the present article. The author is to blame for all such deficiencies.

## APPENDIX A

We shall show that the dependence of the amplitude operator $\hat{M}$ [Eq. (2.16)], on the c.m. coordinate $r_A$ and $r_B$ [Eq. (2.4)], may be factored out, namely, that

$$\hat{M}(\Delta; s^A, s^B) = \hat{M}(\Delta; s^{A'}, s^{B'})\, e^{i(\Delta \cdot r_A - \Delta \cdot r_B)} \tag{A1}$$

The amplitude operator, appearing in Eq. (2.16) is

$$\hat{M}(\Delta; s^A, s^B) = \frac{ik}{2\pi} \int d^2b\, e^{i\Delta \cdot b}\, \hat{\Gamma}(b; s^A, s^B) \tag{A2}$$

where

$$\hat{\Gamma}(b; s^A, s^B) = 1 - \prod_{j}^{A} \prod_{k}^{B} [1 - \gamma_{jk}(b - s_j^A + s_k^B)] \tag{A3}$$

We may, alternately, express $\hat{\Gamma}$ in terms of the amplitudes [compare (2.16)] and get

$$\hat{M}(\Delta; s^A, s^B) = \sum_{j,k} \int d^2\delta_{jk}\, \delta^{(2)}(\Delta - \delta_{jk})\, e^{i\delta_{jk}\cdot(s_j^A - s_k^B)} f_{jk}(\delta_{jk}) - \left(\frac{1}{2\pi i k}\right)$$

$$\times \sum_{j_1 k_1 j_2 k_2} \int d^2\delta_{j_1 k_1}\, d^2\delta_{j_2 k_2}\, \delta^{(2)}(\Delta - \delta_{j_1 k_1} - \delta_{j_2 k_2})\, e^{i\delta_{j_1 k_1}\cdot(s_{j_1}^A - s_{k_1}^B) + i\delta_{j_2 k_2}\cdot(s_{j_2}^A - s_{k_2}^B)}$$

$$\times f_{j_1 k_1}(\delta_{j_1 k_1}) f_{j_2 k_2}(\delta_{j_2 k_2}) + \left(\frac{1}{2\pi i k}\right)^2 \sum_{j_1 k_1 j_2 k_2 j_3 k_3} \int d^2\delta_{j_1 k_1}\, d^2\delta_{j_2 k_2}\, d^2\delta_{j_3 k_3}$$

$$\times \delta^{(2)}(\Delta - \delta_{j_1 k_1} - \delta_{j_2 k_2} - \delta_{j_3 k_3})\, e^{i\delta_{j_1 k_1}\cdot(s_{j_1}^A - s_{k_1}^B) + i\delta_{j_2 k_2}\cdot(s_{j_2}^A - s_{k_2}^B) + i\delta_{j_3 k_3}\cdot(s_{j_3}^A - s_{k_3}^B)}$$

$$\times f_{j_1 k_1}(\delta_{j_1 k_1}) f_{j_2 k_2}(\delta_{j_2 k_2}) f_{j_3 k_3}(\delta_{j_3 k_3}) \cdots \tag{A4}$$

Introducing now the internal coordinates [see Eq. (2.4)]

$$s_j^{A'} = s_j^A - r_A, \qquad s_k^{B'} = s_k^B - r_B$$

we see that the c.m. coordinates appear in the exponent in (A4) as $r_A \cdot (\delta_{j_1 k_1} + \delta_{j_2 k_2} + \delta_{j_3 k_3} + \ldots)$ and $-r_B \cdot (\delta_{j_1 k_1} + \delta_{j_2 k_2} + \delta_{j_3 k_3} + \ldots)$. In view of the $\delta$-function in (A4), these expressions become, upon integration, $r_A \cdot \Delta$ and $-r_B \cdot \Delta$, from which (A1) follows.

## APPENDIX B

In this appendix we illustrate how the scattering of fluctuating objects of Quantum Electrodynamics can be incorporated into the framework of the Glauber model. We do it by presenting *all* the steps in the computation of the Delbrück scattering amplitude. The relations between the normalizations used below and in the main text are not discussed, however.

The Delbrück scattering amplitude is here obtained by taking the matrix element of (2.16) between the initial and the final states of the photon. If we limit ourselves to just positron–electron fluctuation (we could include, in principle, arbitrarily complicated fluctuations, but in our picture of Delbrück scattering they should certainly be less important than the positron–electron fluctuation) we have ($|k| = \omega$):

$$\hat{M} = \frac{i\omega}{2\pi} \int d^2b\, e^{i\Delta \cdot b} [1 - (1 - \gamma_-(b - s_1))(1 - \gamma_+(b - s_2))] \tag{B1}$$

where $\gamma_\mp$ are, respectively, the electron–Coulomb-field and the positron–Coulomb-field profiles. Then, in accordance with equation (2.2), the amplitude is given by the matrix element $\langle \gamma_f | \hat{M} | \gamma_i \rangle$, where $|\gamma_i\rangle$ and $|\gamma_f\rangle$ are the initial and the final states of the photon, respectively. Let us write this amplitude in momentum representation

$$\langle \gamma_f | \hat{M} | \gamma_i \rangle = \int d^3k_1 \, d^3k_2 \, d^3q_1 \, d^3q_2$$
$$\times \langle \gamma_f | \mathbf{k_1 k_2} \rangle \langle \mathbf{k_1 k_2} | \hat{M} | \mathbf{q_1 q_2} \rangle \langle \mathbf{q_1 q_2} | \gamma_i \rangle \qquad \text{(B2)}$$

where $|\mathbf{k_1 k_2}\rangle$ and $|\mathbf{q_1 q_2}\rangle$ are properly normalized plane wave states of the electron–positron pair. Note that, adhering strictly to the model presented in Section 2, the profile (B1) is *two-dimensional*, whereas the wave functions $\langle \mathbf{q_1 q_2} | \gamma_i \rangle$ and $\langle \mathbf{k_1 k_2} | \gamma_f \rangle$ are *three-dimensional*. Besides, since we want to compute the amplitude in the limit of the very high photon momentum, we shall compute $\langle \mathbf{q_1 q_2} | \gamma_i \rangle$ and $\langle \mathbf{k_1 k_2} | \gamma_f \rangle$ in a definite frame where the photon momentum is very large (in the case of Delbrück scattering, where the source of the Coulomb field is infinitely heavy, the obvious choice is the laboratory frame). Then we use a noncovariant perturbation expansion (in our chosen reference system) to express the photon state $|\gamma\rangle$ through the fluctuations (in the case of Delbrück scattering only the electron–positron fluctuation is important; for a general discussion of this expansion compare also Dre 70a):

$$|\gamma\rangle = Z^{1/2} |\gamma_0\rangle + \sum_{k_1 k_2} \frac{\langle \mathbf{k_1 k_2} | H' | \gamma_0 \rangle}{\omega - E(k_1) - E(k_2)} |\mathbf{k_1 k_2}\rangle \qquad \text{(B3)}$$

where $|\gamma_0\rangle$ is the bare photon state and $Z^{1/2}$ is the renormalization constant which assures the right normalization of the dressed photon state $|\gamma\rangle$, $\omega$ is the incident photon energy, $E(k_1)$, $E(k_2)$ are the energies of the electron and the positron, and $k_1$, $k_2$ their momenta. $H'$ is the electron–positron–photon interaction Hamiltonian:

$$H' = e \int d^3x \bar{\psi}(\mathbf{x}) \gamma_\mu \psi(\mathbf{x}) A_\mu(\mathbf{x}) \qquad \text{(B4)}$$

One gets $\langle \mathbf{k_1 k_2} | H' | \gamma_0 \rangle$ from (B4) by replacing the electron, positron, and photon fields by the following plane waves:

$$\psi_{k_1}^{(el)} = \sqrt{\frac{m}{E(k_1)}} \frac{1}{\sqrt{(2\pi)^3}} e^{i\mathbf{k_1} \cdot \mathbf{x}} u(k_1)$$

$$\psi_{k_2}^{(positr)} = \sqrt{\frac{m}{E(k_2)}} \frac{1}{\sqrt{(2\pi)^3}} e^{-i\mathbf{k_2} \cdot \mathbf{x}} v(k_2) \qquad \text{(B5)}$$

$$\chi_K^{(phot)}(i) = \sqrt{\frac{1}{2\omega}} \frac{1}{\sqrt{(2\pi)^3}} e^{i\mathbf{K} \cdot \mathbf{x}} \varepsilon_\mu^{(i)}$$

where $u$, $v$ are the standard electron and positron spinors and $\varepsilon_\mu^{(i)}$ is the photon polarization four-vector (Bjo 64). Inserting (B5) into (B3) and (B4) we get the following expressions for the photon wave functions in the momentum space (the first term in (B3) does not give any contribution):

$$
\begin{aligned}
\langle \mathbf{q}_1 \mathbf{q}_2 \mid \gamma_i \rangle_\mu &= \frac{\langle \mathbf{q}_1 \mathbf{q}_2 \mid H' \mid \gamma_{0i} \rangle}{\omega - E(q_1) - E(q_2)} \\[2mm]
&= \frac{e}{(2\pi)^{3/2}} \sqrt{\frac{m}{E(q_1)}} \sqrt{\frac{m}{E(q_2)}} \sqrt{\frac{1}{2\omega}} \\[2mm]
&\quad \times \frac{[\bar{u}(q_1)\gamma_\mu v(q_2)]}{\omega - E(q_1) - E(q_2)} \delta^{(3)}(\mathbf{K}_i - \mathbf{q}_1 - \mathbf{q}_2)
\end{aligned}
\tag{B6}
$$

$$
\begin{aligned}
\langle \gamma_f \mid \mathbf{k}_1 \mathbf{k}_2 \rangle_\nu &= \frac{\langle \gamma_{0f} \mid H' \mid \mathbf{k}_1 \mathbf{k}_2 \rangle}{\omega - E(k_1) - E(k_2)} \\[2mm]
&= \frac{e}{(2\pi)^{3/2}} \sqrt{\frac{m}{E(k_1)}} \sqrt{\frac{m}{E(q_2)}} \sqrt{\frac{1}{2\omega}} \\[2mm]
&\quad \times \frac{[\bar{v}(k_2)\gamma_\nu u(k_1)]}{\omega - E(k_1) - E(k_2)} \delta^{(3)}(\mathbf{K}_f - \mathbf{k}_1 - \mathbf{k}_2)
\end{aligned}
\tag{B7}
$$

These initial and final photon wave functions can be described by the diagrams shown in Fig. 22.

Using the same electron and positron plane waves (B5) one gets the following expression for $\langle \mathbf{k}_1 \mathbf{k}_2 \mid \hat{M} \mid \mathbf{q}_1 \mathbf{q}_2 \rangle$:

$$
\begin{aligned}
\langle \mathbf{k}_1 \mathbf{k}_2 \mid \hat{M} \mid \mathbf{q}_1 \mathbf{q}_2 \rangle &= \frac{i\omega}{(2\pi)^3} \sqrt{\frac{m^4}{E(k_1)E(k_2)E(q_1)E(q_2)}} \\
&\quad \times \{ (2\pi)^4 \delta^{(2)}(\boldsymbol{\Delta})\, \delta^{(3)}(\mathbf{k}_1 - \mathbf{q}_1)\, \delta^{(3)}(\mathbf{k}_2 - \mathbf{q}_2) \\
&\quad - \delta^{(2)}(\boldsymbol{\Delta} - (\mathbf{k}_{1\perp} - \mathbf{q}_{1\perp}) - (\mathbf{k}_{2\perp} - \mathbf{q}_{2\perp})) \\
&\quad \times \delta(k_{1z} - q_{1z})\, \delta(k_{2z} - q_{2z}) \sum_-(-\mathbf{k}_{1\perp} + \mathbf{q}_{1\perp}) \sum_+(-\mathbf{k}_{2\perp} + \mathbf{q}_{2\perp}) \} \\
&\quad \times (\bar{u}(k_1)\gamma_0 u(q_1))\,(\bar{v}(q_2)\gamma_0 v(k_2)).
\end{aligned}
\tag{B8}
$$

where

$$
\begin{aligned}
\sum_\mp(\boldsymbol{\delta}_\perp) &= \int d^2t\, e^{i\boldsymbol{\delta}_\perp \cdot \mathbf{t}} (1 - \gamma_\mp(t)) \\[2mm]
&= \int d^2t\, e^{i\boldsymbol{\delta}_\perp \cdot \mathbf{t}} \exp\!\left[ \mp ie \int_{-\infty}^{+\infty} dz'\, V_{\mathrm{Coul}}(\mathbf{t}, z') \right]
\end{aligned}
\tag{B9}
$$

In obtaining (B8) we accepted $\boldsymbol{\Delta}$ perpendicular to the $z$-axis. The vectors $\mathbf{k}_{1,2\perp}$ and $\mathbf{q}_{1,2\perp}$ are the components of the three-dimensional vectors $\mathbf{k}_{1,2}$ and $\mathbf{q}_{1,2}$ which are perpendicular to the $z$-axis. Note that the first term of (B8) contributes only in the forward direction ($\boldsymbol{\Delta} = 0$). We shall discuss

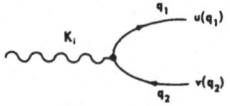

INITIAL STATE, EQ. (B6)

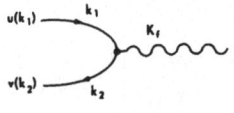

FINAL STATE, EQ. (B7)

**Fig. 22.**

further only $\Delta \neq 0$ contribution. In any case the first term can be easily put back into our formulas if it need be. Putting (B6), (B7), and (B8) into (B2) we obtain the sought expression for Delbrück amplitude. There are 12 integrations in (B2) but there are also many Dirac $\delta$-functions in (B6)–(B8) which reduce considerably the number of integrations. Let us choose the positive $z$-axis to be in the direction of the average momentum of the incoming and the outgoing photons, and let us call the $z$-components of the momenta of these two photons $\omega$. Then their transverse momenta are $-\Delta/2$ and $\Delta/2$, respectively and the momenta of the electron–positron pair can be expressed through some independent variables $\mathbf{p}_\perp$, $\mathbf{q}_\perp$ and $\beta$; the first two being vectors in the $x$, $y$ plane and $\beta = q_{1z}/\omega$. Consequently, from the energy conservation, $1 - \beta = q_{2z}/\omega$, hence $0 < \beta < 1$.

The vectors $q_1$, $q_2$, $k_1$, $k_2$ can then be written as follows:

$$\mathbf{q}_1 = [\beta\omega, \mathbf{p}_\perp], \qquad\qquad \mathbf{q}_2 = [(1 - \beta)\omega, -\mathbf{p}_\perp - \Delta/2]$$
$$\mathbf{k}_1 = [\beta\omega, \mathbf{p}_\perp - \mathbf{q}_\perp + \Delta/2], \qquad \mathbf{k}_2 = [(1 - \beta)\omega, -\mathbf{p}_\perp + \mathbf{q}_\perp] \qquad \text{(B10)}$$

Inserting (B6), (B7), and (B8) into (B2) and summing over the spin orientations we get (after few manipulations):

$$\langle \gamma_f | \hat{M}(\Delta \neq 0) | \gamma_i \rangle = -\frac{1}{4\pi} \delta^{(3)}(\Delta - \mathbf{K}_f + \mathbf{K}_i)$$

$$\times \frac{ie^2\omega}{(2\pi)^5} \int d^2p_\perp \int d^2q_\perp \int_0^1 d\beta \sum_-(-\mathbf{q}_{1\perp} + \mathbf{k}_{1\perp}) \sum_+(-\mathbf{q}_{2\perp} + \mathbf{k}_{2\perp})$$

$$\times \frac{m^4}{E(k_1)E(k_2)E(q_1)E(q_2)}$$

$$\times \sum_{\text{spins}} \frac{(\bar{u}(q_1)\gamma_\mu v(q_2))(\bar{v}(q_2)\gamma_0 v(k_2))(\bar{v}(k_2)\gamma_\nu u(k_1))(\bar{u}(k_1)\gamma_0 u(q_1))}{[\omega - E(q_1) - E(q_2)][\omega - E(k_1) - E(k_2)]} \qquad \text{(B11)}$$

This is virtually the same result as that obtained for Delbrück scattering by Cheng and Wu (Che 69a, Che 70). Let us here make a few comments about it. The first is that the calculations to obtain it are very simple. To obtain the same result by any standard quantum electrodynamical methods (e.g., Feynman graphs) is exceedingly involved [see, e.g., the reference in (Che 69a, Che 70)]. The second point is that (B11) can, of course, be written in a form which [as was done, e.g., in the case of hadron–deuteron scattering; compare equation (2.25)] explicitly exhibits the single-scattering contributions (terms which contain $\gamma_{\mp}$ linearly) and the double-scattering contribution [the term which contains the product $\gamma_-\gamma_+$, for the relations between $\gamma$'s and $\Sigma$'s; see equation (B9)]. For instance, if we want to evaluate the contribution which comes only from the scattering of the electron of the electron–positron fluctuation, we should put $\gamma_+ \to 0$ in (B1). But this implies [compare (B9)] that we should make the following replacement in (B11):

$$\Sigma_+(-\mathbf{q}_{2\perp} + \mathbf{k}_{2\perp}) \to (2\pi)^2\,\delta^{(2)}(-\mathbf{q}_{2\perp} + \mathbf{k}_{2\perp}) = (2\pi)^2\,\delta^{(2)}(\Delta/2 - \mathbf{q}_\perp) \quad \text{(B12)}$$

Equation (B12), together with $\delta(-q_{2z} + k_{2z})$ of equation (B8) gives $\mathbf{q}_2 = \mathbf{k}_2$; hence

$$\bar{v}(q_2)\gamma_0 v(k_2) \to \bar{v}(q_2)\gamma_0 v(q_2) = \frac{E(q_2)}{m} \quad \text{(B13)}$$

Finally, we get the following contribution to Delbrück scattering amplitude coming from just electron scattering:

$$(\langle \gamma_f | \, \hat{M}(\Delta \neq 0) \, | \gamma_i \rangle)_{\text{only electron scatters}} = -\frac{1}{4\pi}\,\delta^{(3)}(\Delta - \mathbf{K}_f + \mathbf{K}_i)$$

$$\times \frac{i\omega e^2}{(2\pi)^3} \int d^2p_\perp \int_0^1 d\beta\, \Sigma_-(\Delta)\, \frac{m^3}{E(q_1)E(q_2)E(k_1)}$$

$$\times \sum_{\text{spins}} \frac{(\bar{u}(q_1)\gamma_\mu v(q_2))(\bar{v}(q_2)\gamma_\nu u(k_1))(\bar{u}(k_1)\gamma_0 u(q_1))}{[\omega - E(q_1) - E(q_2)][\omega - E(k_1) - E(k_2)]} \quad \text{(B14)}$$

This expression also agrees with Cheng and Wu (Che 70). Note that (B14) does not have the $d^2q_\perp$ integration. The vectors $q_1$, $q_2$, and $k_1$ of (B14) are given by (B10) except that, in accordance with (B12), one should everywhere replace $\mathbf{q}_\perp$ by $\Delta/2$.

From the formulas discussed above it is simple to extract the Fourier transform of the electron–positron fluctuation density $\bar{\varrho}^\gamma(\boldsymbol{\delta}_1, \boldsymbol{\delta}_2)$ which was discussed in Section 4. Formula (4.3) defines $\bar{\varrho}^\gamma$. Since $\boldsymbol{\delta}_1$ and $\boldsymbol{\delta}_2$ in (4.3) are the momentum transfers to the electron and positron, we have

$$\boldsymbol{\delta}_1 = -\mathbf{q}_\perp + \boldsymbol{\Delta}/2, \qquad \boldsymbol{\delta}_2 = \mathbf{q}_\perp + \boldsymbol{\Delta}/2 \tag{B15}$$

and from equation (4.3) we get immediately

$$M(\boldsymbol{\Delta} \neq 0) = \frac{i\omega}{(2\pi)^3} \int d^2q_\perp \bar{\varrho}^\gamma(-\mathbf{q}_\perp + \boldsymbol{\Delta}/2, \mathbf{q}_\perp + \boldsymbol{\Delta}/2)$$

$$\times \, \Sigma_-(-\mathbf{q}_\perp + \boldsymbol{\Delta}/2) \, \Sigma_+(\mathbf{q}_\perp + \boldsymbol{\Delta}/2) \tag{B16}$$

One should observe however that, due to the specific form of $\Sigma_-$ and $\Sigma_+$ functions, one can add an arbitrary but $q_\perp$-independent expression to $\bar{\varrho}^\gamma$ and (B16) still holds. Indeed, this is because [compare (B9)]

$$\int d^2q_\perp \, \Sigma_-(-\mathbf{q}_\perp + \boldsymbol{\Delta}/2) \, \Sigma_+(\mathbf{q}_\perp + \boldsymbol{\Delta}/2)$$

$$= (2\pi)^2 \int d^2t \, e^{i\boldsymbol{\Delta}\cdot t} \, e^{-ie \int_{-\infty}^{+\infty} dz' \, V_{\text{Coul}}(t,z')} \, e^{+ie \int_{-\infty}^{+\infty} dz' \, V_{\text{Coul}}(t,z')}$$

$$= (2\pi)^4 \delta^{(2)}(\boldsymbol{\Delta}) = 0, \qquad \text{(we keep } \boldsymbol{\Delta} \neq 0) \tag{B17}$$

From (B16) and (B11) we can extract the following expression for $\bar{\varrho}^\gamma$:

$$\bar{\varrho}^\gamma(-\mathbf{q}_\perp + \boldsymbol{\Delta}/2, \mathbf{q}_\perp + \boldsymbol{\Delta}/2) = -\frac{e^2}{2(2\pi)^3} \int d^2p_\perp \int_0^1 d\beta \, \frac{m^4}{E(k_1)E(k_2)E(q_1)E(q_2)}$$

$$\times \sum_{\text{spins}} \frac{(\bar{u}(q_1)\gamma_\mu v(q_2))(\bar{v}(q_2)\gamma_0 v(k_2))(\bar{v}(k_2)\gamma_\nu u(k_1))(\bar{u}(k_1)\gamma_0 u(q_1))}{[\omega - E(q_1) - E(q_2)][\omega - E(k_1) - E(k_2)]} \tag{B18}$$

As it stands, this expression is not well defined because of the divergence of the the integration over $\mathbf{p}_\perp$. One can however make it well defined by employing the freedom in the choice of $\bar{\varrho}^\gamma$ mentioned above. We shall not go into any details of this problem and into any more detailed computations of Delbrück scattering or any other electrodynamical processes at high energies. One can find many further details in the papers of Cheng and Wu.

We hope that by showing explicitly how one can quickly and very simply get the same results as are obtained by employing more standard methods of quantum electrodynamics, we have indicated how very wide the applicability region of the diffractive multiple scattering model (Glauber model) is.

# REFERENCES

Abe 66    E. S. Abers, H. Burkhardt, V. L. Teplitz, and C. Wilkin, *Nuovo Cimento* 42:365 (1966).

Abu 68    A. Y. Abul-Magd, *Nuclear Phys.* B8:638 (1968).

Abu 69    A. Y. Abul-Magd, G. Alberi, and L. Bertocchi, *Phys. Letters* 30B:182 (1969).

Alb 69    G. Alberi and L. Bertocchi, *Nuovo Cimento* 63A:285 (1969).

All 69a    J. V. Allaby, A. N. Diddens, R. J. Glauber, A. Klovning, O. Kofoed-Hansen, E. J. Sacharidis, K. Schlüpmann, A. M. Thorndike, and A. M. Wetherell, CERN preprint, November (1969).

All 68, 69b    J. V. Allaby, F. Binon, A. N. Diddens, P. Duteil, A. Klovning, R. Meunier, J. P. Peigneux, E. J. Sacharidis, K. Schlüpmann, M. Spighel, J. P. Stroot, A. M. Thorndike, and A. M. Wethrell, *Phys. Letters* 28B:67 (1968); 29B:198 (1969).

All 69c    J. V. Allaby, Yu. B. Bushin, S. P. Denisov, A. N. Diddens, R. W. Robinson, S. V. Donskov, G. Giacomelli, Yu. P. Gorin, A. Klovning, A. I. Petrukhin, Yu. D. Prokoshkin, R. S. Shuvalov, C. A. Stahlbrandt and D. A. Stoyanova, paper submitted to the Lund Intern. Conf. on Elementary Particles, 1969; also *Phys. Letters* 30B:500 (1969).

And 70    D. K. Anderson, D. A. Jenkins, and R. J. Powers, *Phys. Rev. Letters*, 24:71 (1970).

Bas 67    R. H. Bassel and C. Wilkin, *Phys. Rev. Letters* 18:871 (1967).

Bas 68    R. H. Bassel and C. Wilkin, *Phys. Rev.* 174:1179 (1968).

Bell 64    J. S. Bell, *Phys. Rev. Letters* 18:57 (1964).

Bell 68    J. S. Bell, CERN Report TH-887 (1968).

Bel 66    G. Bellettini, G. Cocconi, A. N. Diddens, E. Lillethun, G. Matthiae, J. P. Scanlon, and A. M. Wetherell, *Nucl. Phys.* 79:609 (1966).

Bel 68    G. Bellettini, Raporteur's talk, XIVth Int. Conf. on High-Energy Physics, Vienna (1968).

Ben 67    G. W. Bennett, J. L. Friedes, H. Palevsky, R. J. Sutter, G. J. Igo, W. D. Simpson, G. C. Phillips, R. L. Stearns, and D. M. Corley, *Phys. Rev. Letters* 19:387 (1967).

Bjo 64    J. D. Bjorken and S. D. Drell, *Relativistic Quantum Mechanics and Relativistic Quantum Fields*, McGraw-Hill, New York (1964-1965).

Bos 68    E. T. Boschitz, W. K. Roberts, J. S. Vincent, K. Gotow, P. C. Gugelot, C. F. Pedrisat, and L. W. Swenson, *Phys. Rev. Letters* 20:1116 (1968).

Bos 69    E. T. Boschitz, Paper presented at Symposium on Nuclear Reaction Mechanisms and Polarization Phenomena, Quebec, Canada, September 1–2, 1969, NASA TM X-52673.

Boc 69    G. von Bochman, B. Margolis, and C. L. Tang, *Phys. Letters*, 30B:254 (1969).

Bra 70    F. Bradamante, G. Fidecaro, M. Fideraco, M. Giorgi, P. Palazzi, A. Penzo, L. Piemontese, F. Sauli, P. Schiavon, and A. Vascotto, *Phys. Letters* 31B:87 (1970).

Bro 69    S. J. Brodsky and J. Pumplin, *Phys. Rev.* 182:1794 (1969).

Bug 66    D. V. Bugg, D. C. Salter, G. H. Stafford, R. F. George, K. F. Riley, and R. J. Tapper, *Phys. Rev.* 146:980 (1966).

Bul 69      F. Bules, W. Busza, R. Giese, R. R. Larsen, D. W. G. S. Leith, B. Richter,
            V. Perez-Mendez, A. State, S. H. Williams, M. Beniston, and J. Rettberg,
            *Phys. Rev. Letters* **22**:490 (1969).

Car 68      A. A. Carter, K. F. Riley, R. J. Tapper, D. V. Bugg, R. S. Gilmore, K. M.
            Knight, D. C. Sakter, and G. H. Stafford, *Phys. Rev.* **168**:1457 (1968).

Chan 69     S. J. Chang and S.-k. Ma, *Phys. Rev. Letters* **22**:1334 (1969).

Cha 69      R. C. Chase, E. Coleman, and H. W. J. Courant, *Phys. Rev. Letters* **23**:811
            (1969).

Che 69a     H. Cheng and T. T. Wu, *Phys. Rev. Letters* **23**:670 (1969).

Che 69b     H. Cheng and T. T. Wu, *Phys. Rev. Letters* **22**:666 (1969).

Che 70      H. Cheng and T. T. Wu, *Phys. Rev.*, **D1**:1069 (1970).

Chy 70      Z. Chylinski, *Acta Phys. Polon.* **A37**:97 (1970).

Coc 62      G. Cocconi, A. N. Diddens, E. Lillethun, G. Manning, A. E. Taylor, T. G.
            Walker and A. M. Wetherell, *Phys. Rev.* **126**:277 (1962).

Cho 67      T. T. Chou and C. N. Yang, in *High-Energy Physics and Nuclear Structure*,
            (G. Alexander, ed.), North-Holland, Amsterdam (1967).

Cho 68      T. T. Chou and C. N. Yang, *Phys. Rev. Letters* **20**:1213 (1968); *Phys. Rev.*
            **170**:1591 (1968); **175**:1832 (1968).

Cal 69      D. O. Caldwell, V. B. Elings, W. P. Hesse, G. E. Jahn, R. J. Morrison,
            F. V. Murphy, and D. E. Yount, *Phys. Rev. Letters* **23**:1256 (1969).

Col 67      E. Coleman, R. M. Heinz, O. E. Overseth, and D. E. Pellat *Phys. Rev.* **164**:
            1655 (1967).

Coo 55      T. Coor, D. A. Hill, W. F. Hornyak, L. W. Smith, and G. Snow, *Phys. Rev.*
            **98**:1369 (1955).

Cor 68      D. M. Corley, *Quasi-Free Scattering of 1-BeV Protons from $^{12}C$ and $^{40}Ca$,*
            Thesis, University of Maryland, Department of Physics and Astronomy
            (1968).

Czy 67a     W. Czyż and L. Leśniak, *Phys. Letters* **24B**:227 (1967).

Czy 67b     W. Czyż and L. Leśniak, *Phys. Letters* **25B**:319 (1967).

Czy 67c     W. Czyż and L. Leśniak, in *High-Energy Physics and Nuclear Structure*,
            (G. Alexander, ed.), North-Holland, Amsterdam (1967).

Czy 67d     W. Czyż, *Medium-Energy Nuclear Physics with Electron Linear Accelerator*,
            MIT 1967 Summer Study, p. 359.

Czy 68a     W. Czyż and L. C. Maximon, *Phys. Letters* **27B**:354 (1968).

Czy 68b     W. Czyż and W. D. Simpson, Brookhaven National Lab. Report BNL
            12899 (1968).

Czy 69a     W. Czyż and L. C. Maximon, *Annals of Physics* **52**:59 (1969).

Czy 69b     W. Czyż, L. Leśniak, and H. Wołek, Inst. Nucl. Phys. Report INP No 687/
            PL/PH, *Nucl. Phys.*, **B19**:125 (1970).

Czy 69c     W. Czyż, in *Proceedings of the Third International Conference on High-
            Energy Physics and Nuclear Structure*, Columbia Univ., New York 8–12
            Sept., 1969, Report INP No. 686/PH.

Czy 69d     W. Czyż, in *Proceedings of the Inaugural Meeting of the European Physical
            Society*, Florence 1969, Report INP No. 678/PL/PH.

Czy 69e     W. Czyż and L. C. Maximon, Brookhaven National Lab. Report BNL
            14052.

Czy 69f     W. Czyż, in *Proceedings of the IX Cracow School of Theoretical Physics,
            1969*, INP No. 682/PL/PH.

Czy 69g    W. Czyż, in *International Course of Theoretical Physics*, Trieste 1969, INP No. 679/PH/PL.

Czy 70     W. Czyz and H. Wolek, "Errata and Addendum to (Czy 69b)," *Nucl. Phys.*, to be published.

Dre 66     S. D. Drell and J. S. Trefil, *Phys. Rev. Letters* **16**:552 (1966); **16**:832 (1966).

Dre 69a    S. D. Drell, D. J. Levy, and T. M. Yan, *Phys. Rev. Letters* **22**:744 (1969).

Dre 69b    S. D. Drell, D. J. Levy, and T. M. Yan, *Phys. Rev.* **187**:2159 (1969).

Dre 70a    S. D. Drell, D. J. Levy, and T. M. Yan, *Phys. Rev.* **D1**:1035 (1970).

Dre 70b    S. D. Drell and T. M. Yan, *Phys. Rev. Letters* **24**:181 (1970).

Dur 68     L. Durand and R. Lipes, *Phys. Rev. Letters* **20**:637 (1968).

Eng 68     J. Engler, K. Horn, J. König, F. Mönnig, P. Schludecker, H. Schopper, P. Sievers, and H. Ullrich, *Phys. Letters* **28B**:64 (1968).

Fäl 68     G. Faldt and T. E. O. Ericson, *Nucl. Phys.* **B8**:1 (1968).

Fes 66     H. Feshbach, in *Proceedings of the International School of Physics "Enrico Fermi"*, *Course XXXVIII*, June 27–July 9, 1966, p. 183.

Fes 70     H. Feshbach and J. Hüfner, *Annals of Physics* **52**:268 (1970).

Fel 69     M. Fellinger, E. Gutman, R. C. Lamb, F. C. Paterson, L. S. Schroeder, R. C. Chase, E. Colman, and T. G. Rhoades, *Phys. Rev. Letters* **22**:1265 (1969).

Fol 69     L. L. Foldy and J. D. Walecka, Stanford ITP 330 3/69 and *Annals of Physics*, **54**:447 (1969).

For 67     J. Formanek and J. S. Trefil, *Nucl. Phys.* **B3**:155 (1967).

Fra 66a    V. Franco and R. J. Glauber, *Phys. Rev.* **142**:1195 (1966).

Fra 66b    V. Franco and E. Coleman, *Phys. Rev. Letters* **17**:827 (1966).

Fra 68a    V. Franco, *Phys. Rev.* **175**:1376 (1968).

Fra 68b    V. Franco, *Phys. Rev. Letters* **21**:1369 (1968).

Fra 69     V. Franco and R. J. Glauber, *Phys. Rev. Letters* **22**:370 (1969).

Fri 67     J. L. Friedes, H. Palevsky, R. J. Sutter, G. W. Bennet, G. J. Igo, W. D. Simpson, and D. M. Corley, *Nucl. Phys.* **A104**:294 (1967).

Fro 67     R. F. Frosch, J. S. McCarthy, R. E. Rand, and M. R. Yearian, *Phys. Rev.* **160**:874 (1967).

Gia 68     G. Giacomelli *et al.*, to be published.

Gla 55     R. J. Glauber, *Phys. Rev.* **100**:242 (1955).

Gla 59     R. J. Glauber, "High Energy Collision Theory," in *Lectures in Theoretical Physics*, Interscience Publishers, New York (1959), Vol. I, pp. 315–414.

Gla 60     R. J. Glauber, in *Proceedings of the Conference on Nuclear Forces and the Few Nucleon Problem*, (T. C. Griffith and E. A. Power, ed.), Pergamon Press, London (1960), Vol. I.

Gla 67a    R. J. Glauber, in *High-Energy Physics and Nuclear Structure*, (G. Alexander, ed.), North-Holland, Amsterdam (1967), pp. 311–338.

Gla 67b    R. J. Glauber, and V. Franco, *Phys. Rev.* **156**:1685 (1967).

Gla 69     R. J. Glauber, *Proceedings of the Third International Conference on High Energy Physics and Nuclear Structure*, Columbia Univ., New York, Sept. 8–12, 1969.

Gol 68     A. S. Goldhaber and C. J. Joachain, *Phys. Rev.* **171**:1566 (1968).

Gold 64    M. Goldberger and K. Watson, *Collision Theory*, Wiley-Interscience, New York (1964).

Got 69     K. Gottfried and D. R. Yennie, *Phys. Rev.* **182**:1595 (1969).

Gos 69    A. T. Goshaw, P. J. Oddone, M. J. Bazin, and C. R. Sun, *Phys. Rev. Letters* **23**:990 (1969).

Gri 69    V. N. Gribov, *Zh. Éksp. i Teoret. Fiz.* **57**:1306 (1969).

Gri 69a   V. N. Gribov, *Zh. Éksp. i Teoret. Fiz.* **56**:893 (1969).

Gri 69b   V. N. Gribov, *Yadernaya Fizika* **9**:640 (1969).

Har 64    D. R. Harrington, *Phys. Rev.* **135**:B358 (1964).

Har 68    D. R. Harrington, *Phys. Rev. Letters* **21**:1496 (1968).

Har 69    D. R. Harrington, *Phys. Letters* **29B**:188 (1969).

Har 69a   D. R. Harrington, *Phys. Rev.* **184**:1745 (1969).

Har 69b   D. R. Harrington and A. Pagnamenta, *Phys. Rev.* **184**:1908 (1969).

Hsi 68    H. C. Hsiung, E. Coleman, B. Roe, D. Sinclair, and J. Vander Velde, *Phys. Rev. Letters* **21**:187 (1968).

Igo 67    G. J. Igo, J. L. Friedes, H. Palevsky, R. Sutter, G. Bennett, W. D. Simpson, D. M. Corley, and R. L. Stearns, *Nuclear Phys.* **B3**:181 (1967).

Jon 68    L. Jones, M. J. Longo, B. Gibhard, J. O'Fallon, M. Randall, and M. Kreisle, XIV Int. Conf. on High-Energy Physics, Vienna, 1968.

Ker 69    A. K. Kerman and L. S. Kisslinger, *Phys. Rev.* **180**:1483 (1969).

Kir 66    L. Kirillova *et al.*, *Zh. Éksp. i Teoret. Fiz.* **50**:76 (1966), and unpublished material.

Köl 68    K. S. Kölbing and B. Margolis, *Nucl. Phys.* **B6**:85 (1968).

Kof 69    O. Kofoed-Hansen, *Nuovo Cimento* **60A**:621 (1969).

Kre 68    M. N. Kreisler, L. W. Jones, M. J. Longo, and J. R. O'Fallon, *Phys. Rev. Letters* **20**:468 (1968).

Kuj 68    E. Kujawski, D. Sachs, and J. Trefil, *Phys. Rev. Letters* **21**:581 (1968).

Lee 68    H. K. Lee and H. McManus, *Phys. Rev. Letters* **20**:337 (1968).

Ler 63    T. Leray *et al.*, *Proceedings of the Sienna Conference on Elementary Particles* (1963), Vol. I, p. 102.

Les 69a   L. Leśniak, Doctoral dissertation, Institute of Nuclear Physics, Kraków (1969), unpublished.

Les 69b   L. Leśniak and H. Wołek, *Nucl. Phys.* **A125**:665 (1969).

Les 69c   L. Leśniak and H. Wołek, Private communication.

Mar 68    B. Margolis, *Phys. Letters* **26B**:524 (1968).

Mic 69    C. Michael and C. Wilkin, *Nucl. Phys.* **B11**:99 (1969).

Mol 47    G. Molière, *Z. für Naturforschung* **2A**:133 (1947).

McC 69a   G. McClellan, N. Mistry, P. Mostek, H. Ogren, A. Silverman, J. Swartz, R. Talman, K. Gottfried, and A. I. Lebedev, *Phys. Rev. Letters* **22**:374 (1969).

McC 69b   G. McClellan, N. Mistry, P. Mostek, H. Ogreen, A. Silverman, J. Swartz, and R. Talman, *Phys. Rev. Letters* **22**:377 (1969).

McV 62    K. W. McVoy and L. Van Hove, *Phys. Rev.* **125**:1034 (1962).

Mon 69    E. J. Moniz and G. D. Nixon, *Phys. Letters* **30B**:393 (1969).

Nau 69    M. Nauenberg, *Phys. Rev. Letters* **22**:556 (1969).

Pal 66    H. Palevsky, in *International Nuclear Physics Conference*, Gatlinburg, Tennessee, 1966 (R. L. Becker, ed.), p. 1600.

Pal 67    H. Palevsky, J. L. Friedes, R. J. Sutter, D. M. Corley, N. S. Wall, R. L. Stearns, and B. Gottschalk, *Phys. Rev. Letters* **18**:1200 (1967).

Pan 62    V. S. Pantuev and M. N. Khachaturyan, *Zh. Eksp. i Teoret. Fiz.* **42**:909 (1962).

Pum 68    J. Pumplin and M. Ross, *Phys. Rev. Letters* **21**:1778 (1968).

| | |
|---|---|
| Pum 68a | J. Pumplin, *Phys. Rev.* **173**:1651 (1968). |
| Rem 68 | E. A. Remler, *Phys. Rev.* **176**:2108 (1968). |
| Rem 69 | E. A. Remler, William and Mary Report VM-T6 (1969). |
| Ros 66 | M. Ross and L. Stodolsky, *Phys. Rev.* **149**:1172 (1966). |
| Sax 57 | D. S. Saxon and L. I. Schiff, *Nuovo Cimento* **6**:614 (1957). |
| Sch 56 | L. I. Schiff, *Phys. Rev.* **103**:443 (1956). |
| Sch 68 | L. I. Schiff, *Phys. Rev.* **176**:1390 (1968). |
| Sto 67 | L. Stodolsky, *Phys. Rev. Letters* **18**:135 (1967). |
| Str 70 | N. Straumann and C. Wilkin, *Phys. Rev. Letters* **24**:479 (1970). |
| Tin 68 | S. C. C. Ting, in *Proceedings of the Fourteenth Intern. Conf. on High-Energy Physics*, Vienna (1968). |
| Tre 69a | J. S. Trefil, *Phys. Rev.* **180**:1366 (1969). |
| Tre 69b | J. S. Trefil, *Phys. Rev.* **180**:1379 (1969). |
| Tre 69c | J. S. Trefil, *Phys. Rev. Letters* **23**:1075 (1969). |
| Van 64 | L. Van Hove, *Rev. Modern Phys.* **36**:655 (1964). |
| Vin 70 | J. S. Vincent, W. K. Roberts, E. T. Boschitz, L. S. Kisslinger, K. Gotow, P. C. Gugelot, C. F. Pedrisat, L. W. Swenson, and J. R. Priest, *Phys. Rev. Letters* **24**:236 (1970). |
| Wil 67 | C. Wilkin, in *Medium-Energy Nuclear Physics with Electron Linear Accelerators*, MIT 1967 Summer Study p. 369. |
| Wil 69 | C. Wilkin, "Graphs and Glauber," Lectures given at the Summer School on Diffractive Processes, McGill University (1969). |
| Wu 57 | T. T. Wu, *Phys. Rev.*, **108**:466 (1957). |
| Wu 65 | T. T. Wu and C. N. Yang, *Phys. Rev.* **137**:B708 (1965). |
| Zol 66 | L. S. Zolin, L. F. Kirillova, Lu Ch'ing-Ch'iang, V. A. Nikitin, V. W. Pantuev, V. A. Svindov, L. N. Strunov, M. N. Khachaturyan, M. G. Shafranova, Z. Korbel, L. Rob, P. Devinski, Z. Zlatanov, P. Markov, L. Khristov, Kh. Chernev, N. Dalkhazaz, and D. Tuvdendorzh, *Zh. Éksp. i Teoret. Fiz.-Pis'ma Redakt.* **3**:15 (1966). |

*Chapter 3*

# NUCLEOSYNTHESIS
# BY CHARGED-PARTICLE REACTIONS*

## C. A. Barnes

*W. K. Kellogg Radiation Laboratory*
*California Institute of Technology*
*Pasadena, California*

## 1. INTRODUCTION

Among the many fields of scientific endeavor in which nuclear physics plays a significant role, none is more breathtaking in scope than astrophysics. Indeed, all physical knowledge is clearly relevant in understanding our universe because of the limitless ranges of temperature and density which are believed to have existed in different regions and at different times. Most objects in the universe are, of course, not accessible for experimental analysis in the laboratory and our fragmentary knowledge of these objects can be derived only indirectly through the study of various kinds of radiation. It is therefore essential for physicists to investigate and understand as completely as possible those areas of physics which contribute in an important way to the structure and evolution of our universe, especially those which are verifiable in the laboratory.

Nuclear astrophysics is one such area where the data and understanding can be obtained with continued effort. Nuclear reactions are the source of energy for the vast majority of stars and, at the same time, they produce the rich distribution of nuclides which we observe on the earth and moon, and in meteorites, as well as those we can infer spectroscopically from the

* Supported in part by the National Science Foundation [GP-15911, GP-9114, GP-19887] and the Office of Naval Research [Nonr-220(47)].

solar and stellar atmospheres. In addition, the nuclear abundances in the cosmic radiation contain valuable clues about nucleosynthesis at the sources of the radiation and those nuclear reactions occurring during the traversal of interstellar and interplanetary space by the radiation.

## 1.1. Nuclear and Elemental Abundances

From the standpoint of the nuclear astrophysicist what is really needed is detailed relative abundance information for each nuclide, i.e., each value of $Z$ and $N$, in all objects in the universe. What is available to us in the various tabulations of relative abundances (see, for example, SU 56, GMA 60, Mul 67, Cam 68, Uns 69, Gol 69) falls far short of this for various quite obvious reasons. In the meteoritic samples, and especially in the terrestrial samples, there is considerable fractionation depending on the chemical and physical properties of the material, for which one must attempt some correction. For the solar and stellar spectroscopic data, there are serious problems in line-width interpretation arising from turbulence and collision broadening (Uns 69); in addition many of the absolute atomic transition rates are uncertain, as illustrated by the current discussion about the solar iron abundance (WKM 69, Gar+ 69). There may also be vertical segregation of chemical elements in some stellar atmospheres, arising from the competing effects of gravitation and radiation pressure, if turbulence is suppressed by strong magnetic fields, or is small for other reasons. Except for a few special situations, only elemental rather than nuclear abundances are obtained from spectral data.

In spite of these difficulties, there is quite good agreement on the solar system abundances. We will compare our calculated abundances with those tabulated by (Cam 68), which are derived mainly from analysis of meteorites (carbonaceous chondrites), and thus bypass the difficulty caused by unknown oscillator strengths in determining abundances from optical spectra. It appears, for the most part, that the relative abundances of the elements from carbon upwards are similar in most stars in our galaxy, as well as in the few well-studied stars in other galaxies. Many stellar spectroscopists insist, however, that the frequently reported variations by factors of two to five in relative abundance for various elements are genuine, and not the result of inadequate information about the excitation conditions. Only further careful spectroscopy can settle this question. There do seem to be well-established large differences in the abundances of elements with $A \geq 12$ relative to hydrogen, and to a lesser extent to helium, and this has given rise to the separation of stars into the Population I and Population II classes. Generally speaking, the Population I stars have heavy element to hydrogen ratios one or two orders of magnitude larger than the Population II stars; from this, it is inferred that the Population I stars have been formed more recently than the Population II stars, from material which has already undergone considerable nuclear processing.

A few stars appear to have quite anomalous overabundances of (one or more of) helium, carbon, nitrogen, barium, and other heavy elements. For carbon, there also exist isotopic abundance measurements (from molecular band spectra) for many stars. In the atmospheres of most of these it seems that $^{12}C:^{13}C \sim 100:1$, as in terrestrial and solar carbon; there are stars, however, with $^{12}C:^{13}C$ as low as 5:1 (McK 60, Cli 60).

The general features of the solar system nuclear abundances, as a function of $A$, are shown schematically in Fig. 1 (Bur+ 57). Our aim is the explanation of these relative abundances, and of the observed anomalous cases, by combining our knowledge of nuclear physics with theoretical studies of stellar, galactic, and universal evolution.

## 1.2. Possible Sites for Nucleosynthesis

As soon as one starts to try to understand the observed nuclear abundances in terms of nuclear physics, one immediately encounters the problem that a knowledge of nuclear reaction rates alone is not sufficient to make possible an unambiguous prediction of the resultant abundances. The distribution of nuclei with which one starts, and the temperature and density at which the nuclear reactions occur are also important; these quantities must evolve in accord with accepted physical laws. In other words, it is necessary to construct a model, whether one is discussing nucleosynthesis in stars, or in some primeval universal explosion. For reasons of brevity, stellar, galactic, and universal models cannot be discussed here in great detail, but this subject is clearly of comparable importance with nuclear reaction rates in determining the course of nucleosynthesis. For a review of the remarkable evolutionary paths of stars of different masses and compositions, the reader is referred to the paper by Iben (Ibe 67) and the references listed therein [see also (Ibe 70)].

The cosmological model of the universe favored at the present time is the so-called big-bang hypothesis that the present state of our universe is the result of expansion from a very high-density, high-temperature condition $(T > 10^{12} \, °K)$, as discussed in more detail in Section 7.1 of this paper. Some nucleosynthesis does occur, in this model, at the appropriate temperatures during the expansion, and continues until the rapidly falling

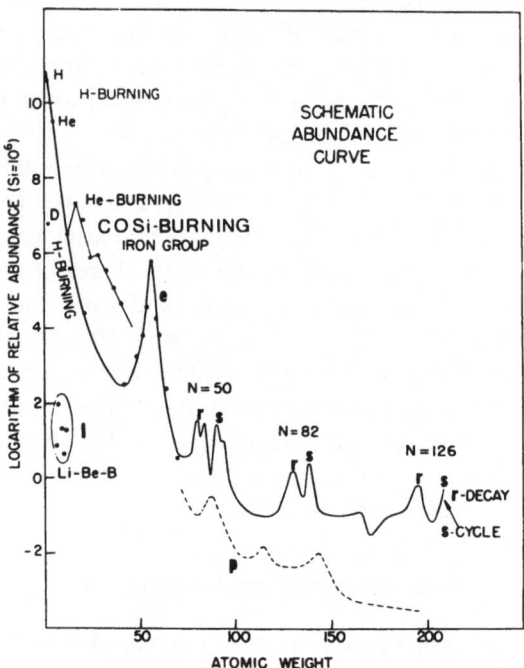

**Fig. 1. Nuclear abundances [adapted from (Bur + 57)].**
The processes believed to be mainly responsible for
the synthesis of nuclei are as follows: hydrogen burning, helium burning, carbon burning, oxygen burning, silicon burning, the equilibrium process (*e*),
neutron capture on a rapid (*r*) or slow (*s*) time scale,
the *p*-process for the low-abundance proton-rich
heavy nuclides, and the *l*-process for the low-abundance highly-reactive light elements lithium, beryllium,
and boron.

temperature cuts off nuclear reactions. Further nucleosynthetic activity can
then not take place until the material has condensed into stars massive
enough to provide from their gravitational energy the requisite high central
temperature to start further nuclear reactions. Perhaps the most important
question we would eventually like to answer is what fraction of the observed
nucleosynthesis occurred during the big bang, and how was the remainder distributed in time. At present, we are not able to give a definitive
answer.

The most obvious objects now present in the universe are the stars,
which are, for the most part, clustered into galaxies. We will adopt the

point of view, in this paper, that the lion's share of nucleosynthesis has taken place in successive generations of stars, with the possible exception of some helium-building during the primeval big bang. Indeed, there is strong evidence from the isotopic constitution of meteoritic xenon, that nucleosynthesis was still occurring locally just $2 \times 10^8$ years prior to the formation of our solar system (HPR 67, SW 70). Further, the discovery in stellar spectra of technetium (Mer 52, Mer 56), whose longest-lived isotope has a half-life of about $2 \times 10^6$ years, provides graphic evidence that nucleosynthesis is a continuing process elsewhere in our galaxy. The necessity of transporting technetium from deep inside a star to the surface within $10^6$ years poses serious but probably manageable problems for stellar models. If the recent identification of promethium in the atmosphere of HR 465 (AC 70) is confirmed, it may be necessary to postulate extensive nucleosynthesis on stellar surfaces, since the longest-lived isotope of promethium has a half-life of 18 years.

The chronologies that can be derived from the heavy neutron-produced radioactive elements, U, Th, and Re, are discussed in the following paper by Allen, Gibbons, and Macklin. Important chronological information can also be derived from many other radioactive nuclei, including some which appear to be mainly produced by charged-particle nucleosynthesis such as $^{40}$K (BLW 66), $^{26}$Al (Cla+ 70),* and $^{10}$Be.

It has been known since the pioneering work of Hertzsprung and Russell that there are strong correlations between the surface color and the luminosity (total energy emitted from the surface) of stars. This correlation is shown schematically in Fig. 2(a). If the birth, evolution, and death of stars is a continuing phenomenon, as we believe it to be, the relative number of stars in a given region of the Hertzsprung–Russell diagram is related inversely to the length of time the stars spend in that configuration. Most of a star's life, one concludes from observation, is spent on the main sequence, and rather shorter periods in the giant phase. The H–R diagram of an associated group of stars, presumably formed about the same time, provides a "snap-shot" of the state of evolution of the stars of different mass. For a young association such as the Pleiades, for example, most of the stars remain on the main sequence, except for the most massive ones, as shown in Fig. 2(c).† The old, highly evolved globular cluster M3, shown in Fig.

---

* The $^{26}$Mg abundance anomaly reported by (Cla+ 70) has not been observed in more recent work by G. J. Wasserburg and D. Schramm (private communication).

† For main sequence stars, the luminosity is approximately proportional to the 3.7th power of the mass.

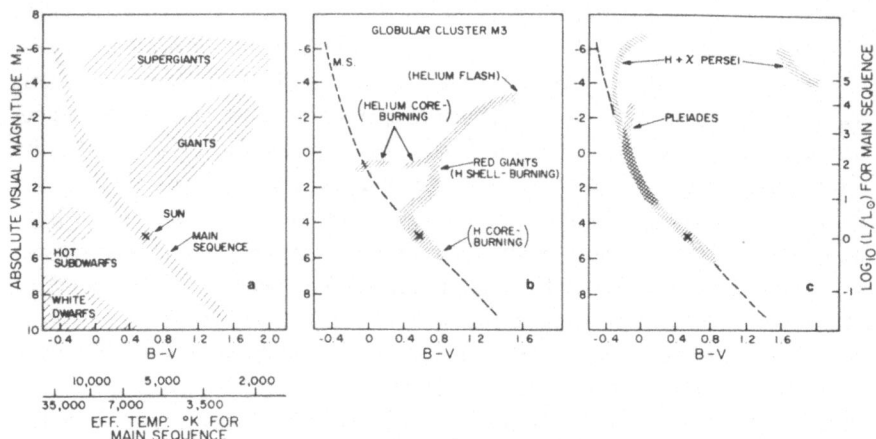

**Fig. 2. Hertzsprung–Russell diagrams (schematic).** Absolute visual magnitude $(M_V)$ is the brightness of a star, corrected to a standard distance (10 parsecs or 32.6 light years), as measured for a band of optical wavelengths centered approximately at 5500 Å. The scale is logarithmic, and increasing magnitude corresponds to decreasing brightness. The color index, B–V, is the difference between the magnitude measured with a band of wavelengths centered at approximately 4500 Å (B), and the visual magnitude (V). The relations between $M_V$ and luminosity (energy output), and between B–V and equivalent black-body temperature are shown on the figure for main sequence stars. Fig. 2(b) shows the distribution of the highly evolved stars in the globular cluster M3, while Fig. 2(c) shows two young clusters of stars. From the points at which the distributions break away from the main sequence, H and χ Persei and the Pleides have been estimated to be only about $10^6$ and $2 \times 10^7$ years old, respectively (San 58).

2(b), has only its least massive stars still on the main sequence; the majority have evolved to the subgiant or giant stages. There appears to be a tendency for the oldest stars (Population II) to be distributed within a large spherical region in our galaxy, while the more recently formed stars are concentrated in the spiral arms, in a disc-shaped region near the equatorial plane of the galaxy (Population I).

## 1.3. Nucleosynthetic Processes

The current picture of stellar nucleosynthesis can be described briefly, as follows. A star condenses from interstellar gas, mainly hydrogen, and heats as a result of the conversion of gravitational to thermal energy. When the temperature at the center becomes high enough, nuclear reactions commence in the most easily burned nuclear fuel. The energy released by the nuclear reactions "stabilizes" the star, i.e., it establishes a balance between thermal pressure and gravitation which persists without a large temperature change until the exhaustion of that nuclear fuel. The star can then contract, converting more gravitational energy to thermal energy until the temperature becomes high enough to ignite the next available fuel. An examination of the abundance data and coulomb barrier penetration considerations led Burbidge, Fowler, and Hoyle (Bur+ 57) to postulate a series of nucleosynthetic processes, which have withstood the succeeding years of experimental and theoretical investigation with little change. These are, with slight modification:

(a) Hydrogen burning (conversion of hydrogen to helium)
(b) Helium burning (conversion of helium to carbon, oxygen, etc.)
(c) Carbon and oxygen burning (production of $16 \lesssim A \lesssim 28$)
(d) Silicon burning (production of $28 \lesssim A \lesssim 60$)
(e) The $s$-process ⎫
(f) The $r$-process ⎬ (production of $A \gtrsim 60$ by neutrons)
(g) The $p$-process (production of the low-abundance proton-rich heavy nuclei)
(h) The $l$-process (production of the reactive light nuclei, D, Li, Be, B)

The $s$- and $r$-processes will be discussed only briefly here; they are discussed in detail, with a survey of the relevant experimental data, in the following paper. The other listed processes, which are induced by charged particles, will be discussed in the following sections of this work.

Following the postulated primeval universal expansion, stars of various masses form from the products of the big bang. The most massive of these, which could be far more massive than any presently known stars, evolve very quickly. They may collapse to very high temperature and density, and then re-expand rapidly ("bounce"), with nucleosynthesis occurring at appropriate temperature and density—the "little bangs" discussed in Section 7.2. Alternatively, if somewhat less massive, they may evolve quickly through the nucleosynthesis processes listed above in a more or less orderly fashion, and then explode as massive supernovae. In either case, partially processed material becomes available even before stars of ordinary mass condense. The smallest of these "original" or primeval stars may still exist due to their slow rates of evolution.

In our model, these early events are followed by successive generations of stars, with a continuing increase in the abundances of the elements heavier than hydrogen, produced by orderly progress through the various nucleosynthesis processes, and in a last final dramatic fling before the star's material is scattered into space. From the observed widespread similarity in relative abundance of the elements heavier than helium, we may conclude that (1) the bulk of the nucleosynthesis in our universe occurred early, before most of the presently observable stars were formed, or (2) that the evolution and death of the majority of stars synthesize nuclei in such a way that generally similar abundances are produced. Otherwise, implausibly effective, continuously operating, mixing processes must be postulated.

Those especially interesting stars which display gross abundance anomalies for a few specific elements must then represent cases where mixing of the atmosphere with inner regions, or loss of the envelope, displays interior material which has gone only part way through the list of nucleosynthesis processes. Even if the bulk of stellar nucleosynthesis should prove to be a product of the last events in a star's history, the study of the earlier nucleosynthesis processes remains important for several reasons: (1) they provide an explanation of anomalous abundances, (2) they represent the energy source for the greater part of the star's life, (3) they provide the "starting conditions" for the final phases of stellar evolution, and (4) they will occur in appropriate concentric layers of the star, on a rapid scale, during the terminal explosive demise of the star.

## 2. CHARGED-PARTICLE REACTION RATES

### 2.1. General Case and Definitions

Before turning to a more detailed discussion of the various nucleosynthesis processes, it is important to examine the temperature dependence to be expected for various kinds of nuclear reaction, since this is relevant in deciding what kind of experimental data should be sought in the laboratory. We restrict ourselves here to reactions with two particles in the initial and final states. The necessary modifications when three or more particles are involved, are given by Fowler and collaborators (FH 64, FV 64, FCZ 67), on whose work this section is based.

For the reaction in which nuclei 1 and 2 form 3 and 4 with energy release $Q$,

$$1 + 2 \rightarrow 3 + 4 + Q \tag{1}$$

assuming nondegenerate, nonrelativistic nuclei, the reaction rate is

$$P_{12} = \frac{n_1 n_2}{1 + \delta_{12}} \langle \sigma v \rangle \qquad \text{reactions per cm}^3 \text{ per second} \tag{2}$$

where $n_1$ and $n_2$ are the number densities of nuclei of types 1 and 2, and

$\delta_{12}$ is the Kronecker delta. The quantity

$$\langle \sigma v \rangle = \frac{(8/\pi)^{1/2}}{M^{1/2}(kT)^{3/2}} \int \sigma(E)E \exp(-E/kT)\, dE \tag{3}$$

is the product of the energy-dependent cross section and the velocity, averaged over the Maxwell–Boltzmann distribution of relative velocities.* $M$ is the reduced mass of particles 1 and 2.

If we express the temperature in units of $10^9$ °K, the kinetic energy in the center-of-mass system in MeV, and $\sigma$ in barns, equation (3) can be written numerically,

$$\langle \sigma v \rangle = 6.197 \times 10^{-14} A^{-1/2} T_9^{-3/2}$$
$$\times \int \sigma_b E_6 \exp(-11.605\, E_6/T_9)\, dE_6 \qquad \text{cm}^3\ \text{sec}^{-1} \tag{4}$$

In equation (4), $A$ is the reduced mass in AMU, given by $A = A_1 A_2 / (A_1 + A_2)$. It is frequently convenient to write the number densities as

$$n_i = \varrho N_A \frac{X_i}{A_i} \qquad \text{cm}^{-3} \tag{5}$$

where $\varrho$ is the density (grams cm$^{-3}$), $N_A$ is Avogadro's number, $X_i$ is called the mass fraction of nuclide $i$, and $A_i$ is the atomic mass of $i$ in AMU (numerically the same as the mass in grams of one gram-atom).

The mean lifetime, $\tau_2(1)$, of nucleus 1 for interaction with nucleus 2 is given by

$$\lambda_2(1) = \frac{1}{\tau_2(1)} = -\frac{1}{n_1}\left(\frac{dn_1}{dt}\right)_2 = -\frac{1}{X_1}\left(\frac{dX_1}{dt}\right)_2$$
$$= n_2 \langle 12 \rangle = \frac{X_2}{A_2}\, [12] \qquad \text{sec}^{-1} \tag{6}$$

In equation (6), $\langle 12 \rangle \equiv \langle \sigma v \rangle_{12}$ and $[12] = \varrho N_A \langle 12 \rangle$ sec$^{-1}$.

Since it may be easier to measure a required nuclear reaction cross section in the reverse direction, use is frequently made of the reciprocity theorem, which follows from the assumption of time-reversal invariance for the nuclear and electromagnetic interactions [see, for example, (BW 52) or (Pre 62)].

$$\frac{\sigma(34)}{\sigma(12)} = \frac{(1 + \delta_{34})g_1 g_2 A_1 A_2 E_{12}}{(1 + \delta_{12})g_3 g_4 A_3 A_4 E_{34}} \tag{7}$$

---

* The astrophysical cross section is usually somewhat greater than the laboratory cross section because of partial screening of the Coulomb barrier by nearby electrons. The enhancement factor for this effect, $f_{12}$, is rarely more than two or three for typical densities and compositions. [See, for example, (Sal 54).]

where $g_i = (2I_i + 1)$ is the statistical weight of nucleus $i$, and $E_{12}$ and $E_{34}$ are the kinetic energies, in the center-of-mass system, on the two sides of the nuclear reaction equation. Since $Q = E_{34} - E_{12}$, it may easily be derived from equation (3) that

$$\frac{\langle 34 \rangle}{\langle 12 \rangle} = \frac{(1 + \delta_{34})}{(1 + \delta_{12})} \frac{g_1 g_2}{g_3 g_4} \left( \frac{A_1 A_2}{A_3 A_4} \right)^{3/2} \exp(-Q/kT) \tag{8}$$

At very high temperatures the reacting nuclei may have an appreciable probability of being in excited states. If the excited nuclei are in thermal equilibrium with their ground states, as is usually the case when excited states are relevant, equation (7) must be modified, as discussed by (FCZ 67). The $g_i$ are replaced by the nuclear partition functions, $G_i = \sum_j g_{ij} \times \exp(-E_j/kT)$, where the sum includes the ground state, and $E_j$ is the excitation energy of the $j$th state. $\langle 12 \rangle$ must be averaged over all combinations of excited states of 1 and 2, and be summed over all excited states of 3 and 4. Similar remarks hold for $\langle 34 \rangle$.

A frequent application of the technique of using an inverse reaction to determine the rate of one harder to measure is the use of a radiative capture reaction to determine a photodisintegration rate. In this case particle 4 is a gamma-ray quantum. The photodisintegration rate is then*

$$\lambda_\gamma(3) = \frac{1}{\tau_\gamma(3)} = \frac{g_1 g_2}{(1 + \delta_{12}) g_3} \left( \frac{M_1 M_2}{M_3} \right)^{3/2} \frac{kT^{3/2}}{h^2} \langle 12 \rangle \exp(-Q/kT)$$

$$= 0.987 \times 10^{10} \frac{g_1 g_2}{(1 + \delta_{12}) g_3} \left( \frac{A_1 A_2}{A_3} \right)^{3/2} \varrho^{-1} T_9^{3/2} \, [12]$$

$$\times \exp(-11.605 Q_6/T_9) \quad \text{sec}^{-1} \tag{9}$$

The best method for evaluating the integral in equation (3) depends on the type of energy dependence exhibited by $\sigma(E)$, and several commonly occurring cases will be considered.

## 2.2. Nonresonant Charged-Particle Reactions

For nonresonant reactions between low-energy charged particles, the steepest energy dependence of $\sigma(E)$ is contained in the penetration factor for the Coulomb and angular momentum barrier. For incident energies small compared with the height of the Coulomb barrier and the angular

---

* See, for example, the discussion by Clayton in *Principles of Stellar Evolution and Nucleosynthesis*, listed among the general references.

momentum barrier $P_l$ is approximately proportional to $\exp(-2\pi Z_1 Z_2 e^2/hv)$. It is therefore convenient to factor out this energy dependence, and an additional factor of $1/E$ (from the $\lambda^2$ factor which always appears in nuclear cross sections). The cross section can thus be written

$$\sigma(E) = \frac{S(E)}{E} \exp(-2\pi\eta) = \frac{S(E)}{E} \exp[-(E_G/E)^{1/2}] \qquad (10)$$

where the Coulomb parameter $\eta$ and the Gamow energy $E_G$ are given by

$$\eta = Z_1 Z_2 e^2/\hbar v \quad \text{and} \quad E_G = (2\pi\alpha Z_1 Z_2)^2 (Mc^2/2) \qquad (11)$$

Numerically, $E_G^{1/2} = 0.989 Z_1 Z_2 A^{1/2}$ MeV$^{1/2}$. The remaining energy dependence resulting from the use of an approximate form for the penetration factor, can be absorbed into $S(E)$.

In most cases of interest it has not been technically feasible to measure the reaction cross sections at the low energies relevant in the astrophysical environment; an extrapolation formula becomes a crucial necessity. The quantity $S(E)$ is expected to have only a weak energy dependence, and a power series expansion in the center-of-mass energy is an adequate way to express $S(E)$.

$$S(E) = S(0) \left[ 1 + \frac{S'(0)}{S(0)} E + \frac{1}{2} \frac{S''(0)}{S(0)} E^2 + \dots \right] \qquad (12)$$

Substitution of (12) and (10) into equation (3) yields (CF 62, Bah 66)

$$\langle \sigma v \rangle = \left( \frac{2}{M} \right)^{1/2} \frac{\Delta E_0}{(kT)^{3/2}} S_{\text{eff}} \exp(-\tau) \qquad (13)$$

where

$$\Delta E_0 = 4(E_0 kT/3)^{1/2} = 0.2368 \, (Z_1^2 Z_2^2 A)^{1/6} T_9^{5/6} \quad \text{MeV} \qquad (14)$$

$$E_0 = [\pi\alpha Z_1 Z_2 kT(Mc^2/2)^{1/2}]^{2/3} = 0.1220 \, (Z_1^2 Z_2^2 A)^{1/3} T_9^{2/3} \quad \text{MeV} \qquad (15)$$

$$\tau = 3E_0/kT = 4.249 \, (Z_1^2 Z_2^2 A)^{1/3} T_9^{-1/3} \qquad (16)$$

and

$$S_{\text{eff}} = S(0) \left[ 1 + \frac{5}{12\tau} + \frac{S'(0)}{S(0)} \left( E_0 + \frac{35}{36} kT \right) \right.$$
$$\left. + \frac{1}{2} \frac{S''(0)}{S(0)} \left( E_0^2 + \frac{89}{36} E_0 kT \right) \right] \qquad (17)$$

In the equations above, $E_0$ is the most effective interaction energy, corresponding to the peak of the integrand in equation (3), and $\Delta E_0$ is the full width of this peak at $1/e$ of the maximum value. Values for the constants to be used in these equations are given for many astrophysically relevant nuclear reactions in (FCZ 67) and a revision of this article which is currently in preparation.

## 2.3. Isolated Narrow Resonances

For an isolated narrow resonance, the energy dependence of the cross section is given by the well-known Breit–Wigner expression,

$$\sigma(E) = \pi \lambdabar^2 \frac{\omega \Gamma_{12} \Gamma_{34}}{(E - E_R)^2 + (\Gamma^2/4)}$$

$$= \frac{0.657}{AE} \frac{\omega \Gamma_{12} \Gamma_{34}}{(E - E_R)^2 + (\Gamma^2/4)} \qquad \text{barn} \qquad (18)$$

where $\lambdabar = \hbar/Mv$, $M$ is the reduced mass of 1 and 2, $v$ is their relative velocity, and $E$ is the center-of-mass energy of 1 and 2, in MeV. The quantities $\Gamma_{12}$ and $\Gamma_{34}$ are the partial widths for the appropriate particle combinations, $\Gamma$ is the total width of the resonance which is equal to $\Gamma_{12} + \Gamma_{34} + \cdots$, and $E_R$ is the resonance energy, all in the center-of-mass system, of course. The quantity $\omega$ is equal to $(2J_r + 1)(1 + \delta_{12})/g_1 g_2$, where $J_r$ is the angular momentum of the resonant state. For the frequently useful case $\Gamma \ll \Delta E_0$, equation (3) can be integrated immediately to yield

$$\langle 12 \rangle = \left(\frac{2\pi\hbar^2}{MkT}\right)^{3/2} \frac{(\omega\gamma)}{\hbar} \exp(-E_R/kT)$$

$$= [2.557 \times 10^{-13} \, A^{-3/2}(\omega\gamma)] T_9^{-3/2} \exp(-11.605 \, E_R/T_9) \qquad \text{cm}^3 \text{ sec}^{-1}$$
$$(19)$$

In equation (19), $(\omega\gamma) = \omega \Gamma_{12} \Gamma_{34}/\Gamma$ which can be determined experimentally from the integrated yield of the resonance under consideration. If several narrow resonances occur within $\Delta E_0$, their contributions to $\langle \sigma v \rangle$ are simply summed.

In the case of a strongly exoergic reaction, $\Gamma_{34}$ will not vary systematically from resonance to resonance, for changes of a few hundred keV incident energy. $\Gamma_{12}$, on the other hand, will vary relatively rapidly as the incident energy is changed, because of changes in the barrier penetration. In the limiting case where $\Gamma_{12} \sim \Gamma$, $(\omega\gamma) \sim \omega \Gamma_{34}$, and the strength of the various resonances will tend to be (very roughly) constant. On the other hand, for low incident energies, $\Gamma_{12} \ll \Gamma$, and the resonance strength $(\omega\gamma) \sim \omega \Gamma_{12}$; in this case, the strength of the resonances will decrease rapidly as the incident energy is lowered, due to the barrier penetration factor contained in $\Gamma_{12}$. From the point of view of the experimentalist, this is a serious problem because it means that some low-energy resonances, which might be of major importance in astrophysical situations, are inaccessible to the experimenter, at least by a direct strength measurement.

In a few cases where direct measurements of $(\omega\gamma)$ have not been feasible, it has been possible to measure $\Gamma_{12}/\Gamma$, or $\Gamma_{34}/\Gamma$, and $\Gamma$ (or $\tau = \hbar/\Gamma$) by other experimental techniques. If the spin of the level is also known, $(\omega\gamma)$ can then be calculated. There remain, however, many energy levels for which neither kind of measurement has been accomplished, and these potentially important energy levels constitute a major challenge to the experimentalist.

It may be of interest to examine the question of how reliably the strength of such difficult-to-measure levels may be estimated from the strengths of levels where direct measurements are possible. Aside from penetration factors, the widths, $\Gamma_{12}$ and $\Gamma_{34}$ should be distributed according to the chi-square distribution (Porter–Thomas distribution) [see, for example, (Pre 62)], where the number of degrees of freedom is just the number of *a priori* equally-likely channels contributing. There are some studies in the heaviest nuclei where these expectations seem to be borne out quite well. However, an examination of some 85 levels studied by (LTS 69) in $^{27}$Al$(p, \gamma)$ is rather disappointing, from this point of view. The average value of $\omega\Gamma_{34}$ is about 7 eV, while the most probable value is only about 0.7 eV; this gives some measure of the very large dispersion of the measured values, or the small "effective" number of degrees of freedom. Although this analysis is very crude because many of the resonance spins and the multipolarities of the $\gamma$-radiations are unknown, it is sufficient to show that estimated resonance strengths may be wrong by orders of magnitude.

If there are many sharp resonances within the effective energy range $\Delta E_0$, one can sum the contributions given by equation (19) for all the resonances, at a series of discrete temperatures covering the range of interest. These summed contributions can then be fitted by least squares to the empirically-suitable simple expression,

$$\langle\sigma v\rangle = A \exp(-B/kT) \tag{20}$$

for use in stellar model calculations. If the range of temperature for which a simplified expression is desired is large, it may sometimes be necessary to use a 4-parameter fit [see, for example, (Lyo+ 70)],

$$\langle\sigma v\rangle = A \exp(-B/kT) + C \exp(-D/kT) \tag{21}$$

## 2.4. Broad Overlapping Resonances

For broad resonances, it is necessary to include the energy dependence of the partial widths and the total width in the Breit–Wigner expression. In cases where there are overlapping resonances of the same spin and parity, the interference between the resonant amplitudes does not integrate to zero when the differential cross section is integrated over angle. It is therefore necessary, when integrating equation (3) for broad overlapping resonances of the same spin and parity to know whether the interference is constructive or destructive (see Section 4.2, for example).

## 2.5. Excited State Reactions

In some situations involving high temperatures and low-lying excited states, an appreciable fraction of nuclei 1 and/or 2 may be in excited states when they react, as remarked earlier. A detailed calculation of $\langle \sigma v \rangle$ for a particular resonance will then require a knowledge of the partial widths for the excited states, as well as for the ground states. These additional partial widths are sometimes available from a study of the inverse reaction, or from inelastic scattering through the resonances in question. Very little work has so far been carried out with the inclusion of the excited states. An example of this kind of analysis is the study of the reactions

$$^{19}\text{F} + p \rightarrow \alpha + ^{16}\text{O} \tag{22}$$

and

$$^{19}\text{F} + p \rightarrow \gamma + ^{20}\text{Ne} \tag{23}$$

carried out by N. A. Bahcall and W. A. Fowler (BF 69, 70). For these reactions most of the requisite data were already available from inelastic proton scattering studies (Bar 55a, Bar 55b). There is always the worry that astrophysically important resonances may be missed in the inelastic scattering measurements because of low cross sections; resonances with large partial widths, $\Gamma_{12*}$, for the excited states, but small widths, $\Gamma_{12}$, for the unexcited nuclei will be weak in inelastic scattering.

## 2.6. High Level Density Region, Continuum Cross Sections

When there is a high density of overlapping compound nucleus energy levels, both in the region available for experimental study and in the region around $E_0$, an approach based on the nuclear optical model may be more suitable [see, for example, (FH 64)]. The cross section, averaged over the closely-spaced overlapping resonances, can again be factored, for the purposes of making extrapolations, as

$$\bar{\sigma} = \frac{\bar{S}(E)}{S} \exp[-(E_G/E)^{1/2}] \tag{24}$$

where the problem now has been transferred to determining a suitable form for the energy-averaged $S$-factor. Fowler and Hoyle (FH 64)* have derived the following form for $\bar{S}$ (logarithms to the base 10):

$$\log \bar{S} \approx -1.184 + \log\langle\beta\Gamma_{34}/\Gamma\rangle + \log Z_1 Z_2 - 0.50 \log A$$
$$+ 0.457(AZ_1Z_2R_f)^{1/2} - 0.053E_6(AR_f^3/Z_1Z_2)^{1/2} \tag{25}$$

In this expression, $R_f$ is the interaction radius in fermis; the quantity $\beta$, which results from averaging over the size resonances of the optical model, is expected to be of order unity. Michaud *et al.* (MSV 70) have suggested that $\beta$ and the radius of interaction should be increased to take account of the diffuseness of the nuclear surface; (FH 64) based their analysis on a square-well potential.

Since equation (25) is the result of several approximations, it should probably be viewed as a parametrization of $\bar{S}$ in terms of $\langle\beta\Gamma_{34}/\Gamma\rangle$, which quantity can then be determined by fitting experimental data. A number of experiments have been analyzed this way [see, for example, (Ada+ 69)] but the validity of extrapolations based on this type of analysis requires further verification.

Considerable theoretical effort is currently being devoted to refining the nuclear optical model, and clarifying its connection with existing statistical treatments of nuclear level densities [see, for example, (MSV 70) and references given therein]. It seems likely that improved ways of estimating continuum nuclear cross sections will result from this work, but a detailed discussion of this important area of research would be too lengthy for inclusion here. No matter how carefully the theoretical work is carried out, however, it will still be imperative for the experimentalist to check theoretically predicted astrophysical cross sections. Low-energy nuclear reaction cross sections are sensitive to the magnitude of the absorptive part of the optical potential at radii much greater than those relevant in the higher energy region where optical potential parameters are usually determined,

---

* A typographical error in this paper has been corrected here.

especially for reactions between heavy ions. In addition, at higher energies direct reactions may contribute strongly to the absorption of the incident particles, while the reaction cross section at energies of astrophysical interest is expected to be dominated by compound nucleus formation; the absorptive part of the optical potential may therefore prove to be strongly energy dependent.

## 3. HYDROGEN BURNING

### 3.1. The $p$–$p$ Chain

Since hydrogen appears to be the most abundant element in the universe, and since it has the lowest $Z$ and therefore the lowest Coulomb barrier, it is natural to consider first the reactions that convert hydrogen into helium. Any traces of deuterium present in the gas from which the star forms will react and become exhausted during the contraction stage, at comparatively low temperatures. When the core temperature and density become high enough, hydrogen will react with itself according to the reaction (Wei 37, Be 38),

$$^1\text{H} + {}^1\text{H} \rightarrow {}^2\text{H} + e^+ + \nu \tag{26}$$

a relatively slow reaction because the weak interaction is involved. This will be followed by the reaction

$$^1\text{H} + {}^2\text{H} \rightarrow {}^3\text{He} + \gamma \tag{27}$$

and the fusion of hydrogen to helium will usually be completed by the reaction (Lau 51, Sch 51)

$$^3\text{He} + {}^3\text{He} \rightarrow {}^4\text{He} + 2\,{}^1\text{H} \tag{28}$$

If there is already considerable $^4$He present in the material from which the star condensed, as there appears to be in most stars ($\sim$25% by mass in the surface of the sun), or as a result of the helium formed by reaction (28), alternative ways of completing the fusion are possible. For example, we may have

$$^3\text{He} + {}^4\text{He} \rightarrow {}^7\text{Be} + \gamma \tag{29}$$

$$^7\text{Be} + e^- \rightarrow {}^7\text{Li} + \nu \tag{30}$$

$$^7\text{Li} + {}^1\text{H} \rightarrow 2\,{}^4\text{He} \tag{31}$$

or alternatively, at higher H-burning temperatures, the sequence

$$^7\text{Be} + {}^1\text{H} \rightarrow {}^8\text{B} + \gamma \tag{32}$$

$$^8\text{B} \rightarrow {}^8\text{Be} + e^+ + \nu \tag{33}$$

$$^8\text{Be} \rightarrow 2\,{}^4\text{He} \tag{34}$$

becomes significant. This alternative, first suggested by Fowler (Fow 58) [see also (Cam 58)], has the important consequence that neutrinos with energies up to 14 MeV are produced [see discussion in Section 3.3]. The proton–proton $\beta$-decay has been reconsidered most recently by Bahcall and May (BM 69a) taking into account the revised value for the Gamow–Teller coupling constant which can be derived from new measurements of the neutron half-life (Chr+ 67) and the Fermi coupling constant, as determined from $J^\pi = 0^+ \rightarrow 0^+$, $\Delta T = 0$ $\beta$-decays (Bar+ 62, Fre+ 69). This reaction appears to be many orders of magnitude too weak to be directly measured in the laboratory,* and it determines the overall rate of the proton–proton chain because it is by far the slowest reaction involved.

The second reaction, equation (27), has been studied down to 16 keV in the center-of-mass system by Griffiths, Lal, and Scarfe (GLS 63). Since the reaction seems to have comparable direct-capture contributions from both $s$- and $p$-waves, at stellar energies, these authors used the measured $\gamma$-ray angular distribution to make separate low-energy extrapolations for the $s$- and $p$-waves. The measurements and extrapolations are in good agreement with a theoretical treatment by Donnelly (Don 67). These extrapolations are shown in Fig. 3.

The reaction given as equation (28) is expected to be the most frequent way of completing the $p$–$p$ chain in main sequence stars like the sum. It has been investigated most recently at low energies by Winkler and Dwarakanath (WD 67) and by Bacher and Tombrello (BT 67). From an experimental point of view, it is interesting to note that the $^3\text{He} + {}^3\text{He}$ reaction appears to proceed mainly to a proton and the unstable nucleus $^5\text{Li}$, at laboratory energies above about 2 MeV. A different mechanism is clearly needed for lower energies, as indicated by a dramatic change in the proton energy spectrum (BT 67). May and Clayton have studied theoretically a reaction mechanism in which a neutron tunnels from one $^3\text{He}$ to the other without the necessity of a coulomb barrier penetration to a radial distance where the nuclei overlap appreciably; a "diproton" remains and this sub-

---

* It has been estimated that a 1-MeV proton beam on a pure hydrogen target would produce about one reaction per ten megawatt-years of integrated beam power (Bur+ 57).

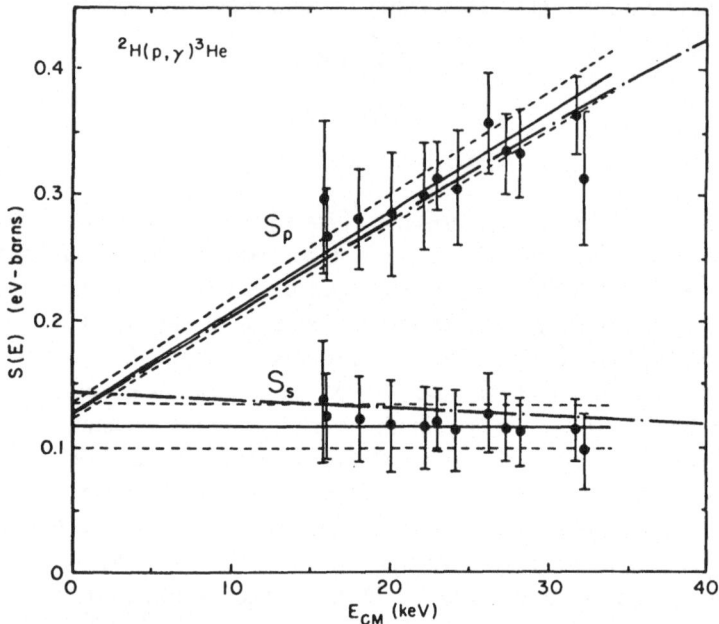

**Fig. 3.** *S*-factor extrapolations for radiative capture of $l = 0$ and $l = 1$ protons by deuterium. The solid lines are fits to the data of (GLS 63), the dashed lines are the estimated uncertainties for these fits, and the dot-dash extrapolations are based on a theoretical study of the reaction by (Don 67).

sequently fissions into two protons (MC 68). The calculated *S*-factor rises more steeply as the energy is lowered than the experimental data, but this may be a result of approximations in the calculation. Experimental *S*-factors from several sources are shown in Fig. 4. The Dwarakanath–Winkler results (Dwa 68) were obtained with a differentially pumped, continuously recirculated gas target, to minimize energy loss and energy straggling of the beam before the beam reaches the target volume. The uncertainty in energy loss and in energy distribution of a low-energy $^3$He beam, passed through the more conventional entrance foil, is intolerable because the cross section is such a steep function of beam energy. In fact, it is a general characteristic of cross section measurements at very low energies, that the energy resolution and the energy of the beam must be rather accurately known in order to extract a meaningful *S*-factor.

The radiative capture of $^3$He by $^4$He has been reexamined most recently by Nagatani, Dwarakanath, and Ashery (NDA 69), with the same differ-

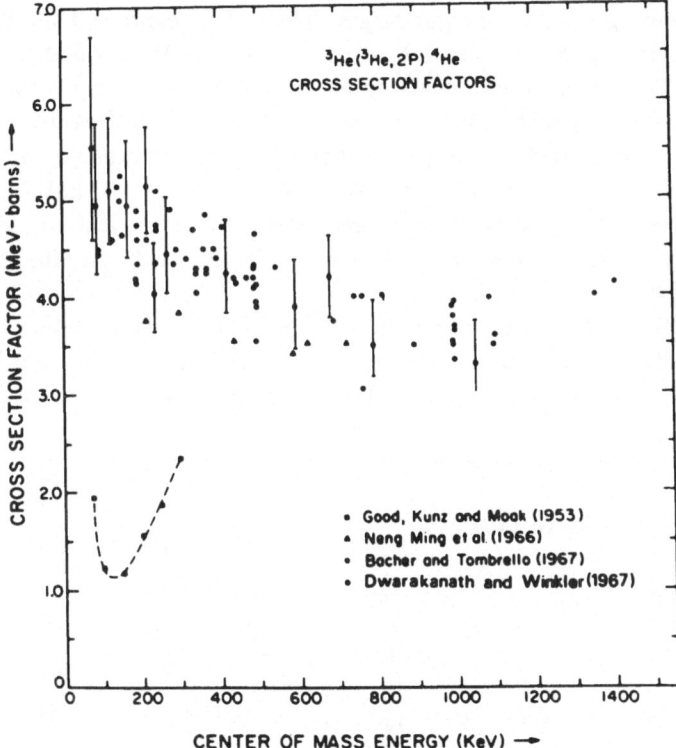

**Fig. 4.** Comparison of cross section measurements for the reaction $^3$He $+$ $^3$He $\rightarrow$ $2p$ $+$ $^4$He (Dwa 68).

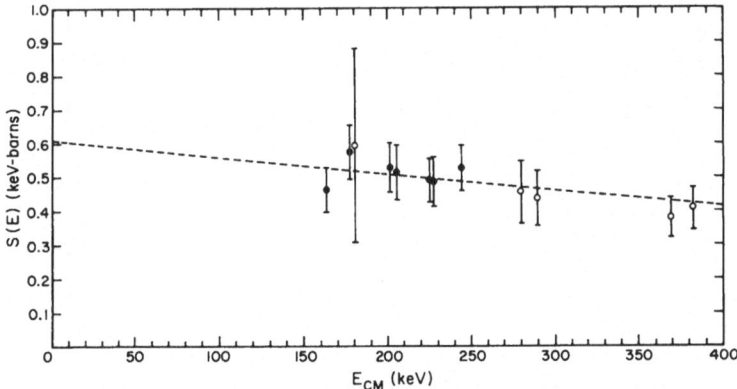

**Fig. 5.** *S*-factor extrapolation for the reaction $^3$He $+$ $^4$He $\rightarrow$ $^7$Be $+$ $\gamma$. The closed circles are from (NDA 69) and the open circles from (PK 63).

ential pumping system and gas target. Their data expressed as $S$-factors are shown in Fig. 5, together with earlier data from Parker and Kavanagh (PK 63). The dashed curve is a second order polynomial fit (in the center-of-mass energy). A theoretical fit based on a direct radiative-capture analysis of this reaction at higher energies by Tombrello and Parker (TP 63), yields a slightly lower value of $S(0)$; we adopt $S(0) = 0.61 \pm 0.07$ keV-barns and $(dS/dE)_0 = (-5.8 \pm 0.3) \times 10^{-4}$ barns from the empirical fit, although there is really no compelling reason to prefer one extrapolation method over the other.

The $^7$Be electron capture rate in stellar environments has been examined theoretically by many authors including, most recently, Bahcall and Moeller (BM 69b). Electron capture and the $^7$Be$(p, \gamma)$ process are competitors,

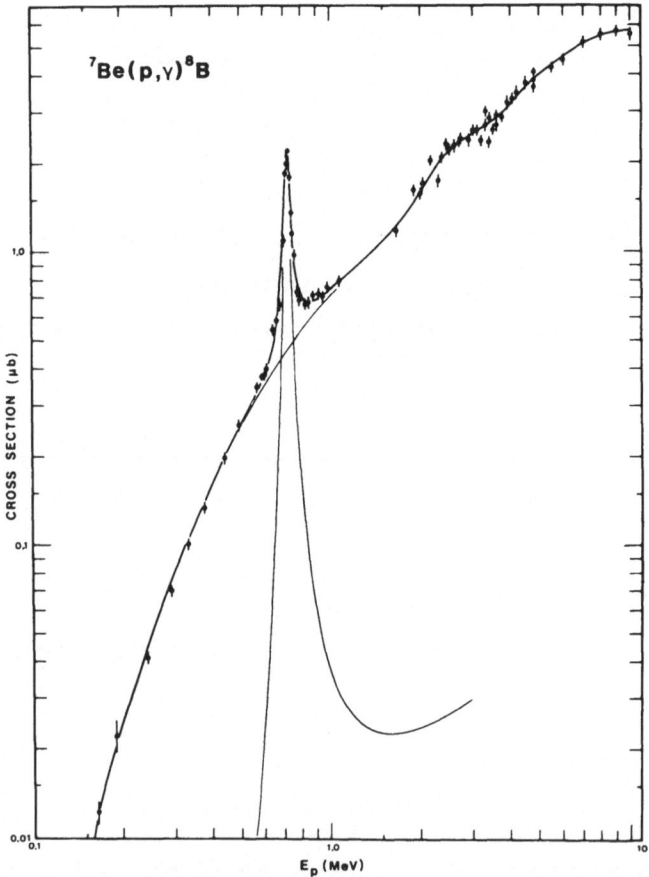

Fig. 6. Cross section for the reaction $^7$Be $+ p \rightarrow {}^8$B $+ \gamma$ (Kav$+$ 69).

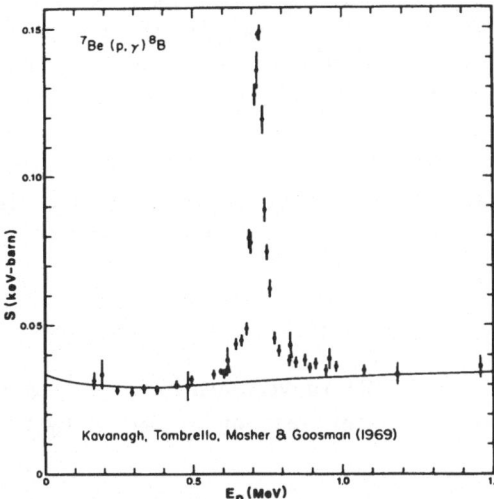

**Fig. 7.** *S*-factor curve for the data of Fig. 6 (Kav+ 69). The solid curve is from a theoretical calculation by (Tom 65) which has been normalized to the experimental data. The narrow resonance at $E_p$(lab) = 0.72 MeV plays no important role at the temperatures relevant for the *p–p* chain.

since they determine whether this branch of the *p–p* chain is terminated by ⁷Li(*p*, α) or by the production of ⁸B. The most recent measurement of the low-energy cross section of reaction (31) has been made by bombarding protons with ⁷Li ions (TSW 70), but the rate is, in any case, much faster than the electron capture rate forming the ⁷Li, and is therefore irrelevant in determining the branching among the various ways of completing the *p–p* chain.

The principal experimental problem in studying the radiative capture of protons by ⁷Be [equation (32)], is that the target is radioactive with a terrestrial half-life of about 55 days. This makes the manufacture of a suitable target a major enterprise, and also leads to a high background of 480-keV *γ*-quanta from the excited state of ⁷Li. In spite of, or perhaps, because of, these difficulties this reaction has been studied several times [see, for example, (Par 66, 68a, Kav+ 69)]. The most recent study (Kav+ 69) is more precise and covers a wider energy range than any of the earlier work; moreover it is in excellent agreement with the data of (Par 68a). Figures 6 and 7 show the cross section and *S*-factor extrapolation of (Kav+ 69) for the reaction ⁷Be(*p*, *γ*)⁸B.

There are, of course, many other reactions involving various combinations of protons, deuterons, britons, and $^3$He which might provide alternate paths for the completion of the $p$–$p$ chain under some circumstances. Parker, Bahcall, and Fowler (PBF 64) have considered these reactions in detail, and have concluded that in most main sequence stars they will play no important role, either because they are in competition with much faster reactions, or because one or both of the respective ingredients never reach a sufficient number density. One of these reactions deserves special mention, however,

$$^3\text{He} + {}^1\text{H} \rightarrow {}^4\text{Li} + \gamma, \qquad {}^4\text{Li} \rightarrow {}^4\text{He} + e^+ + \nu \qquad (35)$$

If $^4$Li were proton-stable, or very nearly so, the main completion of the $p$–$p$ chain would be by this reaction instead of by equations (28) or

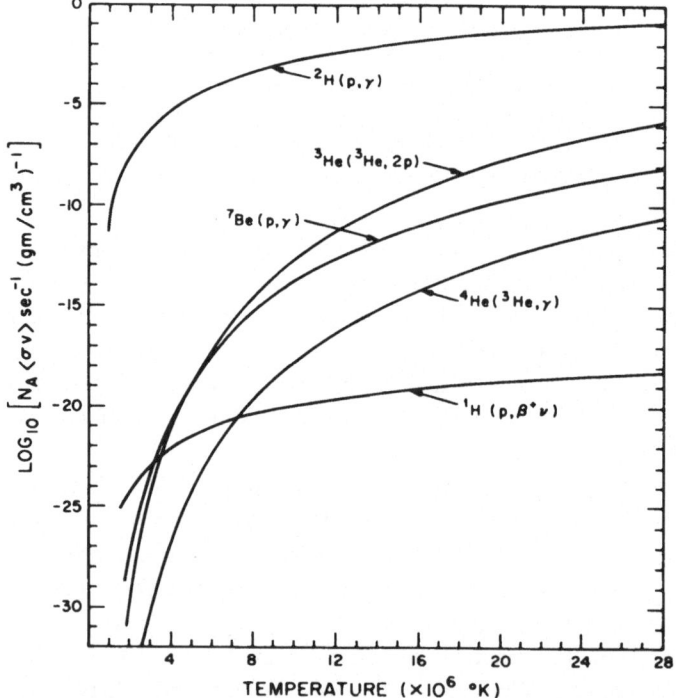

Fig. 8. $N_A\langle\sigma v\rangle$ as a function of temperature for the $p$–$p$ chain reactions. The curves have been plotted from data compiled by W. A. Fowler, G. R. Caughlan, and B. A. Zimmerman. The relation between $N_A\langle\sigma v\rangle$ and the mean time for interaction is given by Equation (6). See Table I for input data.

TABLE I. $p$–$p$ Chain Data Used in Rate Calculations of Fig. 8[a,b]

| Reaction | $S(0)$, MeV-barns | $S'(0)$, barns | $S''(0)$, barns-MeV$^{-1}$ |
|---|---|---|---|
| $^1$H$(p, \beta^+\nu)^2$H | $3.78 \times 10^{-25}$ | $4.23 \times 10^{-24}$ | — |
| $^2$H$(p, \gamma)^3$He | $2.50 \times 10^{-7}$ | $7.90 \times 10^{-6}$ | — |
| $^3$He$(^3$He, $2p)^4$He | $5.10$ | $-2.40$ | $2.00$ |
| $^4$He$(^3$He, $\gamma)^7$Be | $6.10 \times 10^{-4}$ | $-5.80 \times 10^{-4}$ | $5.06 \times 10^{-4}$ |
| $^7$Be$(p, \gamma)^8$B | $3.35 \times 10^{-5}$ | — | — |

[a] These data have been taken from (FCZ 67) and a revised compilation by the same authors.
[b] The mean lifetime for interaction, $\lambda = (\varrho X/A)N_A\langle\sigma v\rangle$; e.g., the mean lifetime for the conversion of $^4$He to $^7$Be by $^3$He nuclei is $(\varrho X_{3\mathrm{He}}/3)$ times $N_A\langle\sigma v\rangle$.

(29). It seems experimentally clear that $^4$Li is unstable (BKP 59), and that the direct reaction,

$$^3\mathrm{He} + {}^1\mathrm{H} \rightarrow {}^4\mathrm{He} + e^+ + \nu \tag{36}$$

is too slow to be relevant (WB 67).

In Fig. 8 we have plotted the logarithm (base 10) of the product $N_A\langle\sigma v\rangle$ for several of the relevant $p$–$p$ chain H-burning reactions, as a function of temperature in millions of degrees K. From equation (6), the mean lifetime for interaction of the target nucleus (designated as 1) with protons or $^3$He (designated as 2), $\tau_2(1)$, is given by the expression

$$\lambda_2(1) = 1/\tau_2(1) = (X_2/A_2)\varrho N_A\langle\sigma v\rangle \qquad \text{sec}^{-1}$$

where $\varrho$ is expressed in grams per cubic centimeter, $X_2$ is the mass fraction of nucleus 2, $A_2$ is the atomic mass of nucleus 2, and $N_A$ is Avogadro's number. In the temperature region of most interest for the $p$–$p$ chain, roughly from 10 to 15 million degrees, it is clear that the rate-determining reaction for the chain is the $p$–$p$ $\beta$-decay, and that the $^3$He + $^3$He reaction is considerably faster than the $^4$He + $^3$He reaction. The $S$-values for these reactions are listed in Table I.

## 3.2. The C–N–O Bi-Cycle

In stars somewhat more massive than the sun, higher temperatures and densities will occur in the core before the star achieves hydrostatic equilibrium. Hydrogen will then be fused to form helium by another, faster chain of reactions, provided at least one of the elements carbon, nitrogen, or oxygen is present to act as a catalyst (Bet 39, Wei 38). These reactions are

$$^{12}C + {}^{1}H \rightarrow {}^{13}N + \gamma \tag{37}$$

$$^{13}N \rightarrow {}^{13}C + e^{+} + \nu \tag{38}$$

$$^{13}C + {}^{1}H \rightarrow {}^{14}N + \gamma \tag{39}$$

$$^{14}N + {}^{1}H \rightarrow {}^{15}O + \gamma \tag{40}$$

$$^{15}O \rightarrow {}^{15}N + e^{+} + \nu \tag{41}$$

$$^{15}N + {}^{1}H \rightarrow {}^{12}C + {}^{4}He \tag{42}$$

together with the weaker, interlocking cycle of reactions

$$^{15}N + {}^{1}H \rightarrow {}^{16}O + \gamma \tag{43}$$

$$^{16}O + {}^{1}H \rightarrow {}^{17}F + \gamma \tag{44}$$

$$^{17}F \rightarrow {}^{17}O + e^{+} + \nu \tag{45}$$

$$^{17}O + {}^{1}H \rightarrow {}^{14}N + \alpha \tag{46}$$

Unlike the reactions of the $p$–$p$ chain, which are predominantly non-resonant, the reaction rates here are mostly controlled by resonances. Most of the relevant cross section measurements were made some years ago, because it appeared at that time that this chain of reactions might be predominant in the sun. Parameters for calculating the rates of these reactions are listed in Table II as compiled by (FCZ 67), which contains references

TABLE II. C–N–O Cycle Data Used in Rate Calculations for Fig. 9[a]

| Reaction | $S(0)$, MeV-barns | $S'(0)$, barns | $S''(0)$, barns-MeV$^{-1}$ |
|---|---|---|---|
| $^{12}C(p, \gamma)^{13}N$ | $1.40 \times 10^{-3}$ | $4.25 \times 10^{-3}$ | $3.75 \times 10^{-2}$ |
| $^{13}C(p, \gamma)^{14}N$ | $5.50 \times 10^{-3}$ | $1.34 \times 10^{-2}$ | $9.87 \times 10^{-2}$ |
| $^{14}N(p, \gamma)^{15}O$ | $2.75 \times 10^{-3}$ | — | — |
| $^{15}N(p, \gamma)^{16}O$ | $2.74 \times 10^{-2}$ | $1.86 \times 10^{-1}$ | — |
| $^{15}N(p, \alpha)^{12}C$ | $5.34 \times 10^{1}$ | $8.22 \times 10^{2}$ | — |
| $^{16}O(p, \gamma)^{17}F$ | $7.45 \times 10^{-3}$ | $-6.47 \times 10^{-3}$ | $5.26 \times 10^{-3}$ |
| $^{17}O(p, \alpha)^{14}N$ | $3.58 \times 10^{-2}$ | — | — |

[a] These data have been taken from (FCZ 67) and a revised compilation by the same authors.

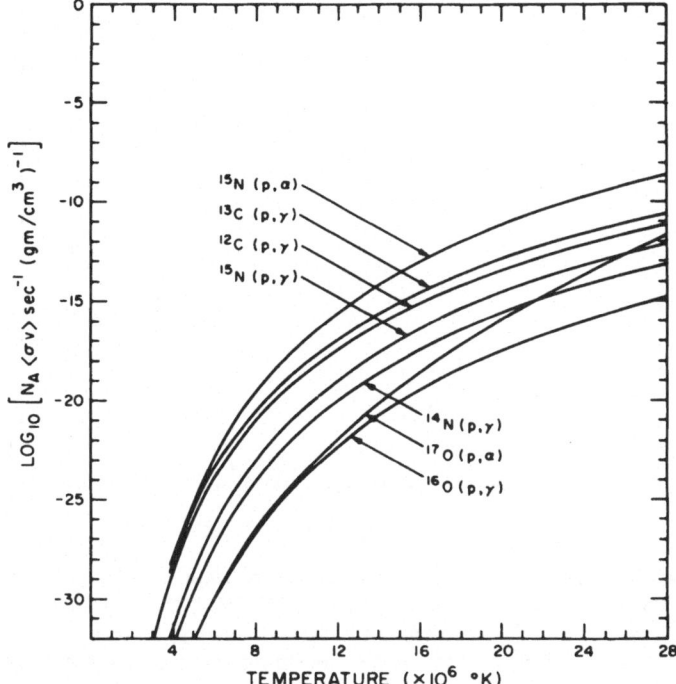

**Fig. 9.** $N_A\langle\sigma v\rangle$ **as a function of temperatures for the C–N–O cycle reactions.** The curves have been plotted from data compiled by W. A. Fowler, G. R. Caughlan, and B. A. Zimmerman. See Table II for input data.

to the original data. The product $N_A\langle\sigma v\rangle$ for these reactions is shown in Fig. 9.

Since none of these reactions can be studied in the astrophysically important energy range, there is always a serious worry that compound-nucleus states, which escape detection in the experimental studies of the various reactions, could nevertheless dominate the rate at lower energies. All one can do is to study the relevant region of the compound nucleus in other reactions, and try to obtain the parameters necessary to calculate the contribution of any states found near threshold. This is well illustrated by the rate-determining reaction, equation (40). A new excited state of $^{15}$O was reported (WOA 65) at an excitation energy of 7284 ± 7 keV, just 8.8 ± 7 keV below the calculated threshold at 7292.8 ± 1.2 keV (MTW 65).

Subsequent measurements placed the state 17.1 ± 1.3 keV (WA 66), or 21.6 ± 1.1 keV (Hen 67), below threshold, as shown in Fig. 10. In addition,

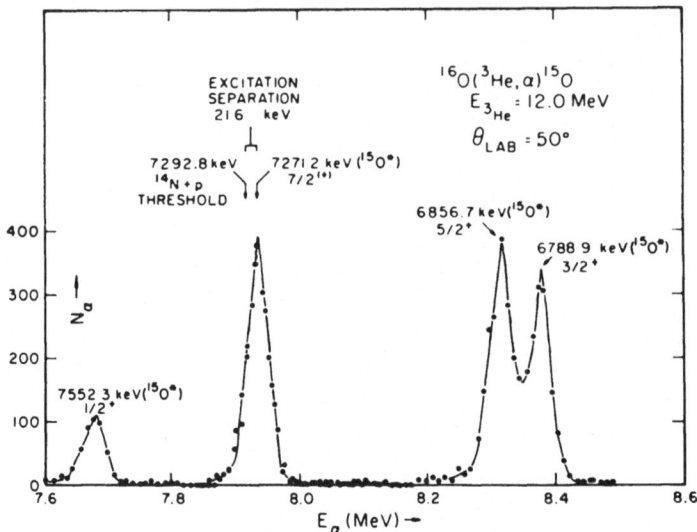

**Fig. 10. Alpha-particle spectrum from the reaction $^{16}O(^3He, \alpha)^{15}O$ at 12.0 MeV incident energy and at $\theta_{Lab} = 50°$.** From the position of the 7552.3 keV level, which occurs as a sharp low-energy resonance in the reaction $^{14}N(p, \gamma)^{15}O$, it can be determined that the recently discovered $7/2^+$ level of $^{15}O$ is $(21.6 \pm 1.1)$ keV below the $^{14}N + p$ threshold; for this reason and because of its high spin value it plays no significant role in determining the astrophysical rate of the $^{14}N(p, \gamma)^{15}O$ reaction (Hen 67).

the latter author identified the spin and parity of the level as $7/2^+$, which means that $d$-wave protons are required to form it from $^{14}N$. The position of the level and its spin and parity prevent it from significantly affecting the astrophysical rate of the reaction. Incidentally, the 4.5 keV discrepancy between the two measurements of the excitation energy could be removed by altering the tabulated mass of $^{15}O$, since one measurement was made with respect to the ground state of $^{15}O$ (WA 66), the other with respect to a resonance at a known bombarding energy above the $^{14}N + {}^1H$ threshold.

Caughlan and Fowler (CF 62) have studied the equilibrium relative abundances of the nuclei in the C–N–O cycle, as a function of temperature. These are, of course, determined by the lifetimes for interaction with protons, for the various nuclei. Near 20 million degrees, for example, $^{12}C/^{14}N$ is predicted to be approximately $8 \times 10^{-3}$, and $^{16}O/^{14}N$ about $2.6 \times 10^{-2}$. $^{12}C/^{13}C$ is predicted to be near 4 for a very wide range of temperatures.

From the solar system relative abundances (Cam 68, Gol 69), which are similar to the spectroscopic abundances derived for most stars, we find $^{12}C/^{14}N$ to be about 5, $^{16}O/^{14}N$ about 10, and $^{12}C/^{13}C$ about 100, in sharp contrast with the C–N–O cycle predictions. There are many possible explanations for the large difference between the predicted and observed abundances: (1) The results of the C–N–O cycle may have been heavily diluted by the addition of $^{12}C$ and $^{16}O$ resulting from subsequent helium burning. (2) If one starts with carbon and oxygen, and adds less than one proton per carbon or oxygen nucleus, equilibrium will not be established for the C–N–O cycle, and $^{12}C/^{14}N$ and $^{16}O/^{14}N$ ratios comparable with solar system ratios may be obtained (Cau 65). (3) Perhaps the most plausible explanation of the discrepancy is that the products of the C–N–O cycle are usually reprocessed by later stages of nucleosynthesis before the material from stellar interiors is distributed through space.

It is not unreasonable to enquire whether there is any visible proof of the operation of the C–N–O cycle. As noted in Section 1.1, there are stars with $^{12}C/^{13}C$ ratios near 5 as predicted for the C–N–O cycle. Certain very hot subdwarf stars display C/N and O/N ratios about two orders of magnitude smaller than solar system values (Pet 70), ratios, in fact, which are characteristic of C–N–O cycle equilibrium at $18 \times 10^6$ °K. These unusual stars have presumably started life with masses of the order of 3.5 solar masses, and somehow lost about 80% of their mass, perhaps as a planetary nebula. The remnant star, with C–N–O cycle products in its surface layer, is not massive enough for further nucleosynthesis and seems doomed to cool slowly into oblivion.

In the model we have adopted for stellar evolution (Section 5.3), the C, N, and O now visible in stellar atmospheres comes partly from the cores of stars in which further nucleosynthesis has taken place, and partly from the outer regions of stars in which the C–N–O cycle has operated on a *rapid* time scale during the final explosion of the star. It is readily apparent that a fast cycle would produce much less $^{14}N$ than the slow equilibrium process, because the $^{13}N$ produced would not have time to $\beta$-decay to $^{13}C$. The rapid C–N–O cycle is being studied currently (Cau 70); it is clear that some truly novel experiments will be required to determine the necessary reaction rates, since reactions like $^{13}N + {}^{1}H \rightarrow {}^{14}O + \gamma$ will be involved.

At higher temperatures than those for quiescent C–N–O hydrogen-burning, other similar cycles for fusing hydrogen into helium are possible, such as the Ne–Na–Mg cycle discussed by Marion and Fowler (MF 57) These will not be discussed here, since their relevance is unclear.

## 3.3. Solar Neutrinos

If the luminosity of the sun is due to the fusion of hydrogen into helium in the central region of the sun, the only "direct" way to verify this is to detect the neutrinos which must accompany the positron decays of the radioactive elements formed. Indeed, an ingenious attempt to accomplish this has been under way for several years by R. Davis and colleagues (DHH 68). In this experiment, the detector is a tank containing $10^5$ gallons of perchlorethylene cleaning fluid, and the relevant detection mechanism is the production of $^{37}Ar$ in the reaction

$$^{37}Cl + \nu \rightarrow {}^{37}Ar + e^- \tag{47}$$

The radioactive $^{37}Ar$ produced is swept out of the tank at intervals of a few months, and the Auger electrons resulting from the electron-capture decay of the $^{37}Ar$ are observed in a heavily shielded, small-volume proportional counter. Because of the $E^2$ dependence of the $\nu$ capture, and the dominance of the matrix element to the 5.1-MeV analogue state in $^{37}Ar$, the experiment is most sensitive to high-energy neutrinos, such as those from the $^8B$ decay (Section 3.1). The $^8B$-producing branch of the $p$–$p$ chain is calculated to account for only about 0.1% of the completions of the chain in the sun; the branching ratio between different modes of completing the $p$–$p$ chain is critically dependent on the temperature of the region producing the solar energy output. Most of the recent theoretical predictions of the solar neutrino flux have been made by J. N. Bahcall and colleagues [see, for example, (BBU 69)]. These predictions involve the temperature-dependent reaction rates of all of the nuclear reactions, as well as a study of the structure and evolution of a solar model.

In the letter cited above (DHH 68), an upper limit of $3 \times 10^{-36}$ neutrino captures per second per $^{37}Cl$ atom was reported, approximately one-half of the rate predicted theoretically, using the best nuclear data and the presently favored solar model. A more recent provisional experimental result, $(2.5 \pm 1.4) \times 10^{-36}$ captures per second per $^{37}Cl$, was presented at the Budapest Cosmic Ray Conference in 1969, which is consistent with the upper limit given earlier. Davis is currently improving some technical features of his experiment and it is expected that the background in the experiment can be improved by a considerable factor.

If the provisional experimental result is confirmed by further work, and especially if a still lower upper limit should be established, it will be necessary to find an explanation for the discrepancy between theory and experiment. From the point of view of the nuclear experimentalist, it appears that the relevant nuclear reaction cross sections have all been checked and rechecked with sufficient precision, and that the source of the discrepancy must lie in the parameters of the solar model. A number of more speculative explanations have been proposed, such as short period oscillations in the central temperature of the sun (which would not show up in the sun's luminosity because of the long diffusion time for energy to reach the sun's surface), and electron-neutrino–muon-neutrino mixing (She 69, Pon 67).

Meanwhile, it is interesting to note that the results already obtained by Davis and his collaborators provide a great deal of useful information. For example, less than 9% of the sun's energy can arise from the C–N–O cycle, without violating the upper limit on the $\nu$-capture rate (DHH 68, BBS 68). If it were not already excluded otherwise, the present results also rule out the existence of a nucleon-stable $^4$Li, since the formation of this high-energy positron emitter would surely be the dominant mode of completing the $p$–$p$ chain, as noted previously.

## 4. HELIUM BURNING

Following the exhaustion of the hydrogen fuel in the core of a star, the central region of the star contracts gradually, increasing the temperature sufficiently for hydrogen burning to continue in a concentric shell around the predominantly helium core. At the same time, the outer regions of the star expand to accommodate the increasing outward heat flux. Stellar model calculations which include the marked nonhomogeneity of the star caused by the helium core, show that the outer regions of the star may expand as much as a factor of fifty or more in radius. Thus, in spite of the increased luminosity (energy output) of the star, the surface becomes cooler. In this phase, the star is called a red giant [see Fig. 2(b)]. The helium core of the star is supported mainly by the pressure contributed by highly degenerate electrons.

The star evolves comparatively rapidly (a few hundred million years) along the red giant branch in the H–R diagram, until the core becomes hot enough to ignite helium burning. This region of the H–R diagram shown in Fig. 2(b) is labeled the helium flash because of the sudden onset of energy production from helium burning. When the thermal pressure in the core exceeds the electron-degeneracy pressure, the core expands, the envelope of the star contracts due to decreasing energy output by the hydrogen-burning shell, and the star enters a period of quiescent helium burning in the core, with hydrogen burning continuing in a shell around the core. This period of the star's evolution probably corresponds to the horizontal branch on the H–R diagram. The core density is typically a few times $10^4$ g cm$^{-3}$ and the temperature is 1 to $2 \times 10^8$ °K.

The reactions generally assigned to this stage of development are

$$3 \, ^4\text{He} \rightarrow \, ^{12}\text{C} + \gamma + \gamma \text{ or } ^{12}\text{C} + e^+ + e^- \tag{48}$$

$$^4\text{He} + \, ^{12}\text{C} \rightarrow \, ^{16}\text{O} + \gamma \tag{49}$$

$$^4\text{He} + \, ^{16}\text{O} \rightarrow \, ^{20}\text{Ne} + \gamma \tag{50}$$

$$^4\text{He} + \, ^{20}\text{Ne} \rightarrow \, ^{24}\text{Mg} + \gamma \tag{51}$$

We will include here also the reactions on any nitrogen which might be present. These reactions will also be discussed in Section 6.2 since they may lead to neutron production.

$$^4\text{He} + \, ^{14}\text{N} \rightarrow \, ^{18}\text{F} + \gamma \tag{52}$$

$$^{18}\text{F} \rightarrow \, ^{18}\text{O} + e^+ + \nu \tag{53}$$

$$^4\text{He} + \, ^{18}\text{O} \rightarrow \, ^{22}\text{Ne} + \gamma \tag{54}$$

The reaction $^4\text{He} + \, ^{13}\text{C} \rightarrow \, ^{16}\text{O} + n$ should also be included with the helium-burning reactions; its rate is included in Fig. 14 (see also the discussion in Section 6.2).

## 4.1. The Triple-α Process

Reaction (48) was originally suggested by Öpik (Opi 51) and Salpeter (Sal 52) as a method of bypassing mass numbers five and eight, for which no stable nuclei exist, in the synthesis of heavier elements from nuclei with $A \leq 4$. It is instructive to consider the experimental data needed for a determination of the rate of the triple-α process in some detail, because no other case shows more clearly how great a variety of nuclear techniques may be involved in pinning down just one astrophysical reaction rate.

The fusion of three alpha particles to form $^{12}\text{C}$ proceeds in two steps —the formation of a small concentration of $^8\text{Be}$ in thermal equilibrium with $^4\text{He}$, and the radiative capture of an additional alpha particle to form $^{12}\text{C}$. The $^8\text{Be}$ concentration depends critically on the difference in masses of the nucleus $^8\text{Be}$ and two alpha particles; the most accurate value for this mass difference, $92.1 \pm 0.05$ keV, comes from the resonant scattering of alpha particles by helium (Ben+ 66). To explain the abundance of $^{12}\text{C}$ in nature, it was postulated by Hoyle that the radiative capture of an alpha particle by $^8\text{Be}$ must also be a resonant process, i.e., that there must be a state of suitable spin and parity in $^{12}$C near the threshold for $^8$Be + $^4$He. Such a state was indeed found at 7.66 MeV excitation energy (Hoy+ 53), and its spin and parity were established as 0$^+$ by Cook *et al.* (Coo+ 57) in a study of the delayed alpha particles following the $\beta$-decay of $^{12}$B. The ratio $(\Gamma_{\text{pair}} + \Gamma_\gamma)/\Gamma$ for the 7.66-MeV state was determined by detecting the formation of $^{12}$C nuclei of the appropriate energy in the reactions, $^{14}$N + $^2$H → $^4$He + $^{12}$C*, $^{12}$C* → $^{12}$C + radiation (SK 63). The ratio $\Gamma_{\text{pair}}/\Gamma_\gamma$ was determined with a pair-spectrometer (Alb 61). Finally, to normalize the partial widths and the total width $\Gamma$, the monopole matrix element between the ground state and the 7.66-MeV state of $^{12}$C was determined by the inelastic scattering of high energy electrons (CG 64, GS 65). The relevant energy levels in $^{12}$C are shown in Fig. 11. It is instructive to speculate how our universe might have looked if the mass of $^8$Be had not been close to the mass of two alpha particles, and if there were no state in $^{12}$C to act as a thermal resonance for the radiative capture of alpha particles by $^8$Be.

Since the rate of capture of alpha particles by $^8$Be is small compared

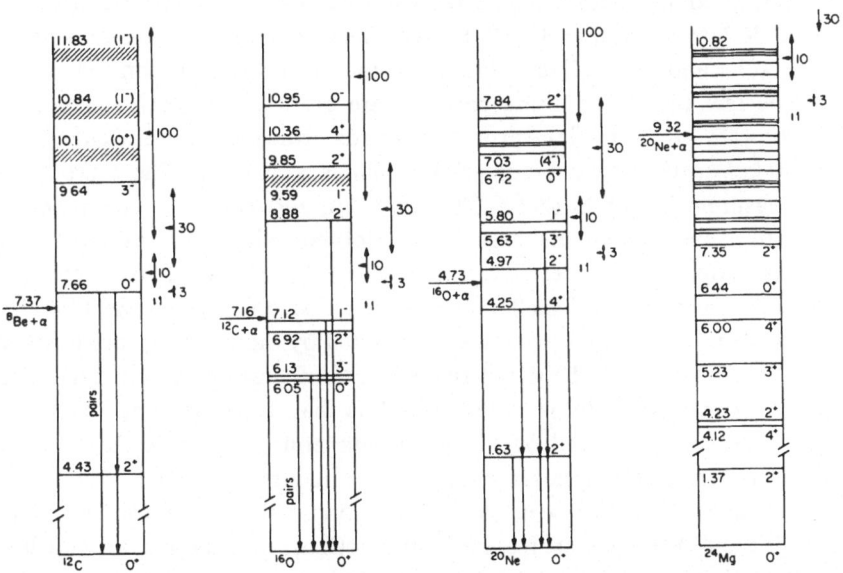

**Fig. 11. Energy-level diagrams for the helium-burning nuclei.** These diagrams have been plotted from data compiled by Lauritsen and Ajzenberg-Selove (La 62), and Endt and Van der Leun (EV 67). The alpha-particle thresholds are shown on the left of each diagram. The most effective energy $E_0$, and the effective range of energies $\Delta E_0$, are shown on the right of each diagram for temperatures of 1, 3, 10, 30, and 100, in units of $10^8$ °K.

with the decay rate of $^8$Be into two alpha particles, and since $\Gamma_{rad}/\Gamma$ is small for the 7.66-MeV state of $^{12}$C, there is very little departure from equilibrium in the reactions $2\ ^4\text{He} \leftrightarrow\ ^8\text{Be}$ and $^8\text{Be} +\ ^4\text{He} \leftrightarrow\ ^{12}\text{C}^*$. Thus the abundance of $^{12}$C* can be calculated from the law of mass action [see, for example, (Sal 57)] to be

$$n_{12_{\text{C}^*}} = n_\alpha^3 \left( \frac{2\pi\hbar^2}{m_\alpha kT} \right)^3 3^{3/2}\, e^{-E_t/kT} \tag{55}$$

where $E_t$ is the energy difference between the excited state of $^{12}$C and three alpha particles. The rate of the $3\alpha \rightarrow\ ^{12}$C reaction, $P_{3\alpha}$ is then just $\Gamma_{rad}/\hbar$ times $n_{12_{\text{C}^*}}$. The mean lifetime for the destruction of $^4$He by the triple-$\alpha$ process is given by

$$\lambda_{3\alpha} = \frac{1}{\tau_{3\alpha}} = 3P_{3\alpha}/n_\alpha = 6.65 \times 10^{-10}(\varrho X_\alpha)^2 T_9^{-3} \exp(-4.433/T_9) \tag{56}$$

This expression depends explicitly only on $\Gamma_{rad}$ and $E_t$ (for which we have used $2.8 \times 10^{-9}$ MeV and 0.382 MeV respectively), and not on the excitation energy or lifetime of the ground state of $^8$Be; these are implicitly involved, of course, because the establishment of an equilibrium concentration of $^{12}$C* requires that $\Gamma_{rad}/\hbar$ be very small compared with the rate of $2\ ^4\text{He} \rightarrow\ ^8\text{Be}$ and $^8\text{Be} +\ ^4\text{He} \rightarrow\ ^{12}\text{C}^*$. The factor of 3 in equation (56) arises from the fact that three alpha particles are destroyed in each reaction.

The rate of the triple-$\alpha$ process depends critically on $E_t$; a change of only 10 keV produces a factor of 3.2 change at $10^8$ °K. There are three new measurements of $E_t$ (ATK 70, MWK 70, and NSB 70) which yield a value 10–12 keV above the previously adopted value, 370 keV (FCZ 67).

For temperatures in excess of $\sim 6 \times 10^9$ °K, which might be relevant during the late stages of stellar evolution to be discussed in Section 5.3, the triple-$\alpha$ process rate is probably dominated by radiative capture through the $3^-$ 9.64-MeV state of $^{12}$C. At these high temperatures and at high densities, the decay rate of this state, as well as that of the state at 7.66 MeV, could be greatly enhanced over their spontaneous decay rates by inelastic collisions with electrons, neutrons, protons, and alpha particles. The importance of these effects has been estimated theoretically by Clayton and Shaw (CS 67, SC 67), for electromagnetic interactions, and by Truran and Koslovsky (TK 69) for nuclear interactions. Morgan and Weisser have measured the inelastic scattering of alpha particles by $^{12}$C to the 7.6-MeV and 9.6-MeV states, extending the measurements as close to threshold as possible to determine the temperature and densities under which this kind of enhancement might be important (MW 70). Davids and Bonner (DB 70) have reported a new energy level at 10.32 MeV excitation energy in $^{13}$N, which is an $s$-wave resonance for $^{12}C(p, p'')$. Thus deexcitation of the 7.66-MeV state of $^{12}$C could also be enhanced by inelastic proton scattering if enough protons escape capture by other nuclei present. At the very high temperatures and densities where such collisional deexcitation processes are relevant, there are many other kinds of nuclear reaction going on, including very rapid photodisintegrations; as our understanding of the nuclear and structural problems improves, enhancement processes like that just discussed may assume an important role.

## 4.2. The $^{12}C(\alpha, \gamma)^{16}O$ Reaction

For temperatures $\lesssim 5 \times 10^8$ °K, the dominant contribution to the $^{12}C(\alpha, \gamma)^{16}O$ cross section is expected to be made by the high-energy tail of the $1^-$ 7.115-MeV state of $^{16}O$, which lies just 46 keV below the $^{12}C + {}^4$He threshold. $\Gamma_\gamma$ for this state is $0.057 \pm 0.005$ eV, a weighted average of the resonant gamma-ray scattering measurements of (Eve+ 68) and (Swa 70). The dimensionless reduced alpha-particle width of this state, $\theta_\alpha{}^2$, cannot, of course, be measured directly in an elastic scattering experiment because the state lies below threshold for $^4$He + $^{12}$C.

An attempt has been made by Loebenstein *et al.* to measure the reduced alpha width indirectly in the reaction (Loe+ 67),

$$^{12}C + {}^6Li \rightarrow {}^{16}O^* + {}^2H \tag{57}$$

In this work, it was assumed that the compound-nucleus contribution to the cross sections for the reactions leading to the 7.11-MeV and 9.58-MeV $1^-$ states could be calculated from the measured cross section for producing the $2^-$ 8.8-MeV $^{16}O$ state, which is assumed to be formed only by a compound-nucleus process. [This assumes that a two-step direct reaction involving an inelastic scattering as one step, for example, has a negligible cross section—an assumption for which there is little experimental evidence either way.] After subtracting the compound-nucleus contributions, to get the direct-reaction cross sections for the two $1^-$ states, the dependence of the direct $\alpha$-transfer cross section on $\theta_\alpha{}^2$ was then normalized to the known value for the 9.58-MeV state. In this way $\theta_\alpha{}^2$ was determined to be in the range from 0.06 to 0.14 for the 7.11-MeV state.

A theoretical calculation by Stephenson (Ste 66), based on a model in which the odd-parity states of $^{16}O$ are members of two intermixed odd-parity rotational bands, yields $\theta_\alpha{}^2(7.11 \text{ MeV}) = (0.085 \pm 0.040)$. Although the agreement between theory and experiment is impressive, it should be kept in mind, on the theoretical side, that the prediction of small particle widths from nuclear models is precarious, and on the experimental side, that the subtracted compound-nucleus cross section was approximately three times as large as the residual direct-reaction cross section for the 7.11-MeV state. A recently published $R$-matrix fit to the $p$-wave phase shifts for elastic alpha-scattering from $^{12}C$ finds $\theta_\alpha^2(7.11 \text{ MeV}) = 0.71^{+0.37}_{-0.17}$, in strong disagreement with the value, quoted above (Cla 69). A recent direct reaction study of $^{12}C(^7Li, ^3H)^{16}O$ yields a much smaller value, $\theta_\alpha^2 = 0.025$, with an unknown normalizing factor (Puh+ 70).

A more direct approach to the determination of $\theta_\alpha^2$ is to measure the $^{12}C(\alpha, \gamma)^{16}O$ cross section to a sufficiently low energy that one can determine the magnitude and phase (constructive or destructive) of the interference between the broad 9.58-MeV state and the 7.11-MeV state. From the magnitude of the interference, one can then determine $\theta_\alpha^2$ for the 7.11-MeV state and, perhaps more relevantly, one can predict the $^{12}C(\alpha, \gamma)^{16}O$ cross section in the vicinity of a few hundred keV $\alpha$-energy. A comprehensive theoretical and computational $R$-matrix analysis, using the available $(\alpha, \alpha)$ and $(\alpha, \gamma)$ data, has been made by Weisser, Morgan, and Thompson (WMT 70). This study indicates clearly the difficulties with this approach. The predicted $(\alpha, \gamma)$ cross sections are very small at energies low enough to be useful—of the order of 0.1 nanobarn at $E_\alpha \sim 1.5$ MeV. In addition to the interference between the 9.58 and 7.11 MeV states, there can be interference with the tails of higher $1^-$ states, the effect of which was simulated in the study by including one higher state whose position and partial widths were taken as free parameters, to be determined by the elastic scattering and $(\alpha, \gamma)$ cross-section data. The possibility that $^{16}O$ states with spins and parities $0^+$ or $2^+$ might contribute significantly requires further examination.

There are a few early $(\alpha, \gamma)$ cross-section measurements near the 9.58-MeV state ($E_\alpha \sim 3.2$ MeV) [see, for example, (BTW 57, LS 64)], but the neutrons produced from the reaction $^{13}C(\alpha, n)^{16}O$, even in the best available $^{13}C$-depleted targets, produced such a serious background in the $\gamma$-ray detectors that these workers were prevented from extending their measurements to lower $\alpha$-particle energies. At some loss in average beam current and solid angle, Adams *et al.* introduced time-of-flight separation of neutrons and gamma rays (Ada+ 68), by using a pulsed beam, a 10-cm flight path, and fast-timing techniques, which at the same time reduces room and cosmic-ray background. This technique has been employed by Jaszczyak, Gibbons, and Macklin (JGM 70), along with pulse-rise-time discrimination, to obtain cross sections down to $E_\alpha = 1.86$ MeV; similar measurements are currently underway in our laboratory (Wei + 70).

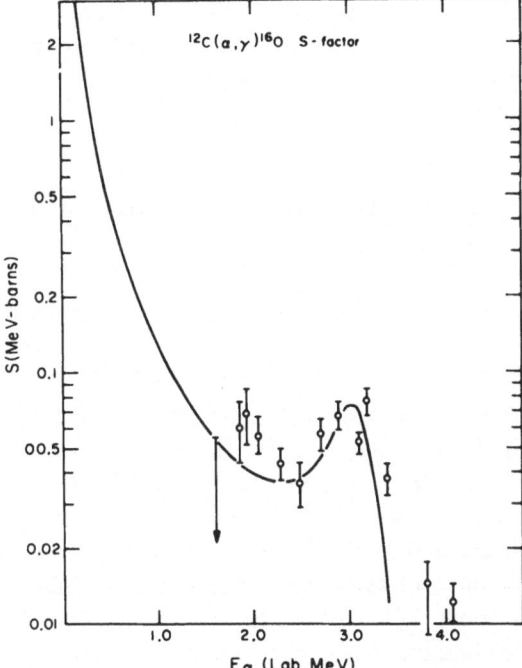

**Fig. 12. The $^{12}C(\alpha, \gamma)^{16}O$ S-factor as a function of alpha-particle energy.** This plot is from a three-level R-matrix analysis (WMT 70) of published data for the elastic scattering of alpha particles by $^{12}C$, and the $^{12}C(\alpha, \gamma)^{16}O$ data of Jaszczak *et al.*, shown here as open circles (JGM 70). This analysis yields an astrophysical S-factor with an uncertainty of about 200%. The large uncertainty is the extrapolated value underlines the importance of data of higher precision.

Other suggestions to minimize background troubles have been proposed, such as the inversion of the reaction, i.e., bombarding helium gas with $^{12}C$ ions, and attempts to measure the cross section by this technique are also being made in several laboratories. At present, the available data fall far short of what is required to make an unambiguous prediction of the astrophysical $^{12}C(\alpha, \gamma)^{16}O$ cross section.

Figure 12 shows a fit to the data of (JGM 70) with the three-level R-matrix formulation of (WMT 70). Because of the large extrapolation and the number of parameters to be determined from the data, data of very high precision are needed. Achieving the necessary precision constitutes a major challenge to the experimental physicist; however, the relative abundances of carbon and oxygen are almost completely determined by the rates of the triple-$\alpha$ process and the $^{12}C(\alpha, \gamma)$ reaction. A major experimental effort is clearly justified.

## 4.3. Further Helium-Burning Reactions and Comparison with Abundances

The cross section for the reaction $^{16}O(\alpha, \gamma)^{20}Ne$ should be dominated by the resonances at $E_\alpha(\text{lab}) = 1.12, 1.32,$ and $2.49$ MeV corresponding to excited states of $^{20}Ne$ at 5.62 (3$^-$), 5.79 (1$^-$), and 6.72 (0$^+$) MeV (VSS 65). The possibility that the more favorably located state at an excitation energy of 4.97 MeV might constitute a stellar thermal resonance was eliminated when the spin and parity were shown to be 2$^-$ (CGL 61). Recent measurements by Toevs have confirmed the previously measured yields from the three resonances, and have established stringent upper limits on the thick-target yield below the 1.12-MeV resonance (Toe 70). Since the listed resonances are so far above $E_0$ for the temperature region expected to be relevant during helium burning, the $^{16}O(\alpha, \gamma)^{20}Ne$ reaction is expected to be relatively slow, and indeed even slower than the succeeding reaction $^{20}Ne + \alpha \rightarrow {}^{24}Mg + \gamma$, due to the presence of several favorably located resonances in the latter reaction (Smu 65). Thus, at the temperatures required to yield an appreciable rate for the reaction $^{16}O(\alpha, \gamma)^{20}Ne$, helium burning would yield much larger amounts of $^{24}Mg$ than $^{20}Ne$, while the observed abundances favor neon over magnesium by a factor estimated by different authors to be between two and ten. In any case, at the temperature required for appreciable conversion of oxygen to neon by helium burning, it is probable that other nuclear processes to be described later are already operative.

It is perhaps more interesting to consider whether helium burning can account for the observed abundance ratio of oxygen to carbon which is near 2:1. This question has been examined by Cameron (Cam 57), by Deinzer and Salpeter (DS 64), and others. Deinzer and Salpeter concluded that $^{12}C$ and $^{16}O$ would indeed be produced in comparable amounts in the great majority of helium-burning stars, namely those with helium-core masses less than about 10 solar masses; more massive helium cores would result mainly in $^{16}O$ production, or magnesium production for the most massive objects with $M \gtrsim 1000$ $M_\odot$. Their estimates were based on a $[^{12}C(7.6 \text{ MeV}) - 3\alpha]$ mass difference of 375 keV, and $\theta_\alpha{}^2 \sim 0.1$ for the 7.1-MeV state of $^{16}O$. The value of $\theta_\alpha{}^2$, to which the $^{12}C(\alpha, \gamma)^{16}O$ rate is directly proportional, should really be considered as uncertain at present by at least an order of magnitude either way. There is thus at present no incompatibility between a helium-burning origin for $^{12}$C and $^{16}$O, and the observed abundances; however, the abundances resulting from helium burning are related simply to the relative rates of the two reactions, and better nuclear data will lead to confirmation or conflict.

## 4.4. $^{14}$N$(\alpha, \gamma)^{18}$F and $^{18}$O$(\alpha, \gamma)^{22}$Ne

When hydrogen is entirely converted to helium by the C–N–O cycle of reactions in the core of a star, the next most abundant nuclide, after $^{4}$He, will be $^{14}$N, as discussed in Section 3.2. There is a possibility that, when the temperature rises subsequent to hydrogen burning, the reaction $^{14}$N$(\alpha, \gamma)^{18}$F may occur before the triple-$\alpha$ process. Indeed with the $S$-value listed for this reaction by (FCZ 67), $8.73 \times 10^6$ MeV-barns, the $^{14}$N$(\alpha, \gamma)^{18}$F reaction would be considerably faster than the triple-$\alpha$ process at the temperatures and densities given by most models for the onset of helium burning. The careful study of four resonances for the $^{14}$N$(\alpha, \gamma)$ reaction at $E_\alpha$(lab) = 1.14, 1.40, 1.53, and 1.62 MeV by Parker (Par 68b) has shown that the $S$-value given above may be a gross overestimate. In order to estimate the effect of the low-energy tails of these resonances in the stellar thermal region, it is necessary to know the cross section and width for each of the resonances, and also their relative phases should any two of them have the same spin and parity. The result of the available experimental measurements, however, is the integrated yield from each resonance

$$Y = \int_R (\sigma/\varepsilon)\, dE = 2\pi^2 \lambda^2 (\omega\gamma)/\varepsilon \tag{58}$$

where $\lambda$, $\omega$, $\gamma$, and $\varepsilon$ are the reduced de Broglie wavelength, the statistical weight $(2J + 1)/[(2S + 1)(2I + 1)]$, the ratio $\Gamma_{12}\Gamma_{34}/\Gamma$, and the stopping cross section per $^{14}$N atom. In the experiment cited, the resonances were too narrow for the cross section to be measured as a function of energy over the resonance.

If we assume that $\Gamma_\alpha \ll \Gamma$ for the four resonances, a crude approximation for the energy-averaged $S$-factor can be made as follows (Par 68c):

$$\bar{S} = \int S(E)\, dE/\Delta E = \frac{1}{\Delta E} \int \sigma(E) E \exp[(E_G/E)^{1/2}]\, dE$$

$$\approx \frac{1}{\Delta E} \sum_R \left( E_R \exp[(E_G/E_R)^{1/2}] \int_R \sigma dE \right) \tag{59}$$

The values of $(\omega\gamma)$ obtained experimentally are $0.028 \pm 0.001$, $0.009 \pm 0.001$, $1.60 \pm 0.13$, and $0.43 \pm 0.05$ eV, for the resonances at $E_\alpha(\text{lab}) = 1.14$, $1.40$, $1.53$, and $1.62$ MeV, respectively. Taking $\Delta E$ very crudely as 1 MeV, one finds $\bar{S} \sim 2 \times 10^4$ MeV-barns. If we now assume that $\bar{S}$ is independent of energy, we can calculate the lifetime for the destruction of $^{14}$N by the $(\alpha, \gamma)$ reaction, from equations (6) and (13). We find

$$\tau_\alpha(^{14}\text{N}) = \frac{4}{\varrho X_\alpha N_A} \left(\frac{M}{2}\right)^{1/2} \frac{(kT)^{3/2}}{\Delta E_0} \frac{\exp(\tau)}{S(0)}$$

$$= \frac{5.41 \times 10^{-14}}{\varrho X_\alpha} T_9^{2/3} \exp(36.03/T_9^{1/3}) \tag{60}$$

where $\tau$ and $\Delta E_0$ have been defined in Section 2.2. If we use $\varrho X_\alpha = 2 \times 10^4$ g cm$^{-3}$ and $T = 2 \times 10^8$ °K, for purposes of illustration, we obtain $\tau_\alpha(^{14}\text{N}) \approx 15$ years.

In order to compare this value with the lifetime of the triple-$\alpha$ process, some modifications of the relations given in Section 2.1 are necessary because three alpha particles are required in the incident channel. The appropriate expression is just the reciprocal of equation (56)

$$\tau_{3\alpha}(^4\text{He}) = 1.51 \times 10^9 T_9^3 \exp(4.433/T_9)/(\varrho X_\alpha)^2 \tag{61}$$

Numerical evaluation for $\varrho X_\alpha = 2 \times 10^4$ g cm$^{-3}$ and $T = 2 \times 10^8$ °K leads to $\tau_{3\alpha}(^4\text{He}) \approx$ four years, which means that the triple-$\alpha$ process occurs more rapidly than the $^{14}$N$(\alpha, \gamma)$ reaction. Of course, the ratio of the two rates depends on the temperature chosen and also on $(\varrho X_\alpha)$, since there is an extra power of $(\varrho X_\alpha)$ in the denominator of (61).

The important point, however, is that the available experimental data, and the wide variation of $(\varrho X_\alpha)$ in various stellar models, still leave open the question of which reaction occurs most rapidly during the helium flash at the onset of helium burning. The lower states in the compound nucleus $^{18}$F could greatly augment the rate of the $^{14}$N$(\alpha, \gamma)$ process. There are four of these, at excitation energies of 4.96, 4.84, 4.74, and 4.65 MeV, corresponding to $E_\alpha(\text{c.m.}) = 0.56$, 0.44, 0.34, and 0.25 MeV, respectively, as shown in Fig. 13. There are also states at 4.40 and 4.36 MeV, close to the threshold for $^{14}$N + $^4$He. Of these six states, the most likely one to make a large contribution to the reaction under study is the 4.84-MeV state, which has isospin 0 and spin 1, because it is so close to the effective energy at helium-burning temperatures. The other states are less likely contri-

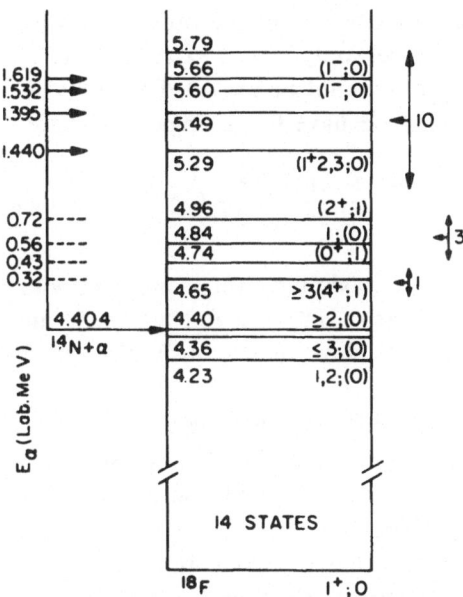

**Fig. 13. Energy-level diagram for $^{18}$F (LA 62).** The alpha-particle threshold is shown on the left, and the effective energy regions for temperatures of 1, 3, and $10 \times 10^8$ °K are shown on the right. The relevance of the $^{18}$F energy level at $E_{exc} = 4.84$ MeV to the prediction of the astrophysical rate of the reaction, $^{14}$N($\alpha,\gamma$)$^{18}$F, is discussed in Section 4.4 of the text.

butors, either because they are farther away from $E_0$, or because of unfavorable isospin or spin values.

The difficulty in being sure that a reliable $S$-value has been obtained from laboratory measurements at energies above those relevant in the stellar situation is brought sharply into focus by considering the 4.84-MeV state of $^{18}$F. Since $\Gamma_\alpha$ is much smaller than $\Gamma_\gamma$ at low $E_\alpha$, the integrated yield from a resonance is proportional to $\Gamma_\alpha$, which quantity is varying rapidly with energy because of the coulomb barrier penetrability. In the $^{14}$N($\alpha, \gamma$) work, an upper limit for the thick-target yield of 0.01 captures per $10^{12}$ incident $\alpha$-particles was established just below the $E_\alpha$(lab) $= 1.14$-MeV resonance. If we attribute this yield to the 4.84-MeV state, we can derive $(\omega\gamma) \lesssim 4 \times 10^{-4}$ eV. Since $\Gamma_\alpha \ll \Gamma$ for the 4.84-MeV state (Gor+ 67), it follows that $\Gamma_\alpha \lesssim 4 \times 10^{-4}$ eV (taking $\omega \sim 1$). From this number we obtain a limit for the reduced width, $\gamma_\alpha^2 \lesssim 0.06$ MeV. Since the Wigner single-particle limit for an $\alpha$-particle is of the order of 1 MeV, a value of $\gamma_\alpha^2$ as large as 0.06 MeV can certainly not be excluded, and thus a value of $(\omega\gamma)$ as large as $4 \times 10^{-4}$ eV cannot be excluded for the 4.84-MeV state. From equations (6) and (19) we have for a single resonance

$$\tau_\alpha(^{14}N) = \frac{4}{\varrho X_\alpha} \frac{A^{3/2} T_9^{3/2}}{1.54 \times 10^{11}} (\omega\gamma)^{-1} \exp(11.605\, E_R/T_9) \qquad (62)$$

With $(\omega\gamma)$ equal to the maximum value consistent with the upper limit of (Par 68b), $4 \times 10^{-10}$ MeV, $\varrho X_\alpha = 2 \times 10^4$ g cm$^{-3}$, and $T = 2 \times 10^8$ °K, we obtain $\tau_\alpha(^{14}N) \approx 1.7$ days, enormously shorter than the other two estimates.

Measurements of the $^{14}N(\alpha, \gamma)^{18}F$ reaction are currently in progress in our laboratory by T. A. Tombrello, H. M. Spinka, and R. G. Couch. Preliminary data suggest that a yield considerably below Parker's upper limit will be measurable.

Even if $(\omega\gamma)$ is very much smaller than the upper limit $4 \times 10^{-4}$ eV discussed above, it is very likely that the 4.84-MeV level will contribute enough to the reaction rate to make the $^{14}N(\alpha, \gamma)$ reaction occur before the triple-$\alpha$ process. The $^{18}F$ produced will then $\beta$-decay to $^{18}O$, and the question immediately arises as to whether $^{18}O(\alpha, \gamma)^{22}Ne$ also proceeds before the triple-$\alpha$ process. The available information on this reaction is even less satisfactory than that for the $^{14}N(\alpha, \gamma)$ reaction. The cross section has been measured for $E_\alpha(\text{lab}) = 2.15$ to 3.70 MeV by using the neutron–$\gamma$ time-of-flight difference to separate the effects of the neutrons and gamma rays in the detector (Ada+ 69). While this energy range is of interest in the explosive burning processes to be described later, $E_0$ for $^{18}O(\alpha, \gamma)^{22}Ne$ is about 0.39 MeV for $T = 2 \times 10^8$ °K, and extrapolation from the measured energy region to the energies of interest during helium burning is likely to be grossly in error. The basic reason for this is that the threshold for the $^{18}O(\alpha, n)$ reaction occurs at 0.7 MeV in the center-of-mass system. Above this energy the resonance widths will be dominated by $\Gamma_n$, while they will dominated by $\Gamma_\gamma$ below this energy. The low-energy rate for $^{18}O(\alpha, \gamma)^{22}Ne$ can thus be expected to exceed that which one would get by extrapolation from energies well above the neutron threshold, by a large factor. Indeed, it appears likely that $^{18}O(\alpha, \gamma)^{22}Ne$ would also proceed before the triple-$\alpha$ process, even though the rate for this process is still uncertain by orders of magnitude. The rates of several of the helium burning reactions are shown in Fig. 14.

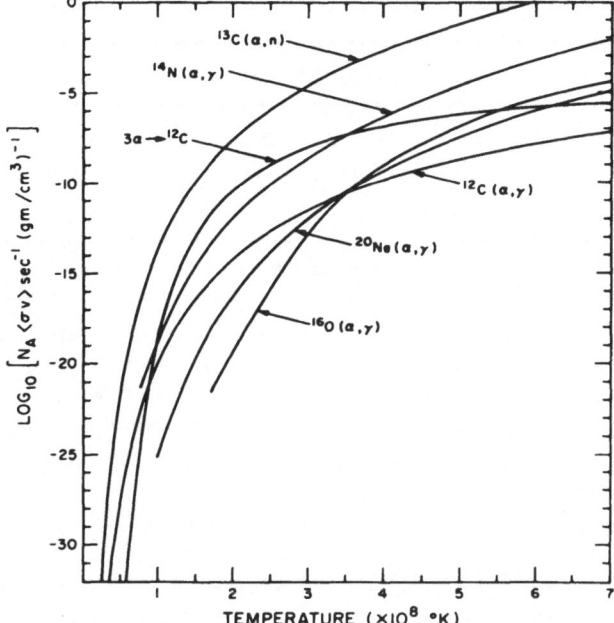

**Fig. 14.** $N_A \langle \sigma v \rangle$ **for several helium-burning reactions as a function of temperature.** The curves have been plotted from data compiled by W. A. Fowler, G. R. Caughlan, and B. A. Zimmerman. See Table III.

# 5. SYNTHESIS OF THE NUCLEI WITH $20 \lesssim A \lesssim 60$

## 5.1. Quiescent Carbon and Oxygen Burning

As the helium-burning phase comes to an end with the exhaustion of the helium supply, the core of the star must contract again while the temperature rises perhaps to 6 or $7 \times 10^8$ °K. At these temperatures $^{12}$C begins to react with $^{12}$C, and the energy output may once again stabilize the star by balancing the gravitational forces. Of course, quiescent carbon burning, the burning of carbon at relatively constant temperature, can only occur if the thermal pressure generated by the energy output is capable of balancing gravity. If the pressure is insensitive to temperature, as would be the case at a density high enough to produce a dominant electron degeneracy pressure, quiescent carbon-burning does not occur, as discussed later.

Because of the higher coulomb barrier, the reaction of $^{16}$O with $^{16}$O is not expected to occur until higher temperatures are reached, $\sim 10^9$ °K,

**TABLE III. Helium-Burning Reaction Data Used in Rate Calculations for Fig. 14[a]**

| Reaction | $S(0)$, MeV-barns | $S'(0)$, barns |
|---|---|---|
| $^{13}C(\alpha, n)^{16}O$ | $5.48 \times 10^5$ | $1.21 \times 10^6$ |
| $^{16}O(\alpha, \gamma)^{20}Ne$ | $1.00 \times 10^{-1}$ | — |
| $^{14}N(\alpha, \gamma)^{18}F^b$ | $1.50 \times 10^3$ | — |

| | Special Cases |
|---|---|
| $3\alpha \rightarrow {}^{12}C$ | The quantity plotted in Fig. 14 is $\lambda_{3\alpha}(\alpha)(\varrho X_4/4)^{-1} = 2.66 \times 10^{-9}$ $(\varrho X_4) \exp(-4.294/T_9)$, with $(\varrho X_4) = 5.6 \times 10^4$ g cm$^{-3}$. The curve should be lowered by a factor of 4.4 at $10^8$ °K, and a factor of 1.2 at $7 \times 10^8$ °K, because of the new value of $E_t$ [see Section 4.1]. |
| $^{12}C(\alpha, \gamma)^{16}O$ | Dominated by a resonance below threshold (FCZ 67). $$S_{\text{eff}} = 3.14 \times 10^6 \frac{\theta_\alpha{}^2\Gamma_\gamma}{T_9{}^{4/3}(1 + 0.050T_9^{-2/3})^2}$$ In Fig. 14, $(\theta_\alpha{}^2\Gamma_\gamma)$ has been taken to be $5.6 \times 10^{-9}$ MeV. |
| $^{20}Ne(\alpha, \gamma)^{24}Mg$ | Extrapolated from $N_A\langle\sigma v\rangle$ curve for the region $1 \leq T_9 \leq 5$. (Insufficient experimental data available at low energies.) |

[a] These data have been taken from (FCZ 67) and a revised compilation by the same authors
[b] See Section 4.4 for a discussion of the rate of this reaction.

after the exhaustion of the $^{12}C$. The reaction of $^{12}C$ with $^{16}O$ is not expected to be an important contributor to energy production or nucleosynthesis for the same reason. The rate of the $^{12}C + {}^{16}O$ reaction is, however, being checked in the laboratory to be sure that it may be neglected (Sym 70).

The $^{12}C + {}^{12}C$ reaction proceeds through many exit channels; alpha particles, protons, neutrons, or gamma rays may be emitted at low $^{12}C$ energies, with the alpha-particle and proton channels by far the most prolific. In addition to earlier experimental and theoretical study of the $^{12}C + {}^{12}C$ reactions over the energy range $5 \leq E_{\text{cm}} < 12.5$ MeV (ABK 60, Vog+ 64), these reactions have been studied recently for $3.23 \leq E_{\text{cm}} \leq 8.75$ by Patterson, Winkler, and Zaidins (PWZ 69). Strong resonance-like variations in the cross section occur down to the lowest energies studied, as shown in Fig. 15. This feature of the reactions poses a serious new problem, i.e., how to extrapolate the cross section to the still lower energies desired for astrophysical calculations. The approach taken by Patterson *et al.* was to average the observed total cross sections over a center-of-mass energy interval of 0.63 MeV, a procedure which finds its justification in the thermal

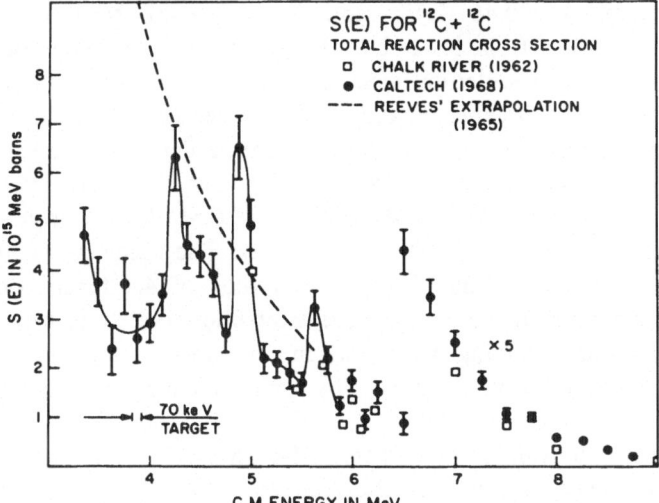

**Fig. 15.** *S*-factor for the $^{12}C + ^{12}C$ reactions as a function of center-of-mass energy. The open squares are data obtained by the Chalk River Laboratories [see, for examples, (Vog+ 64)], and the closed circles are data taken by (PWZ 69) at Caltech. The dashed curve is an optical model extrapolation by (Ree 66a), from higher energy data.

averaging which will take place in any case, in the stellar environment. When this somewhat smoother cross section was fitted to the expression,

$$\bar{\sigma} = \frac{\overline{S(E)}}{E} \exp[-(E_G/E)^{1/2}] \tag{10}$$

it was found that the resulting $\overline{S(E)}$ was still a steep function of energy. The cross section was then fitted to the expression,

$$\bar{\sigma} = \frac{\overline{S(E)}}{E} \exp[-(E_G/E)^{1/2} - gE] \tag{63}$$

where $g$ is expected from Coulomb barrier penetration theory to be given by

$$g = 0.122 \, (AR_f^3/Z_1Z_2)^{1/2} \qquad \text{MeV}^{-1} \tag{64}$$

The quantities $A$, $Z_1$, and $Z_2$ have been defined earlier, and $R_f$ is the interaction radius expressed in fermis. With $g$ and $\bar{S}$ taken as free parameters, but with the constraint that $\bar{S}$ should be independent of energy, Patterson *et al.* obtained $g = 0.46$ MeV$^{-1}$ and $\bar{S} = 2.9 \times 10^{16}$ MeV-barns. These values yield cross sections lower by a factor of 3.5 than those obtained by an optical-model extrapolation from the earlier high-energy data (Ree 66a), even though the newer experimental work includes several significant alpha-particle groups which were not included in the earlier work. Patterson *et al.* have estimated that the quoted values of $g$ and $\bar{S}$ describe the energy-averaged cross section from $2$ MeV $\leq E_{cm} \leq 6$ MeV with an uncertainty of not more than a factor of two either way. The value $g = 0.46$ MeV$^{-1}$ yields the surprisingly small interaction radius of 4.4 fermi; one could readily understand an unusually large interaction radius as being due to nuclear reactions occurring when only the extreme outer edges of the two $^{12}$C nuclei overlap, but it is unclear why such a small radius is found in this work.

At carbon-burning temperatures, the alpha particles, protons, and neutrons will react rapidly with the other products of the reactions, $^{20}$Ne, $^{23}$Na, $^{23}$Mg, and $^{24}$Mg, as well as with unburned carbon and oxygen. In particular the protons will interact most readily with $^{12}$C, producing $^{13}$N, and thus $^{13}$C. The $^{13}$C can then react with liberated alpha particles by the reaction $^{13}$C$(\alpha, n)^{16}$O to produce more oxygen, and additional neutrons. It therefore seems likely that the result of quiescent carbon burning is a spread of nuclides extending up to $^{24}$Mg, with some $^{28}$Si probably being formed also [see, for example, (AT 69) and references listed therein].

Possibly at the temperature required for carbon burning, and certainly at the higher temperatures required for oxygen burning, energy loss in the form of neutrinos will be the dominant mode of energy loss for a star.* While normal energy transport to the surface of a star is by electromagnetic radiation, which has a very short mean path, and to a lesser extent by conduction and convention, neutrinos produced in the core escape unimpeded. The consequences of a large neutrino luminosity are very rapid evolution of the star and a correspondingly small probability of observing stars in this phase of their existence. The significance of neutrinos in this context depends on the existence of $(\nu, e)^{\dagger}(\nu, e)$, the self-interaction term, in the interaction Lagrangian for the weak interaction [see, for example, (FG 58)]. As yet there is no experimental evidence for the existence of this term, but the observation of a weak parity-violating component in nuclear forces confirms the existence of a self-interaction term of the form $(p, n)^{\dagger}(p, n)$ (Hen 69). This does not necessarily prove the existence of the

---

* Neutrinos play a significant role in several other astrophysical situations. [See, for example, the review by Ruderman (Rud 69).]

$(\nu, e)^+(\nu, e)$ term, but the present theoretical consensus favors the existence of the self-terms. Direct evidence for the electron–neutrino term is currently being searched for in experiments to detect electron–neutrino scattering.

The various reactions which follow the interaction of two $^{16}$O nuclei have been studied most recently by Patterson, Winkler, and Spinka for $7.25 \leq E_{\mathrm{cm}} < 12$ MeV (PWS 68). Here the cross sections vary smoothly as a function of energy, in contrast with the $^{12}$C + $^{12}$C reactions. Proton emission is the dominant reaction, but significant contributions are made by alpha-particle emission, neutron emission, or pairs of nucleons. When fitted with the expression given as Equation (63), in the same way as the $^{12}$C + $^{12}$C reaction was fitted, the parameters obtained are $\bar{S} = 6 \times 10^{27}$ MeV-barns and $g = 0.72$ MeV$^{-1}$, with a factor of two uncertainty in $\bar{S}$. From the nucleosynthetic point of view, the higher temperature for oxygen burning ($T \sim 10^9$ °K) produces another problem. In addition to the extensive rearrangement of nuclei by the $^{16}$O + $^{16}$O reactions and their products, the high-energy tail of the thermal radiation spectrum is intense enough to cause the liberation of alpha particles and nucleons which will participate in altering the nuclear distribution. It is, in fact, not clear whether quiescent oxygen burning forms a separate phase of stellar evolution or merges with the following silicon-burning phase. There are probably stellar objects or, at least, shell regions in some stars, where oxygen burning does form the last nucleosynthetic stage before the material is distributed into space and becomes available for the formation of other stars. To calculate the nucleosynthetic results of oxygen burning requires the simultaneous consideration of a large number of nuclear reactions because of the photo-disintegration processes which accompany oxygen burning. Such calculations show that there will be abundant production of nuclides up to $^{28}$Si, with small quantities of nuclides extending up to $A \sim 40$. Because of the high temperatures involved, and the high neutrino luminosity, the oxygen-burning phase (if it exists at all) will be very short-lived.

## 5.2. Quiescent Silicon Burning

The principal result of quiescent carbon and oxygen burning is a spread of nuclei from oxygen to silicon, with $^{24}$Mg and $^{28}$Si the most abundant nuclides. With the exhaustion of yet another nuclear fuel, or perhaps merging with the later stages of oxygen burning, the star will contract until the core temperature is high enough to initiate the so-called silicon-burning stage. As presently envisaged, this very short-lived stage of stellar activity builds the nuclides up to the abundance peak near iron, by successive captures of alpha particles, protons, and neutrons. The alpha particles, protons, and neutrons are in turn produced by photodisintegration of the nuclides, principally silicon and magnesium, resulting from earlier nucleosynthesis. Although this stage of nucleosynthesis is not understood in all details at present, the general course of nucleosynthesis seems to be fairly clear. Truran, Cameron, and Gilbert (TCG 66) have examined the problem using experimental cross sections where available, and supplementing these with cross sections based on theoretically estimated level densities. In most cases the reaction rates are high enough that a quasi-equilibrium exists among alpha particles, protons, neutrons, and the nuclides for $A > 28$. This fact has been exploited in a calculation by Bodansky, Clayton, and Fowler (BCF 68), in which the $^{24}$Mg and $^{28}$Si come into a slowly changing equilibrium, and the rate at which alpha particles, neutrons, and protons are made available for building heavier nuclei is controlled by the net photodisintegration rate of $^{24}$Mg. This quasi-equilibrium is made possible by the less refractory nature of the nuclides below $A = 24$ as far as photodisintegration is concerned. As the buildup of the nuclides continues, the quasi-equilibrium approaches true statistical equilibrium, and an abundance peak develops in the region of iron, where the binding energy per nucleon reaches a maximum, reminiscent of the $e$-process of (Bur+ 57).

To the extent that true equilibrium exists, a detailed knowledge of nuclear reaction rates is unnecessary; all that is required are nuclear masses and statistical weights, including those for low-lying excited states. Rather impressive agreement can be obtained between the observed elemental abundances for $A < 60$, and those predicted from the silicon-burning process, especially if the radial zones of the star at increasing distance from the center process decreasing fractions of the $^{28}$Si initially present, as has been demonstrated by Michaud (Mic 69). Figure 16 from Michaud's work compares the calculated abundances for $28 \leq A \leq 60$ with the solar system abundances of Cameron (Cam 68).

Although there seems little doubt that the theorists will eventually ask for the rates of all nuclear reactions leading up to $A \sim 60$ at least, if past history is any guide, it appears at present that the most important, rate-determining, nuclear reactions in silicon burning are those in the vicinity of $^{20}$Ne, $^{24}$Mg, and $^{28}$Si. The photodisintegration rates of $^{28}$Si have been reexamined recently by Lyons et al. (LTS 69, Lyo 69, Lyo+ 70). These rates have been calculated from the inverse reaction rates, i.e., the rates of

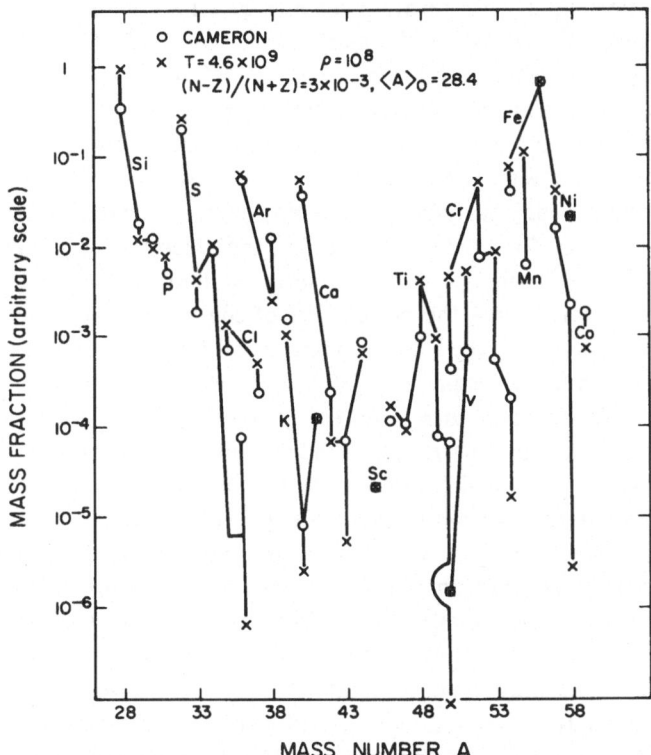

**Fig. 16. Nuclear abundances produced by silicon burning (Mic 69).** The open circles are the solar system abundances as tabulated by Cameron (Cam 68), and the crosses are the result of a calculation in which various concentric zones of the star burned different fraction of the $^{28}$Si present before silicon burning. In this particular example, the maximum temperature reached at the center was $4.6 \times 10^9$ °K and the central density was $10^8$ g cm$^{-3}$. The initial neutron excess was $3 \times 10^{-3}$, where $N$ and $Z$ include both bound and free neutrons and protons, and the average value of $A$ was 28.4 at the beginning of silicon burning. The solid lines in the figure connect the isotopes of each element, for identification purposes.

$^{27}$Al$(p, \gamma)^{28}$Si, and $^{24}$Mg$(\alpha, \gamma)^{28}$Si. Even for these well-studied reactions large discrepancies have long existed in the published literature for the strengths, $\omega\Gamma_{12}\Gamma_{\gamma}/\Gamma$, of the various resonances. The desiderata for theoretical use are the resonance strengths to an absolute precision of perhaps 20%, a measurement of the nonresonant yield, if significant, and a search for new

resonances, particularly those at lower energy where experimental measurement is difficult because of the small values of $\Gamma_{12}/\Gamma$. Such resonances
are the crucial ones at temperatures near $10^9$ °K. As part of a continuing
program, we have also reexamined the reaction $^{28}\text{Si}(\alpha, \gamma)^{32}\text{S}$ recently (Toe
70), and measurements of the reaction rates of $^{20}\text{Ne}(\alpha, \gamma)^{24}\text{Mg}$ and $^{32}\text{S}(\alpha, \gamma)$
are currently under way (CS 70).

For use in astrophysical photodisintegration-rate calculations, the quan-
tity that is needed, for each resonance, is the strength $\omega\Gamma_{12}\Gamma_\gamma/\Gamma$, where
$\Gamma_\gamma = \Gamma_{\gamma_0} + \Gamma_{\gamma_1} + \Gamma_{\gamma_2} + \cdots$. It might seem, at first glance, that only $\Gamma_{\gamma_0}$
would be important, since the population of the ground state of the nu-
cleus will normally far exceed the population of the excited states because
of the Boltzmann factor, $\exp(-E_{\text{exc}}/kT)$. However, for photodisintegration
through a particular resonance level, the probability is proportional to the
product, (initial state population) $\times$ ($\Gamma_\gamma$) $\times$ (probability of finding a photon
of the requisite energy in the high-energy tail of the blackbody spectrum).
The latter probability is proportional to $\exp[-(E_R - E_{\text{exc}})/kT]$, where $E_R$
is the photon energy required to reach the resonance from the ground state
of the nucleus. The excitation energy does not appear in the product of the
two exponentials, which makes the excited states of the nucleus equal in
importance to the ground state, as far as photodisintegration is concerned.
(The excited states may even be somewhat more important because they
will frequently have higher $J$-values, and hence larger statistical weights
than the ground state.) It can thus be seen that detailed information on the
various ways gamma-rays de-excite a particular resonance is not really
necessary; all that is needed is the total gamma-ray width, $\Gamma_\gamma$. This can be
determined by a detector which weights all possible gamma-ray cascades
equally; an approximation to this approach, combined with high detection
efficiency, has been employed by (LTS 69).

The calculations cited above (BCF 68, Mic 69) suggest that the reaction
rates in the vicinity of $A \approx 45$ are only marginally fast enough to establish
quasi-equilibrium, and will be very important in determining the relative
abundances of the nuclides above and below $A \approx 45$ at "freeze-out," i.e.,
when, because of the expansion and dispersal of the star, the temperature
drops low enough to stop further nuclear reactions. These reaction rates
are therefore prime candidates for measurement; unfortunately, the most
desired cross sections involve radioactive target nuclei such as $^{44}\text{Ti}$. In any
case, there is a crucial need for more experimental studies of the systematics
of nuclear rates for intermediate mass nuclei, to determine the validity of
theoretical estimates for the densities of levels of different $J$-values, and to
check the reaction rates calculated from them. It is very clear, also, that

further progress will require a more careful study of the dynamics of these late stages of stellar evolution prior to and during the dispersal of the star's contribution to nucleosynthesis.

## 5.3. Explosive Carbon, Oxygen, and Silicon Burning: Supernovae

As already discussed in Section 1.3 we do not have any conclusive evidence identifying the process (or processes) by which the products of nucleosynthesis are exposed to view in stellar atmospheres and in the solar system. It is clear, both from stellar models and from observation, that a slow steady mass loss occurs from the surface of some stars. It is conceivable that this might proceed to the point where the regions in which nucleosynthetic activity has taken place are exposed to view. Similarly, one could imagine stars in which some instability has caused more rapid mass loss from the outer layers of the star, and occasional examples of such stars are seen—notably the planetary nebulae. Other mechanisms which have been suggested include the transfer of the envelope of one star to a companion star in a binary or multiple star system, and the transfer of the surface layers of a star to another star in a close collision. Although any of these processes might contribute to the list of stars which exhibit truly anomalous isotopic or elemental abundances, it is difficult to attribute the widely distributed, general uniformity of the products of nucleosynthetic activity to such apparently accidental circumstances.

It seems at present more plausible to assume that the general similarity in composition of most stellar objects is the result of universal nucleosynthesis, to be discussed later in Section 7.1, or the result of many supernovae events. In the latter case, it would have to be shown that most stellar explosions produce generally similar nuclidic abundances, for which there is in fact some theoretical evidence, or that these catastrophic stellar events mostly took place at an early stage of universal history and the products have been thoroughly mixed on a possibly intergalactic scale. If the observed nuclidic abundances are to be attributed to supernova activity, the role of the various stages of stellar evolution discussed in earlier sections of this paper is to provide the stellar energy production during most of the life of the star, and the stellar composition at the onset of the explosive phase. It must be anticipated that nucleosynthesis occurring during the explosion will extensively alter the presupernova nuclear composition.

Fowler and Hoyle (FH 64) have considered a 30-solar-mass star whose core region has evolved through silicon burning to iron. With the exhaustion of nuclear fuels in the core, the core contracts and the temperature rises rapidly in an ineffectual effort to supply the energy carried away from the star by neutrinos, produced mainly in the reaction $e^+ + e^- \rightarrow \nu + \bar{\nu}$. This refrigerating reaction would probably be sufficient to cause the eventual collapse of the core, but the endoergic photoconversion of iron to helium causes the core regions of the star to collapse in free fall even before the neutrinos can accomplish the same result. Although the dynamical details are by no means clear at this time, the shock waves from such a collapse must surely heat the outer layers of the star to high enough temperature to ignite explosively any unburned nuclear fuels. Some, as yet poorly determined, fraction of the star will achieve escape velocity and the results of this very rapid nucleosynthesis will be distributed throughout the surrounding space.

For less massive stars, Arnett and Truran have considered the evolution of a core consisting of comparable amounts of oxygen and carbon [see, for example, (Arn 69a, TA 70)]. Energy loss in the form of neutrinos should be sufficient to carry away the nuclear energy output while the core is enlarged by further helium burning. In their models, the pressure supporting the core of the star arises mainly from electron degeneracy, and is therefore only weakly related to the temperature, especially as the electrons become relativistic. Eventually high enough temperatures and pressures are achieved in the core to ignite explosively the $^{12}C + {}^{12}C$ and $^{16}O + {}^{16}O$ reactions, with silicon burning probably occurring in the center of the core. The resultant energy production overwhelms the neutrino energy loss and sends a detonation wave throughout the star, igniting helium burning and a rapid C–N–O cycle at the appropriate levels in the star. The total energy output is calculated to be sufficient to give most or all of the material in the star escape velocity. The combination of these nuclear processes yields an elemental abundance curve agreeing closely with that of the solar system, given appropriate fractions of each of the main processes—fast C–N–O cycle, helium burning, and combined carbon, oxygen, and silicon burning, as shown in Fig. 17. In this work nuclear reaction rates were taken from experiment, where available, and calculated theoretically for the other cases, as in the earlier work of Truran, Cameron, and Gilbert (TCG 66). Electron captures and the predetonation value of the average nuclear $N/Z$ ratio play an important role in determining the resulting abundances of the more neutron-rich products.

There remains considerable uncertainty regarding the dynamics of such explosions. For example, it has been suggested recently (BBW 70) that electron captures may reduce the pressure sufficiently to prevent detonation

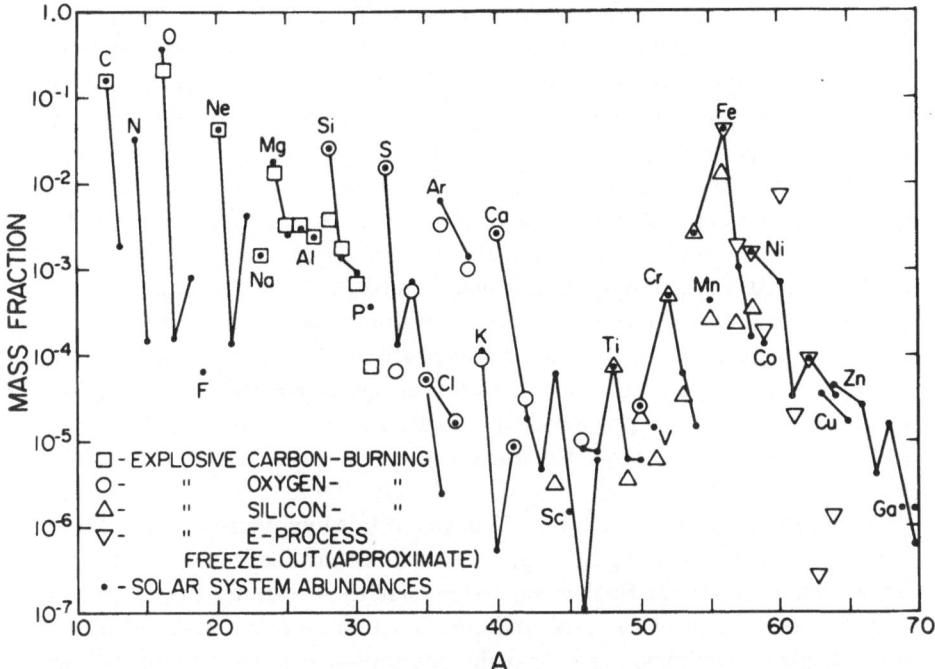

Fig. 17. Nuclear abundances produced by explosive carbon, oxygen, and silicon burning, in different concentric zones of an exploding star. (W. D. Arnett and J. W. Truran, to be published). The solar system abundances of (Cam 68) are shown as black dots. The isotopes of each element are joined for identification purposes. In this model, the nuclei from carbon to magnesium are produced mainly in the carbon-burning zone, those from silicon to calcium in the oxygen-burning zone, those from scandium to chromium by silicon burning, and the iron abundance peak by an equilibrium process, following silicon burning, as in the earlier work of (Bur+ 57).

of the core, in some stars. On the other hand, electron captures provide yet another refrigerating mechanism for inducing the collapse and subsequent detonation of a stellar core which has evolved to carbon, oxygen, and smaller amounts of elements in the magnesium region (WH 70). It is clear that supernova models will provide a challenge to the best efforts of astrophysicists for some time to come. At the same time, it is urgent that nuclear physicists provide better information on the reaction rates of protons, neutrons, and alpha particles with intermediate mass nuclei.

If the remnant mass after a supernova explosion is below the Chandrasekhar limit, about 1.4 solar masses, a white dwarf is expected to be formed. This kind of star is supported by electron degeneracy, and is expected to cool slowly into a black dwarf with negligible further nucleosynthesis. If the remnant mass is greater than perhaps two solar masses, there is no known physical law which could prevent the star from collapsing to a Schwarzschild singularity, a so-called black hole, detectable only by its gravitational effect on other objects, or by the gravitational and other radiation emitted during its collapse and subsequent accretion of matter. In the narrow range between 1.4 and 2 solar masses, a neutron star may be formed [see, for example, (Arn 69b)], as hypothesized for the pulsar remnant of the 1054 A.D. supernova which formed the Crab Nebula.

It seems unavoidable that the temperatures and densities reached in supernovae must result in major nucleosynthesis. Since $^{56}$Ni seems likely to be a prolific product of the conversion of lighter elements to the region of the iron peak, it was proposed by Colgate and McKee (CM 69) that the dominant sources of energy for heating the expelled material, during the first few weeks of expansion, are the $\beta$-decays, $^{56}$Ni$(e^-, \nu)^{56}$Co $(T_{1/2} = 6.1$ days$)$, and $^{56}$Co$(e^-, \nu)^{56}$Fe $(T_{1/2} = 77$ days$)$. If this hypothesis is correct, the gamma rays accompanying these decays should be produced with observable intensities. It has been further suggested that these gamma rays form an appreciable fraction of the observed cosmic gamma-ray background in the appropriate energy region, and that the intensity–red shift relation for these gamma rays could even provide a history of the supernovae frequency throughout the age of our expanding universe (CS 69).

One of the mechanisms proposed for the production of cosmic rays is the acceleration of nuclei in the intense time-dependent magnetic fields which must exist during and after a supernova. The nuclear composition of the galactic cosmic rays should then reflect the advanced stages of nucleosynthesis expected in such objects. Shapiro, Silverberg, and Tsao have corrected for the effects of traversal of interstellar matter, to derive the elemental composition of galactic cosmic rays at their source (SST 70). Unlike the composition of the particles from solar flares, which closely resembles the composition of the solar photosphere, the derived source composition for galactic cosmic rays differs in the following respects from the solar system composition (normalized for carbon): H and He are underabundant by a factor of 10 to 20, N and O are underabundant by about a factor 2, Mg and Si are overabundant by a factor of 3, Fe and Pb are overabundant by a factor of about 12. These abundances strongly favor the association of cosmic ray production with supernovae events, as discussed in this and the preceding sections. [See also (Bla+ 70).]

# 6. SYNTHESIS OF THE NUCLEI WITH $A \gtrsim 60$ AND THE $l$-PROCESS

## 6.1. The $s$- and $r$-Processes

Since the binding energy per nucleon decreases with increasing $A$ for nuclides beyond the iron peak, fusion of such nuclei would be an endoergic process, even if it were not so severely limited by the Coulomb barrier. Even the addition of protons or alpha particles is very strongly hindered by the rising Coulomb barrier, as $Z$ becomes greater than about thirty. It thus seems likely that the major fraction of the synthesis of nuclei heavier than iron is due to successive neutron captures, much as in early models of universal nucleosynthesis in which all nucleosynthesis beyond hydrogen was attributed to neutron capture.

If the neutrons are captured at time intervals greater than or comparable with beta-decay lifetimes near the curve of stability in the $N$–$Z$ plane, the nuclides produced will lie on or very near the stability curve. If, on the other hand, extremely high fluxes of neutrons are available, nuclides with very large neutron excess will be constructed, limited only by the small binding energies and short $\beta$-decay lifetimes for nuclei far removed from stability. These two neutron-capture modes have been named the $s$ (for slow) and $r$ (for rapid) processes (Bur+ 57, Cla+ 61, SFC 65). The seed nuclei for the process are postulated to be those in the iron abundance peak, but it must be expected that neutron captures will also effect some alteration of the nuclear abundances for lighter nuclei. The latter point has not been investigated yet in much detail, probably because it is only recently that a consistent picture of the charged-particle reactions in this mass region has been developed.

A characteristic feature of the $s$-process hypothesis is that the abundances of neighboring nuclides should be approximately inversely proportional to their capture cross sections for neutrons in the keV region. Thus the product $\sigma n$ should be a smoothly decreasing function of $A$ for those nuclides which are made only by the $s$-process. That this expectation seems to be rather well fulfilled is discussed most recently in the paper by Allen, Macklin, and Gibbons in this volume. Perhaps the most persuasive proof of the necessity for postulating the $r$-process is the occurrence in nature of such long-lived nuclides as $^{235}$U, $^{238}$U, and $^{232}$Th, which are separated from the stable nuclides with $A \leq 209$ by many short-lived nuclides, and could therefore not be made by the $s$-process. Indeed, in a supernova a copious supply of neutrons is made available in a very short time and there seems to be no way to prevent the $r$-process from asserting itself.

## 6.2. Possible Neutron Sources

While we are far from being able to identify with certainty the nuclear reactions which provide the neutrons for the *s*- and *r*-processes, and even less able to be sure of the stage of stellar evolution at which these processes occur, several likely neutron sources have been considered.

The reaction, $^{13}C + {}^4He \rightarrow {}^{16}O + n$, was first suggested by Cameron (Cam 54) and Greenstein (Gre 54), since it might be expected to follow immediately after the C–N–O hydrogen-burning phase. As pointed out in Section 3.2, however, the principal result of the C–N–O cycle is $^{14}N$ which has a large cross section for absorbing neutrons by the reaction

$$^{14}N + n \rightarrow {}^1H + {}^{14}C \tag{65}$$

Even though the released protons could produce more $^{13}N$, and thus more $^{13}C$, the concentration of $^{12}C$ is too small compared with the other elements present, and the $^{14}C$ eventually decays to $^{14}N$ in any case. It thus seems very difficult to assign an important role to the $(\alpha, n)$ reaction on $^{13}C$, where the $^{13}C$ is present as the result of the C–N–O cycle, unless the $^{14}N(\alpha, \gamma)^{18}F$ reaction occurs even before the $^{13}C(\alpha, n)^{16}O$ reaction. From the data available at present this seems very unlikely [see discussion in Section 4.4 and Fig. 14]. In a red giant model, discussed by Schwarzschild, Härm, and Sanders (SH 67, San 67), convective instability mixes limited amounts of fresh hydrogen into a shell region where the 3 $^4He \rightarrow {}^{12}C$ reaction is adding $^{12}C$ to a core consisting essentially of $^{12}C$. The hydrogen reacts rapidly with $^{12}C$ to form $^{13}C$, which in turn reacts with $^4He$ to produce neutrons. The important feature of this model is that there is such an excess of $^{12}C$ and $^4He$ that the hydrogen is prevented from establishing an appreciable $^{14}N$ concentration. [See also (CF 64) and (CFT 64) for combined hydrogen and helium burning in the core of a Population II red giant.]

The $^{13}C(\alpha, n)$ reaction has been studied experimentally down to a center-of-mass energy of 360 keV by Davids (Dav 68), who found the *S*-factor to be given by the expression

$$S(E) = [(5.48 \pm 1.77) + (12.05 \pm 3.91)E] \times 10^5 \quad \text{MeV-barns} \tag{66}$$

After the recognition that the presence of $^{14}N$ probably rules out appreciable neutron production by the $^{13}C(\alpha, n)$ reaction in the products of the C–N–O cycle, the following sequence of nuclear reactions was suggested by Cameron (Cam 60):

$$^{14}\text{N} + {}^4\text{He} \rightarrow {}^{18}\text{F} + \gamma \tag{67}$$

$$^{18}\text{F} \rightarrow \beta^+ + \nu + {}^{18}\text{O} \tag{68}$$

$$^{18}\text{O} + {}^4\text{He} \rightarrow {}^{22}\text{Ne} + \gamma \tag{69}$$

$$^{18}\text{O} + {}^4\text{He} \rightarrow n + {}^{21}\text{Ne} \qquad Q = -0.705 \text{ MeV} \tag{70}$$

$$^{22}\text{Ne} + {}^4\text{He} \rightarrow n + {}^{25}\text{Mg} \qquad Q = -0.48 \text{ MeV} \tag{71}$$

These reactions could, of course, be accompanied by $(\alpha, n)$ reactions on $^{21}\text{Ne}$, $^{25}\text{Mg}$, or on $^{26}\text{Mg}$ formed by radiative capture of $\alpha$-particles by $^{22}\text{Ne}$. It is expected that the dominant course of events would be the radiative capture of $\alpha$-particles by $^{18}\text{O}$, followed by neutron production by $(\alpha, n)$ on $^{22}\text{Ne}$, because of the high threshold for the $(\alpha, n)$ process on $^{18}\text{O}$. The rate for the $(\alpha, \gamma)$ reaction on $^{18}\text{O}$ has been studied experimentally by Adams *et al.* (Ada+ 69). The rate for the $^{22}\text{Ne}(\alpha, n)$ reaction was measured by Ashery (Ash 69) over the range $1.6 < E_{cm} < 4.0$ MeV; his extrapolated $S$-values at threshold, $S_{th} = 2.5 \times 10^9$ MeV-barns, is in good agreement with optical model estimates of Reeves (Ree 66b).

As always, one must treat extrapolations with caution, since one favorably-placed resonance could dominate the astrophysical rate. Indeed, Berman *et al.* (BHB 69) have shown that there is a prominant resonance in the reaction $^{26}\text{Mg}(\gamma, n){}^{25}\text{Mg}$ just 54 keV above the neutron threshold. If that level should also have sufficient $\alpha$-particle width it could completely invalidate an extrapolation of the cross section for $^{22}\text{Ne}(\alpha, n){}^{25}\text{Mg}$ from higher energies. The role of the $^{22}\text{Ne}(\alpha, n)$ reaction in the $s$-process has been investigated by Peters (Pet 68), using the cross section estimate of Reeves.

While we have just argued that $^{18}\text{O}$ would be effectively removed by radiative capture, rather than producing $^{21}\text{Ne}$ by an $(\alpha, n)$ reaction, $^{21}\text{Ne}$ could also be produced by the Ne–Na–Mg cycle, following the C–N–O cycle (MF 57). The reaction $^{21}\text{Ne}(\alpha, n)$ might therefore be a significant neutron source. The cross section for this reaction has been studied by Tanner (Tan 65), but current unpublished work in our laboratory by D. Ashery and Hay Boon Mak disagrees strongly with the earlier data.

At temperatures higher than the 1 to $2 \times 10^8$ °K values relevant for the $\alpha$-processes just described, carbon burning, oxygen burning, and silicon burning all produce significant numbers of neutrons per iron nucleus, even though neutrons are produced in only a few of the multitude of reactions occurring. These can build heavy elements in either quiescent (*s*-process) or explosive (*r*-process) burning. In the final collapse of the cores of very massive stars, triggered by the Fe → He conversion, approximately four neutrons are produced per disintegrated iron nucleus. However, this occurs near the center of the star, and it is not obvious how these neutrons could be effectively used. It is nevertheless clear that supernovae explosions will in general be accompanied by massive neutron production, both by charged-particle reactions and by photodisintegration; supernovae are thus the probable regimes for the *r*-process.

## 6.3. The *p*-Process

A detailed examination of the naturally-occurring nuclides with $A > 60$ shows that there are some, usually low-abundance, nuclides on the proton-rich side of the stability valley, which are by-passed by both the *s*- and *r*-processes. Formation of these nuclides is attributed to the *p*-process, originally believed to be radiative proton capture. Other possible candidates for the *p*-process include $(p, n)$, $(\gamma, n)$, and $(n, 2n)$ reactions, and positron capture. [See, for example, (Ito 61, RS 65, Mac 70) and references listed therein.] A more detailed experimental and theoretical investigation of the *p*-process would provide some new critical tests of nucleosynthesis theory.

## 6.4. The *l*-Process: Synthesis of D, Li, Be, and B

The light nuclides $^2$H, $^6$Li, $^7$Li, $^9$Be, $^{10}$B, $^{11}$B constitute another group of nuclei which require special attention, because they are so readily destroyed by proton bombardment. Hence, their presence in stellar spectra, in meteorites, and on the earth is likely to be due to a nonthermal process (see, however, Section 7.2). Since the sun and many other stars exhibit strong surface magnetic activity, it seems reasonable to attribute the formation of these low-abundance nuclides to spallation of the more abundant products of nucleosynthesis, such as carbon, nitrogen, and oxygen, by protons (FBB 55). Solar flares are indeed observed to be sources of high-energy protons, and it is likely that newly-formed stars are much more magnetically active than the sun is at present. Spallation will alter the abundances of other nuclei as well, of course, but the fractional changes are expected to be small except for those nuclides produced with small abundance in other nucleosynthetic processes.

For the laboratory study of spallation the principal techniques have been activity measurements and on-line mass spectrometry for short-lived radioactive products, mass spectrometry for long-lived or stable products, and on-line time-of-flight mass identification (Ber+ 65, Yio+ 68, AER 67, DLA 69).

Although much experimental work still remains, the cross sections available at this time indicate that a quantitative understanding of the origin of the $l$-nuclides is likely to be provided by the hypothesis of spallation on stellar surfaces, especially if one postulates deep enough convection to allow some reprocessing of the spallation products by low-energy protons. An earlier hypothesis (FGH 62, BFH 65) attributed the relative abundances of the $l$-nuclei to spallation and subsequent slow neutron capture in meter-sized planetesimals containing a large fraction of ice, and thus required very high-energy protons. In the light of our present knowledge of the steep energy spectrum of solar flare protons ($\sim E^{-4}$), it is somewhat easier to accept the main production of the spallation products as being due to protons with energies below $\sim100$ MeV. This puts the measurement of most currently interesting spallation cross sections within the energy range of many of the newer cyclotrons. A promising alternative hypothesis for the synthesis of the $l$-nuclides is that they are produced by spallation in interstellar space—spallation of ambient heavy nuclei by cosmic-ray protons or $\alpha$-particles, or spallation of heavy cosmic rays by interstellar hydrogen (RFH 70).

# 7. NUCLEOSYNTHESIS IN MASSIVE EXPLODING OBJECTS

## 7.1. Universal Nucleosynthesis

There is abundant evidence that our universe is expanding. Various cosmological models have been developed to describe this situation; steady-state universes with continuous (or intermittent) creation of matter, expanding universes with hot (or cold) singular origins, and pulsating universes have been considered, among others. The principal evidence for expansion is the distance–red-shift relation for stars and galaxies; the radio-source count as a function of brightness (or distance) suggests an evolutionary universe with a singular origin. Perhaps the most compelling evidence for a hot singular origin is the recently discovered background microwave radiation which can be interpreted as the 2.7 °K, red-shifted blackbody radiation of an earlier high-temperature–high-density state of the universe (PW 65, Dic+ 65).

The study of nucleosynthesis in expanding universes can be traced back to the models of Alpher, Bethe, and Gamow (ABG 48), Fermi and Turkevich (FT 50), and Alpher and Herman (AH 48, 49). In these early models, as in several others which are omitted for reasons of brevity [see, for example, (Tay 69)], the primeval matter was in the form of neutrons, and protons were made available only by free neutron decay. It was found that implausibly narrow limits had to be placed on the baryon density to produce helium/hydrogen mass-fraction ratios of order unity; outside these limits either no helium was produced or all matter was processed to helium. Hayashi (Hay 50) pointed out an important omission in these models; at the higher temperatures existing before expansion cools the universe to the nucleosynthesis temperature range, the weak interactions are fast enough to keep neutrons and protons in thermal equilibrium. Providing the net electron lepton number is near zero, the proton and neutron mass fractions will be given by

$$X_p/X_n = \exp[(M_n - M_p)c^2/kT] \tag{72}$$

With this correction to the model, several authors have verified that comparable mass fractions of helium and hydrogen are produced by a hot expanding universe, over a wide range of baryon density [see, for example (HT 64, Pee 66, WHF 67)]. Since $^4$He is such a strongly bound nucleus, it is not easily destroyed in subsequent nucleosynthesis, and the existence of a minimum, universal helium mass fraction becomes a critical test for models of the evolution of the universe. The helium mass fractions reported range from near zero in certain old stars (SS 68, GM 66), about 0.27 in the sun, 0.4 in some planetary nebulae, and up to 0.9 in some hot subdwarfs (Pet 70). It is possible that the occasional very low values reported are due to gravitational settling of helium in the stellar atmosphere (GTC 67), or to unfavorable excitation conditions in the stellar atmosphere. Before a definitive answer can be given on the existence or nonexistence of universal nucleosynthesis, the occurrence of a universal minimum helium mass fraction will have to be more surely demonstrated or disproved.

Wagoner, Fowler, and Hoyle (WFH 67, Wag 67, Fow 67) have studied the relative abundances of helium and heavier elements that would be produced by the expansion of a universal "fireball" under various assumed conditions. In addition to assumptions about the degree of isotropy and homogeneity of the expansion, the rate of expansion, and the validity of general relativity, it is necessary to specify several other parameters. The relation between baryon number and radiation density must be spelled out; a convenient way to express this is to make use of the fact that adiabatic expansion relates the baryon density to the radiation temperature in a very simple way,

$$\varrho_B = hT_\gamma^3 \tag{73}$$

The quantity $h$ can then be fixed by the present-day mean baryon density, $3 \times 10^{-31}$ g cm$^{-3} \lesssim \varrho_0 \lesssim 3 \times 10^{-29}$ g cm$^{-3}$, and the present blackbody temperature, $T_0$. For $T_\gamma$ in units of $10^9$ °K, $h$ lies in the range $10^{-3}$ to $10^{-5}$. The density of the various constituents of the universe, and the temperature are plotted against the expansion factor, $V/V_0$, in Fig. 18 for $\varrho_0 = 2 \times 10^{-29}$ g cm$^{-3}$ and $T_0 = 3$ °K.

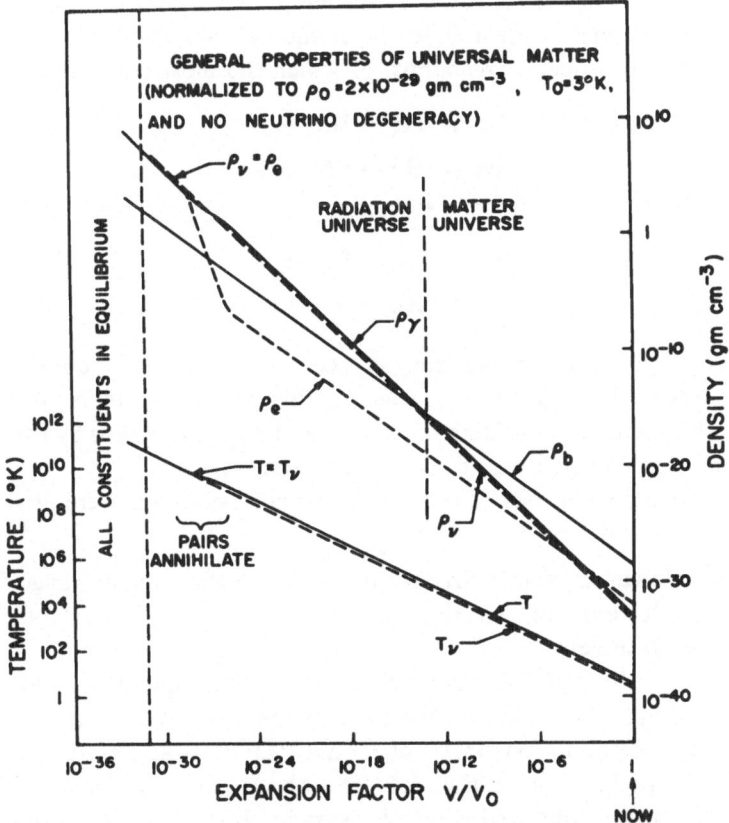

**Fig. 18. Expansion of the universe from a high-temperature, high-density origin (WFH 67).** Shown here are the energy densities of neutrinos, electrons, photons, and matter as a function of the expansion factor $V/V_0$. In this example, the matter density and temperature have been normalized to present values of $2 \times 10^{-29}$ g cm$^{-3}$ and $3$°K, respectively.

In addition, the muon and electron lepton numbers have a small effect on the rate of expansion, which we will ignore; the electron lepton number, however, has a critical effect on nucleosynthesis because a large excess of neutrinos or antineutrinos will drive the neutron–proton equilibrium to protons or neutrons, respectively. In either extreme, nucleosynthesis is greatly hindered, even for the synthesis of helium. The sensitivity of nucleosynthesis to lepton number is shown in Fig. 19.

By constructing a chain of some sixty-three nuclear reactions, and their inverse reactions, Wagoner *et al.* followed the evolution in time of the abundances of the nuclides with $A \leq 24$, as the universal temperature falls through the nucleosynthesis region, roughly from $10^{10}$ °K to $10^8$ °K. The well-known gaps in the sequence of nuclei at $A = 5$ and 8, which were a considerable embarrassment in early studies of nucleosynthesis, can be bridged by several nuclear reactions, of which the most relevant are

$$^3\text{He} + {}^4\text{He} \rightarrow {}^7\text{Be} + \gamma \tag{74}$$

$$^7\text{Be} + {}^4\text{He} \rightarrow {}^{11}\text{C} + \gamma \tag{75}$$

$$3\,{}^4\text{He} \rightarrow {}^{12}\text{C} + \gamma \tag{76}$$

$$2\,{}^4\text{He} + n \rightarrow {}^9\text{Be} + \gamma \tag{77}$$

$$^9\text{Be} + {}^4\text{He} \rightarrow n + {}^{13}\text{C} \tag{78}$$

The $^7\text{Be}(\alpha, \gamma){}^{11}\text{C}$ reaction has not yet been studied in the laboratory; although technically difficult to measure, this reaction rate is important because this method of bridging the $A = 5$ and 8 gaps proceeds by two-body reactions, and thus at lower densities than the others.

The main conclusions of this study of nucleosynthesis were as follows (Fow 67, Fow 70):

(1) $^4\text{He}$ can be synthesized with a mass fraction in the range 0.2 to 0.3 for a wide range of densities, provided the electron lepton number is not too far from zero.

(2) $^2\text{H}$, $^3\text{He}$, and $^7\text{Li}$ may be produced in small quantities, with mass fractions comparable to those observed in the solar system, as shown in Fig. 20. This method of synthesis of $^2\text{H}$ and $^7\text{Li}$ offers an alternative to the spallation hypothesis of Section 6.4. The boron isotopes, $^{10}\text{B}$ and $^{11}\text{B}$, do not appear to be synthesized with solar system abundances. A more serious problem in producing the *l*-elements by universal nucleosynthesis is that these elements are so easily destroyed by thermonuclear reactions that it may not prove feasible to construct models in which the *l*-elements could be preserved until the present epoch.

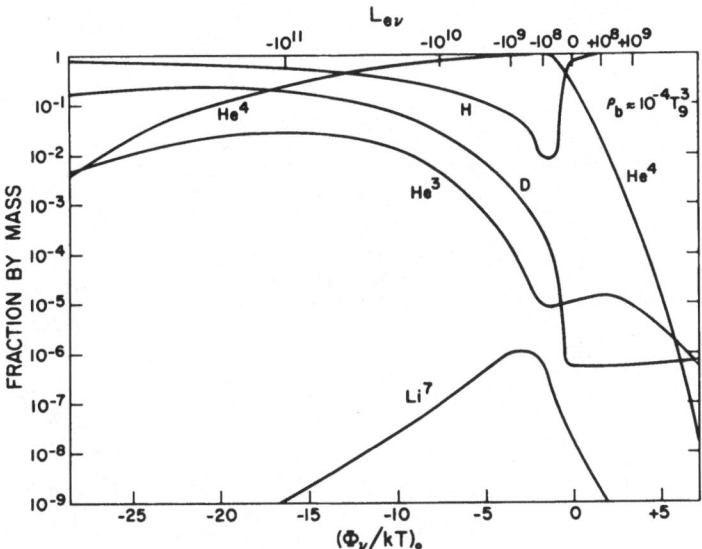

**Fig. 19. The dependence of universal nucleosynthesis on the electron–neutrino lepton number.** $L_{e\nu}$ is approximately $(n_\nu - n_{\bar{\nu}})/n_\gamma$, where $n_\nu$, $n_{\bar{\nu}}$, and $n_\gamma$ are the number densities of electron neutrinos, electron antineutrinos, and photons respectively. $L_{e\nu}$ can also be expressed as the present value of the ratio of the electron-neutrino fermi energy to $kT$, as shown on the lower scale. Only for a narrow range near $L_{e\nu} = 0$ is helium synthesized with a mass fraction of about 0.3 (Fow 70).

(3)  In spite of the bridges over $A = 5$ and $A = 8$, the constraints imposed by the present universal density and temperature do not permit high enough densities during the nucleosynthesis temperature range to produce nuclides with $A \geq 4$ (except for $^6$Li, $^7$Li) in abundances comparable with those observed even in very old stars. Thus we must apparently attribute the synthesis of essentially all of the nuclides with $A > 4$ to later processing.

## 7.2. Nucleosynthesis in Supermassive Objects

Nucleosynthesis in very massive objects which expand from a high-density–high-temperature state has been investigated by methods very similar to those employed in the universal nucleosynthesis study discussed in the previous section (WFH 67, Wag 69). Masses ranging from about $10^3$ solar masses to galactic mass were considered; these objects could reasonably be conjectured to be a class of stars existing for a short time early in

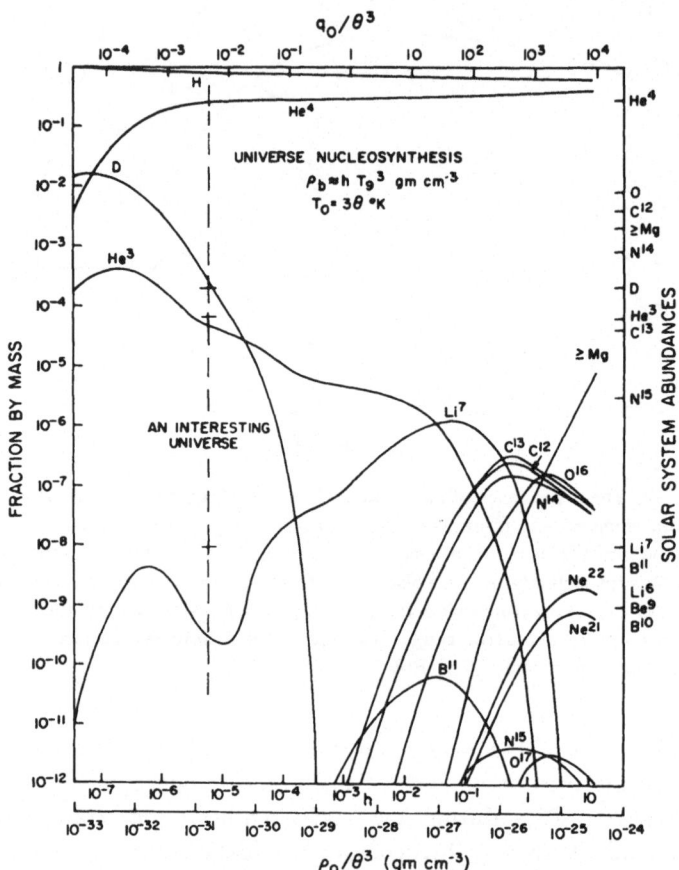

**Fig. 20. Synthesis of the light elements in a universal expansion (WFH 67).** The mass fractions of various light nuclei are shown as a function of the parameter $h$ in the relation $\varrho_B = hT_9^3$, where $\varrho_B$ is the baryon density and $T_9$ the universal blackbody temperature. The parameter $h$ also determines $q_0$, the present value of the deceleration parameter of the expansion [see, for example, (WFH 67)]. Constraints imposed by the present-day density and temperature rule out the production of significant amounts of the elements heavier than lithium in such a universal process.

galactic history, or even the galaxies themselves, which have "bounced" after contraction to high density, and to temperatures above $10^{10}$ °K. At these high temperatures both strong and weak reactions are in statistical equilibrium, and, as the temperature falls with expansion, the higher $Z$-elements successively drop out of equilibrium while nuclear reactions continue to build up the abundances of the lower $Z$-elements.

Wagoner *et al.* assumed that nucleosynthesis proceeded close to the stability curve, not because $\beta$-decays are fast enough to compete effectively with the rapid expansion rates assumed, but because exoergic nuclear reactions, such as $(p, n)$ and $(n, p)$ processes, limit the excursions of $N$ and $A$ away from stability. The nuclear reactions included in the calculations were similar to those for the universal study, and the calculations were limited to $A \leq 28$, by the size of computer available.

By invoking higher densities and faster expansion rates during the nucleosynthesis temperature range than in the universal study, greatly enhanced production of nuclides with $A > 4$ was achieved. During the early high-temperature phase some production of nuclides up to the iron peak was achieved, and the nucleosynthesis continuing after the freeze-out of the iron peak elements enhances the abundances of the nuclides up to about silicon or sulfur.

The kind of agreement between calculated abundances and the solar system abundances is illustrated in Fig. 21. In this figure the abundances are expressed as mass fractions relative to $^4$He. It is clear that these models produce too little of the heavy elements relative to helium as compared with the solar system (Population I) abundances, but the C, N, and O abundances appear to be comparable with the minimum values found for the extreme Population II stars, believed to have formed early in galactic history. Wagoner has conjectured that it might be possible to achieve a mass-fraction curve resembling the solar system by increasing both the density and the expansion rate, while reducing the bounce temperature to prevent forming unreasonably large mass fractions for $A > 28$. An interesting point of the calculations is that there are still about $10^5$ neutrons per iron peak nucleus when these nuclei freeze out, so that the r-process could produce even the heaviest nuclei. Another interesting feature of the calculations, pointed out by Wagoner, is that the nuclides $^{25}$Mg, $^{26}$Mg, and $^{27}$Al are produced by charged-particle nucleosynthesis in relative abundances which satisfy the condition that the product of the $(n, \gamma)$ reaction cross section and the abundance is approximately constant. This was just the criterion employed by Burbidge *et al.* (Bur+ 57) in assigning the production of these nuclides to the s-process, and should perhaps serve as a reminder that we

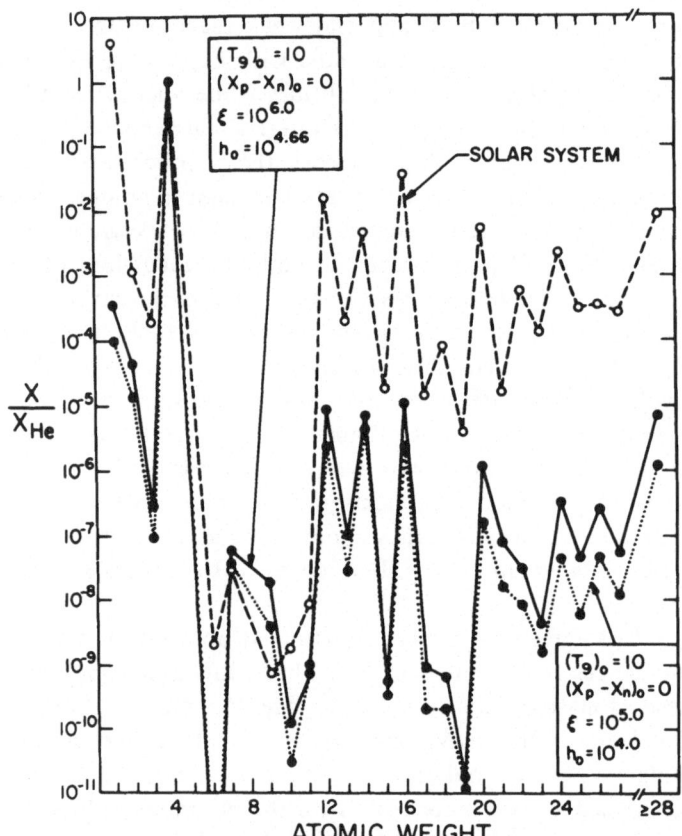

**Fig. 21. Nuclear abundances produced by the rapid expansion of a hot massive object (Wag 69).** The conditions assumed in this particular model were (1) initial temperature, $10^{10}$ °K, (2) neutron–proton equality during the nucleosynthesis stage of the expansion, (3) $\xi = 10^5$ or $10^6$, as shown, where $\xi$ characterizes the expansion rate $[V^{-1}\,dV/dt = \xi\,\sqrt{24\pi G\rho}]$, and (4) $h_0 = 10^{4.0}$ or $10^{4.66}$, as shown, where $h_0$ is the parameter relating the baryon density and temperature at the beginning of the expansion. The observed and calculated abundances have been normalized to unity for $^4$He. Although the trend of the calculated abundance curve resembles that of the solar system, the abundances are too low by a factor of about $10^4$ for $A > 9$. The calculated abundances may not be very different from those of extreme Population II stars, however.

are far from being able to produce a unique route map through the nucleo-synthesis process. Since there is no known mechanism which can produce the enormous expansion rates necessary to produce Population-I abundances, we must conclude that *most* of our nuclidic heritage is the result of stellar nucleosynthesis, with the possible exception of helium.

From the point of view of the nuclear physicist, the models employed by Wagoner and collaborators have an important feature in common with the explosive carbon, oxygen, and silicon-burning models of Arnett, Truran, and others. They require reaction rates for virtually all nuclear reactions, at least up to the iron peak. Of course the rates of many of the relevant nuclear reactions have been measured (FCZ 67), often, however, with inadequate precision, or at energies so far removed from the relevant energies that extrapolation is risky. Wagoner (Wag 67) has considered the charges in the calculated abundances which would be produced by altering the assumed reaction rates; many of the reaction rates critically affect the calculated abundances. In particular, the reaction rates for bridging the $A = 5$ and 8 gaps produce nearly proportional changes in the abundances of all elements with $A \geq 12$, as might be guessed. To make possible more critical tests of the massive-object nucleosynthesis model, the nuclear physicist will have to supply rate information for many of the reactions listed by Wagoner (Wag 69).

## ACKNOWLEDGMENTS

Except for the hydrogen-burning reactions, for which the experimental and theoretical work seems reasonably complete, we have raised more questions than we have settled; conclusions are premature, to say the least. This paper would not be complete, however, without grateful acknowledgment of the many informative discussions the author has had with W. D. Arnett, G. R. Caughlan, W. A. Fowler, R. W. Kavanagh, T. Lauritsen, T. A. Tombrello, J. W. Truran, R. W. Wagoner, W. Whaling, H. C. Winkler, B. A. Zimmerman, and others. The expert assistance of Jan Rasmussen, in the preparation of the manuscript, has been most valuable. Figures 18, 20, and 21 have been reproduced (with slight modifications) from *The Astrophysical Journal*, by permission of the University of Chicago Press (Copyrights 1967, 1969 by the University of Chicago).

# BIBLIOGRAPHY

*Astrophysics*, by R. J. Tayler, W. Davidson, J. V. Narlikar, and M. A. Ruderman, (reprinted from *Reports on Progress in Physics*), W. A. Benjamin, Inc., New York (1969).

*New Uses for Low-Energy Accelerators*, Prepared by the Ad Hoc Panel of the National Research Council, W. A. Fowler, Chairman, National Academy of Sciences, Washington, D.C. (1968).

*Nuclear and Neutrino Processes in Stars and Supernovae*, by W. A. Fowler and J. L. Vogl, *Lectures in Theoretical Physics*, Vol. 6, University of Colorado Press, Boulder, Colorado (1964).

*Nuclear Physics—A Bibliographic Survey*, by B. Kuchowicz, Gordon and Breach Science Publishers, New York (1967).

*Nucleosynthesis* (W. D. Arnett, C. J. Hansen, J. W. Truran, and A. G. W. Cameron, eds.), Gordon and Breach Science Publishers, New York (1968).

*Physical Processes in Stellar Interiors*, by D. Frank-Kamenetskii (1958). English translation available from the Office of Technical Services, U. S. Department of Commerce, Washington, D.C.

*Principles of Stellar Evolution and Nucleosynthesis*, by D. D. Clayton, McGraw-Hill Book Co., New York (1968).

*Stellar Evolution and Nucleosynthesis*, by H. Reeves, Gordon and Breach Science Publishers, New York (1968).

*Stellar Evolution* (R. F. Stein and A. G. W. Cameron, eds.), Plenum Press, New York (1966).

# REFERENCES

(ABG 48)   R. A. Alpher, H. A. Bethe, and G. Gamow, *Phys. Rev.*, **73**:803 (1948).

(ABK 60)   E. Almqvist, D. A. Bromley, and J. A. Kuehner, *Phys. Rev. Letters*, **4**:515 (1960).

(AC 70)    M. F. Aller and C. R. Cowley, *Astrophys. J.*, **162**:L145 (1970).

(Ada+ 68)  A. Adams, M. H. Shapiro, C. A. Barnes, E. G. Adelberger, and W. M. Denny, *Bull. Amer. Phys. Soc.*, **13**:698 (1968).

(Ada+ 69)  A. Adams, M. H. Shapiro, W. M. Denny, E. G. Adelberger, and C. A. Barnes, *Nucl. Phys.*, **A131**:430 (1969).

(AER 67)   J. Adouze, M. Epherre, and H. Reeves, in *High-Energy Nuclear Interactions in Astrophysics* (B. S. P. Shen, ed.), W. A. Benjamin, Inc., New York (1967).

(AH 48)    R. A. Alpher and R. Herman, *Nature*, **162**:774 (1948).

(AH 49)    R. A. Alpher and R. Herman, *Phys. Rev.*, **75**:1089 (1949).

(Alb 61)   D. E. Alburger, *Phys. Rev.*, **124**:193 (1961).

(Arn 69a)  W. D. Arnett, *Astrophys. J.*, **157**:1369 (1969); *Astrophys. and Space Sci.*, **5**:180 (1969).

(Arn 69b)  W. D. Arnett, *Nature*, **222**:359 (1969).

(Ash 69)   D. Ashery, *Nucl. Phys.*, **A136**:481 (1969).

(AT 69)    W. D. Arnett and J. W. Truran, *Astrophys. J.*, **157**:339 (1969).

(ATK 70)   S. M. Austin, G. F. Trentelman, and E. Kashy, *Bull. Amer. Phys. Soc.*, **15**, No. 12 (1970).

(Bah 66)   J. N. Bahcall, *Astrophys. J.*, **143**:259 (1966).

(Bar 55a)   C. A. Barnes, *Phys. Rev.*, **97**:1226 (1955).

(Bar 55b)   E. U. Baranger, *Phys. Rev.*, **99**:145 (1955).

(Bar+ 62)   R. K. Bardin, C. A. Barnes, W. A. Fowler, and P. A. Seeger, *Phys. Rev.*, **127**:583 (1962).

(BBS 68)    J. N. Bahcall, N. A. Bahcall, and G. Shaviv, *Phys. Rev. Letters*, **20**:1201 (1968).

(BBU 69)    J. N. Bahcall, N. A. Bahcall, and R. K. Ulrich, *Astrophys. J.*, **156**:559 (1969).

(BBW 70)    Z. Barkat, J. R. Buchler, and J. C. Wheeler, *Astrophys. Letters* **6**:117 (1970).

(BC 38)     H. A. Bethe and C. L. Critchfield, *Phys. Rev.*, **54**:248 (1938).

(BCF 68)    D. Bodansky, D. D. Clayton, and W. A. Fowler, *Astrophys. J. Suppl.* No. 148, **16**:299 (1968).

(Ben+ 66)   J. Benn, E. B. Dally, H. H. Müller, R. E. Pixley, H. H. Staub, and H. Winkler, *Phys. Letters*, **20**:43 (1966); and *Nucl. Phys.*, **A106**:296 (1968).

(Ber+ 65)   R. Bernas, M. Epherre, E. Gradsztajn, R. Klapisch, and F. Yiou, *Phys. Letters*, **15**:147 (1965).

(Bet 39)    H. A. Bethe, *Phys. Rev.*, **55**:103 (1939); and *Phys. Rev.*, **55**:434.

(BF 69)     N. A. Bahcall and W. A. Fowler, *Astrophys. J.*, **157**:645 (1969).

(BF 70)     N. A. Bahcall and W. A. Fowler, *Astrophys. J.*, **161**:119 (1970).

(BFH 65)    D. S. Burnett, W. A. Fowler, and F. Hoyle, *Geochim. Cosmochim. Acta*, **29**:1209 (1965).

(BHB 69)    B. L. Berman, R. L. Van Hemert, and C. D. Bowman, *Phys. Rev. Letters*, **23**:386 (1969).

(BKP 59)    S. Bashkin, R. W. Kavanagh, and P. D. Parker, *Phys. Rev. Letters*, **3**:518 (1959).

(Bla+ 70)   G. E. Blanford, Jr., R. L. Fleischler, P. H. Fowler, M. W. Friedlander, J. Klarman, J. M. Kidd, G. E. Nichols, P. B. Price, R. M. Walker, J. P. Wefel, and W. C. Wells, *Acta Physica Hungarica* **28**, Suppl. 1 (1970).

(BLW 66)    D. S. Burnett, H. J. Lippolt, and G. J. Wasserburg, *J. Geophys. Res.*, **171**: 1249 (1966).

(BM 69a)    J. N. Bahcall and R. M. May, *Astrophys. J.*, **155**:501 (1969).

(BM 69b)    J. N. Bahcall and C. P. Moeller, *Astrophys. J.*, **155**:511 (1969).

(BT 67)     A. D. Bacher and T. Tombrello, quoted by T. A. Tombrello, in *Nuclear Research with Low-Energy Accelerators* (J. B. Marion and D. M. Van Patter, eds.), Academic Press, Inc., New York (1967).

(BTW 57)    S. D. Bloom, B. J. Toppel, and D. H. Wilkinson, *Phil. Mag.*, **2**:57 (1957).

(Bur+ 57)   E. M. Burbidge, G. R. Burbidge, W. A. Fowler, and F. Hoyle, *Rev. Mod. Phys.*, **29**:547 (1957).

(BW 52)     J. M. Blatt and V. F. Weisskopf, *Theoretical Nuclear Physics*, John Wiley and Sons, Inc., New York (1952).

(Cam 54)    A. G. W. Cameron, *Phys. Rev.*, **93**:932 (1954).

(Cam 57)    A. G. W. Cameron, Atomic Energy of Canada Ltd. Report AECL 454 (1957).

(Cam 58)    A. G. W. Cameron, *Ann. Rev. Nucl. Phys.*, **8**:299 (1958).

(Cam 60)    A. G. W. Cameron, *Astron. J.*, **65**:485 (1960).

(Cam 68)    A. G. W. Cameron, in *Origin and Distribution of the Elements* (L. H. Ahrens, ed.), Pergamon Press, London and New York (1968).

(Cau 65)    G. R. Caughlan, *Astrophys. J.*, **141**:688 (1965).

(Cau 70)    G. R. Caughlan, private communication.

(CF 62)    G. R. Caughlan and W. A. Fowler, *Astrophys. J.*, **136**:453 (1962).

(CF 64)    G. R. Caughlan and W. A. Fowler, *Astrophys. J.*, **139**:1180 (1964).

(CFT 64)   G. R. Caughlan, W. A. Fowler, and R. J. Talbot, *Astrophys. J.*, **140**:380 (1964).

(CG 64)    H. L. Crannell and T. A. Griffy, *Phys. Rev.*, **136**:B1580 (1964); also private communication quoted in (FCZ 67).

(CGL 61)   M. A. Clark, H. E. Gove, and A. E. Litherland, *Canadian J. Phys.*, **39**:1241, 1243 (1961).

(Chr+ 67)  C. J. Christensen, A. Nielsen, A. Bahnsen, W. K. Brown, and B. M. Rustad, *Phys. Letters*, **26B**:11 (1967).

(Cla 69)   G. J. Clark, *Australian J. Phys.*, **22**:289 (1969).

(Cla+ 61)  D. D. Clayton, W. A. Fowler, T. C. Hull, and B. A. Zimmerman, *Ann. Phys.*, **12**:331 (1961).

(Cla+ 70)  W. B. Clarke, J. R. de Laeter, H. P. Schwarcz, and K. C. Shane, *J. Geophys. Res.*, **75**:448 (1970).

(Cli 60)   J. L. Climenhaga, *Publ. Dominion Astrophys. Obs.*, Victoria, B.C., **11**:307 (1960).

(CM 69)    S. A. Colgate and C. McKee, *Astrophys. J.*, **157**:623 (1969).

(Coo+ 57)  C. W. Cook, W. A. Fowler, C. C. Lauritsen, and T. Lauritsen, *Phys. Rev.*, **107**:508 (1957).

(CS 67)    D. D. Clayton and P. B. Shaw, *Astrophys. J.*, **148**:301 (1967).

(CS 69)    D. D. Clayton and J. Silk, *Astrophys. J.*, **158**:143 (1969).

(CS 70)    R. Couch and K. Shane, private communication.

(Dav 68)   C. N. Davids, *Nucl. Phys.*, **A110**:619 (1968); and *Astrophys. J.*, **151**:775 (1968).

(DB 70)    C. N. Davids and T. I. Bonner, *Bull. Amer. Phys. Soc.*, **15**, No. 12 (1970).

(DHH 68)   R. Davis, K. C. Hoffman, and D. S. Harmer, *Phys. Rev. Letters*, **20**:1205 (1968).

(Dic+ 65)  R. H. Dicke, P. J. E. Peebles, P. G. Roll, and D. T. Wilkinson, *Astrophys. J.*, **142**:144 (1965).

(DLA 69)   C. N. Davids, H. Laumer, and S. M. Austin, *Phys. Rev. Letters*, **22**:1388 (1969); *Phys. Rev.* **C1**:270 (1970). (1969); and to be published.

(Don 67)   T. W. Donnelly, Ph.D. Thesis, University of British Columbia (1967).

(DS 64)    W. Deinzer and E. E. Salpeter, *Astrophys. J.*, **140**:499 (1964).

(Dwa 68)   M. R. Dwarakanath, Ph.D. Thesis, California Institute of Technology (1968).

(EV 67)    P. Endt and C. Van der Leun, *Nucl. Phys.*, **A105**:1 (1967).

(Eve+ 68)  D. Evers, G. Flügge, J. Morgenstern, T. W. Retz-Schmidt, H. Schmidt, and S. J. Skorka, *Phys. Letters*, **27B**:423 (1968).

(FBB 55)   W. A. Fowler, G. R. Burbidge, and E. M. Burbidge, *Astrophys. J. Suppl.* No. 17, **2**:167 (1955).

(FCZ 67)   W. A. Fowler, G. R. Caughlan, and B. A. Zimmerman, *Ann. Rev. Astron. and Astrophys.*, **5**:525 (1967). An updated list of reaction rates is in preparation.

(FG 58)    R. P. Feynman and M. Gell-Mann, *Phys. Rev.*, **109**:193 (1958).

(FGH 62)   W. A. Fowler, J. L. Greenstein, and F. Hoyle, *Geophys. J. Roy. Astron. Soc.*, **6**:148 (1962).

(FH 64)    W. A. Fowler and F. Hoyle, *Astrophys. J. Suppl.* No. 91, **9**:201 (1964).

(Fow 58)  W. A. Fowler, *Astrophys. J.*, **127**:551 (1958).

(Fow 67)  W. A. Fowler, in *High-Energy Physics and Nuclear Structure*, North-Holland Publishing Co., Amsterdam (1967).

(Fow 70)  W. A. Fowler, in *Proceedings of The Meeting on "Astrophysical Aspects of the Weak Interactions"* at Cortona, Italy, June 12, 1970.

(Fre+ 69)  J. M. Freeman, J. G. Jenkin, G. Murray, and W. E. Burcham, *Nucl. Phys.*, **A132**:593 (1969).

(FT 50)  E. Fermi and A. Turkevich, quoted in R. A. Alpher and R. C. Herman, *Rev. Mod. Phys.*, **22**:153 (1950).

(FV 64)  W. A. Fowler and J. L. Vogl, *Lectures in Theoretical Physics*, VI, University of Colorado Press, Boulder (1964).

(Gar+ 69)  T. Gary, M. Koch, J. Richter, B. Baschek, H. Holweger, and A. Unsöld, *Nature*, **223**:1254 (1969).

(GLS 63)  G. M. Griffiths, M. Lal, and C. D. Scarfe, *Canadian J. Phys.*, **41**:724 (1963).

(GM 66)  J. L. Greenstein and G. Münch, *Astrophys. J.*, **146**:618 (1966).

(GMA 60)  L. Goldberg, E. A. Müller, and L. H. Aller, *Astrophys. J. Suppl.* No. 45, **5**:1 (1960).

(Gol 69)  G. G. Goles, *Handbook of Geochemistry*, Vol. 1, Springer-Verlag, Berlin (1969), p. 116.

(Gor+ 67)  S. Gorodetzky, R. M. Freeman, A. Gallman, F. Hass, and B. Heusch, *Phys. Rev.*, **155**:1119 (1967).

(Gre 54)  J. L. Greenstein, in *Modern Physics for the Engineer* (L. N. Ridenour, ed.), McGraw-Hill Book Co., New York (1954).

(GS 65)  F. Gudden and P. Strehl, *Z. Physik*, **185**:111 (1965).

(GTC 67)  G. S. Greenstein, J. W. Truran, and A. G. W. Cameron, *Nature*, **213**: 871 (1967).

(Hay 50)  C. Hayashi, *Progr. Theoret. Phys.*, **5**:224 (1950).

(Hen 67)  D. C. Hensley, *Astrophys. J.*, **147**:818 (1967).

(Hen 69)  E. M. Henley, *Ann. Revs. Nucl. Sci.*, **19**:367 (1969).

(Hoy+ 53)  F. Hoyle, D. N. F. Dunbar, W. A. Wenzel, and W. Whaling, *Phys. Rev.*, **92**:1095 (1953).

(HPR 67)  C. M. Hohenberg, F. A. Podosek, and J. H. Reynolds, *Science*, **156**:202 (1967).

(HT 64)  F. Hoyle and R. J. Tayler, *Nature*, **203**:1108 (1964).

(Ibe 67)  I. Iben, Jr., *Ann. Rev. Astron. and Astrophys.*, **5**:571 (1967).

(Ibe 70)  I. Iben, Jr., *Scientific American*, **223**:26 (1970).

(Ito 61)  K. Ito, *Progr. Theoret. Phys.*, **26**:990 (1961).

(JGM 70)  R. J. Jaszczak, J. H. Gibbons, and R. L. Macklin, *Phys. Rev.*, **C2**:63 (1970); *Phys. Rev.*, **C2**:2452 (1970).

(Kav+ 69)  R. W. Kavanagh, T. A. Tombrello, J. M. Mosher, and D. R. Goosman, *Bull. Amer. Phys. Soc.* **14**, 1209 (1969); and to be published.

(LA 62)  T. Lauritsen and F. Ajzenberg-Selove, *Nuclear Data Sheets*, Sets 5 and 6 (1961), National Academy of Sciences-National Research Council, Washington, D.C.

(Lau 51)  C. C. Lauritsen, quoted in W. A. Fowler, J. L. Greenstein, and F. Hoyle, *Am. J. Phys.*, **29**:393 (1961).

(Loe+ 67)  H. M. Loebenstein, D. W. Mingay, H. Winkler, and C. S. Zaidins, *Nucl. Phys.*, **91**:481 (1967).

(LS 64)    J. D. Larson and R. H. Spear, *Nucl. Phys.*, **56**:497 (1964).
(LTS 69)   P. B. Lyons, J. W. Toevs, and D. G. Sargood, *Nucl. Phys.*, **A130**:1 (1969).
(Lyo 69)   P. B. Lyons, *Nucl. Phys.*, **A130**:25 (1969).
(Lyo+ 70)  P. B. Lyons, J. W. Toevs, C. A. Barnes, W. A. Fowler, and D. G. Sargood, *Astrophys. J.*, **159**:913 (1970).
(Mac 70)   R. L. Macklin, *Astrophys. J.*, **162**:353 (1970).
(MC 68)    R. May and D. Clayton, *Astrophys. J.*, **153**:855 (1968).
(McK 60)   A. McKellar, *J. Roy. Astron. Soc. Can.*, **54**:97 (1960).
(Mer 52)   P. W. Merrill, *Astrophys. J.*, **116**:21 (1952).
(Mer 56)   P. W. Merrill, *Publ. Astron. Soc. Pacific*, **68**:70 (1956).
(MF 57)    J. B. Marion and W. A. Fowler, *Astrophys. J.*, **125**:221 (1957).
(Mic 69)   G. Michaud, Ph.D. Thesis, California Institute of Technology (1969).
(MMK 70)   S. J. McCaslin, F. A. Mann, and R. W. Kavanagh, to be published.
(MSV 70)   G. Michaud, L. Scherk, and E. Vogt, *Phys. Rev.*, **C1**:864 (1970).
(MTW 65)   J. H. E. Mattauch, W. Thiele, and A. H. Wapstra, *Nucl. Phys.*, **67**:1 (1965).
(Mul 67)   E. A. Müller, *Symposium on the Origin and Distribution of the Elements*, Paris (1967).
(MW 70)    J. F. Morgan and D. C. Weisser, *Nucl. Phys.*, **A151**:561 (1970).
(NDA 69)   K. Nagatani, M. R. Dwarakanath, and D. Ashery, *Nucl. Phys.* **A128**:325 (1969).
(NSB 70)   D. B. Nichols, K. Shane, and C. A. Barnes, to be published.
(Opi 51)   E. J. Öpik, *Proc. Roy. Irish Acad.*, **A54**:49 (1951).
(Par 66)   P. D. Parker, *Phys. Rev.*, **150**:851 (1966); *Astrophys. J.*, **145**:960 (1966).
(Par 68a)  P. D. Parker, *Astrophys. J.*, **153**:L85 (1968).
(Par 68b)  P. D. Parker, *Phys. Rev.*, **173**:1021 (1968).
(Par 68c)  P. D. Parker, private communication (1968).
(PBF 64)   P. D. Parker, J. N. Bahcall, and W. A. Fowler, *Astrophys. J.*, **132**:602 (1964).
(Pee 68)   P. J. E. Peebles, *Phys. Rev. Letters*, **16**:410 (1966); and *Astrophys. J.*, **146**:542 (1966).
(Pet 68)   J. G. Peters, *Astrophys. J.*, **154**:225 (1968).
(Pet 70)   A. V. Peterson, Ph.D. Thesis, California Institute of Technology (1970).
(PK 63)    P. D. Parker and R. W. Kavanagh, *Phys. Rev.*, **131**:2578 (1963).
(Pon 67)   B. Pontecorvo, *Zh. Eskp. i Teor. Fiz.*, **53**:1717 (1967).
(Pre 62)   M. A. Preston, *Physics of the Nucleus*, Addison-Wesley, Reading, Massachusetts (1962).
(Puh+ 70)  Pühlhofer, H. G. Ritter, R. Bock, G. Brommundt, H. Schmidt, and K. Bethge, *Nucl. Phys.*, **A147**:258 (1970).
(PW 65)    A. A. Penzias and R. W. Wilson, *Astrophys. J.*, **142**:419 (1965).
(PWS 68)   J. R. Patterson, H. Winkler, and H. M. Spinka, *Bull. Amer. Phys. Soc.*, **13**:1465 (1968); and to be published.
(PWZ 69)   J. R. Patterson, H. Winkler, and C. S. Zaidins, *Astrophys. J.*, **157**:367 (1969).
(Ree 66a)  H. Reeves, in *Stellar Evolution*, Plenum Press, New York (1966); and *Astrophys. J.*, **146**:447 (1966).
(Ree 66b)  H. Reeves, *Astrophys. J.*, **146**:447 (1966).
(RFH 70)   H. Reeves, W. A. Fowler, and F. Hoyle, *Nature* **226**:727 (1970).
(RS 65)    H. Reeves and P. Stewart, *Astrophys. J.*, **141**:1432 (1965).
(Rud 69)   M. A. Ruderman, in *Astrophysics*, W. A. Benjamin, Inc., New York (1969).

(Sal 52)    E. E. Salpeter, *Phys. Rev.*, **88**:547 (1952).
(Sal 54)    E. E. Salpeter, *Australian J. Phys.*, **7**:373 (1954).
(Sal 57)    E. E. Salpeter, *Phys. Rev.*, **107**:516 (1957).
(San 58)    A. R. Sandage in *Stellar Populations* (D. J. K. O'Connell, S. J., ed.) North-Holland Publishing Co., Amsterdam (1958).
(San 67)    R. H. Sanders, *Astrophys. J.*, **150**:971 (1967).
(SC 67)     P. B. Shaw and D. D. Clayton, *Phys. Rev.*, **160**:1193 (1967).
(Sch 51)    E. Schatzman, *Compt. Rend.*, **232**:1740 (1951).
(SFC 65)    P. A. Seeger, W. A. Fowler, and D. D. Clayton, *Astrophys. J. Suppl.* No. 97, **11**:121 (1965).
(SH 67)     M. Schwarzschild and R. Härm, *Astrophys. J.*, **150**:961 (1967).
(She 69)    W. R. Sheldon, *Nature*, **221**:650 (1969).
(SK 63)     P. A. Seeger and R. W. Kavanagh, *Nucl. Phys.*, **46**:577 (1963).
(Smu 65)    P. J. M. Smulders, *Physica*, **31**:973 (1965).
(SS 68)     W. L. W. Sargent and L. Searle, *Astrophys. J.*, **145**:652 (1966).
(SST 70)    M. M. Shapiro, R. Silberberg, and C. H. Tsao, *Acta Physica Hungarica*, **28**, Suppl. 1, 479 (1970).
(Ste 66)    G. J. Stephenson, Jr., *Astrophys. J.*, **146**:950 (1966).
(SU 56)     H. E. Suess and H. C. Urey, *Revs. Mod. Phys.*, **28**:53 (1956).
(SW 70)     D. N. Schramm and G. J. Wasserburg, *Astrophys. J.*, **161** (1970), to be published.
(Swa 70)    C. P. Swann, *Nucl. Phys.*, **A150**:300 (1970).
(Sym 70)    G. Symonds, private communication.
(TA 70)     J. W. Truran and W. D. Arnett, *Astrophys. J.*, **160**:181 (1970), and to be published.
(Tan 65)    N. W. Tanner, *Nucl. Phys.*, **61**:297 (1965).
(Tay 69)    R. J. Tayler, in *Astrophysics*, W. A. Benjamin, Inc., New York (1969).
(TCG 66)    J. W. Truran, A. G. W. Cameron, and A. Gilbert, *Canadian J. Phys.*, **44**: 563 (1966). See also: J. W. Truran, C. J. Hansen, A. G. W. Cameron, and A. Gilbert, *Canadian J. Phys.*, **44**:151 (1966).
(TK 69)     J. W. Truran and B.-Z. Kozlovsky, *Astrophys. J.*, **158**:1021 (1969).
(Toe 70)    J. W. Toevs, Ph.D. Thesis, California Institute of Technology (1970), and to be published.
(Tom 65)    T. A. Tombrello, *Nucl. Phys.*, **71**:459 (1965).
(TP 63)     T. A. Tombrello and P. D. Parker, *Phys. Rev.*, **131**:2582 (1963).
(TSW 70)    T. A. Tombrello, H. M. Spinka, and H. Winkler, to be published.
(Uns 69)    A. O. J. Unsöld, *Science*, **163**:1015 (1969).
(Vog+ 64)   E. W. Vogt, D. McPherson, J. Kuehner, and E. Almqvist, *Phys. Rev.*, **136**:B99 (1964).
(VSS 65)    C. Van der Leun, D. M. Shephard, and P. J. M. Smulders, *Phys. Letters*, **18**:134 (1965).
(WA 66)     E. K. Warburton and D. E. Alburger, private communication.
(Wag 67)    R. V. Wagoner, *Science*, **155**:1369 (1967).
(Wag 69)    R. V. Wagoner, *Astrophys. J. Suppl.* No. 162, **18**:247 (1969).
(WB 67)     C. Werntz and J. G. Brennan, *Phys. Rev.*, **157**:759 (1967); and private communication.
(WD 67)     H. C. Winkler and M. R. Dwarakanath, *Bull. Amer. Phys. Soc.*, **12**:16 (1967).

(Wei 37)    C. F. von Weizsäcker, *Phys. Zeits.*, **38**:176 (1937).

(Wei 38)    C. F. von Weizsäcker, *Phys. Zeits.*, **39**:633 (1939).

(Wei+ 70)   D. C. Weisser, J. F. Morgan, P. Dyer, and C. A. Barnes, to be published.

(WFH 67)    R. V. Wagoner, W. A. Fowler, and F. Hoyle, *Astrophys. J.*, **148**:3 (1967).

(WH 70)     J. C. Wheeler and C. J. Hansen, to be published.

(WKM 69)    W. Whaling, R. B. King, and M. Martinez-Garcia, *Astrophys. J.*, **158**:389 (1969).

(WMT 70)    D. C. Weisser, J. F. Morgan, and D. R. Thompson, *Bull. Amer. Phys. Soc.*, **15**:805 (1970), and to be published.

(WOA 65)    E. K. Warburton, J. W. Olness, and D. E. Alburger, *Phys. Rev.*, **140**:B1202 (1965).

(Yio+ 68)   F. Yiou, M. Baril, J. DuFaure de Citres, P. Fontes, E. Gradsztajn, and R. Bernas, *Phys. Rev.*, **166**:968 (1968).

Chapter 4

# NUCLEOSYNTHESIS AND NEUTRON-CAPTURE CROSS SECTIONS*

## B. J. Allen,† J. H. Gibbons, and R. L. Macklin

*Oak Ridge National Laboratory*
*Oak Ridge, Tennessee*

## ABSTRACT

The role of neutron capture reactions in the nucleosynthesis of heavy elements in stars is reviewed. Maxwellian-averaged capture cross sections at 30 keV are presented for most isotopes across the periodic table. Unmeasured cross sections are obtained by interpolations based on the systematic variation of cross sections with neutron number for odd- and even-$Z$ nuclides. The empirical correlation of capture cross sections and $s$-process abundances is shown, and a consistent set of $r$-process abundances derived.

## 1. INTRODUCTION

Since his earliest beginnings, man has striven to understand his environment, though his tools have been his senses and his laboratory the earth and the stars and skies above. We have not yet departed far from that same condition although we now have extensions for our hands and our eyes that have enabled us, particularly during the last 50 years, to perceive our environment with an incredible sensitivity that links infinitesimal nu-

* Research sponsored by the U. S. Atomic Energy Commission under contract with the Union Carbide Corporation.
† On assignment from Australian Atomic Energy Commission.

clear dimensions with the vastness of the universe. In this paper we review the results of laboratory measurements of nuclear cross sections and their correlations with observed elemental and isotopic abundances, and derive from this some remarkably quantitative information about the specific processes that occurred in the creation of the elements.

The historical route to understanding of heavy element nucleosynthesis has several converging branches. During the 1930's the discovery of explicit nuclear burning processes as the energy source for stars led to stellar models which in turn delineated conditions such as internal temperature. It was quickly apparent that in ordinary stars the Coulomb barrier precluded significant element creation beyond medium weight nuclei. Concurrently the slowly amassing body of data on terrestrial and solar elemental abundances gave conclusive evidence that abundances, with a prominent peak near iron, are not correlated with chemical properties and therefore must be correlated with some physical properties.

The explosive growth in nuclear-structure physics during the war years of the early 1940's resulted in a wealth of empirical data on nuclear reaction rates. In particular, and for the first time, cross sections for both slow and fast neutrons on a variety of heavy nuclei became available (Hug 46, HGL 53). These were activation cross sections for a large variety of elements exposed to fission-produced neutrons ($E_n \approx 1$ MeV). Alpher (Alp 48), in collaboration with Gamow, immediately observed that there is an approximately inverse relationship between neutron-capture cross sections and the relative elemental abundances found in the solar system (Fig. 1). From this there emerged a nonequilibrium model for element formation, involving a neutron gas, at high temperature, from which all of the elements were formed in an interval of about 15 minutes by successive neutron captures. The assumption was made that essentially the entire process of element

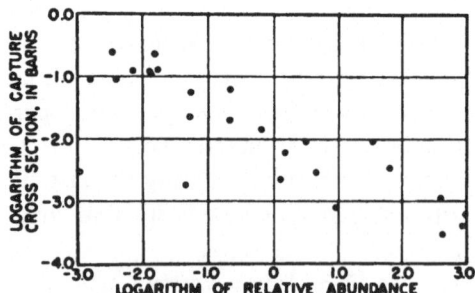

Fig. 1. First correlations between fast capture
cross sections and abundance (Alp 48).

formation was completed in the primordial "big bang" at the beginning of the pregalactic phase of the universe. It was established that the average neutron temperature was probably greater than $10^3$ eV, since correlations between capture cross sections and abundance do not appear for "thermal" neutrons (energy of $\sim 0.025$ eV) or neutrons in the 1–1000 eV range. Alpher estimated that the average neutron energy probably corresponded to about $10^9$ °K (or about 100,000 eV). He admitted that there were some weak points in this theory. For example, he pointed out that some isotopes are "shielded" by other stable isotopes with equal weight but lower $Z$, and thus could not have been produced in such a rapid chain capture process. Also, the lack of a stable nucleus at mass 5 or mass 8 meant that simple neutron-capture chains could not have been solely responsible for element building. In short, the assumption of element formation by a purely "rapid" process of neutron capture was not entirely satisfactory, but *an intimate connection between neutron capture and element synthesis was clearly established.*

The elemental abundances used in examining these correlations were admittedly uncertain by large factors (up to an order of magnitude) and cross section values were relatively crude, so the challenges for further experimental work were obvious.

The monumental work of Goldschmidt (1937) in solar-system elemental abundances (Gol 37) was followed by Brown in 1949 (Bro 49) and by Suess and Urey in 1956 (SU 56), based on measurements in the earth's crust and also on data from meteorites and the atmosphere of the sun, which revealed subtle relationships between elemental abundances and atomic weight. It became increasingly clear that the elements were formed on the basis of their nuclear properties, and that our world bears clear signs of being the collective ashes of what Suess and Urey labeled a "cosmic nuclear fire," in which the matter of our sun and solar system was created.

Two observations in the 1950's helped set the stage for the next advance in our knowledge of the role of neutron capture in stellar nucleosynthesis. In 1952, Merrill (Mer 52) discovered atomic absorption lines of technetium in the atmosphere of type-S stars (red giants). Since technetium has a half-life of less than a million years, this observation proved conclusively that significant neutron capture does occur in these stars and that nucleosynthesis by neutron capture is a dynamic continuing stellar process. In 1956, a rough correspondence between the spontaneous fission half-life (56 days) of californium-254 and the characteristic (approximately exponential) decay time (50 to 60 nights) of light from type-I supernova explosions was noted (Bur+ 56). This suggested that rapid multiple neutron capture occurs in stellar explosions such as supernovae. Even though the hypothesis of $^{254}$Cf spontaneous fission as the main energy source for supernovae decay characteristics is probably inadequate (MS 66), we know that some such process *must* have occurred in the case of our solar-system material, since we have a significant natural abundance of thorium and uranium. These elements could only have been produced in some rapid process, since rather short-lived elements (for example, astatine, whose longest-lived isotope has a half-life of 1 minute) lie between them and their stable progenitors.

## 2. DEVELOPMENT OF NEUTRON BUILDUP THEORY

In 1957, Burbidge, Burbidge, Fowler, and Hoyle (Bur+ 57) and also Cameron (Cam 57) integrated all of the new ideas and information on element formation into a coherent picture. With regard to heavy-element synthesis, they incorporated Gamow's basic idea of neutron capture but made at least three important modifications:

1) The location of element synthesis was placed in stellar interiors and in violent stellar explosions.

2) Charged-particle reactions were recognized as primarily responsible for production of elements up to iron; for production of elements heavier than iron, neutron capture was considered the predominant mechanism.

3) Two quite different and independent neutron capture processes were assumed. The first of these, the rapid process, or r-process, occurs on a rapid time scale (up to 100 neutron captures in 1 to 100 sec) and builds up neutron-rich isotopes which subsequently undergo beta decay until they achieve stability. The site of r-process isotope production was thought to be supernovae or some other cataclysmic explosion. In the second process, called the slow- or s-process, the end products of charged-particle reactions near maximum nuclear binding are the assumed starting material for the neutron-capture synthesis mechanisms. The most abundant is $^{56}$Fe. In s-process synthesis, the $^{56}$Fe "raw material" is immersed in a low-density sea of neutrons whose average energy is 2 to $3 \times 10^8$ °K (several tens of keV). The neutron-production mechanisms are discussed later in this paper. Neutrons are captured at intervals long ($10^3$–$10^5$ years) compared to radioactive lifetimes of neutron-rich isotopes and thus the capture chain follows the valley of $\beta$ stability. Beyond $^{209}$Bi, alpha decay terminates the process.

The r-process path lies far to the neutron-rich side of the stable elements. While most nuclides are formed by both of these processes, certain ones are formed solely by one process. There are some heavy stable isotopes whose next-higher isotope (containing one more neutron) has a relatively short half-life ($\ll 10^3$ years). According to the s-process hypothesis, in these cases the next-heavier isotope cannot be produced since beta decay would occur before the second neutron was absorbed. Likewise, some stable nuclides exist that are "shielded" against r-process production because a stable nucleus with the same atomic weight but lower $Z$ terminates the chain of beta decays after completion of the r-process for that particular atomic weight (see Fig. 2).

That different processes have been operative in the nucleosynthesis of the heavy elements is apparent from Fig. 3 (Cam 68). Solar-system abundance distributions are shown for s- and r-process nuclides. Also included is the distribution of proton-rich (p) nuclides which are bypassed by both s and r neutron capture reactions and are much less abundant. These nuclides are probably formed by proton capture, positron capture, or photodisintegration reactions and subsequent decay (ITO 61).

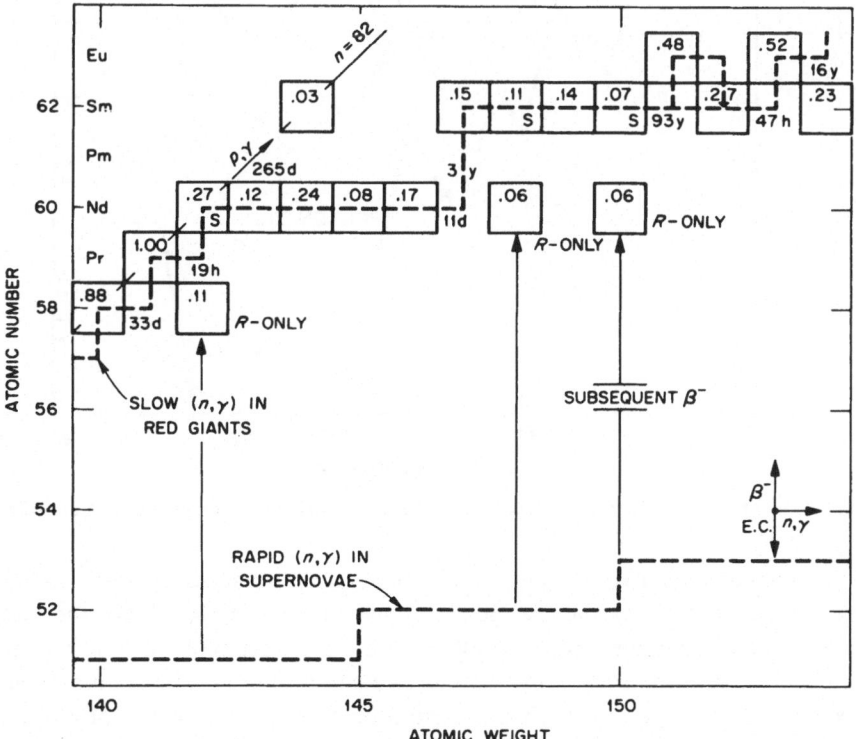

Fig. 2. The s-process path in samarium. ¹⁴⁸Sm and ¹⁵⁰Sm are due only to the s-process as they are shielded from r-process contributions by the corresponding isotopes of Nd.

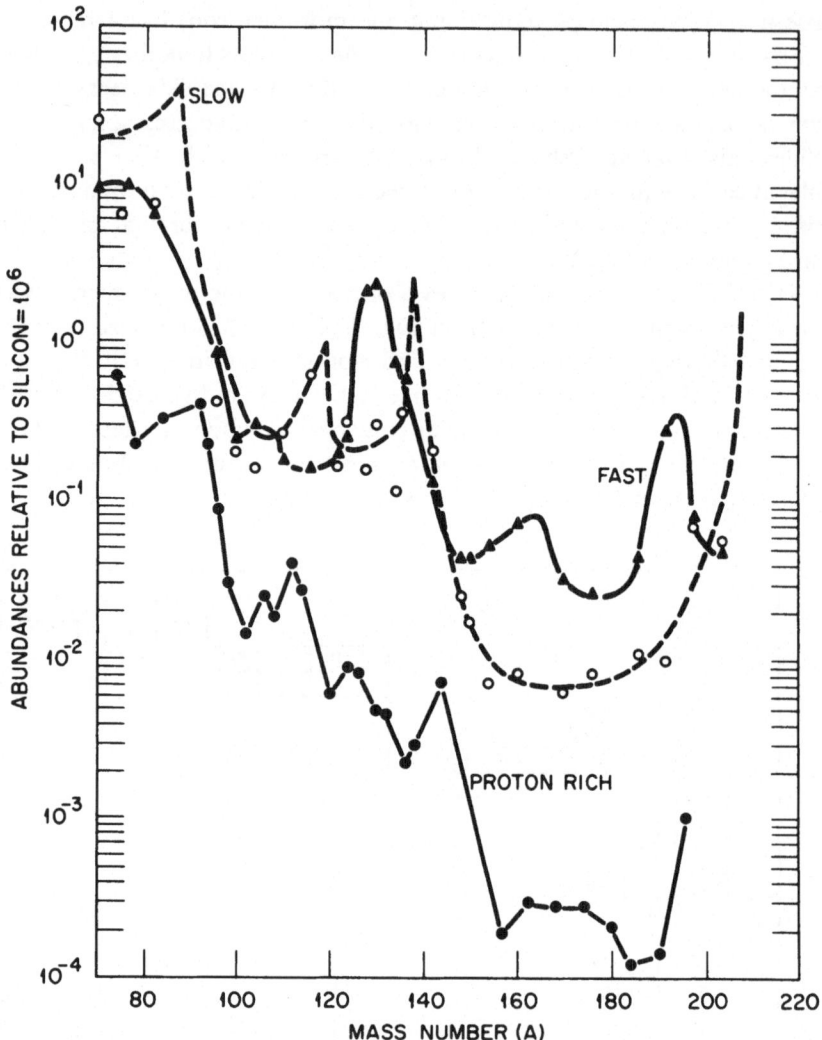

**Fig. 3. Solar system abundance distributions for *s*-, *r*- and *p*-process nuclides (Cam 68).**

The curve drawn through the *r* isobar abundances is remarkably smooth, in contrast to the scatter about the *s* isobar curve. Hence the fast time scale abundances are not affected by the variability of individual neutron-capture cross sections. Abundance smoothing processes must have been operative, either through a contribution from several values of atomic number *Z* to each mass number, or through frequent neutron emission following the high-energy beta decays of the neutron-rich final products of the capture process. In addition, the neutron-capture process must have terminated quite abruptly; otherwise fast beta decays would have produced only one capture product per mass number, and the final abundances would have been affected by cross section variations.

The publication of (Bur+ 57) gave strong impetus to experimentalists to measure neutron-capture cross sections in the energy range corresponding to a Maxwellian temperature of 10 to 30 keV because of the prediction of a specific inverse proportionality between cross section and abundance for certain neighboring isotopes. Such a correlation was first considered by Cameron (Cam 55) and Fowler, Burbidge, and Burbidge (FBB 55). These authors further predicted that the product of cross section times abundance should be a smoothly varying function of atomic weight. The existing cross section data were relatively scarce because there were few strong sources of neutrons in the keV energy range and capture detection techniques (activation and spherical-shell transmission) were not widely applicable. A significant advance occurred by the dual development of pulsed-beam time-of-flight neutron techniques and of large fast liquid scintillators that detected a neutron-capture event by the promptly ensuing gamma rays (DTH 60, Gib+ 61). The list of measured capture cross sections in the keV range thus grew extensively in the late 1950's and early 1960's, but mostly consisted of natural element cross sections since large ($\sim$one mole) samples were normally required.

Even on the basis of such measurements of natural elements, the s-process correlations between abundance ($N_s$) and cross section gave a clear indication (Cla+ 61) that this hypothesis should be taken seriously and that measurements more crucial to the hypothesis should be undertaken (Fig. 4). The curve of Fig. 4(a), composed mainly of nuclei produced by the s-process (s-process nuclei), shows a general clustering of points along an empirical curve, but the curve of Figure 4(b), composed mainly of r-process nuclei, shows no evidence of any such correlation.

The degree of scatter of data points in Fig. 4(a) is small enough to be readily attributed to uncertainties in relative elemental abundances (due to chemical and physical fractionation of the primitive solar-system material) and also to the fact that specific isotopic cross sections were not available. Burbidge et al. (Bur+ 57) pointed out that highly quantitative testing of the s-process hypothesis would be possible only when cross sections of individual isotopes could be measured. Nevertheless the empirical correlation [Fig. 4(a)] was quite revealing and was used by (Cla+ 61) to develop a model for the neutron-flux distribution required to generate the observed correlation.

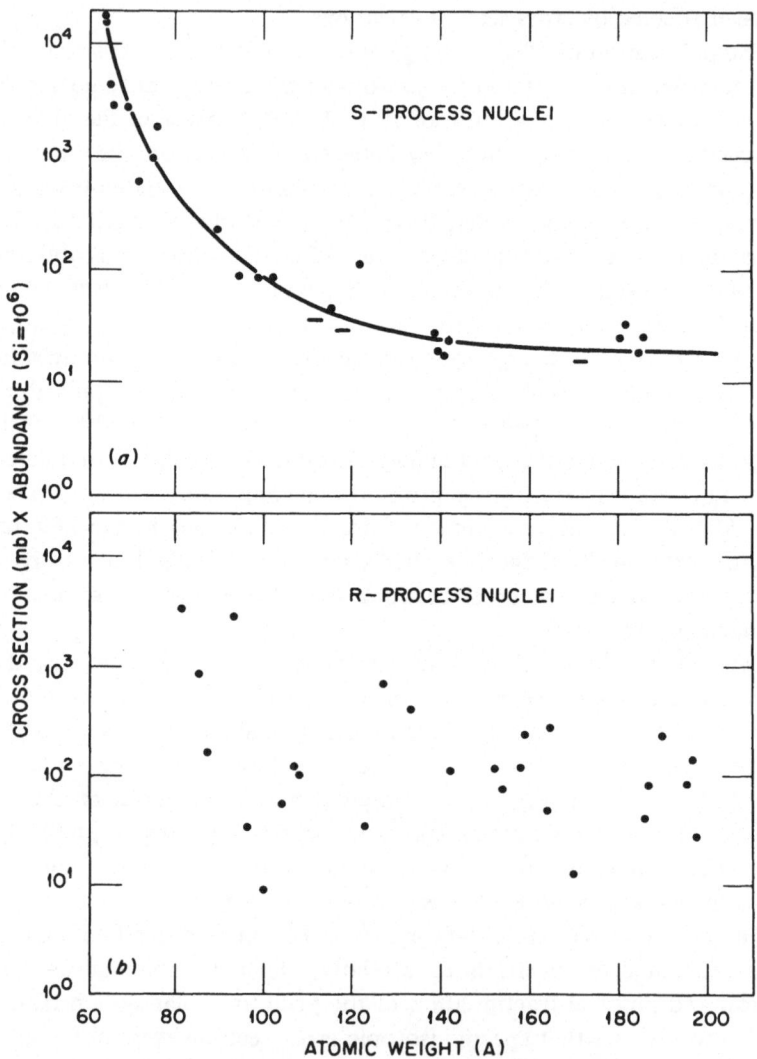

Fig. 4. Correlations between cross sections and abundance (a) for s-process nuclei as a function of atomic weight and (b) r-process nuclei. This figure shows the state of knowledge 13 years ago (Bur 57).

The rate of change of the number of $s$-process nuclei of weight $A$, $N_A(t)$, depends upon its capture rate $\langle \sigma v \rangle_A$ (see Section 4.2), the neutron density $n(t)$, and the abundance and capture rate of its lighter neighbor $(A - 1)$. That is,

$$\frac{dN_A(t)}{dt} = \langle \sigma v \rangle_{A-1} n(t) N_{A-1}(t) - \langle \sigma v \rangle_A n(t) N_A(t)$$

If we define the average capture cross section for weight $A$ at temperature $T$ as $\sigma_A = \langle \sigma v \rangle_A / v_T$ and define the integrated neutron flux time $\tau$ (a measure of the total accumulated neutron bombardment per unit area) as

$$\tau = \int_0^t n(t') v_T \, dt' \quad \text{and} \quad v_T \equiv (2kT/m)^{1/2}$$

then on change of variable we have

$$\frac{dN_A}{d\tau} = \sigma_{A-1} N_{A-1} - \sigma_A N_A$$

This equation assumes that the $\beta$-decay lifetimes of unstable isobars at $A$ are short compared to the mean time between neutron captures. Clayton, Fowler, Hull, and Zimmerman (Cla+ 61) showed that $N_s \sigma(A)$ (the subscript henceforth denoting $s$-process abundances) should be a smooth slowly varying function of atomic weight. These authors also showed that the observed distribution of $N_s \sigma$ products could not have been produced by a uniform exposure of $^{56}$Fe to a single neutron flux. Instead an exposure to a distribution of flux histories was called for by the data. Let the $N_s \sigma$ product per initial $^{56}$Fe nucleus that results from a neutron exposure $\tau$ be

$$\psi(\tau) \equiv N_s \sigma(\tau)/N_0$$

where $N_0$ is the total initial nuclei. Then, if $\varrho(\tau) \, d\tau$ is the number of Fe nuclei (per $10^6$ Si atoms) exposed to an integrated flux between $\tau$ and $(\tau + d\tau)$ the $N_s \sigma$ products produced are

$$N_s \sigma(A) = \int_0^\infty \varrho(\tau) \psi(\tau) \, d\tau$$

Thus the function $N_s \sigma(A)$ is a measure of the integrated flux-time to which seed nuclei have been exposed. Given capture cross sections and abundances one should be able to deduce $\varrho(\tau)$, the "history" of solar-system $s$-process material synthesis.

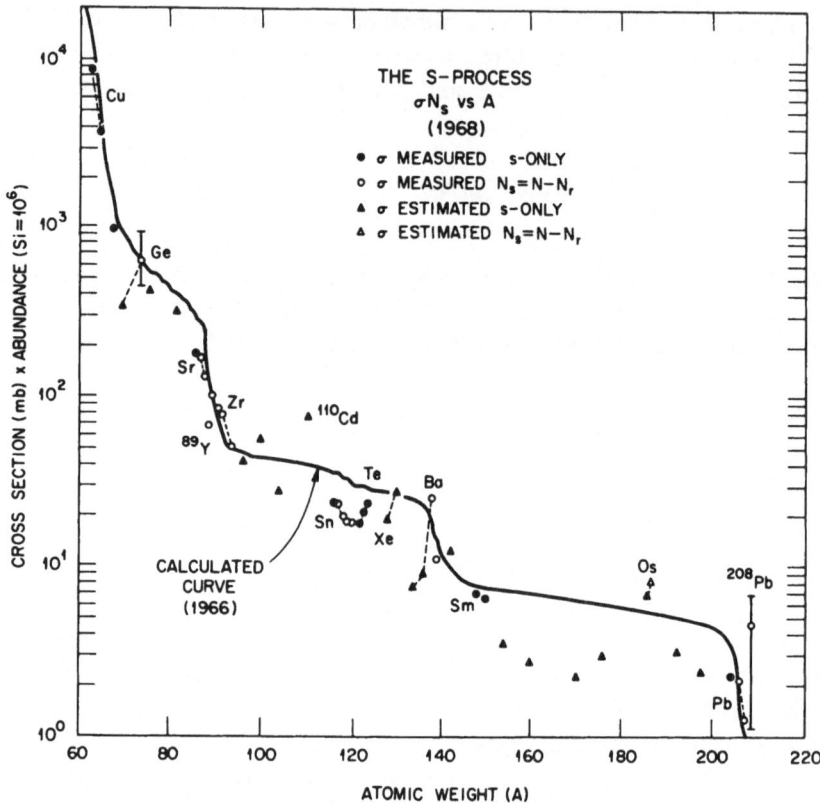

**Fig. 5.** $N_s\sigma$ **Correlations at early 1968.** The curve is that calculated for exposure to an integrated neutron flux given by $\varrho(\tau)\alpha\tau^{-3/2}$ with a cutoff at $\tau_{max} = 1.35 \times 10^{27}$ neutrons cm$^{-2}$.

Since capture cross sections have low minima for certain regions of atomic weight, the calculated shapes of $N_s\sigma(A)$ as a function of atomic weight are characterized by a ledge–precipice structure (see for example, Fig. 5). This structure is clearly present for any continuously decreasing flux distribution. A distribution of integrated neutron flux exposures, found to give a remarkably good fit to the data, was a simple power law (SFC 65). The calculation of Seeger and Fowler (SF 66) incorporates a cutoff term and is compared in Fig. 5 with the best data (Fow 68) available prior to this publication (see Section 3 for a more detailed discussion). Uncertainties in relative abundances, combined with a persistent paucity of cross section data permitted few quantitative conclusions about the s-process, but the panorama of the origins of heavy elements was beginning to unfold. The characteristic breaks in $N_s\sigma(A)$ at neutron magic numbers are a direct consequence of nuclear-shell effects on capture cross sections and their appearance in the observed distribution is a clear "fingerprint" of the synthesizing mechanism.

The $N_s\sigma$ data in Fig. 5 are relevant only to our own solar system; Danziger (Dan 66) has pointed out that elemental distributions in other parts of the galaxy can differ significantly. These abundance distributions have been measured by observing atomic absorption lines from stars. The $N_s\sigma(A)$ distributions (Fig. 6) for these stars reveal different s-process histories, as is evident from the substantial variation in the ledge–precipice structure. Those stars with the largest discontinuity, near $A = 140$, are found to be the most deficient in the heaviest elements, in agreement with s-process expectations. The recent observation (JM 70) of stars rich in the heavy elements (Os, Pt, Au, and U) in the 73 Draconis group indicates that different r-process histories can also occur elsewhere in the galaxy. It is therefore evident that the conditions for nucleosynthesis can vary widely across the galaxy.

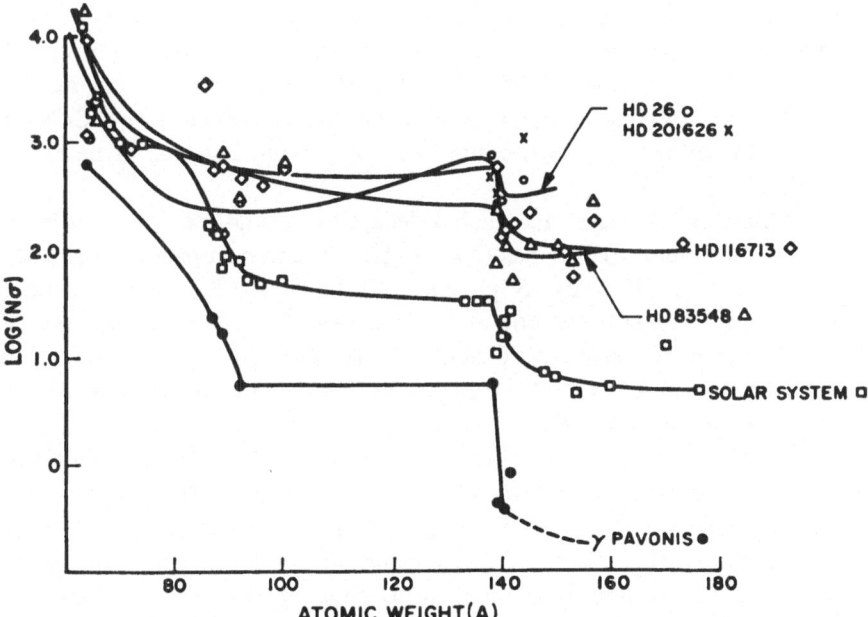

Fig. 6. $N_s\sigma$ **Correlation in other stars.** Elemental abundances differ considerably, but the characteristic ledge–precipice structure, with breaks corresponding to shell effects on neutron capture, are clearly evident. $\gamma$ Pavonis, which is the most deficient in the heaviest elements, shows the largest discontinuity near $A{\sim}140$, as predicted by s-process theory.

## 3. STELLAR NEUTRON SOURCES

Neutrons are produced by charged-particle reactions at most stages of stellar evolution. However significant fluxes are available only under certain conditions of temperature and class of star. In this section we consider the probable neutron sources that contribute to the s-process.

In the initial interstellar gas, the deuterium component ($\sim 10^{-4}$ relative to $^1$H) can produce neutrons via the D($d, n$)$^3$He reaction. However other reactions compete, i.e., D($p, \gamma$)$^3$He and D($d, p$)$^3$H and only 0.1 neutrons per deuterium nucleus are produced. In terms of the abundance of iron seed nuclei, the foundation stone of the s-process, one neutron per Fe seed is available. However the production of $^3$He acts as a neutron poison via $^3$He($n, p$)$^3$H and further reduces the neutron flux, which becomes effectively negligible.

Neutrons are not produced in the initial hydrogen-burning phase ($p, p$) but the subsequent C–N–O bi-cycle (Bur+ 57) yields $^{13}$C which produces neutrons after $\alpha$-capture, i.e., $^{13}$C($\alpha, n$)$^{16}$O, at a temperature of $10^8$ °K (i.e., $T_8 = 1$). However $^{14}$N acts as a neutron poison in this instance as neutrons are lost in the reaction $^{14}$N($n, p$)$^{14}$C, and the net production rate is again estimated to be negligible. At the beginning of He burning ($T_8 > 2$), all $^{14}$N has been transformed to $^{18}$O, and should a helium flash occur at this time (causing a mixing of outer H with the He core) then a new C–N–O cycle would commence which could become a prolific source of neutrons (Ree 68).

He burning produces $^{18}$F, which $\beta$-decays to $^{18}$O and $\alpha$-captures to $^{22}$Ne. The reaction $^{22}$Ne($\alpha, n$)$^{25}$Mg has $Q = -482$ keV and occurs at a sufficiently low temperature ($T_8 > 2$) to prevent $^{18}$O burning. This reaction rate, calculated by Reeves (Ree 66) and experimentally confirmed by Ashery (Ash 69) can produce about 10 neutrons per Fe seed.

Carbon burning, $^{12}$C($^{12}$C, $n$)$^{32}$Mg, is estimated to be a major source of neutrons at higher temperatures ($T_8 = 9$–11), producing 5 to 25 neutrons per Fe seed (Ree 68). Oxygen burning, $^{16}$O($^{16}$O, $n$)$^{31}$S, is more prolific still and yields approximately 100 neutrons per Fe seed at $T_8 = 14$ to 20.

Other processes are possible which yield high neutron fluxes but occur at temperatures which would destroy heavy nuclei at the same time (e.g., supernovae). A summary of neutron producing reactions is shown in Table I.

Further information pertinent to the neutron source for the s-process can be gained from the $N_s\sigma$ correlations. Clayton et al. (Cla+ 61) recognized that it was not possible to fit the $N_s\sigma$ data by a single exposure of the Fe

**TABLE I. Stellar Neutron Sources**

| Temperature, $T_8 = 10^8 \,°K$ | Reaction | Yield, neutrons/Fe seed |
|---|---|---|
| | $D(d, n)$ | $< 1$ |
| 1 | $C^{13}(\alpha, n)$ | $\begin{cases} < 1 \text{ (no mixing)} \\ \leq 2000 \text{ (mixing)}^a \end{cases}$ |
| $>2$ | $^{14}N \rightarrow {}^{22}Ne(\alpha, n)$ | 10 |
| $>3.5$ | $^{14}N \rightarrow {}^{18}O(\alpha, n)$ | — |
| 9–12 | $^{12}C({}^{12}C, n)^{23}Mg$ | 5–25 (Pop I) |
| 14–20 | $^{16}O({}^{16}O, n)^{31}Si$ | $\sim 100$ |
| 10–30 | Supernovae | |

$^a$ (San 67).

seed to a fixed neutron flux. A superposition of exposures was required:

$$N_s\sigma(A) = 4 \times 10^{-3}\, \psi(n_c = 3) + 10^{-3}\, \psi(n_c = 7) + 10^{-4}\, \psi(n_c = 35) + 10^{-5} \times \psi(n_c = 100)$$

where $\psi(n_c)$ is the $N_s\sigma$ distribution per Fe seed for $n_c$ neutron captures per Fe seed. This result remains relevant today (although the $N_s\sigma$ data have been updated), in that (a) the probability for high neutron exposures is low and (b) less than 1% of the Fe seed nuclei are involved in the s-process.

Seeger et al. (SFC 65) found the $N_s\sigma$ curve could be fitted by a continuous distribution of exposures of the form:

$$\varrho(\tau) = 0.02 \exp(-\tau/0.17)$$

where $\varrho(\tau)\, d\tau$ is the number of Fe seed nuclei exposed to an integrated neutron flux between $\tau$ and $d\tau$. With improved $N_s\sigma$ data, Seeger and Fowler (SF 66) obtained a better fit (shown in Fig. 5) for $A \sim 200$ with a maximum exposure cutoff:

$$\varrho(\tau)\alpha\tau^{-3.2}, \qquad \tau_{\text{max}} = 1.35 \times 10^{27} \text{ neutrons cm}^{-3}$$

Whereas the $N_s\sigma$ curve indicates the occurence of multiple neutron exposures, there is no evidence as to where and when they took place. The s-process requires a stable period of stellar evolution, and the observation of technetium in S-type stars (Mer 52) at temperatures of $T_8 \sim 2$ suggests that either the $^{13}C(\alpha, n)$ or $^{22}Ne(\alpha, n)$ reaction is the major neutron source (see Table I).

Peters (Pet 68) has attempted to calculate different exposures within a single star, using as the neutron source the $^{22}Ne(\alpha, n)$ reaction which has been studied by Davids (Dav 68). Different neutron fluxes are postulated at the central core, where a high exposure results from the s-process occurring at longer times and higher temperatures, and at the outer shell of the core, where a low exposure results from a short time period and low temperature. Different exposures could be obtained by mixing outer sections of the s-processed core with the stellar envelope. However, the various exposure distributions obtained could not reproduce the $N_s\sigma$ distribution, and it appears unlikely that the $^{22}Ne(\alpha, n)$ reaction is the sole neutron source during He burning. It was possible to obtain a satisfactory fit by requiring an initial mass system which decreases rapidly with $A$. The implication then is that a small fraction of high-exposure material is present in the interstellar medium prior to condensation of the star. Thus, the $N_s\sigma$ correlations may result from second or third generation s-processes.

Sanders (San 67) has proposed that the $^{13}C(\alpha, n)$ reaction is the major neutron source, as a result of mixing of the He-burning shell (surrounding an inert $^{12}C$ core) with the hydrogen envelope. The s-process mechanism then operates on the Fe seed already present when the star condensed from the interstellar medium. However, it is the young Population I stars, like the sun, which utilize the Fe seed ejected by the much older Population II stars in slow or rapid explosions. The more massive members of the population II class ($M \gg M_0$) burn fast and end their days as supernovae, leaving less massive and slower burning stars which characterize the observed population.

# 4. NEUTRON CAPTURE CROSS SECTIONS

## 4.1. Techniques in the Measurement of keV Capture Cross Sections

A variety of methods are available for the measurement of capture cross sections: activity, prompt gamma rays, absorption (shell-transmission method), modification of pile reactivity, and mass spectrometry of product nuclei. However, the bulk of measurements have been made by either the activity or prompt gamma-ray technique. Unfortunately, agreement is poor in many cases, and well outside experimental errors; and it is clear that problems of normalization and background are responsible for many of these discrepancies.

### 4.1.1. *Sources of Fast Neutrons*

Many activity and shell-transmission measurements have been made with Sb-$\gamma$-Be photoneutron sources which yield primarily neutrons with energy 22.8 $\pm$ 1.0 keV (RB 67, Sch 60), and with varying degrees of energy degradation depending on source construction. A small higher-energy neutron group is also present.

The slowing-down-time spectrometer is also used as a source of keV neutrons. An 8-m³ lead block serves to moderate a pulse of fast neutrons from the T($d, n$) reaction (IPS 60). However, the use of time-of-flight methods with ($p, n$) reactions in pulsed charged-particle accelerators or from ($\gamma, n$) production with linacs yields the most reliable keV neutron flux with the simultaneous measurement of low-energy backgrounds. The pulsed Van de Graaff accelerator has been used extensively as a neutron source employing the Li($p, n$) reaction to generate bursts of keV neutrons, as well as a continuous source of neutrons with a variable sharp cutoff at both upper and lower energies. For protons incident on a lithium target just above the ($p, n$) threshold, the center-of-mass motion throws the total reaction yield into a forward cone, resulting in an intense kinematically collimated source of 30 keV neutrons, the most prolific monoenergetic source in the keV range.

### 4.1.2. *Prompt Gamma-Ray Detectors*

In recent years, there has been a continuing development in methods for the detection of prompt capture gamma rays. Because this technique is used in conjunction with time-of-flight analysis of neutron energy, fast timing as well as high efficiency is called for. A further requirement is for the detection efficiency to be independent of the capture gamma-ray spectrum as this can vary substantially with neutron energy. The large liquid scintillator tank, while highly efficient, suffers from the problem of the spectrum fraction below the relatively high discriminator level (1 to 3 MeV) required to reject background. Uncertainties with regard to the use of small ($\sim$1/2 m diameter) tanks (e.g., sensitivity to changes in capture $\gamma$-ray spectrum) have been discussed recently in the literature (Kom 69, Fri +68, Mac 61).

The Moxon–Rae detector (MR 63) was devised to yield an efficiency proportional to the energy of the incident gamma ray. For a capture gamma-ray cascade, then, the efficiency per capture is proportional to the neutron binding energy (plus the neutron center-of-mass energy), and independent of the capture spectrum. The most recent development has been the total-energy detector (TED) where the low efficiency of the Moxon–Rae converter is exchanged for a computer weighting of pulse heights from a fast nonhydrogenous liquid scintillator (MG 67a). The discriminator level can be set very low ($\sim$150 keV) and all events are weighted according to the energy deposited in the detector. The method, therefore, necessitates a two-parameter experiment, with either analog or digital on-line computer weighting of capture events. The efficiency, including solid angle, of two 5 inch detectors in close proximity to a target can be as high as 20%, and a time resolution of about 1 nsec can be achieved by the use of fast cross-over timing.

Earlier methods of measuring cross sections, such as the large liquid scintillator (Gib+ 61), required about one mole of the material under study but samples with good isotopic purity are very expensive to produce and consequently are normally available in small fractions of a mole. The challenge, therefore, was to increase the detection sensitivity. This was done by making the neutron source more intense (klystron bunching of the Van de Graaff proton beam) and improving the time-of-flight resolution of the capture gamma-ray detector to permit a much shorter neutron flight path (MGI 63a). Together with use of the total-energy detector sufficient sensitivity became available to obtain satisfactory results with 0.1 mole samples of low cross section material. A typical experimental arrangement is sketched in Fig. 7.

## 4.2. Maxwellian-Averaged Cross Section at 30 keV

Stellar neutrons, produced by the various charged particle reactions, can reach kinetic equilibrium at the local temperature through elastic scattering before capture. The effective temperature of a red giant star, estimated at 2 to $3 \times 10^8$ °K (i.e., 20–30 keV), results in a thermalization time for a nascent neutron of $10^{-11}$ sec, and is short compared to the mean lifetime for capture.

While there is some uncertainty as to the actual stellar temperature, averaged capture cross sections are relatively independent of Maxwellian temperature for most nuclides between 10 and 100 keV. From the viewpoint of a compilation it is therefore sufficient to choose the most convenient and common energy, and this is found at 30 keV. Neutrons with this energy are generated in the Li($p, n$) reaction at threshold, and a large number of measurements have been made using this reaction with a pulsed Van de Graaff accelerator. Many activation measurements have also been made at

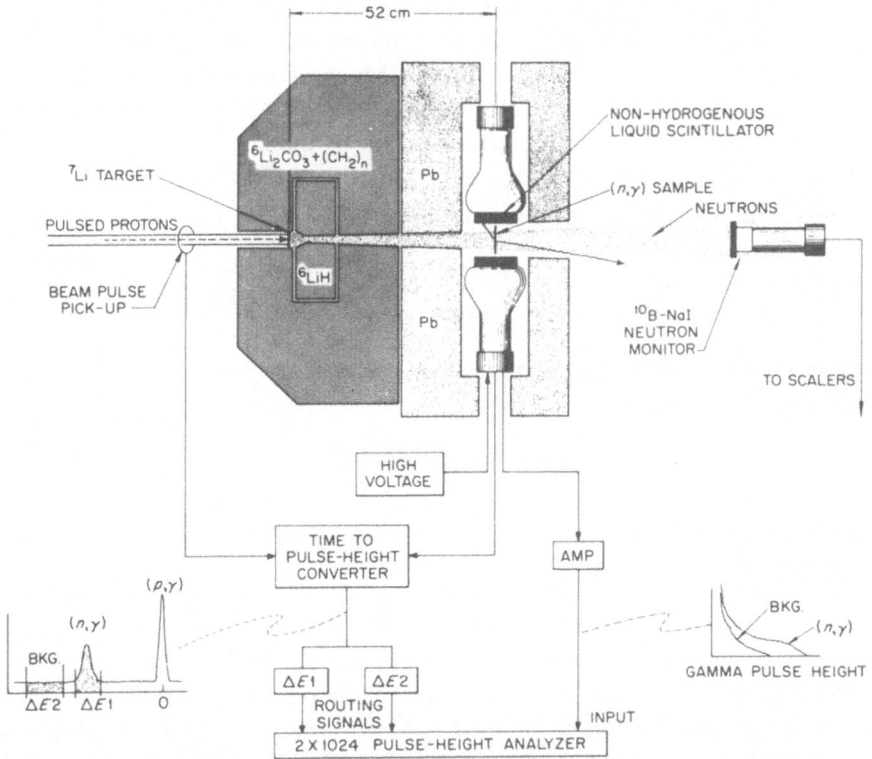

**Fig. 7. Experimental arrangement for small sample capture cross section measurements.** Nanosecond bursts of neutrons from the $Li(p, n)$ reaction permit discrimination of capture events in time and consequent reduction in background.

23 keV with Sb–$\gamma$–Be sources, and these results have been extrapolated to 30 keV by a $\sim 15\%$ reduction in capture cross section. This reduction is typical of the observed energy dependence for a number of capture cross sections and is close to a $1/v$ dependence. The uncertainty introduced by this extrapolation is generally less than the experimental errors of the activation measurements.

Actually, the energy dependence of neutron capture is so closely uniform that the $s$-process model is insensitive to the precise average neutron energy chosen, as long as it is consistent (with the exception of a few nuclei with large resonance spacing). This is unfortunate since significantly different dependences could lead, through cross section–abundance correlations, to a determination of the average neutron temperature that must have prevailed during the nucleosynthesis of the heavy nuclides.

Under the conditions of temperature ($3 \times 10^8$ °K) and density ($10^4$ g-cm$^{-3}$) as found in the stellar interior, the relative velocity between the neutron and target is determined by the Maxwell–Boltzmann distribution and the reaction rates are therefore weighted averages of a product of the relative velocity $v$, the capture cross section $\sigma$, the abundance $N$ for each target isotope, and the local free neutron density (Cla+ 61).

The Maxwellian weighting function is

$$\Phi(v)\, dv = \frac{4}{\pi^{1/2}} \left(\frac{v}{v_T}\right)^2 \exp\left[-\left(\frac{v}{v_T}\right)^2\right] \frac{dv}{v_T}$$

where $v_T = (2kT/m)^{1/2}$ and $m$ is the reduced mass. The Maxwellian-averaged capture rate is defined as

$$\langle \sigma v \rangle = \int_0^\infty \sigma v \Phi(v)\, dv$$

and the Maxwellian-averaged cross section is given by

$$\sigma = \langle \sigma v \rangle / v_T$$

Capture cross section data are categorized on the basis of whether or not individual resonances can be experimentally resolved. In the latter case, an averaged cross section is obtained at 30 keV which is virtually equivalent to the Maxwellian averaged cross section for $kT = 30$ keV. However in the case of resolved resonance data it is necessary to integrate the resonance capture areas (in barn-eV) over the Maxwellian neutron distribution. As the resonances are very narrow compared with $kT$, their contribution to capture can be treated as a series of delta functions superimposed on a small $1/v$ component.

In terms of resonance parameters the capture area of an isolated resonance is

$$A_c = (2\pi^2/k^2)g_J(\Gamma_n\Gamma_\gamma/\Gamma) \text{ barn-eV}$$

where $k^2 = 0.004818\,[A/(A + 1.008956)]E_r b^{-1}$, $A$ is the atomic weight of the target nucleus, $E_r$ is the resonance neutron energy (lab. keV), and $g_J = (2J + 1)/2(2I + 1)$ with $J$ the compound state spin and $I$ the target spin. The resonance width $\Gamma$(eV) is generally just the sum of the radiative width $\Gamma_\gamma$ and neutron width $\Gamma_n$ in the energy region of interest. As the peak total cross section of a resonance is

$$\sigma_0 = (4\pi/k^2)g_J\Gamma_n/\Gamma$$

then,

$$A_c = \frac{\pi}{2} \sigma_0 \Gamma_\gamma$$

The $1/v$ components attributable to the combined effect of distant resonances are evaluated from the thermal (0.0253 eV) capture cross section ($\sigma_{\mathrm{th}}$). The numerical evaluation procedure at $kT = 30$ keV (derived from the preceding integral equation) is

$$\frac{\langle \sigma v \rangle}{v_T} = \sigma_{\mathrm{th}} \left( \frac{25.3 \times 10^{-6}}{kT} \right)^{1/2} + \frac{2}{\pi^{1/2}} \sum_r A_c(r) \frac{E_r}{(kT)^2} \exp - \left( \frac{E_r}{kT} \right)$$

$$= 9.18 \times 10^{-4} \sigma_{\mathrm{th}} + 1.128 \left[ \sum_r A_o(r)(E_r/900) \exp(-E_r/30) \right] \quad \text{millibarns}$$

for $\sigma_{\mathrm{th}}$ in mb, $kT$ and $E_r$ in keV, and $A_c$ in barn-eV.

## 4.3. Comments on Cross-Section Data

It is apparent from the capture cross sections shown in Fig. 8 and 9 that the data show separate systematic trends with neutron number for even-and odd-$Z$ nuclei. The odd-$Z$ nuclei, generally limited to two isotopes, exhibit pronounced minima at the magic neutron numbers $N = 50$, 82, and 126. A similar effect is observed for the even-$Z$ nuclei, though the situation is more complex as a result of the large number of isotopes. Many cross section measurements have been made across the periodic table, and most odd-$Z$ nuclides have been measured at least once. However there are many gaps in the even-$Z$ isotopes, and semiempirical estimates are therefore required. The capture cross sections, listed in Table II, were compiled up to January, 1970.

### 4.3.1. *Odd-Z Nuclei*

Isotopic cross sections are plotted where possible, and many of these are the average of a number of measurements. The cross sections are seen to vary quite smoothly over about three orders of magnitudes. Results of measurements (mainly activation) for the isotopes [63,65]Cu; [71]Ga; [185,187]Re; elements Ho and Re appear to be in substantial disagreement with more accurate data and have not been included in any average result nor in the figure. However these data are shown in the table, and are bracketed beside the accepted values.

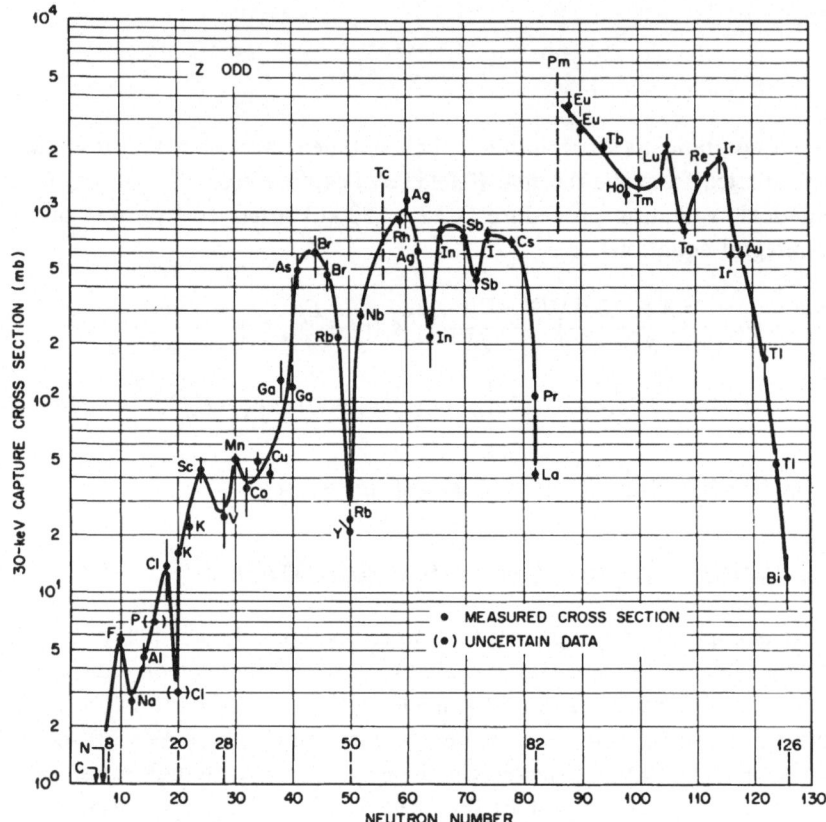

**Fig. 8. 30-keV Maxwellian-averaged capture cross sections for odd-$Z$ nuclides.** Neutron shell effects are clearly apparent.

In contrast to the marked minima at $N = 50$, 82, and 128, the magic neutron number of 28 appears to have little effect on keV cross sections. Secondary minima at $Z = 49$ and 51 presumably relate to the magic proton number $Z = 50$.

Estimates of the cross sections of the unstable isotopes of Tc and Pm can be made and are seen to be exceptionally high. Cross sections for a number of light nuclei, often dominated by one or two resonances, are very low ($< 1$ mb) and are not shown.

## 4.3.2. Even-Z Nuclei

Isotopic cross sections have been measured for many elements and these indicate a systematic effect not only within a particular element, but also relative to the distribution of cross sections with neutron number. These effects are:

a) slowly varying ratio for the cross sections of odd and even neutron isotopes,

b) systematic variations of isotopic cross sections with neutron number for many elements, and

c) systematic variations of the isotopic cross section gradient (i.e., the rate of change of cross section with neutron number) for each element which shows to a marked extent the importance of the magic neutron numbers at 50, 82, and 128.

All open circle data points shown in the figure are based on cross section measurements, and on the assumption of systematic behavior with respect to the smooth dependence on $N$ and a slowly varying odd–even ratio. In many cases the elemental cross section [$\sigma(\text{Nat})$] is known, and from this, assuming the appropriate odd–even ratio ($\alpha$), the average odd-isotope

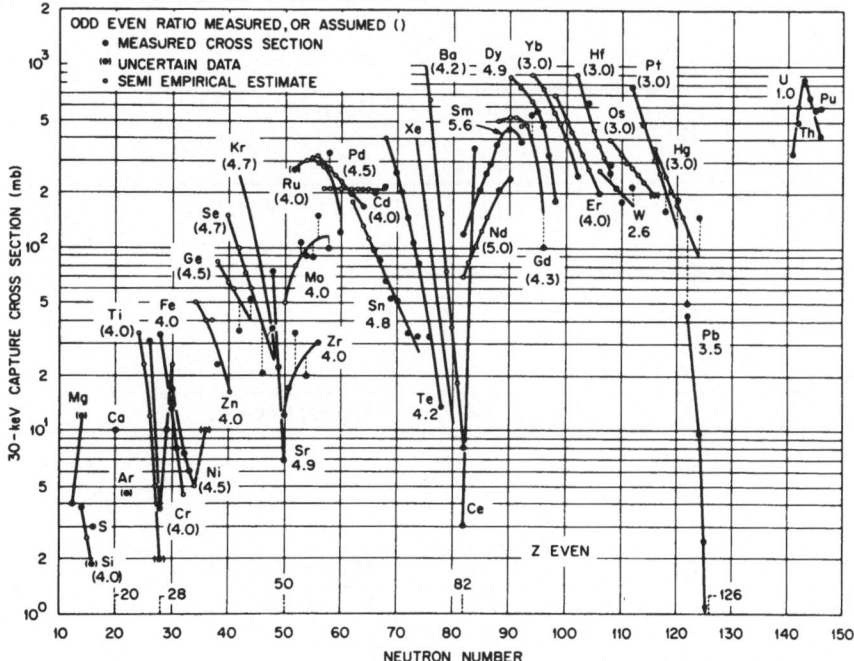

**Fig. 9. 30-keV Maxwellian-averaged capture cross section for even-$Z$ nuclides.** Isotopic cross sections are smoothly related, and their gradients follow a pattern strongly influenced by the magic neutron numbers.

TABLE II. 30-keV Maxwellian-Averaged Capture Cross Sections

| | |
|---|---|
| *: Semiempirical estimates | $p$: proton-rich nuclides |
| ( ): Uncertain experimental value | $s(r)$: dominant $s$ ($> 80\%$) |
| $r$: rapid process only | $s, r$: comparable $s$ and $r$ contributions |
| $s$: slow process only | $r(s)$: dominant $r$ ($> 80\%$) |

| $Z$ | Element | $A$ | Class | Cross section, millibarns | References |
|---|---|---|---|---|---|
| 1 | H | | | | |
| | | 1 | | | |
| | | 2 | | | |
| 2 | He | | | | |
| | | 3 | | | |
| | | 4 | | | |
| 3 | Li | | | | |
| | | 6 | | | |
| | | 7 | | | |
| 4 | Be | 9 | | | |
| 5 | B | | | | |
| | | 10 | | | |
| | | 11 | | | |
| 6 | C | | | $0.2 \pm 0.4$ | Gol+ 66 |
| | | 12 | | | |
| | | 13 | | | |
| 7 | N | | | | |
| | | 14 | | | |
| | | 15 | | | |
| 8 | O | | | | |
| | | 16 | | | |
| | | 17 | | | |
| | | 18 | | | |
| 9 | F | 19 | | $5.6 \pm 0.4$ | MG 65, Hoc+ 65 |
| 10 | Ne | | | | |
| | | 20 | | | |
| | | 21 | | | |
| | | 22 | | | |
| 11 | Na | 23 | | $2.7 \pm 0.4$ | MG 65, Hoc+ 69 |
| 12 | Mg | | | $4.0 \pm 1.0$ | MG 65, Hoc+ 68 |
| | | 24 | | | |
| | | 25 | | | |
| | | 26 | | (12) | Gol+ 66 |
| 13 | Al | 27 | | $4.6 \pm 0.8$ | MG 65, Hoc+ 69 |

**TABLE II** (*Continued*)

| Z | Element | A | Class | Cross section, millibarns | References |
|---|---------|-----|-------|---------------------------|------------|
| 14 | Si |    |   | $3.8 \pm 1.0$ | MG 65 |
|    |    | 28 |   | $3.8 \pm 1.0$ | *d* |
|    |    | 29 |   | 10.4* | |
|    |    | 30 |   | (1.9) | Gol+ 66 |
| 15 | P | 31 |   | (7) | Gol+ 66 |
| 16 | S |    |   | $3.0 \pm 0.6$ | MG 65, Hoc+ 68 |
|    |    | 32 |   | $3.0 \pm 0.6$ | *d* |
|    |    | 33 |   | | |
|    |    | 34 |   | | |
|    |    | 36 | *r* | | |
| 17 | Cl |    |   | $11 \pm 4$ | Gol+ 66 |
|    |    | 35 |   | $13.5 \pm 5$ | *d* |
|    |    | 37 |   | (3) | Gol+ 66, Tol+ 67, HCS 68 |
| 18 | Ar |    |   | | |
|    |    | 36 |   | | |
|    |    | 38 |   | | |
|    |    | 40 | *r* | (4.5) | Gol+ 66 |
| 19 | K |    |   | $16 \pm 2$ | Gol+ 66 |
|    |    | 39 |   | $16 \pm 2$ | *d* |
|    |    | 40 |   | — | |
|    |    | 41 |   | $22 \pm 3$ | Gol+ 66, CS 66, Stu+ 68 |
| 20 | Ca |    |   | $10 \pm 1$ | Gol+ 66 |
|    |    | 40 |   | | |
|    |    | 42 |   | | |
|    |    | 43 |   | | |
|    |    | 44 |   | | |
|    |    | 46 | *r* | | |
|    |    | 48 | *r* | | |
| 21 | Sc | 45 |   | $44 \pm 6$ | Gol+ 66 |
| 22 | Ti |    |   | $20 \pm$ | Gol+ 66 |
|    |    | 46 |   | 34* | |
|    |    | 47 |   | 92* | |
|    |    | 48 |   | 12* | |
|    |    | 49 |   | 20* | |
|    |    | 50 |   | (2) | Gol+ 66 |
| 23 | V |    |   | $25 \pm 8$ | Gol+ 66, CS 66, MG 65 |
|    |    | 50 |   | | |
|    |    | 51 |   | $25 \pm 8$ | *d* |

**TABLE II** (*Continued*)

| Z | Element | A | Class | Cross section, millibarns | References |
|---|---------|---|-------|---------------------------|------------|
| 24 | Cr | | | $6.2 \pm 2$ | Gol+ 66 |
| | | 50 | | $31 \pm 4$ | Gol+ 66 |
| | | 52 | | $3.8 \pm 1.0$ | Gol+ 66 |
| | | 53 | | $40 \pm 5$ | Gol+ 66 |
| | | 54 | | 23* | |
| 25 | Mn | 55 | | $50 \pm 2$ | Gol+ 66, Stu+ 68, HCS 68 |
| 26 | Fe | | | $18 \pm 8$ | Gol+ 66 |
| | | 54 | | $34 \pm 10$ | MG 65 |
| | | 56 | | $13.5 \pm 2.0$ | Mg 65, Hoc+ 69 |
| | | 57 | | $30 \pm 5$ | Mg 65, Hoc+ 69 |
| | | 58 | | 4.5* | |
| 27 | Co | 59 | | $35 \pm 10$ | Gol+ 66, Blo 68 |
| 28 | Ni | | | $12.4 \pm 2$ | MG 65 |
| | | 58 | | $17 \pm 3$ | Hoc+ 69, *a* |
| | | 60 | | $7.5 \pm 2$ | Hoc+ 69, *a* |
| | | 61 | | (30) | Hoc+ 69 |
| | | 62 | | 6* | *a* |
| | | 64 | | (10) | Gol+ 66 |
| 29 | Cu | | | $47 \pm 7$ | Gol+ 66, Abr+ 67 |
| | | 63 | *s(r)* | $49 \pm 14$ (92) | *d* |
| | | 65 | *s(r)* | $42 \pm 7$ (18) | Gol+ 66, CS 66 |
| 30 | Zn | | | $41 \pm 10$ | Gol+ 66, Abr+ 67 |
| | | 64 | *s* | 50* | HCS 68 |
| | | 66 | *s(r)* | 40* | HCS 68 |
| | | 67 | *s(r)* | 160* | |
| | | 68 | *s(r)* | $23 \pm 3$ | CS 66 |
| | | 70 | *r* | 16* | |
| 31 | Ga | | | $115 \pm 20$ | Gol+ 66 |
| | | 69 | *s, r* | $130 \pm 30$ | Gol+ 66, CS 66 |
| | | 71 | *s, r* | $120 \pm 30$ (60) | Gol+ 66, CS 66 |
| 32 | Ge | | | $74 \pm 7$ | Gol +66 |
| | | 70 | *s* | 84* | |
| | | 72 | *s, r* | 65*, 40 | *e* |
| | | 73 | *s, r* | 270* | |
| | | 74 | *s, r* | $35 \pm 20$, 20 | CS 66, TKK 67a, *e* |
| | | 76 | *r* | $53 \pm 10$ | CS 66, CS 68 |
| 33 | As | 75 | *s, r* | $490 \pm 100$ | Gol+ 66, HCS 68, |

**TABLE II** (*Continued*)

| Z | Element | A | Class | Cross section, millibarns | References |
|---|---------|---|-------|---------------------------|------------|
| 34 | Se | | | 94 ± 8 | Gol+ 66 |
| | | 74 | p | 160* | |
| | | 76 | s | 100* | |
| | | 77 | s, r | 340* | |
| | | 78 | s, r | 60* | |
| | | 80 | s, r | 20 ± 12 | CS 66, TKK 67b |
| | | 82 | r | 36 ± 15 | CS 66 |
| 35 | Br | | | 600 ± 60 | Gol+ 66 |
| | | 79 | r | 600 ± 150 | CS 66, Gol+ 66 |
| | | 81 | s, r | 460 ± 80 | CS 66, Gol+ 66 |
| 36 | Kr | | | — | |
| | | 78 | p | 250*   500 | e |
| | | 80 | p | 140*   280 | e |
| | | 82 | s | 80*   200 | e |
| | | 83 | s, r | 225*   670 | e |
| | | 84 | s, r | 28*   60 | e |
| | | 86 | r | 9*   20 | e |
| 37 | Rb | | | 160 ± 20 | Gol+ 66 |
| | | 85 | s, r | 215 ± 20 | Stu+ 68, Gol+ 66 |
| | | 87 | r | 24 ± 4 | Gol+ 66, HCS 68, Tol+ 67 |
| 38 | Sr | | | 120 ± 40 | Gol+ 66, c |
| | | 84 | p | 330* | |
| | | 86 | s | 74 ± 7 | MG 65 |
| | | 87 | s(r) | 109 ± 9 | MG 65 |
| | | 88 | s(r) | 6.9 ± 2.5 | MG 65 |
| 39 | Y | 89 | s, r | 21 ± 4 | Gol+ 66, HCS 68, Stu+ 68 |
| 40 | Zr | | | 25 ± 10 | Gol+ 66 |
| | | 90 | s(r) | 12 ± 2 | Gol+ 66 |
| | | 91 | s, r | 68 ± 8 | Gol+ 66 |
| | | 92 | s, r | 34 ± 6 | Gol+ 66 |
| | | 94 | s, r | 20 ± 2 | Gol+ 66 |
| | | 96 | r | 30 ± 12 | Gol+ 66 |
| 41 | Nb | 93 | s, r | 285 ± 30 | Gol+ 66, Kom 69 |
| 42 | Mo | | | 160 ± 20 | Gol+ 66, Kom 69 |
| | | 92 | p | 50* | |
| | | 94 | p(s) | 80* | |
| | | 95 | s, r(p) | 430 ± 50 | Gol+ 66 |

**TABLE II** (*Continued*)

| Z | Element | A | Class | Cross section, millibarns | References |
|---|---------|---|-------|---------------------------|------------|
| 42 | Mo (*cont.*) | | | | |
| | | 96 | *s* | 90 ± 10 | Gol+ 66 |
| | | 97 | *s, r* | 350 ± 50 | Gol+ 66 |
| | | 98 | *s, r* | 150 ± 40, 110 | Gol+ 66, HCS 68, Stu+ 68, *e* |
| | | 100 | *r* | 100 ± 40 | Gol+ 66 |
| 43 | Tc | 99 | *s, r* | 800* | |
| 44 | Ru | | | (550) | Gol+ 66 |
| | | 96 | *p* | 270 ± 60 | Gol+ 66 |
| | | 98 | *p* | 300* | |
| | | 99 | *s, r* | 1240* | |
| | | 100 | *s* | 290* | |
| | | 101 | *s, r* | 1120* | |
| | | 102 | *s, r* | 330 ± 50 | Gol+ 66 |
| | | 104 | *r* | 120 ± 60 | Gol+ 66, CS 66 |
| 45 | Rh | 103 | *s, r* | 900 ± 100 | Gol+ 66 |
| 46 | Pd | | | 440 ± 40 | Gol+ 66, Kom 69 |
| | | 102 | *p* | 320* | |
| | | 104 | *s* | 270* | |
| | | 105 | *r(s)* | 1130* | |
| | | 106 | *s, r* | 230* | |
| | | 108 | *s, r* | 200 ± 60 | Gol+ 66, CS 66, CS 68 |
| | | 110 | *r* | 170 ± 70 | CS 66 |
| 47 | Ag | | | 920 ± 100 | Gol+ 66, Kom 69, SBB 68, Abr+ 67, Kon+ 67 |
| | | 107 | *s, r* | 1150 ± 150 | Gol+ 66, Kon+ 67 |
| | | 109 | *s, r* | 620 ± 50 | Kon+ 67, CS 66 |
| 48 | Cd | | | 340 ± 50 | Gol+ 66, Abr+ 67, Kom 69 |
| | | 106 | *p* | 210* | |
| | | 108 | *p* | 210* | |
| | | 110 | *s* | 210* | |
| | | 111 | *s, r* | 840* | |
| | | 112 | *s, r* | 210* | |
| | | 113 | *s, r* | 840* | |
| | | 114 | *s, r* | 200 ± 40 | Gol+ 66, CS 65 |
| | | 116 | *r* | 220 ± 40 | CS 66 |
| 49 | In | | | 760 ± 80 | Gol+ 66, Abr+ 67, Kom 69, SBB 68 |
| | | 113 | | 220 ± 70 | CS 66 |
| | | 115 | *s, r* | 800 ± 100 | Gol+ 66, *d* |

**TABLE II** (*Continued*)

| Z | Element | A | Class | Cross section, millibarns | References |
|---|---------|---|-------|---------------------------|------------|
| 50 | Sn | | | 95 ± 15 | Gol +66 |
| | | 112 | p | 180* | |
| | | 114 | p | 130* | |
| | | 115 | p | 550* | |
| | | 116 | s | 100 ± 15 | Gol+ 66 |
| | | 117 | s, r | 420 ± 30 | Gol+ 66 |
| | | 118 | s, r | 63 ± 5 | Gol+ 66 |
| | | 119 | s, r | 260 ± 40 | Gol+ 66 |
| | | 120 | s, r | 50 ± 15 | Gol+ 66, HCS 68 |
| | | 122 | r | 23 ± 5 (165) | Gol+ 66, Tol+ 68 |
| | | 124 | r | 23 ± 4 (180) | Gol+ 66, Tol+ 68 |
| 51 | Sb | | | 490 ± 50 | Gol+ 66, Abr+ 67, SBB 68 |
| | | 121 | s, r | 740 ± 100 | HCS 68, Tol+ 68 |
| | | 123 | r | 440 ± 50 | HCS 68, Tol+ 68 |
| 52 | Te | | | 97 ± 9 (204) | Gol+ 66 |
| | | 120 | p | 400* | |
| | | 122 | s(p) | 270 ± 30 | MG 67b, BR 67 |
| | | 123 | s(p) | 820 ± 30 | MG 67b, BR 67 |
| | | 124 | s | 150 ± 20 | MG 67b, BR 67 |
| | | 125 | s, r | 430 ± 30 | MG 67b, BR 67 |
| | | 126 | s, r | 82 ± 8 | MG 67b, BR 67 |
| | | 128 | r | 32.5 ± 5 | MG 67b, BR 67 |
| | | 130 | r | 13.5 ± 2.0 | MG 67b, BR 67 |
| 53 | I | 127 | r(s) | 760 ± 50 | Gol+ 66 |
| 54 | Xe | | | | |
| | | 124 | p | 1200* | |
| | | 126 | p | 800* | |
| | | 128 | s | 300* | |
| | | 129 | r(s) | 760* | |
| | | 130 | s | 100* | |
| | | 131 | r(s) | 250* | |
| | | 132 | s, r | 36* | |
| | | 134 | r | 13* | |
| | | 136 | r | 5* | |
| 55 | Cs | 133 | r(s) | 700 ± 40 | Kom 69 |
| 56 | Ba | | | 61 ± 5 | Gol+ 66 |
| | | 130 | p | 2000* | |
| | | 132 | p | 650* | |

**TABLE II** (*Continued*)

| Z | Element | A | Class | Cross section, millibarns | References |
|---|---------|---|-------|---------------------------|------------|
| 56 | Ba (*cont.*) | | | | |
| | | 134 | *s* | 155* | |
| | | 135 | *s, r* | 315* | |
| | | 136 | *s* | 37* | |
| | | 137 | *s, r* | 76* | |
| | | 138 | *s, r* | 8 ± 2, 5 | Gol+ 66, CS 66, *e* |
| 57 | La | | | 44 ± 4 | Gol+ 66, CS 66, Stu+ 68 |
| | | 138 | *p* | | |
| | | 139 | *s(r)* | 44 ± 4, 48 | *d, e* |
| 58 | Ce | | | 35 ± 5 | Gol+ 66 |
| | | 136 | *p* | 100* | |
| | | 138 | *p* | 30* | |
| | | 140 | *s(r)* | 3 ± 3, 12 | *d, e* |
| | | 142 | *r* | 360 ± 60 (450) | CS 66 |
| 59 | Pr | 141 | *s, r* | 110 ± 20 | Gol+ 66, CS 66, Stu+ 68 |
| 60 | Nd | | | — | |
| | | 142 | *s* | 70* | |
| | | 143 | *s, r* | 425* | |
| | | 144 | *s, r* | 100* | |
| | | 145 | *s, r* | 600* | |
| | | 146 | *s, r* | 150* | |
| | | 148 | *r* | 210 ± 80 | HCS 68 |
| | | 150 | *r* | 240 ± 150 | HCS 68 |
| 61 | Pm | 147 | | 2000* | |
| 62 | Sm | | | 920 ± 50 | Gol+ 66 |
| | | 144 | *p* | 120 ± 55 | Gol+ 66 |
| | | 147 | *r(s)* | 1150 ± 190 | Gol+ 66 |
| | | 148 | *s* | 260 ± 50 | Gol+ 66 |
| | | 149 | *r(s)* | 1620 ± 280 | Gol+ 66 |
| | | 150 | *s* | 370 ± 70 | Gol+ 66 |
| | | 152 | *s, r* | 450 ± 50 | Gol+ 66, CS 66 |
| | | 154 | *r* | 380 ± 60 | Gol+ 66, CS 66 |
| 63 | Eu | | | 3350 ± 150 | Gol+ 66, MG 67a |
| | | 151 | *r(s)* | 3600 ± 500 | Gol+ 66 |
| | | 153 | *r(s)* | 2700 ± 300 | Gol+ 66 |
| 64 | Gd | | | 940 ± 50 | Gol+ 66 |
| | | 152 | *s, p* | 500* | |
| | | 154 | *s* | 520* | |
| | | 155 | *r(s)* | 2280* | |

**TABLE II** (*Continued*)

| Z | Element | A | Class | Cross section, millibarns | References |
|---|---------|---|-------|---------------------------|------------|
| 64 | Gd (*cont.*) | | | | |
| | | 156 | r(s) | 470* | |
| | | 157 | r(s) | 2070* | |
| | | 158 | s, r | 540 ± 70 | Gol+ 66, CS 66, Stu+ 68 |
| | | 160 | r | 100 ± 30 | CS 66 |
| 65 | Tb | 159 | r(s) | 2200 ± 200 | Gol+ 66 |
| 66 | Dy | | | 730 ± 40 | Gol+ 66 |
| | | 156 | p | 870* | |
| | | 158 | p | 770* | |
| | | 160 | s | 650* | |
| | | 161 | r(s) | 2800 ± 300 | Kon+ 67 |
| | | 162 | r(s) | 470 ± 50 | Kon+ 67 |
| | | 163 | r(s) | 1600 ± 300 | Kon+ 67 |
| | | 164 | r(s) | 180 ± 40 | Gol+ 66, CS 68, CS 65 |
| 67 | Ho | 165 | s, r | 1250 ± 150 (2000) | ACM 68, Gol+ 66 |
| 68 | Er | | | 750 ± 50 | Gol+ 66 |
| | | 162 | p | 900* | |
| | | 164 | p | 750* | |
| | | 166 | r(s) | 560* | |
| | | 167 | r(s) | 2000* | |
| | | 168 | s, r | 400* | |
| | | 170 | r | 250 ± 30 | Hoc+ 69, HCS 68 |
| 69 | Tm | 169 | r(s) | 1500 ± 200 | Gol+ 66 |
| 70 | Yb | | | 600 ± 50 | Gol+ 66 |
| | | 168 | p | 700* | |
| | | 170 | s | 510* | |
| | | 171 | r(s) | 1320* | |
| | | 172 | s, r | 380* | |
| | | 173 | r(s) | 990* | |
| | | 174 | s, r | 275* | |
| | | 176 | r | 200 ± 50 | Gol+ 66 |
| 71 | Lu | | | 1400 ± 300 (3700) | Gol+ 66 |
| | | 175 | r(s) | 1460 ± 110 | CS 76, MG 67a |
| | | 176 | s ⌐ | 2250 ± 200 | MG 67a |
| 72 | Hf | | | 600 ± 50 | Gol+ 66, Kom 69 |
| | | 174 | p | 800* | |
| | | 176 | s ⌐ | 640 ± 160 | Ber+ 65 |
| | | 177 | r(s) | 110* | |

**TABLE II** (*Continued*)

| Z | Element | A | Class | Cross section, millibarns | References |
|---|---------|---|-------|---------------------------|------------|
| 72 | Hf (*cont.*) | | | | |
| | | 178 | s, r | 370* | |
| | | 179 | s, r | 960* | |
| | | 180 | s, r | 290 ± 80 | Gol+ 66 |
| 73 | Ta | | | 800 ± 80 | Gol+ 66, Kom 69, Kon+ 67, Brz+ 68 |
| | | 180 | p | — | |
| | | 181 | s, r | 800 ± 80 | d |
| 74 | W | | | 290 ± 30 | Gol+ 66, Kom 69, Bar+ 68, Kon+ 67 |
| | | 180 | p | 270* | |
| | | 182 | s, r | 260 ± 30 | Gol+ 66, Bar+ 68 |
| | | 183 | s, r | 550 ± 50 | Gol+ 66, Bar+ 68 |
| | | 184 | s, r | 180 ± 20 | Gol+ 66, Bar+ 68, Kon+ 67 |
| | | 186 | r | 220 ± 20 | Gol+ 66, CS 66, Bar+ 68, Kon+ 67 |
| 75 | Re | | | 1420 ± 100 (950) | Gol+ 66, Kom 69, Kon+ 67 |
| | | 185 | r(s) | 1530 ± 200 (2200) | Fri+ 68, CS 65 |
| | | 187 | r(s)⌐ | 1570 ± 100 (780) | Fri+ 68, CS 65 |
| 76 | Os | | | 300 ± 40 | Gol+ 66 |
| | | 184 | p | 400* | |
| | | 186 | s | 330* | |
| | | 187 | s(r)◄ | 900* | |
| | | 188 | s, r | 275* | |
| | | 189 | r(s) | 765* | |
| | | 190 | r(s) | 230* (750) | Gol+ 66 |
| | | 192 | r | (200) | TKK 67a |
| 77 | Ir | | | 1120 ± 200 | Gol+ 66 |
| | | 191 | r(s) | 1900 ± 300 | d |
| | | 193 | r(s) | 600 ± 80 | CS 66, Tol+ 67 |
| 78 | Pt | | | 470 ± 60 | Gol+ 66 |
| | | 190 | p | 770* | |
| | | 192 | s | 490* | |
| | | 194 | r(s) | 310* | |
| | | 195 | r(s) | 780* | |
| | | 196 | r(s) | 160 ± 40 | Gol+ 66, CS 66 |
| | | 198 | r | 185 ± 20 | CS 66 |

**TABLE II** (*Continued*)

| Z | Element | A | Class | Cross section, millibarns | References |
|---|---------|---|-------|---------------------------|------------|
| 79 | Au | 197 | r(s) | 600 ± 50 | Gol+ 66, Abr+ 67, Kom 69, SBB 6ⁿ |
| 80 | Hg |     |      | 250 ± 60 | Gol+ 66, Abr+ 67 |
|    |    | 196 | p    | 360* |  |
|    |    | 198 | s    | 250*, 125 | e |
|    |    | 199 | s(r) | 630* |  |
|    |    | 200 | s, r | 175* |  |
|    |    | 201 | s, r | 450* |  |
|    |    | 202 | s, r | 50 ± 15 | Gol+ 66 |
|    |    | 204 | r    | 150 ± 50 | CS 66 |
| 81 | Tl |     |      | 70 ± 5 | Gol+ 66 |
|    |    | 203 | s, r | 170 ± 30 | Gol+ 66 |
|    |    | 205 | s, r | 48 ± 10 | Gol+ 66, HCS 68 |
| 82 | Pb |     |      | 4.6 ± 1.5 | Gol+ 66, d |
|    |    | 204 | s    | 43 ± 5 | MG 67a |
|    |    | 206 | s, r | 9.6 ± 3.0 | Gol+ 66 |
|    |    | 207 | s, r | 8.7 ± 3.0 | Gol+ 66 |
|    |    | 208 | s, r | 0.33 ± 0.07 | MG 69 |
| 83 | Bi | 209 | s, r | 12.1 ± 4.0 | MG 65 |
| 90 | Th | 232 | r    | 500 ± 100 | Gol+ 66, CS 66 |
| 91 | Pa | 231 | r    | — |  |
| 92 | U  |     |      |  |  |
|    |    | 233 | r    | 330 ± 40 | BCD 68, Gol+ 66 |
|    |    | 234 | r    | 610* |  |
|    |    | 235 | r    | 860 ± 80 | BHC 68, Gol+ 66 |
|    |    | 236 | r    | (680) | Car 68 |
|    |    | 238 | r    | 415 ± 50 | BHC 68, Gol+ 66, Mox 68 |
| 94 | Pu |     |      |  |  |
|    |    | 239 | r    | 580 ± 60 | Gol+ 66, Gwi+ 69, HBC 68 |
|    |    | 240 | r    | 600 ± 100 | HBC 68 |

ᵃ Disagreement with relative cross sections in (AKS 68).
ᵇ Agreement with relative cross sections in (All 70).
ᶜ Disagreement with isotopic cross sections.
ᵈ Cross section is deduced from the natural and/or isotopic values.
ᵉ Alternative value required by $N_s\sigma(A)$ and/or $N_r(A)$.

cross section ($\bar{\sigma}_0$) can be determined. This then provides a data point which, if at least one other isotopic cross section has been measured, can determine the gradient and permit semiempirical estimates of other isotopic cross sections to be made. With the exception of Kr and Xe, all gradients have been determined independently (from each other) in this manner:

$$\sigma(\text{Nat}) = N_e \bar{\sigma}_e + N_0 \bar{\sigma}_0 \qquad N_e, N_0\text{—even and odd abundances}$$
$$= \bar{\sigma}_e (N_e + \alpha N_0) \qquad \text{where } \bar{\sigma}_0 = \alpha \bar{\sigma}_e$$

and $\alpha$ is the odd–even cross section ratio.

In the figure the capture cross sections of all odd neutron number isotopes have been reduced by the odd–even ratio to preserve the systematic effect. Evidence for the systematic dependence and odd–even ratio is found in those elements where three or more data points are available with less than 20% deviation from a smooth neutron number dependence, e.g., Fe(4.5,) Sr(4.9), Mo(4.0), Sn(4.8), Te(4.2), Dy(4.9), W(2.6), and Pb(3.5). In addition Cr(4) and Sm(5.6) show a turn over for the heaviest isotopes in concordance with the local gradient trends.

Twenty-six elements show the gradient variation consistent with minima at the magic numbers 28, 50, 82, and 126. The gradient changes are sharp at the magic numbers and slow for nuclides in between. Major exceptions are Fe, Ni, and Zn which show the opposite gradient to that expected at $N = 28$. In fact $^{54}$Fe, with $N = 28$, has the highest isotopic capture cross section among the iron isotopes.

There are, however, a number of cases where evidence in favor of a smooth neutron number dependence is poor. To achieve systematic behavior, errors of 50 to 100% are implied for isotopes in Ni, Ge, Zr, Ru, and Hg, and the measured cross section for $^{190}$Os is too high by a factor of 3. Some immediate comments on the data can be made:

(a) $\sigma(\text{Nat})$ for Nd is not available,

(b) $\sigma(\text{Nat})$ for Sr is much greater than the weighted sum of its isotopes,

(c) results of measurements (mainly activation) for $^{122,124}$Sn, Te, $^{142}$Ce, and $^{190}$Os have not been indicated in the figure because of substantial disagreement with other data, and

(d) Kr and Xe are the significantly discordant elements as in both cases the gradients must be "guessed" from a comparison of nearest neighbor gradients. In both cases neighboring elements have very steep and well determined gradients (with the exception of Ba).

Because such estimates and gradients are not well based interpolated data points for Kr and Xe are not shown. They are however included in Table II.

On the basis of well-established systematic behavior in the 30-keV capture cross sections it is possible to make semiempirical estimates for many unmeasured isotopes and these are designated by an * in the cross section table. No attempt has been made to smooth the actual experimental data, though estimates are naturally based on best fit extrapolations of these data.

From this compilation estimates of unmeasured s-process capture cross sections can be obtained. In addition, fission-product isotopic cross sections can be derived, and it is of interest to compare our semi-empirical estimates with those calculations by Benzi (BB 65), Musgrove (Mus 69), and Shorin (SKK+ 66) based on capture theory and resonance parameters. The need for estimates of fission-product cross sections has been pointed out by Benzi who considers that 50% of the effective fission-product capture cross sections has not been and, generally, cannot be measured. Table III

### TABLE III. Comparative Cross Section Estimates (mb)

| Element | Class | Semiempirical | Theory | | |
|---------|-------|---------------|--------|--------|--------|
| | | | Benzi | Musgrove | Shorin |
| $^{83}$Kr | FIS | 225 | 580 | 350 | — |
| $^{93}$Zr | FIS | 18 | 140 | 57 | — |
| $^{99}$Tc | FIS | 800 | 870 | 770 | — |
| $^{100}$Ru | S | 290 | — | 530 | 160 |
| $^{110}$Cd | S | 210 | — | 900 | 620 |
| $^{113}$Cd | FIS | 840 | 590 | 470 | — |
| $^{128}$Xe | S | 300 | — | 250 | 280 |
| $^{130}$Xe | S | 100 | — | 195 | 140 |
| $^{134}$Xe | FIS | 13 | 70 | 15 | — |
| $^{134}$Ba | S | 155 | — | 240 | 140 |
| $^{135}$Cs | FIS | 520 | 300 | 100 | — |
| $^{142}$Nd | S | 70 | — | 170 | 60 |
| $^{143}$Nd | FIS | 425 | 310 | 280 | — |
| $^{170}$Yb | S | 510 | — | 620 | 800 |
| $^{186}$Os | S | 330 | — | 390 | 340 |
| $^{198}$Hg | S | 250 | — | 310 | 150 |

gives a comparison of estimates for a representative number of $s$-process and high yield fission product nuclei (denoted by FIS).

The average behavior of resonances has been intensively studied and is well-formulated in terms of neutron strength functions ($S_l$), Porter–Thomas distributions, and of particular importance to capture cross sections, the radiative strength function $S_\gamma = \Gamma_\gamma / \langle D \rangle$, where $\Gamma_\gamma$ is the radiative width and $\langle D \rangle$ the average resonance spacing.

A problem in the calculation of average capture cross sections is the number and uncertainty of the parameters involved. While $\Gamma_\gamma$, $\langle D \rangle$, and $S_0$ are often well known from low-energy resonance measurements, in many cases it is necessary to interpolate to obtain these parameters. $S_1$ and $S_2$, the $p$- and $d$-wave neutron strength functions, are only poorly known and as a result calculated cross sections are subject to large errors.

All the above parameters show systematic variations with mass number. However the neutron strength functions are not correlated with the magic neutron numbers but rather with the zero binding of the $s$, $p$, and $d$ neutron shells. The radiative widths vary slowly with mass number but the resonance spacings are markedly dependent on the position of the closed neutron shells. Consequently the radiative strength function $S_\gamma$ exhibits deep minima at $N = 50$, 82, and 126 as well as a systematic difference in magnitude between odd- and even-$A$ nuclei (Mus 69).

# 5. ABUNDANCE–CROSS-SECTION CORRELATIONS

With the set of 30-keV Maxwellian-averaged neutron capture cross sections, data are now available for a comprehensive investigation of correlations with $s$-process abundances. From the empirical curve $(N_s\sigma(A))$ interpolations can then be made to yield the $r$-process abundances by subtraction. However, it is first necessary to obtain a consistent set of elemental solar system abundances.

## 5.1. Abundances

Since the early $s$-process correlations of Clayton et al. (Cla+ 61) additional measurements have improved the quantity and quality of the abundance data. Cameron (Cam 63) and Seeger et al. (SFC 64) presented revised distributions which have been further improved by more recent measurements. The latest compilations of Cameron (Cam 68) and Amiet and Zeh (AZ 68) are based primarily on type-I carbonaceous chondrites, which are assumed to be the most representative source of solar-system abundances. However, data are also taken from ordinary chondrites, solar atmospheric abundances, and solar cosmic-ray abundances (solar wind). Additional results have been obtained by interpolation on a basis of nuclear regularities.

However there are a number of discrepancies between the two most recent distributions, the most outstanding being Hg ($\times$ 20) and then V, Sr, and Ag ($\times$ 2). It is apparent that considerable uncertainty is associated with the Hg estimate. The abundances of Hg in carbonaceous chondrites are anomalously high and greatly variable. Cameron obtained the Hg estimate by interpolation of $^{199}$Hg and $^{201}$Hg between $^{197}$Au and $^{203}$Te, with as high a value as seemed reasonable.

Volatile elemental abundances are obtained by normalization methods, and Cameron notes that substantial uncertainties (several tens of per cent) may result. In the following work, abundances taken from (Cam 68) have been used exclusively, and these are tabulated in full in that reference.

## 5.2. Solar System $N_s\sigma(A)$

The $N_s\sigma(A)$ results were first obtained by restricting data to $s$-process nuclei only. A smooth correlation was apparent with only one point, $^{198}$Hg, showing anomalous behavior. Generally the scatter of points about a smooth curve is in the range of $\pm 20\%$, and well within the expected uncertainties of abundances and cross sections. About two-thirds of the cross section data were derived from semiempirical estimates.

However, it was found that interpolation by a smooth curve through the $N_s\sigma$ data was not sufficient in itself to derive a consistent set of $r$-process abundances ($N_r(A)$). In a number of instances the $r$-process abundances set constraints on possible values for $N_s\sigma$, and it was necessary to feed back $N_r$ data to modify $N_s\sigma$ data. The final $N_s\sigma$ data, as shown in Fig. 10, therefore include a number of nuclides for which $r$-process abundance subtractions have been made. In these cases only nuclides with measured cross sections have been used.

In addition, new abundance estimates for Zr, Xe, Nd, and Hg have been made. These were suggested from the empirical $r$-process distribution (Fig. 11) and are discussed later. (See Section 5.4.)

As the abundances for Kr are completely anomalous and measured cross sections are not available, $^{86}$Kr has not been included in the $N_s\sigma$ data. For reasons of uncertainty in $s$-process branching, $^{176}$Lu and $^{176}$Hf, have also been excluded (see Section 5.4.)

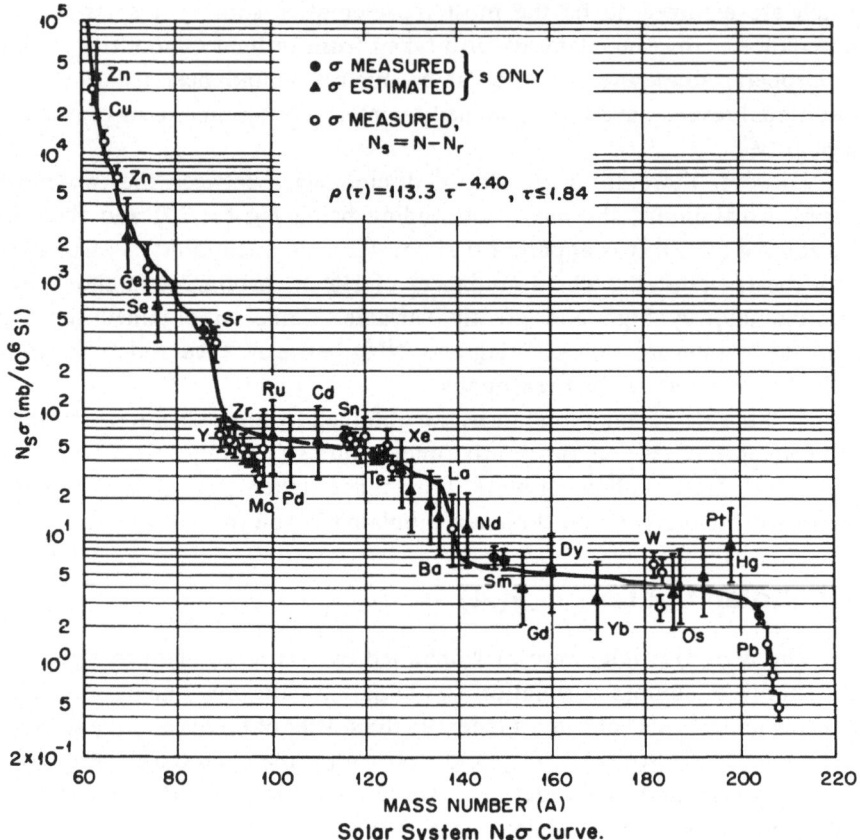

Solar System $N_s\sigma$ Curve.

**Fig. 10. Correlations of solar system $s$-process abundances and capture cross sections at 30 keV.** The product $N_s\sigma$ is shown as a function of mass number. Estimated cross sections are given an uncertainty of a factor of two. The curve is a weighted, least squares fit to the data by Seeger (See 70) and corresponds to the integrated neutron exposure $\varrho(\tau) = 113.3\tau^{-4.4}$, with a cutoff at $\tau_{max} = 1.84 \times 10^{27}$ neutrons cm$^{-2}$.

The $N_s\sigma$ data as such represents a significant advance over earlier data (*cf.* Fig. 5). W. A. Fowler has recently discussed cross sections needed for the clarification of the $s$-process (Fow 68). Further information was required on the following nuclides: Fe–Cr isotopes, Ge, Cd, Ba, and Os isotopes, and $^{208}$Pb. At that time such data were either not available or were anomalous in terms of the $N_s\sigma$ correlations.

Estimates are now available for the Fe–Cr isotopes which are required for Cu–Ni synthesis. The new values for Ge and Ba eliminate the reverse trends in the 1968 $N_s\sigma$ curve, and high values of Cd and Os have been substantially reduced. Semiempirical estimates for Os can now be used with more confidence in cosmochronology calculations.

The $N_s\sigma$ products for the lead isotopes show a marked decrease over earlier estimates. This is the result of a recent measurement of the $^{208}$Pb capture cross section (MG 69) which has reduced its value by an order of magnitude (2.9 mb → 0.33 mb). Clayton and Rassbach (CR 67) had earlier emphasized that the determination of the $s$-process contributions to Pb becomes independent of $\sigma(^{208}$Pb) if this cross section is very small. While the actual value of $\sigma(^{208}$Pb) is therefore an insensitive parameter, its low value satisfies a requirement by Clayton (Cla 68) for a consistent Pb cosmochronology.

In previous $N_s\sigma$ correlations the evidence for a ledge–precipice struc-

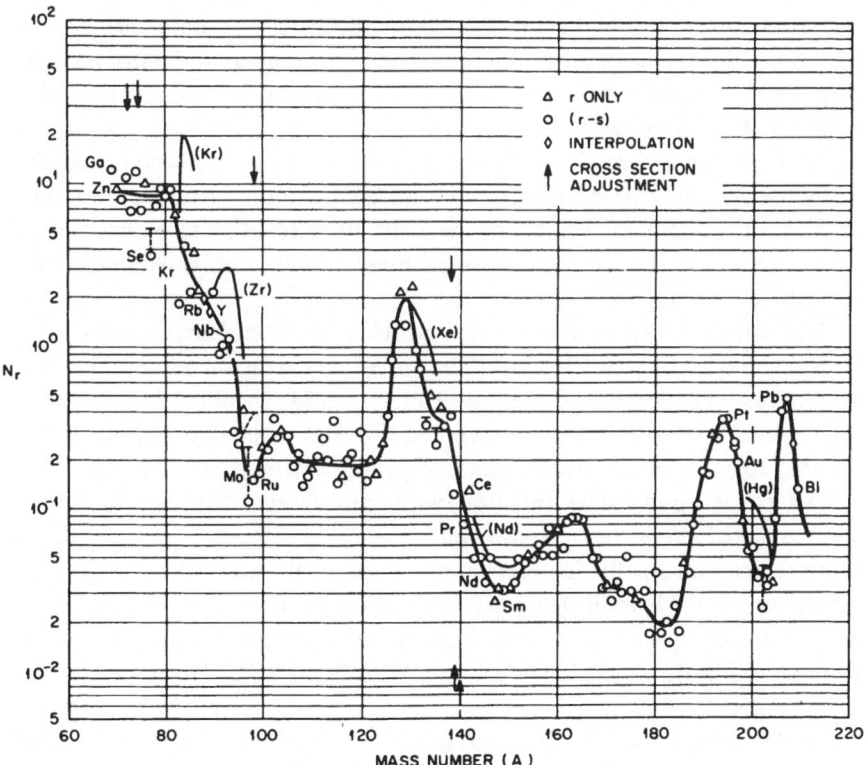

Solar System r−Process Abundances.

**Fig. 11. Solar system r-process abundances.** The solid curve is a best fit to the data, while curves at Kr, Zr, Xe, Nd, and Hg are the original distributions as derived from (Cam 68).

ture at $A \sim 88$ has never been explicitly determined. Clayton (Cla 64b) has discussed this feature and noted that the data could be equally well fitted by a smoothly decreasing function instead of the structured curve of the type shown in Figs. 5 and 10. Such structure is however an important feature of the $s$-process theory, or rather $s$-process boundary conditions such as the Fe-seed and the decreasing-exposure distribution $\varrho(\tau)$. According to Seeger (See 70) there is no single exposure or monotonic distribution of exposures which does not produce a discontinuity at $A \sim 88$, save for a uniform exposure with $N_s\sigma = $ constant. This is because of the proximity to the magic neutron number $N = 50$ and the consequent low capture cross sections as indicated in Figs. 8 and 9.

As a result of these considerations it was necessary for Clayton to postulate a minimum in the integrated neutron exposure $\varrho(\tau)$ so as to compensate for the precipice tendency and obtain a smoothly decreasing $N_s\sigma(A)$. Such a minimum could be interpreted in terms of a dominance of two distinct astrophysical neutron sources, e.g., $^{13}C(\alpha, n)$ and $^{22}Ne(\alpha, n)$. The sharp ledge–precipice structure at $A \sim 88$ as shown in Fig. 10 therefore precludes this argument.

Less apparent though is the break at $A \sim 138$ which seems smeared out by the $N_s\sigma$ data based mainly on estimated capture cross sections. The existence of a dominant ledge–precipice structure remains to be verified by measurements of the capture cross sections of the $s$-only isotopes $^{134}Ba$ and $^{136}Ba$.

The curve shown in Fig. 10 represents a weighted least-squares fit to the $N_s\sigma$ data by Seeger (See 70). The integrated neutron exposure called for is $\varrho(\tau) = 113.3\,\tau^{-4.4}$, with a cutoff at $\tau = 1.84$. In obtaining this fit, a factor of two uncertainty was assumed for the estimated capture cross sections, 10% for $s$-only abundances, 20% for $s(r)$, and 30% for $s, r$.

An earlier power-law function with cutoff term (SF 66) had been found unsatisfactory by Clayton and Rassbach (CR 67) because of its failure to account for the isotopic composition of Pb. In particular the high abundance of $^{208}Pb$ indicates a substantial equilibrium component as the low-capture cross section of that isotope has a trapping effect for nuclei in the recycling through alpha-decay to $^{206}Pb$ from $^{210}Po$. However the new fit by Seeger as shown in Fig. 10 accounts readily for the $N_s$ abundances of the Pb isotopes and in fact substantially exceeds $N_s(^{208}Pb)$. The calculated result $N_s\sigma(^{208}Pb) = 1.00$, yields $N_s(^{208}Pb) = 1.00/0.33 = 3.0$ for $\sigma(^{208}Pb) = 0.33$ mb, and can be compared with the empirical result $N_s(^{208}Pb) = 1.45$. Thus a significant equilibrium component can result at large $\tau$ with a suitably truncated power law.

## 5.3. Isotopic Tests of $N_s\sigma(A)$

The $N_s\sigma$ correlations, shown in Fig. 10, depend to a large extent on the relative elemental abundances which can be distorted by indeterminable amounts of physical and chemical fractionation over billions of years. In addition many data points depend on estimates of $r$-process abundances, which in fact are derived from interpolations of $N_s\sigma(A)$. The data used in the preparation of Fig. 10 are given in Table IV.

The only true and independent tests of the $s$-process are the correlations exhibited by the $s$-only isotopes of a particular element, where it is only required that the $N_s\sigma$ values should correspond to the local equilibrium approximation (Cla 64a), $N_s\sigma(A-1) = N_s\sigma(A)$ as modified by the general trend of the $N_s\sigma(A)$ curve. Two such cases are Te and Sm.

### 5.3.1. *Tellurium*

This highly interesting element is unique in that it has three $s$-only isotopes, shielded from the $r$-process by the heavy tin isotopes and by antimony (see Fig. 12). The natural abundances are quite low (e.g., the isotopic abundance of $^{123}$Te is only 0.87%), but several-gram samples of relatively pure isotopes were available for the measurements from the AEC's Isotopes Pool.

The shape of the observed cross section versus energy (MG 67a) was somewhat unusual in that it indicated a shoulder in the 30–50 keV range, possibly due to $p$-wave neutron effects but also possibly due to an undetected experimental error. The individual isotopic cross section results, when weighted for normal isotopic abundances, give a capture cross section at 30 keV for natural tellurium of 85 millibarns. This is in reasonable agreement with an earlier semiindependent measurement (MGI 63b) of $(97 \pm 9)$ millibarns. The difference is well within experimental uncertainties. The cross section results for $^{124}$Te are in excellent agreement with those of Bergman, Kapchigashev, Popov, and Romanov (Ber+ 65). In addition Bergman and Romanov (BR 67) later reported measurements on all of the relevant tellurium isotopes, which are in good agreement with the Oak Ridge data.

The averaged experimental results are summarized in Table IV. The $p$-process contributions, quite minor except for $^{123}$Te, have been estimated from the $p$-only $^{120}$Te and the neighboring $^{112}$Sn, $^{114}$Sn, and $^{115}$Sn abundances. The tellurium cross sections are relatively small, mostly due to the small level density associated with the proximity of the closed shells ($Z = 50$,

## TABLE IV. s-Process Data

| Element | Class | Abundance[a] | $N_r(A)$[d] | $N_s$ | $\langle\sigma\rangle$[e] | $N_s\sigma$ |
|---|---|---|---|---|---|---|
| Cu 63 | s(r) | 635 | 0 | 635 | 49 ± 14 | 31,100 |
| 65 | s(r) | 284 | 0 | 284 | 42 ± 7 | 11,900 |
| Zn 64 | s | 732 | — | 732 | 50* | 36,600 |
| 68 | s(r) | 278 | 0 | 278 | 23 ± 3 | 6,400 |
| Ge 70 | s | 25.8 | — | 25.8 | 84* | 2,170 |
| 74 | s, r | 46.0 | 10.0 | 36.0 | 35 ± 20 | 1,260 |
| Se 76 | s | 6.32 | — | 6.32 | 100* | 630 |
| Sr 86 | s | 5.76 | — | 5.76 | 74 ± 7 | 426 |
| 87 | s(r) | 4.10 | 0.22 | 3.8 | 109 ± 9 | 414 |
| 88 | s(r) | 48.2 | 2.0 | 46.2 | 6.9 ± 2.5 | 319 |
| Y 89 | s, r | 4.6 | 1.6 | 3.0 | 21 ± 4 | 63 |
| Zr 90 | s(r) | 7.6[b] | 2.2 | 5.4 | 12 ± 2 | 65 |
| 91 | s, r | 1.7 | 0.9 | 0.8 | 68 ± 8 | 56 |
| 92 | s, r | 2.6 | 1.0 | 1.6 | 34 ± 6 | 54 |
| 94 | s, r | 2.7 | 0.3 | 2.4 | 20 ± 2 | 48 |
| Mo 95 | s, r | 0.40 | 0.3 | 0.1 | 430 ± 50 | 43 |
| 96 | s | 0.42 | — | 0.42 | 90 ± 10 | 38 |
| 97 | s, r | 0.24 | 0.16 | 0.08 | 350 ± 50 | 28 |
| 98 | s, r | 0.60 | 0.16 | 0.44 | 110[c] | 48 |
| Ru 100 | s | 0.202 | — | 0.202 | 290* | 59 |
| Pd 104 | s | 0.164 | — | 0.164 | 270* | 44 |
| Cd 110 | s | 0.262 | — | 0.262 | 210* | 55 |
| Sn 116 | s | 0.60 | — | 0.60 | 100 ± 15 | 60 |
| 117 | s, r | 0.32 | 0.18 | 0.14 | 420 ± 30 | 59 |
| 118 | s, r | 1.01 | 0.18 | 0.83 | 63 ± 5 | 52 |
| 119 | s, r | 0.36 | 0.18 | 0.18 | 260 ± 40 | 47 |
| 120 | s, r | 1.39 | 0.18 | 1.21 | 50 ± 15 | 61 |
| Te 122 | s(p) | 0.17 | — | 0.164 | 270 ± 30 | 44 |
| 123 | s(p) | 0.059 | — | 0.054 | 820 ± 30 | 44 |
| 124 | s | 0.31 | — | 0.31 | 150 ± 20 | 46 |
| 125 | s, r | 0.47 | 0.35 | 0.12 | 430 ± 30 | 52 |
| 126 | s, r | 1.27 | 0.85 | 0.42 | 82 ± 8 | 34 |
| Xe 128 | s | 0.11[b] | — | 0.11 | 300* | 32 |
| 130 | s | 0.21 | — | 0.21 | 100* | 21 |
| Ba 134 | s | 0.114 | — | 0.114 | 155* | 17.7 |
| 136 | s | 0.367 | — | 0.367 | 37* | 13.6 |
| La 139 | s(r) | 0.36 | 0.12 | 0.24 | 48[c] | 11.5 |
| Nd 142 | s | 0.16 | — | 0.16 | 70* | 11.2 |
| Sm 148 | s | 0.026 | — | 0.026 | 260 ± 50 | 6.8 |
| 150 | s | 0.017 | — | 0.017 | 370 ± 70 | 6.3 |
| Gd 154 | s | 0.0073 | — | 0.0073 | 520* | 3.8 |
| Dy 160 | s | 0.0083 | — | 0.0083 | 650* | 5.4 |

**TABLE IV** (*Continued*)

| Element | Class | Abundance[a] | $N_r(A)$[d] | $N_s$ | $\langle\sigma\rangle$[e] | $N_s\sigma$ |
|---|---|---|---|---|---|---|
| Yb 170 | s | 0.0064 | — | 0.0064 | 510* | 3.2 |
| W  182 | s, r | 0.0042 | 0.018 | 0.0024 | 260 ± 30 | 6.2 |
|    183 | s, r | 0.023 | 0.018 | 0.005 | 550 ± 50 | 2.8 |
|    184 | s, r | 0.049 | 0.019 | 0.030 | 180 ± 20 | 5.4 |
| Os 186 | s | 0.0113 | — | 0.0113 | 330* | 3.7 |
|    187 | s, r | 0.0116 | 0.0071 | 0.0045 | 900* | 4.1 |
| Pt 192 | s | 0.0099 | — | 0.0099 | 490* | 4.9 |
| Hg 198 | s | 0.035 | — | 0.035 | 250* | 8.8 |
| Pb 204 | s | 0.057 | — | 0.057 | 43 ± 5 | 2.45 |
|    206 | s, r | 0.546 | 0.40 | 0.15 | 9.6 ± 3.0 | 1.44 |
|    207 | s, r | 0.596 | 0.50 | 0.10 | 8.7 ± 3.0 | 0.87 |
|    208 | s, r | 1.70 | 0.25 | 1.45 | 0.33 ± 0.07 | 0.48 |

* Semiempirical cross section estimates.
[a] Relative to Si = $10^6$, taken from (Cam 68).
[b] Revised abundances from Table V.
[c] Upper or lower limit of cross section used.
[d] Smooth interpolations of r-process abundances.
[e] Maxwellian-averaged capture cross sections at 30 keV (mb).

$N = 82$). For example, the resonance spacing is several keV for some of the isotopes and cross section is inversely proportional to spacing. This means that much more experimental sensitivity is needed than for the samarium isotopes, for example, which have level spacings of a few eV. The Oak Ridge results are not detailed (six neutron energy bands from 0.30 to 125 keV were used to obtain the Maxwellian-averaged cross section for $kT = 30$ keV) but agree with predictions of the s-process theory to within better than two statistical standard deviations.

The average ratios of the $N_s\sigma$ products for both sets of data, $N_s\sigma(122):N_s\sigma(123):N_s\sigma(124) = 1.0:1.08:1.04$, define the extent of the ledge indicated in the empirical $N_s\sigma$ curve for $A = 95$–124, and confirm to ± 5% the validity of the s-process theory.

## 5.3.2. *Samarium*

$^{148}$Sm and $^{150}$Sm are both s-only isotopes (see Fig. 2). Thus, the $N_s\sigma$ correlation can be tested in detail for two isotopes without any correction for r-process. Neighboring $N_s\sigma$ products should be very close to equal in this region of atomic weight since it is far from any closed nuclear shells (see Fig. 10).

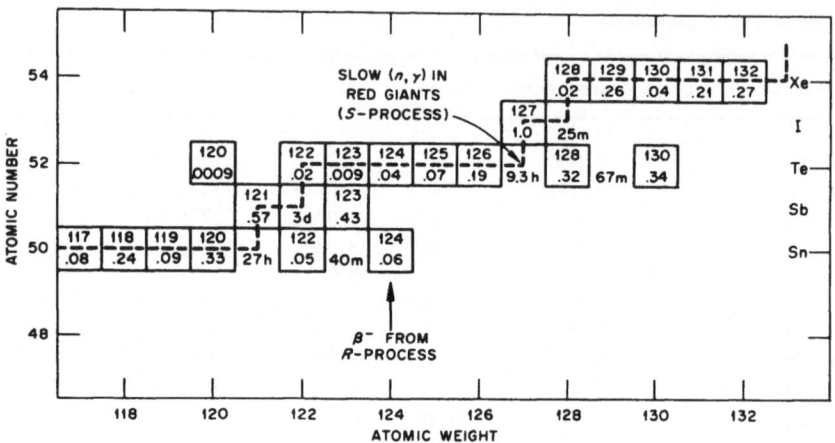

**Fig. 12. The s-process path near tellurium.** This element is unique in that three isotopes are shielded from the r-process, permitting the best quantitative test of s-process theory.

The results are summarized in Table IV. The $N_s\sigma$ products of ($^{148}$Sm + $n$) and ($^{150}$Sm + $n$) are identical, within experimental error. An even more sensitive test was obtained as follows: the cross section ratio, subject to considerably less error than separate absolute values, was determined for the two isotopes. The resultant ratio of the $N_s\sigma$ products is $N_s\sigma(148)/N_s\sigma(150) = 0.98 \pm 0.06$, and provides excellent confirmation of the local equilibrium approximation for the s-process.

If we accept the s-process hypothesis as valid, the close correlation observed allows an additional conclusion to be drawn (MG 67b). $^{149}$Sm has a huge capture cross section, even when compared to $^{147}$Sm, for neutron energies below several hundred electron volts. From this fact and the equality of the $N_s\sigma$ products we can conclude that (supported by other $N_s\sigma$ correlation results) solar system $^{149}$Sm has not been depleted by more than 4 percent by means of post s-process exposure to neutrons of energy $\lesssim 500$ eV such as might have resulted from high-energy solar flares in the early history of the planetary system.

Te and Sm are true isotopic tests of the $N_s\sigma$ correlation, in that these results are independent of r-process and elemental abundance assumptions. There are a number of "secondary" isotopic tests which require estimation and subtraction of $N_r$. The s, r isotopes of Te and Sm can also be included in this category. However, it is important to realize that to obtain the $N_r$ estimates, the r-process abundance distributions are derived in part from the smoothed empirical $N_s\sigma(A)$ curve, and the r-only abundances. Then individual values of $N_r$ are taken from the best fit empirical $N_r(A)$ curve. Results for Sr, Zr, Mo, Sn, Te, Sm, W, and Pb are included in Table IV. However, when $N_s \ll N_r$, as in the case of the Sm isotopes, the errors in $N_s$ forbid a direct test of $N_s\sigma(A)$. These data are discussed in detail in the following section. The complete list of abundances and cross sections of all data plotted in Fig. 10 is also given in Table IV.

## 5.4. Comments on $N_s\sigma(A)$ and $N_r(A)$

The interdependence of the $N_s\sigma(A)$ correlations and the $r$-process abundances has already been discussed. In this section the cross sections and abundances of various isotopes are considered in the framework of these two curves.

With reference to $N_r(A)$ it is assumed that odd–even effects should be absent, and if present, result from residual $s$-process abundance contributions or just bad data. The validity of this argument is based on the general smoothness of $N_r(A)$, particularly for the heavier elements where no significant odd–even effect is observed (see Section 1). Considerable success is achieved in a number of cases by varying capture cross sections (mostly within the experimental errors) to obtain a better fit to a smooth $N_r(A)$ curve. Such values as required are noted in the cross section table.

The greatest odd–even effect was in the $70 < A < 100$ region. While certain cross section modifications accounted for this, the abundances of Kr and Zr remain outstanding. Other abundance effects are at Xe, Nd, and Hg. Often $N_r(A)$ is extremely sensitive to $N_s\sigma(A)$ and effectively feeds back to give improved values for the parameters $N_s$, $\sigma$, and $N_r$. Specific features of the data follow in order of mass number:

(a)   The upper limit of $N_s\sigma$ for Cu and Ge (assuming $N = N_s$) requires that $N_s\sigma(A)$ be lower than that indicated by the $s$-only isotopes $^{64}$Zn and $^{70}$Ge. Consequently Cu and Ge are included, with $N_r = 0$, to improve the accuracy of $N_s\sigma(A)$ in this mass region.

(b)   The odd–even effect in $N_r$(Ge) was substantially reduced by lowering cross sections for $^{72}$Ge and $^{74}$Ge within the limits of experimental errors. Note however that $N_r(^{77}$Se) must remain unusually low because of its low isotopic abundance ($N$).

(c)   $^{86}$Kr is not plotted in the $N_s\sigma$ diagram as it appears anomalous in the $N_r$ diagram, and because measured cross sections are not available. A reduction in abundance to approximately $N/4$ is required to fit $N_r(A)$. A consequent increase in $\sigma(^{82}$Kr) is required ($\times 2.5$) to fit $N_s\sigma(A)$. Cross

TABLE V. Comparative Elemental Abundances

| Element | This work | (Cam 69) | (SFC 64) | (SU 56) | Notes |
|---|---|---|---|---|---|
| *Volatile elements* | | | | | |
| Kr | ~16 | 64.4 | 30 | 51.3 | interpolated |
| Xe | 5.0 | 7.10 | 3.2 | 4.0 | interpolated |
| Hg | ~ 0.38 | 0.75 | 0.15 | 0.28 | interpolated |
| *Nonvolatile elements* | | | | | |
| Zr | ~15 | 30 | 16 | 54.5 | chondrites |
| Nd | 0.57 | 0.77 | 0.64 | 1.44 | chondrites |

section values for the Kr isotopes as required from this analysis are given in Table II. The original $r$-process abundances for Kr are also shown in Fig. 11, and a new elemental abundance is recommended in Table V.

(d)  $^{87}Rb$ decays to $^{87}Sr$ with a half-life of $5 \times 10^{10}$ years. It is therefore necessary to modify abundances to that expected at the end of nucleo-synthesis of solar-system material. The value of $t \sim 10^{10}$ years used here is not critical and results in $N(^{87}Rb) = N_r(^{87}Rb) = 1.88$, and $N(^{87}Sr) = N_s(^{87}Sr) = 3.78$. Good agreement is obtained for $N_s\sigma(^{86}Sr)$ and $N_s\sigma(^{87}Sr)$, which define the extent of the ledge in $N_s\sigma(A)$. $N_r(^{88}Sr)$ is interpolated from $N_r(A)$ and indicates the sharp drop in $N_s\sigma(A)$ to $^{89}Y$.

(e)  The low abundance of $^{89}Y$ requires that $N_s\sigma(^{89}Y) < 100$, and inter-polation of $N_r(^{89}Y)$ fixes $N_s\sigma(^{89}Y)$. However this value is in conflict with that for the Zr isotopes.

(f)  It is apparent from consideration of $N_s\sigma(A)$ and $N_r(A)$ that the abundance of the Zr isotopes has been overestimated. Odd–even oscilla-tion in $N_r$ depends solely on $N(Zr)$, and a reduction to $0.5N(Zr)$ results in good agreement with all other data and improved agreement with Y in $N_s\sigma(A)$. The original Zr $r$-process abundances are shown in Fig. 11, and a new elemental abundance for Zr is suggested in Table V.

(g)  $^{93}Nb$ is mostly bypassed by the $s$-process, and the low abundance of this isotope sets an upper limit for $N_r(A)$.

(h)  To eliminate a gross odd–even effect in $N_r$ in the Mo isotopes, $\sigma(^{98}Mo)$ was reduced to its lower limit. This reduction is required to achieve positive values for $N_s$ in this region, and is suggested by the trend in isotopic cross sections for Mo (see Fig. 9).

(i)   The Sn isotopes show good agreement in $N_s\sigma(A)$ to within 10%, a satisfactory result in terms of errors in cross sections ($\pm 15\%$) and $N_r(A)$ (10%).

(j)   The Te isotopes, discussed earlier, define the edge of the plateau at $A \sim 125$.

(k)   A markedly improved fit to $N_r(A)$ is obtained for Xe by a reduction in $N(\text{Xe})$ to $0.7N(\text{Xe})$ (see Table V). $N_r(^{123}\text{Cs})$ and $N_r(^{135}\text{Ba})$ are both limited by their natural abundances. A consequence of the reduction in $N(\text{Xe})$ is an improved fit for $N_s\sigma(A)$ of $^{128}\text{Xe}$ and $^{130}\text{Xe}$.

(l)   The odd–even effect in $N_r$ for Nd and Sm isotopes can be eliminated by a change in $N$ for either isotope. However for positive values of $N_s$, as determined from $N_s\sigma(A)$, $N_r(A)$ must pass through Sm. Consequently $N(\text{Nd})$ is reduced to $0.75N(\text{Nd})$. Improved agreement is also obtained for $^{142}\text{Nd}$ with $N_s\sigma(A)$.

(m)   $^{110}\text{Cd}$, $^{142}\text{Nd}$, $^{160}\text{Dy}$, and $^{170}\text{Yb}$ reveal an average trend in $N_s\sigma(A)$ to an accuracy of 20%—well within the expected errors of the semiempirical cross section estimates.

(n)   $^{138}\text{Ba}$ and $^{139}\text{La}$ originally showed an unrealistic discontinuity in $N_r$, which could be eliminated by variation of capture cross sections of the order of the experimental errors.

(o)   For $^{176}\text{Lu}/^{176}\text{Hf}$, uncertainty as to the interpretation of these abundances (due to $s$-process branching) forbids the inclusion of these "$s$" process nuclides in the $N_s\sigma$ curve.

(p)   $\sigma(^{140}\text{Ce})$ needs to be increased to 12 mb to obtain a positive value for $N_r(^{140}\text{Ce})$. This result implies that either $\sigma[\text{Ce(Nat)}]$ or $\sigma(^{142}\text{Ce})$ are too low or high, respectively.

(q)   W values show a considerable spread in $N_s\sigma$, due mainly to errors in the estimation of $N_r$.

(r)   $N_s\sigma(^{187}\text{Os})$ shows good agreement with $N_s\sigma(A)$ for $t \sim 10^{10}$ year.

(s)   A clear case for overabundance is established for Hg from $N_r(A)$. Pt, Au, and Tl establish $N_r(A)$ for this region, and a reduction in $N(\text{Hg})$ to $0.5N(\text{Hg})$ eliminates the original Hg shoulder. A consequent reduction in $N_s\sigma(^{198}\text{Hg})$ also results, though a peak still remains. Further reduction in $N_r(\text{Hg})$ eliminates this peak but introduces excessive odd–even effects between Hg and Te. A reduction in $\sigma(\text{Hg}^{198})$ to one half the original semiempirical estimate eliminates the peak in $N_s\sigma(A)$.

Abundance data for the volatile elements Kr, Xe, Hg can be modified without fear of reprisal from the "geocosmologists," as various normalization factors and interpolations are used in their abundance estimations. While the small reduction in Nd abundance should not cause too much alarm, Zr is another case. Zr is a stable element whose abundance has been determined from carbonaceous chondrites (Ure 64). However Seeger *et al.* used a value comparable to that deduced from our $N_r(A)$ curve, and equal to that determined by Schmitt *et al.* (SBC 64) from chondrites. On this basis the new value for $N(Zr)$ becomes less disputable. Amiet and Zeh (AZ 68) have also questioned the Zr abundance.

## 5.5. *r*-Process Abundances—$N_r(A)$

The final *r*-process abundance data are plotted in Fig. 11. In those cases where major elemental abundance changes have been made, the original $N_r$ data are also indicated, and the new elemental abundances are given in Table V, together with Cameron's original estimates (Cam 68). All other abundance data remain as tabulated in that reference.

The major features of $N_r(A)$ correspond to those first determined by Seeger *et al.* (SFC 65), though in this paper the data are more complete. However, the peak at Pb, including in its wings Tl and Bi, has not been previously observed. This peak probably results from the *r*-process production of unstable transbismuth isotopes. Thus $N_r(^{206}Pb)$ is the sum total of *r*-process synthesis at mass numbers 206, 210, 214, 218, 222, 226, 230, and 234, and $N_r(^{207}Pb)$ at 207, 211, 215, 219, 223, 227, and 231. Larger mass numbers in these 4*n*-chains result in the production of $^{238}U$ and $^{235}U$ respectively which contribute also by their cosmoradiogenic decay (Cla 64a). Burbidge *et al.* (Bur+ 57) calculated, on the basis of extrapolated nuclear systematics, the abundances expected from the *r*-process at each of these atomic weights. The sum total of abundances for the progenitors of $^{206}Pb$ and $^{207}Pb$ are 0.48 and 0.37 (relative to Si $= 10^6$), a result which is in agreement with that deduced from $N_s\sigma(A)$ in this paper, remembering that a cosmoradiogenic correction has not been made (see next section). $N_r(^{209}Bi)$ is also the sum of various *r*-process products, but not so $^{205}Tl$, which defines the lower side of the Pb peak.

The peaks at $A \sim 75$, 130, and 195 correspond to neutron numbers just below the magic values of $N = 50, 82$, and 128. Seeger *et al.* (SFC 65) have indicated the possible *r*-process path, which on reaching the low capture cross sections at the magic neutron numbers, decays by fast $\beta$-emissions to less neutron-rich nuclei, until capture again becomes more probable than $\beta$-decay (Fig. 13).

The broad bump at $A \sim 160$ is possibly due to deformation energy in

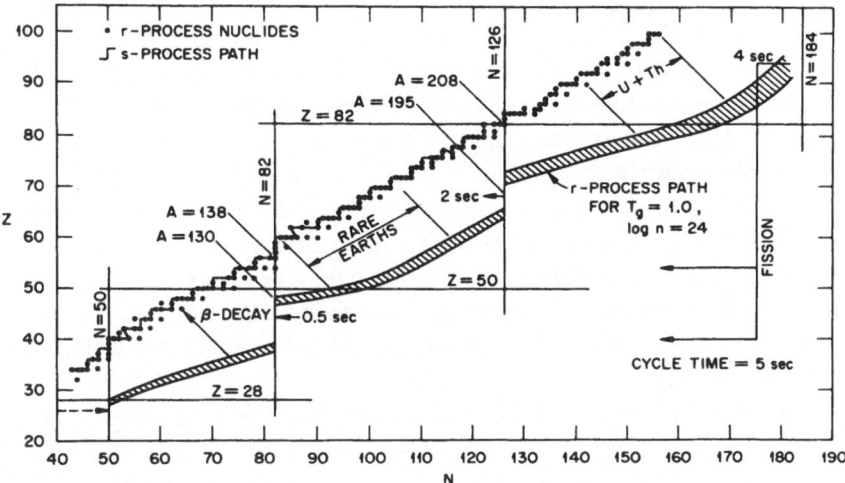

**Fig. 13. Neutron capture paths.** The s-process follows a path in the $N$–$Z$ plane near the line of beta-stability (represented by the stepped line). The r-process progenitor nuclei occur in a neutron rich area of the $N$–$Z$ plane (see shaded band), as calculated for $T_8 = 10$ and $10^{24}$ neutrons $cm^{-3}$. These progenitors subsequently beta-decay to the stable nuclei (dots). The times shown indicate the speed of the r-process buildup, till heavy nuclei undergo fission and double the number of nuclei every five seconds.

the progenitor nuclei. Seeger *et al.* could only reproduce this peak with a mass law including deformation terms based on the Nilsson model.

The overall observed structure requires two different fast neutron irradiations, for Seeger *et al.* found that production of the $A = 195$ peak would definitely imply the destruction of the lower mass peaks in a single exposure. Thus a short time scale ($\sim$4 sec) is required for the bulk of the r-process to form the high abundances at $A \sim 75$ and most of the $A = 130$ peak, while a second time scale long enough to cause fission cycling is required to account for the $A = 195$ peak.

Large numbers of neutrons are required for the formation of the various peaks ($A \sim 75$, 25 neutrons per Fe seed; $A = 130$, 75 neutrons; $A = 195$, $\sim$140 neutrons and more than 220 neutrons are required for cycling after fission). Truran *et al.* (Tru+ 68) have suggested that the base of an ejected supernova envelope is a probable source of such neutron fluxes which result in the production of peaks at $A = 130$ and 195. The passage of a supernova shock wave through the helium shell in a presupernova structure could induce rapid ($\alpha$, $n$) reactions which would result in the formation of the $A \sim 75$ peak.

## 5.6. Chronology

The study of the distant past has had quite a long history itself. Dating specific events like the formation of the earth and the sun has moved from the realm of religious insight to quantitative scientific inference based on laboratory measurements.

Some of the chemical elements and isotopes are, as far as we can measure, stable and thus might have existed forever. Others that we find in the earth and meteorites are constantly changing (back?) into the stable nuclides with a dissipation of energy. This leads us to infer a series of cataclysmic events in the distant past capable of incidentally producing these unstable isotopes.

Since the discoveries of Becquerel and the Curies, the stepwise decays of uranium and thorium to lead isotopes have been studied extensively. By measuring the lead isotopes in a rock one can get a rough measure of its age, and some still not washed away into the sea are as much as 3400 million years old. By measuring the lead isotopes in lead ore bodies formed at different times from the hot fluid interior parts of the earth it has been possible to set an age of 4600 million years for the earth with a precision of one percent or so (ORR 63). The formation of the sun and its planets (the solar system) occurred only slightly earlier. These results are fully consistent with the ages reported for lunar samples (Moo 70).

The temperatures, pressures, and other effects accompanying the solar-system condensation, however, are clearly insufficient to produce uranium and thorium. Only in conditions approximating a nuclear explosion (Bur+ 56) can we find ways to make the heaviest elements. The only observed astronomical candidate is the supernova. Its interior must get hot enough to promote interchange of parts among the lighter elements and may thereby produce enough free neutrons to build up the trace of very heavy elements we find in nature. The supernova explosion, of course, provides a convenient way to spew the newly formed elements out into space where they can later be gathered up into a new star such as the sun.

Other lines of evidence, particularly the radio waves from all over the universe, point to a universe some 2.5 to 4 times as old as the earth, formed by a primordial explosion of all matter. The temperatures and densities involved then, however, were too great for the heavy elements to form. Mostly hydrogen plus some 20% helium would result from that event, to later condense into the galaxies we see. Within each galaxy, stars later formed, glowed for a longer or shorter time, and finally burned out or exploded.

The heaviest stars burn fastest and explode. They should also form in the highest concentrations of gas and dust. Since over half of a burned out star is left behind as a "cinder" after the explosion, the gas and dust in a galaxy available for forming further stars is progressively reduced by those that have formed earlier in the history of a galaxy. On the average, then, we can expect the rate of supernova explosions and heavy elements synthesis to have been high soon after the formation of a galaxy and to have progressively decreased thereafter. While our solar system may have partaken of this average production of heavy elements it has also a particular history of its own, which is discussed in the next section.

## 5.6.1. *Lead–Uranium–Thorium Chronology*

The rapid fall off in $N_s\sigma(A)$ at Pb is determined by $N_s\sigma(^{204}\text{Pb})$ and $N_s\sigma(^{208}\text{Pb})$, the latter deriving from an interpolation of $N_r'$ which indicates $N_s \gg N_r'$. $N_r'$ is used here to indicate that

$$N(A) - N_s(A) = N_r'(A) = N_r(A) + N_c(A)$$

where $N_r$ is the sum of all progenitor nuclei in the $4n - 2$, $4n - 1$ and $4n$ series up to but not including $^{238}\text{U}$, $^{235}\text{U}$, and $^{232}\text{Th}$, for $^{206}\text{Pb}$, $^{207}\text{Pb}$, and $^{208}\text{Pb}$, respectively. $N_c$ refers to the cosmoradiogenic contribution of these long-lived isotopes (and their progenitors) to the corresponding Pb isotopes.

In order to determine the contributions to the Pb isotopic abundances, $N_r(A)$ and $N_c(A)$, estimates of the initial $r$-process abundances of the two groups (i.e., transbismuth and transuranium) of heavy isotopes are required. Clayton (Cla 64a) has obtained estimates in terms of two different $r$-process normalization factors $r$ and $r'$ referring to transbismuth and transuranium progenitors respectively.

In terms of these factors values of $N_r(A)$ and $N_c(A)$ are given below (Cla 64a):

|       | $^{206}\text{Pb}$ | $^{207}\text{Pb}$ | $^{208}\text{Pb}$ |
|-------|---------|---------|---------|
| $N_r$ | $0.48\,r$ | $0.37\,r$ | $0.27\,r$ |
| $N_c$ | $0.18\,r'$ | $0.60\,r'$ | $0.12\,r'$ |

Clayton's estimates are $r = r' = 0.4 \pm 0.1$; however these values do not result in agreement with our values of $N(A) - N_s(A)$, which require $r = 0.6$, $r' = 0.5$.

From the ratio of $N_c(^{206}\text{Pb})/N_c(^{207}\text{Pb})$ Clayton has devised a chronometric diagram for various exponential models for the nucleosynthesis of heavy nuclei. Clayton derived the following relation:

$$\frac{N_c(^{206}\text{Pb})}{N_c(^{207}\text{Pb})} = \frac{0.189(1 - f_s) - 0.48\,r/N}{0.206 - N_s(207)/N - 0.37\,r/N}$$

where $f_s = N_s(^{206}\text{Pb})/N(^{206}\text{Pb})$. Inserting our value of $f_s = 0.27$, and the primordial isotopic ratios (MS 63)

$$N(206) : N(207) : N(208) :: 0.189 : 0.206 : 0.585$$

we obtain

$$\frac{N_c(^{206}\text{Pb})}{N_c(^{207}\text{Pb})} = 0.41$$

Reading from Clayton's chronometric diagram this value leads to the conclusion that nucleosynthesis began 8 to $20 \times 10^9$ years before solar system formation,[*] and continued at an approximately uniform rate (i.e., negligible exponential falloff in rate).

An alternative model, first suggested by Cameron (Cam 62) and updated by Hohenburg (Hoh 69) and Wasserburg *et al.* (WSH 69), calls for an initial prompt synthesis ($A$), a continuous synthesis ($B$), and a last minute final synthesis ($C$) occurring before solar system formation. On the basis of the abundances of U and Th isotopes and the evidence of decay products from $^{129}$I and $^{244}$Pu, the relative contributions were determined by Hohenburg to be $A \sim 0.88$, $B \sim 0$, $C \sim 0.12$. The time interval between the last $r$-process burst and the onset of xenon retention was estimated at $\sim 170 \times 10^6$ years with a total galactic age of $\sim 8.7 \times 10^9$ years, with nucleosynthesis beginning only $\sim 4.1 \times 10^9$ years before solar system formation.

These values are also in reasonable agreement with our results above. Assuming $B = 0$, we find $A \simeq 0.94$, $C \sim 0.06$ with nucleosynthesis beginning $\sim 3.4 \times 10^9$ years before solar system formation or $\sim 8.2 \times 10^9$ years before the present.

With the exponential plus last event model (Fow 69), we obviously can fit a wider range of parameters. If we take Hohenburg's 12% for the fraction of solar system $r$-process production coming from the last event

---

[*] Estimated time of solar system formation is $4.6 \times 10^9$ years, and is defined by the beginning of Xe retention in meteorites.

for instance, the beginning of the exponential stage reaches back 11 to 14 × 10⁹ years ago with an exponential fall off in average $r$-process rate of about 60% by the time of the final event (local supernova?). The information available in 1969 has been analyzed in a detailed review by W. A. Fowler (Fow 70) who finds the duration of presolar system synthesis to be $(6.9 \pm 2) \times 10^9$ years, with a galactic age of $(11.7 \pm 2) \times 10^9$ years.

## 5.6.2. *Rhenium–Osmium Chronology*

The radioactive nucleus $^{187}$Re with a half-life of $5 \times 10^{10}$ years provides an example of the type of information not quite yet available to further elucidate the history of nucleosynthesis prior to the formation of the solar system.

During various stages of nucleosynthesis, $^{185}$Re and $^{186,187,188}$Os were formed in slow neutron capture chains, while $^{185,187}$Re and $^{188}$Os received additions (via subsequent beta-decay) from explosively rapid neutron capture events. While $^{186}$Os is shielded from the last sort of contribution, $^{187}$Os receives $r$-process contributions through the very slow beta-decay of the $^{187}$Re.

If we knew the solar Re/Os ratio, the $^{187}$Re half-life, and the fraction of the $^{187}$Os attributable to the $r$-process we could calculate "how long the $^{187}$Re had to decay" since its production in $r$-process explosions. With many such explosions contributing before the isolation of the solar system some $0.48 \times 10^{10}$ years ago in various amounts, we could obviously only test models, such as an exponentially decreasing frequency of $r$-process explosions throughout the galaxy followed by a single local one just proceding the formation of the solar system (Fow 69). Clayton (Cla 69) has recently pointed out that the $^{187}$Re half-life can have been substantially reduced from the laboratory value by any recycling through stellar interiors, as the excited-state decays induced at high temperatures are orders of magnitude faster.

To determine the fraction of $^{187}$Os attributable to the $r$-process, we must find the fraction attributable to slow neutron capture (the $s$-process). To do this we can use the local equilibrium approximation (Cla 64a)

$$N_s\sigma(187) = N_s\sigma(186) = N\sigma(186)$$

and yet to be performed measurements of the neutron-capture cross sections of $^{186}$Os and $^{187}$Os.

## 6. CONCLUSION

From the intensive knowledge gained from laboratory nuclear physics in this century science has become able to push back the veil of past history some ten billion years. With some understanding of how the stars keep shining we can see how gold and the other heavy elements could have been made from lesser things. Each new insight also entails quantitative relations that can be checked in our laboratories and observatories. In regard to the chemical elements beyond iron, neutron capture probabilities and beta-decay rates play a decisive role. We have here reviewed and summarized the relevant data, finding that a coherent and consistent picture can be obtained.

The neutron-capture cross sections, solar-system abundances, etc., are presented in detail so that further interpretation and inference can be applied in later analyses. The conclusions that emerge based on present interpretations lead to a time scale of roughly ten billion years for our galaxy.

The chemical elements formed in a series of nuclear reactions starting with hydrogen in stars which finally exploded, scattering some of the newly formed material back into the galactic gas and dust. Thus, new star systems could form incorporating heavy elements at their start. The solar system was one of these, forming about 4.6 billion years ago. An intriguing finding, recently receiving more experimental support, is a fresh addition of heavy elements to the solar system material just 170 million years or so before the earth formed and melted. We will probably see many studies in the near future of the possibility that this supernova-like event had more than an accidental connection with the mechanics of the formation of our sun and the rest of our solar system.

## ACKNOWLEDGMENTS

The authors are grateful to Dr. P. A. Seeger for his $N_s\sigma(A)$ calculations, and to Professors D. D. Clayton and W. A. Fowler for their enlightening comments on the manuscript.

## REFERENCES

(Abr+ 67)   A. I. Abramov, A. A. Vankov, V. N. Kononov, A. V. Malishev, and H. Stavisskii, A. V. Waparo, *Nucl. Data for Reactors*, **1**:459 (1967).
(ACM 68)   M. Ashgar, C. M. Chaffey, and M. C. Moxon, *Nucl. Phys.*, **A108**:535 (1968).
(AKS 68)   B. J. Allen, M. J. Kenny, and R. J. Sparks, *Nucl. Phys.*, **A122**:220 (1968).

(All+ 70)   B. J. Allen, and A. R. de L. Musgrove, to be published AAEC/TM (1970).

(Alp 48)   R. A. Alpher, *Phys. Rev.*, **74**:1577 (1948).

(Ash 69)   D. Ashery, *Nucl. Phys.*, **A135**:481 (1969).

(AZ 68)   J. P. Amiet and H. D. Zeh, *Z. Physik*, **217**:485 (1968).

(Bar+ 68)   Z. M. Bartolome, R. W. Hockenbury, W. R. Moyer, and J. R. Tatarczuk, R. C. Block, RPI-328-142, 11 (1968).

(BB 65)   V. Benzi and V. V. Bortonlani, *Nuovo Cim.*, **38**:216 (1965).

(BCD 68)   G. L. Boroughs, C. R. Cravens, and M. K. Drake, GA-8854 (1968).

(Ber+ 65)   A. A. Bergman, S. P. Kapchigashev, Yu. P. Popov, and S. A. Romanov, *Proc. Conf. Study Nuclear Structure with Neutrons, Antwerp (1965)*, North-Holland, Amsterdam; also BNL-325 (1966).

(BHC 68)   J. J. H. Berlijn, R. E. Hunter, and C. C. Cremer, LA-3527 (1968).

(BR 67)   A. A. Bergman and S. A. Romanov, *Lebedev Inst. Report*, No. 70, Moscow (1967).

(Bro 49)   H. S. Brown, *Rev. Mod. Phys.*, **21**:625 (1949).

(Bur+ 56)   G. R. Burbidge, F. Hoyle, E. M. Burbidge, R. F. Christy, and W. A. Fowler, *Phys. Rev.*, **103**:1145 (1956).

(Bur+ 57)   E. M. Burbidge, G. R. Burbidge, W. A. Fowler, and F. Hoyle, *Rev. Mod. Phys.*, **29**:547 (1957).

(Brz+ 68)   J. S. Brzosko, E. Gierlik, A. Soganik, A. Sollan, and Z. Wilhelmi, PAS-INR-940/I/P1 (1968).

(Cam 55)   A. G. W. Cameron, *Astrophys. J.*, **121**:144 (1955).

(Cam 57)   A. G. W. Cameron, AECL-454 (1957).

(Cam 62)   A. G. W. Cameron, *Icarus* **1**:31 (1962).

(Cam 63)   A. G. W. Cameron, *Nuclear Astrophysics*, Unpublished lectures (1963) [see (Cam 68)].

(Cam 68)   A. G. W. Cameron, in *Origin and Distributions of the Elements* (L. H. Ahrens, ed.), Pergamon, London (1968), p. 125.

(Car 68)   A. D. Carlson, CA-9057 (1968).

(Cla+ 61)   D. D. Clayton, W. A. Fowler, T. E. Hull, and B. A. Zimmerman, *Ann. Phys.*, **12**:331 (1961).

(Cla 64a)   D. D. Clayton, *Astrophys. J.*, **139**:637 (1964).

(Cla 64b)   D. D. Clayton, *J. Geophys. Research*, **69**:5081 (1964).

(Cla 69)   D. D. Clayton, *Nature*, **224**:56 (1969).

(CR 67)   D. D. Clayton and M. E. Rassbach, *Astrophys. J.*, **148**:69 (1967).

(CS 65)   A. K. Chaubey and M. L. Sehgal, *Nucl. Phys.*, **66**:267 (1965).

(CS 66)   A. K. Chaubey and M. L. Sehgal, *Phys. Rev.*, **152**:1055 (1966).

(CS 68)   A. K. Chaubey and M. L. Sehgal, *Nucl. Phys.*, **A117**:545 (1968).

(Dan 66)   I. J. Danziger, *Astrophys. J.*, **143**:527 (1966).

(Dav 68)   C. N. Davids, *Nucl. Phys.*, **A110**:619 (1968).

(DTH 60)   B. C. Diven, J. Terrel, and A. Hemmendinger, *Phys. Rev.*, **120**:556 (1960).

(FBB 55)   W. A. Fowler, E. M. Burbidge, and G. R. Burbidge, *Astrophys. J.*, **122**:271 (1955).

(Fow 68)   W. A. Fowler, *Proc. Conf. Neutron Cross Sections and Technology, Washington, D. C.*, NBS Special Publ. No. 299, **1**:1 (1968).

(Fow 69)   W. A. Fowler, *Bull. Am. Phys. Soc.*, **11, 14, 12** (Boulder Meeting) (1969).

(Fow 70)    W. A. Fowler, OAP-197 (1970).

(Fri+ 68)   S. J. Friesenhan, D. A. Gibbs, E. Haddad, F. H. Frohner, and W. N. Lopez, *J. Nucl. Energy*, **22**:191 (1968).

(Gib+ 61)   J. H. Gibbons, R. L. Macklin, P. D. Miller, and J. H. Neiler, *Phys. Rev.*, **122**:182 (1961).

(Gol 37)    V. M. Goldschmidt, *Skrifter Norske Videnskaps-Akad, Oslo I: Mat. Naturu*, K1 No. 4 (1937).

(Gol+ 66)   M. D. Goldberg, S. F. Mughabghab, S. N. Purobit, B. A. Magurno, and V. M. May, BNL-325 (1966).

(Gwi+ 69)   R. Gwin, L. W. Weston, G. de Saussure, R. W. Ingle, J. H. Todd, F. E. Gillespie, R. W. Hocknbury, and R. C. Block, ORNL-TM-2598 (1969).

(HBC 68)    R. E. Hunter, J. J. H. Berlijn, and C. C. Cremer, LA-3528 (1968).

(HCS 68)    S. S. Hasan, A. R. Chaubey, and M. L. Sehgal, *Nuovo Cim.*, **58B**:402 (1968).

(HGL 53)    D. J. Hughes, R. C. Garth, and J. S. Levin, *Phys. Rev.*, **91**: 1423 (1953).

(Hoc+ 68)   R. W. Hockenbury, Z. M. Bartolome, W. R. Moyer, J. R. Tatarczuk, and R. C. Block, RPI-328-142, 18 (1968).

(Hoc+ 69)   R. W. Hockenbury, Z. M. Bartolome, W. R. Moyer, J. R. Tatarczuk, and R. C. Block, *Phys. Rev.*, **178**:1746 (1969).

(Hoh 69)    C. M. Hohenburg, *Science*, **166**:212 (1969).

(Hug 46)    D. J. Hughes, *Phys. Rev.*, **70**:106A (1946).

(IPS 60)    A. I. Isakov, Yu P. Popov, and F. L. Shapiro, *Zh. Eksp. i Teoret. Fiz.*, **38**:989 (1960).

(ITO 61)    K. Ito, *Prog. Theor. Phys.*, **26**:990 (1961).

(JM 70)     M. Jaschek and S. Malaroda, *Nature*, **225**:246 (1970).

(Kom 69)    D. Kompe, *Nucl. Phys.*, **A133**:513 (1969).

(KPS 63)    V. A. Konks, Yu P. Popov, F. L. Shapiro, *Zh. Eksp. i Teoret. Fiz.*, **46**:80 (1963).

(Kon+ 67)   V. N. Kononov, Yu. Ya. Stavisskii, S. R. Chistozvonov, V. S. Shorin, *Nuc. Data Reactors Proc. Conf., Paris*, **1**:469 (1967).

(Mac 61)    R. L. Macklin, private communication (1961).

(Mer 52)    P. W. Merrill, *Science*, **115**:484 (1952).

(MG 65)     R. L. Macklin and J. H. Gibbons, *Rev. Mod. Phys.*, **37**:166 (1965).

(MG 67a)    R. L. Macklin and J. H. Gibbons, *Phys. Rev.*, **159**:1007 (1967).

(MG 67b)    R. L. Macklin and J. H. Gibbons, *Astrophys. J.*, **149**:577 (1967).

(MG 69)     R. L. Macklin and J. H. Gibbons, *Phys. Rev.*, **181**:1639 (1969).

(Moo 70)    Moon issue, Age Measurements, *Science*, **167**:3918, 461 (1970).

(MGI 63a)   R. L. Macklin, J. H. Gibbons, T. Inada, *Nucl. Phys.*, **43**:353 (1963).

(MGI 63b)   R. L. Macklin, J. H. Gibbons, T. Inada, *Phys. Rev.*, **129**:2695 (1963).

(Mox 68)    M. C. Moxon, AERE-R6074 (1968).

(MR 63)     M. C. Moxon and E. R. Rae, *Nucl. Instr. and Methods*, **24**:445 (1963).

(Mus 69)    A. R. de L. Musgrove, AAEC-E198 (1969).

(MS 63)     V. R. Murthy and R. A. Schmitt, *J. Geophys. Res.*, **68**:911 (1963).

(MS 66)     P. Morrison and L. Sartori, *Phys. Rev. Letters*, **16**:414 (1966).

(ORR 63)    R. G. Ostic, R. D. Russell, and P. H. Reynolds, *Nature*, **199**:1160 (1963).

(Pet 68)    J. G. Peters, *Astrophys. J.*, **154**:225 (1968).

(RB 67)     T. B. Rynes and D. W. Beale, *Int. J. Appl. Rad. Isotopes*, **18**:204 (1967).

(Ree 66)    H. Reeves, *Astrophys. J.*, **146**:447 (1966).

(Ree 68)    H. Reeves, *Stellar Evolution and Nucleosynthesis*, Gordon and Breach (1968).

(San 67)    R. H. Sanders, *Astrophys. J.*, **150**:971 (1967).

(SBB 68)    L. M. Spitz, E. Barnard, and F. D. Brooks, *Nucl. Phys.*, **A121**:655 (1968).

(SBC 64)    R. A. Schmitt, E. Bingham, and A. A. Chodos, *Geochim. et Cosmochim. Acta*, (1964).

(Sch 60)    H. W. Schmitt, *Nucl. Phys.*, **20**:220 (1960).

(See 70)    P. A. Seeger, private communication (1970).

(SFC 65)    P. A. Seeger, W. A. Fowler, and D. D. Clayton, *Astrophys. J. Suppl.*, **97**:121 (1965).

(SF 66)     P. A. Seeger and W. A. Fowler, *Astrophys. J.*, **144**:822 (1966).

(SKK+ 66)   V. S. Shorin, S. P. Kapchigashev, and V. E. Kolesov, Fiziko-Energeticheskii Institut, FE1-64 (1966).

(Stu+ 68)   D. C. Stupegia, M. Schmidt, C. R. Keedy, and A. A. Madson, *J. Nucl. Energy*, **22**:267 (1968).

(SU 56)     H. E. Snell and H. C. Arey, *Rev. Mod. Phys.*, **28**:53 (1956).

(TKK 67a)   V. A. Tolstikov, V. P. Koroleva, and V. E. Kolesov, BNL-tr-240 (1967).

(TKK 67b)   V. A. Tolstikov, V. P. Koroleva, and V. E. Kolesov, UDC-539-172-4 (1967).

(Tol+ 67)   V. A. Tolstikov *et al.*, *Atomnaya Énergiya* (*USSR*), **23**:151 (1967).

(Tol+ 68)   V. A. Tolstikov *et al.*, *Atomnaya Énergiya* (*USSR*), **24**:576 (1968).

(Tru+ 68)   J. W. Truran, W. D. Arnett, S. Tsuruta, and A. G. W. Cameron, *Origin and Distribution of the Elements*, (L. H. Ahrens, ed.) Pergamon, New York (1968), p. 77.

(Ure 64)    H. C. Urey, *Rev. Geophys.*, **2**:1 (1964).

(WSH 69)    G. J. Wasserburg, D. N. Schramm, and J. C. Huneke, *Astrophys. J.*, **157**: L91 (1969).

*Chapter 5*

# NUCLEAR STRUCTURE STUDIES IN THE $Z=50$ REGION*

## Elizabeth Urey Baranger

*Laboratory of Nuclear Science and Physics Department*
*Massachusetts Institute of Technology*
*Cambridge, Massachusetts*

and

*Department of Physics*
*University of Pittsburgh*
*Pittsburgh, Pennsylvania*

## 1. INTRODUCTION

### 1.1. The Tin Region

The purpose of this article is to present briefly the state of our knowledge of nuclei in a particular region of the periodic table, namely those nuclei with proton number near fifty. It is natural to group these together. According to our simplest models, their structures are intimately related. They are spherical nuclei, or at least not strongly deformed. The tin isotopes, with $Z = 50$, are classified as "single-closed-shell nuclei," since 50 is a magic number, and the differences in their structure are determined completely

* This work supported in part through funds provided by the U. S. Atomic Energy Commission under Contract No. AT(30-1) 2098 at Massachusetts Institute of Technology, Cambridge, Massachusetts, and by the National Science Foundation at the University of Pittsburgh, Pittsburgh, Pennsylvania under Grant #GP-9330.

by differences in the numbers of neutrons in the valence shells. The indium and antimony isotopes (with $Z = 49$ and $51$) have a proton particle or hole in addition to the fifty in the closed shell. There is a close relationship between their structure and that of the tin nuclei as well as relationships between their various isotopes. The shell-model description of nuclei in this region is thus fairly simple, and makes possible detailed microscopic calculations. On the other hand, the even-mass cadmium and tellurium isotopes ($Z = 48$ and $Z = 52$, respectively) look experimentally quite different from the single-closed-shell tin isotopes. It is also much more difficult to calculate their structure. For these reasons, the tin, antimony and indium isotopes make an island of nuclei, about fifty in number, whose properties can profitably be discussed as a unit.

This region near the single-closed shell at $Z = 50$ deserves to be studied both theoretically and experimentally. The variety of experiments which can now be done leads to a wealth of information; this can be used in a straightforward way to confirm or disprove basic shell-model ideas. A large number of stable isotopes exist; this makes possible the measurement of the change of various quantities with neutron number. And finally, this region is one of the very few places among heavy nuclei where detailed shell-model calculations can be made and tested by comparison with experiment.

In this article the discussion is restricted to the odd- and even-mass tin isotopes and the odd-mass antimony and indium isotopes. The even-mass antimony and indium isotopes have much more complicated spectra and there is not sufficient data to warrant a review at the present time.

In order to understand the experimental results, it is necessary to know the shell-model description of these nuclei and for this reason a brief discussion of it is given first, in Section 1.2. In Section 2, I survey what has been learned from experiment. Types of data which are available are listed, a discussion of what each reveals about the structure is given, examples are shown, facts which remain unexplained are pointed out, but no detailed, thorough comparison of all existing data is made. The hope is that someone unfamiliar with these nuclei can gain an understanding of what the experiments tell us about the structure without getting lost in detail.

After this discussion of experiment, Section 3 presents the successes and failures of the theoretical calculations. This will be restricted to shell-model calculations which use "realistic" residual interactions, i.e., interactions which fit the free nucleon–nucleon scattering data, and calculations which use more phenomenological approaches will not be discussed. At the end of the article, there is an extensive list of papers in which experimental results concerning these nuclei have been reported.

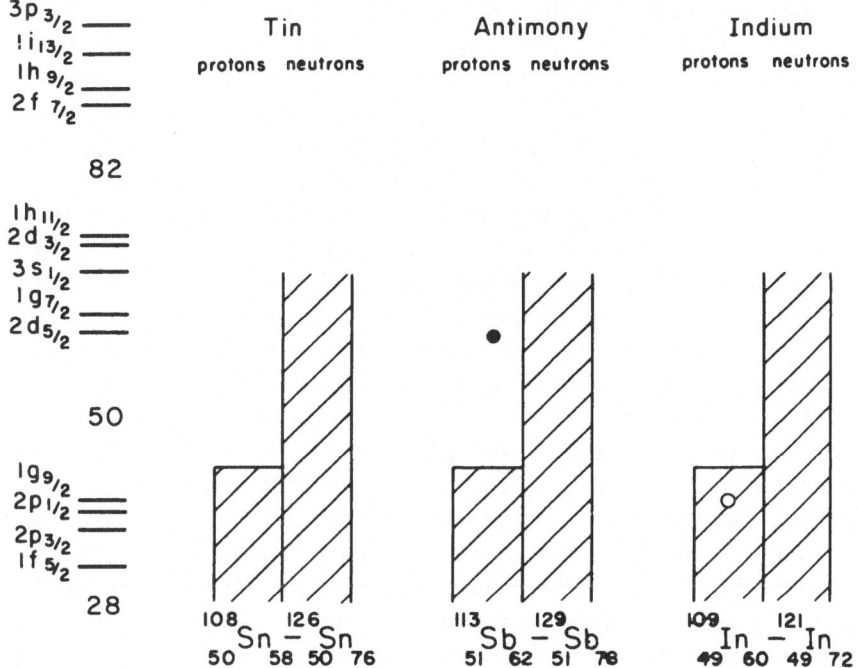

**Fig. 1. Shell-model description of the tin, antimony, and indium isotopes.** The levels which are being filled are plotted on the left. At the bottom of each column are listed those isotopes about whose spectra there is some information.

## 1.2. Shell-Model Description

The shell model predicts the single-particle levels which are occupied in any given nucleus; the appropriate ones for this region of the periodic table are listed in Fig. 1. The tin isotopes have fifty protons and therefore, according to the basic shell model, the lowest states of these nuclei have a closed proton shell. The proton levels up to and including the $1g_{9/2}$ are filled and the $2d_{5/2}$, $1g_{7/2}$, etc., are empty. There is experimental information for $^{108}_{50}\text{Sn}_{58}$ to $^{126}_{50}\text{Sn}_{76}$. The shell model says that in these there are 8 to 26 neutrons distributed among the $2d_{5/2}$, $1g_{7/2}$, $3s_{1/2}$, $3d_{3/2}$, and $1h_{11/2}$ levels or 6 to 24 neutron holes if one works down from the 82 closed shell. Thus all isotopes are far from a closed neutron shell. The antimony isotopes with $Z = 51$ have, in their lowest states, an extra proton in one of the levels $2d_{5/2}$, $1g_{7/2}$, etc., and indium, with $Z = 49$, has a proton hole in one of the levels $1g_{9/2}$, $2p_{1/2}$, and below. Both of them have neutrons distributed among the five levels $2d_{5/2}$ and $1h_{11/2}$.

Since the single-particle levels lie quite close together there is an enormous number of states very close in energy which will be mixed by the residual interaction neglected in the simplest shell model. So far this problem has been too large to be tackled with the conventional, or exact, shell-model programs (Fre+ 69, Coh+ 69). These programs have reached the stage where they could calculate either the heavy or light tin isotopes; their use should be extended as far as possible. The method used so far in these nuclei is the BCS or quasi-particle approximation* and even with exact shell-model calculations this will still be a useful way to think about them.

The most important property of the residual interaction between identical nucleons is to favor coupling them to angular momentum zero. In order to get the lowest possible energy, the particles in a given level must all be paired off, each pair coupled to zero. An odd-mass nucleus, with one unpaired particle, has therefore a higher energy. The BCS theory applied to nuclei, which is borrowed from the BCS theory of superconductivity, is designed to treat well these pairing properties of the interaction, in a simple, approximate way. It is applicable to single-closed shell nuclei, such as the tin isotopes, where there is only one kind of particle outside the closed shells. In this method the ground state of an even number of identical particles is presumed to be a linear combination of seniority zero wave functions, i.e., wave functions in which pairs of particles are coupled to angular momentum zero. The particular linear combination which is lowered in energy is determined by solving the "gap equations." These determine the occupation probabilities of the single-particle levels. The lowest states of the odd isotopes are certain seniority one states, called one-quasi-particle (1QP) states. In the formalism there is a 1QP state for each single-particle state. By solving the gap equations, the 1QP energies can be calculated; the smallest 1QP energy is the odd–even mass difference, the others give the excitation energy of the 1QP levels. The excited states of the even isotopes are 2QP, 4QP states; the excited states of the odd are 3QP, 5QP states. These occur at an excitation energy which is the sum of the appropriate 1QP energies. Thus the lowest excited states in the even isotopes occur at twice the lowest 1QP energy, and there is a gap in the energy spectrum which depends on this energy. To get more accurate results, interactions between the quasi-particles are then included. The low levels of the odd antimony and indium isotopes are described as single-proton particle or hole states with the neutrons in the BCS ground state; in the higher levels the neutrons are in 2QP states.

---

* Some review articles which discuss this approximation are (Bar 63, Lan 64, Jea 67, BS 69).

Calculations of this type were originally done with a pairing-plus-quadrupole force by Kisslinger and Sorensen (KS 59, KS 63), and since by others. It is however quite possible to use any form of residual interaction, and calculations with potentials of Gaussian shape were made quite early (Arv 63, Arv+ 63). Section 3 of this article reviews work in which "realistic" interactions were used with the BCS approximation.

This simple description of these nuclei is sufficient to start a discussion of the experimental data. As the article proceeds, a more sophisticated picture should emerge.

## 2. EXPERIMENTAL DATA

### 2.1. Even-Mass Tin Isotopes

In Fig. 2 the known levels of the even tin isotopes up to 3.2 MeV are shown. In Table I the energies, spins and, parities of some of the low levels are listed. The nuclei in the center of the figure are accessible with a wide variety of reactions because they and nuclei in their immediate vicinity are stable. A recent article (Bee+ 70) lists the different types of experiments which have been done to study the isotopes $^{116}$Sn to $^{124}$Sn. Tables of excitation energies together with spin and parity assignments are given and these

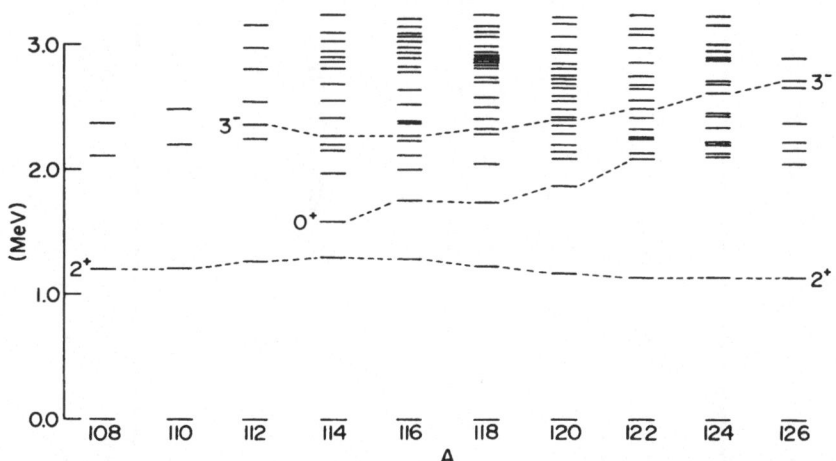

**Fig. 2. Levels below 3.2 MeV in the even-mass tin isotopes.** The data is taken from the following references: *A* = 108, 110 (Ye 69), *A* = 112 (Mak+ 68, Ye 69, KC 66), *A* = 114 (Bje+ 69), *A* = 116-124 (Bee+ 69), and *A* = 126 (Bje+ 69, Fly 69).

TABLE I. Experimental Excitation Energies of Levels in the Even Tin Isotopes. The lowest levels for $J$ = 2+, 3-, 4+, 5-, 6-, 6+, 7- and the two lowest for $J$ = 0+ are listed. The assignment of levels in parentheses is uncertain.

| $J$ | 108 | 110 | 112 | 114 | 116 | 118 | 120 | 122 | 124 | 126 |
|---|---|---|---|---|---|---|---|---|---|---|
| | | | | | $A$ | | | | | |
| 2+ | 1.205[b] | 1.211[a] | 1.257[a] | 1.299[b,c] | 1.293[a] | 1.230[a] | 1.172[a] | 1.140[a] | 1.131[a] | 1.164[c] |
| 3- | | | 2.35[h] | 2.274[b,c] | 2.268[d] | 2.328[d] | 2.408[d] | 2.492[d] | 2.613[d] | (2.720)[f] |
| 4+ | (2.111)[b] | (2.199)[b] | (2.251)[b] | 2.189[b,c] | 2.393[d] | 2.285[d] | 2.180[d] | 2.150[d] | (2.100)[d,i] | |
| 5- | | | | 2.817[b,c] | 2.368[d] | 2.326[d] | 2.280[d] | 2.245[d] | 2.213[d] | 2.054[f] |
| 6- | | | | | 2.777[d] | | | | | |
| 6+ | (2.365)[b] | (2.480)[b] | (2.552)[b] | | | | | | | (2.167)[f] |
| 7- | | | | | 2.913[d] | 2.580[d] | 2.490[d] | (2.400)[d] | 2.335[d] | (2.222)[f] |
| 0+ | | | | (1.58)[g] | 1.76[g] | 1.75[g] | 1.88[g] | 2.08[e] | | |
| | | | | 1.95[g] | 2.03[g] | 2.05[g] | 2.17[g] | | | |

a (Ste + 68).    g (SPC 67).
b (YE 69).    h (Mak + 68).
c (Bje + 69).    i 5-, according to (FB 70).
d (Bee + 70).
e (SHC 67).
f (FBB 70).

were used in composing the figure. Bjerregaard *et al.* (Bje+ 69) make a similar summary for $A = 114$. The level density appears less for $A = 108$, 110, 112, and 126, probably because fewer experiments have been done there. The levels in $A = 108$ and 110 were excited by the Cd($\alpha$, $xn$) reaction (YE 69), those in $A = 112$ by this reaction as well as inelastic scattering of protons and deuterons (Mak+ 68, KC 66), those in $A = 125$ by the $(t, p)$ reaction on $^{124}$Sn (Bje+ 69, FBB 70). As Fig. 2 emphasizes, there is a tremendous range of isotopes. This means there is a real opportunity to study properties of both ground and excited states as a function of neutron number.

The sudden increase in level density above 2 MeV can be explained using the quasi-particle picture. The lowest 2QP states should occur at an excitation energy approximately twice the odd–even mass difference which is about 1.1 MeV for these isotopes. Thus some of the increase in level density can be attributed to the many 2QP states lying above 2.2 MeV. For certain spins and parities there should be a number of 2QP states close together in energy and we can expect large changes due to the residual interaction. The lowest $2^+$ level is most affected; it is pulled down into the gap to an excitation energy of about 1.2 MeV. The position of the lowest 4QP states can then be estimated by assuming them to consist of two of the $2^+$ states weakly-coupled together; the excitation energy of these "two-phonon states" would be double that of the lowest $2^+$ level. Thus at 2.2 MeV and above there should be some 4QP states of even spin and parity in addition to a variety of 2QP states.

The energy of the lowest $2^+$ level is surprisingly constant with neutron number, considering that its wavefunction is expected to contain quite different quasi-particle components as the number of neutrons is changed. Its collectivity is evident in several experiments. The $(p, p')$ cross sections are strongly enhanced; for instance, for 25 MeV incident protons the cross sections in the isotopes $^{116}$Sn to $^{124}$Sn are four to ten times greater than to any other level excepting the $3^-$, which is also collective (Bee+ 70). The $B(E2)$ transition probabilities to the ground state are 12–15 times the single-particle value (KS 63). In (p, $t$) reactions, it is the strongest level after the very strongly excited ground state (Fle+ 70).

The radiative transition probabilities between these levels and the ground states have been accurately measured using Coulomb excitation. They decrease from $B(E2)_{0\to2} = (2.56 \pm 0.06) \times 10^{-49}$ cm$^4$ for $^{112}$Sn to $(1.61 \pm 0.04) \times 10^{-49}$ cm$^4$ for $^{124}$Sn (Ste+ 70). The fact that the $2^+$ states radiate is the first indication that the naive picture of a closed inert proton core is not valid. The extra neutrons polarize the core; the crudest way to include this in our simple model is to give a state-independent effective charge to the neutrons. Coulomb excitation techniques were also used to investigate the static quadrupole moments of the first 2+ states of the isotopes $^{112}$Sn and $^{116}$Sn to $^{124}$Sn (Ste+ 70). The resulting values are not large (for $^{120}$Sn, 0.09 $\pm$ 0.10 barn) and are, on the whole, consistent with zero (Ste+ 70, Kle+ 70). This is in contrast to the cadmium isotopes, $Z = 48$, where large values have been found.

Quasi-particle states of spin 0+ will also be shifted by the residual interaction. There have been observed low excited 0+ states, whose energies are given in Table I, and whose nature is not well understood. None of the available experimental information determines whether these are mainly 2QP states or 4QP (2-phonon) states. It would be very interesting to know the position of 0+ states in the light isotopes, since Fig. 2 indicates they might be very low. However, even in $^{114}$Sn their position is not certain. The level at 1.58 MeV was excited weakly in the $(d, t)$ experiments of (SPC 67), but only the 0+ level at 1.95 MeV was observed in the $(t, p)$ reaction (Bje+ 69).

It was early recognized that there exists in most nuclei a 3$^-$ collective level which is primarily a core-excited state (LP 60). The 3$^-$ levels shown in Fig. 2 are of this type. They are collective levels, in the sense that they have enhanced proton and deuteron inelastic scattering cross sections. However, since there are only two 2QP states in our model space which can couple to 3$^-$ (these are an $h_{11/2}$ coupled to a $d_{5/2}$ or $g_{7/2}$ quasi-particle) one is led to the supposition that these collective states lie outside the model space and are primarily particle–hole excitations of the core. Then the constancy of the excitation energy with changing neutron number is not surprising.

Figure 2 is somewhat misleading because it implies that every tin nucleus is like every other, while in fact the properties of the upper levels vary from nucleus to nucleus. There is information accumulating about them such as spins, a few magnetic moments and lifetimes, and extensive $(p, p')$, $(t, p)$, $(p, d)$, and $(d, p)$ data. The $(p, p')$ experiments were done with a resolution better than 20 keV (Bee+ 70); the $(t, p)$ with 15–20 keV (Bje+ 69, FBB 70). This kind of resolution, or better, is needed before one can be reasonably sure that all levels have been found. References to relevant experimental papers are given at the end of this article. Some of them will be discussed in Section 3.

Besides such studies of the region below 3 MeV, experiments have been done which pick out states at higher energies, outside the shell-model space described thus far. We will discuss three reactions of this kind; i.e.,

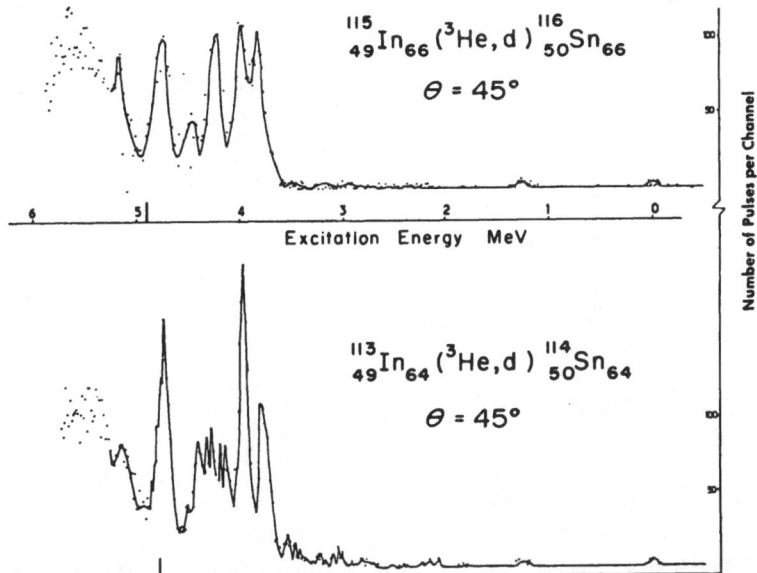

**Fig. 3. Spectrum of deuterons from In($^3$He, $d$) reactions with 18 MeV $^3$He particles.** The resolution is ~100 keV [from (Har 68)]. The line at 4.9 MeV marks the energy of particle–hole states estimated from proton separation energies.

the ($^3$He, $d$) reaction on the odd-mass indium isotopes, the ($t, p$) reaction on the heavy tin isotopes and a classic example, the ($\gamma, n$) reaction.

Since the ground state of indium consists predominantly of a $g_{9/2}$ proton hole configuration, a ($^3$He, $d$) reaction on an odd indium target is expected to excite, in addition to the ground state, certain proton particle–hole states in the resulting even tin nucleus. Figure 3 shows the spectrum obtained from ($^3$He, $d$) on $^{113}$In and $^{115}$In resulting in $^{114}$Sn and $^{116}$Sn. The resolution is not good, i.e., about 100 keV. This type of experiment has been done at Saclay (CHP 66, BCH 67, Har 68), Oak Ridge (Big+ 67), Colorado (Lin+ 69), and Florida (SFV 69). It is clear that the low states are weakly excited; that the strong peaks start at 3.8 MeV in both nuclei showing that these particle–hole states are indeed at high excitation. Also shown is the energy at which they should occur if there were no residual particle–hole interaction and if the single-particle and hole energies were taken in the usual way from experimental proton separation energies. This energy is 1.1 MeV above the first of the peaks and 0.6 MeV above the centroid (SFV 69). Shell-model calculations which include the particle–hole interaction and include mixing of various particle–hole states show that the

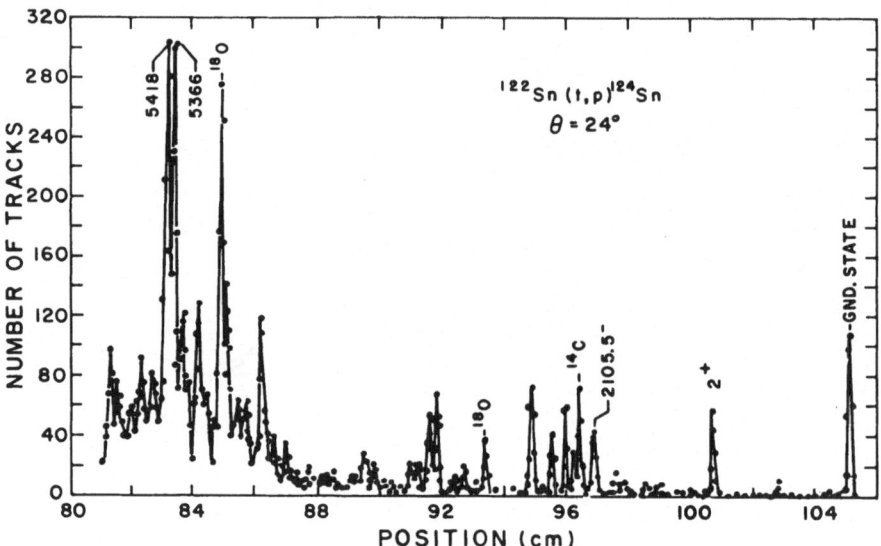

Fig. 4. Spectrum of protons from the $^{122}$Sn($t$, $p$)$^{124}$Sn reaction at 20 MeV. The resolution is 18 keV [from (FB 70)].

nuclear force yields a repulsive contribution to the energies, the Coulomb force, an attractive one, and that the result is close to the estimate from separation energies (CB 68). The procedure for taking particle–hole energies from experiment thus seems less successful in this single-closed shell nucleus than in the double-closed-shell nucleus lead (BT 67, Bje+ 68b, ASS 68). It is difficult to get a detailed understanding of these states. The Florida group (with 35 keV resolution) finds sixteen $l = 2$ peaks instead of the six expected from the $g_{9/2}^{-1}d_{5/2}$ configuration. This indicates that interaction is occurring between these simple particle–hole states and other states in this region, for instance, with states which are rearrangements of the neutrons in the valence shells or with two-particle, two-hole proton states which, from experimental binding energies, are expected to start at 5.2 MeV. (These should occur almost as low as the $1p1h$ states because of the pairing attraction between the two identical particles as well as between the two holes.)

The ($t$, $p$) reaction on the heavier isotopes has been found to excite strongly sharp states at high excitation energy. The experiments were done at Aldermaston (Bje+ 69) and Los Alamos (FB 70, FBB 70). In Fig. 4 the $^{122}$Sn($t$, $p$) spectrum of (FB 70) is plotted. The 0$^+$ ground state is strongly excited, as is usual in ($t$, $p$) reactions. The dominance of the ground-state cross section is in accord with BCS theory; it demonstrates physically that the ground states of the even-mass tin isotopes consist of correlated pairs of neutrons coupled to angular momentum zero. The higher states in the 2–3 MeV region are weakly excited, as expected. Then at ~5.3 MeV, there are two distinct strong peaks. The usual explanation (Boh 68, BRS 68) is that they are states consisting of the ground state of $^{122}$Sn plus two neutrons in the next shell—in single-particle levels not occurring in the ground state configuration. They would thus lie at high excitation energy. One particular $0^+$ state would be somewhat lower than the others due to the attractive nucleon–nucleon interaction and it is just such a coherent state, in which the attractive interaction brings the particles closer together, which would be strongly excited in $(t, p)$. This explanation is very reasonable, but unfortunately the angular distributions do not confirm a $0^+$ assignment but instead it is claimed (FB 70, FBB 70) they have $5^-$ spin, and that these states consist of a quasi-particle coupled to a particle in one of the levels in the next shell. Peaks similar to these are seen in $^{122}$Sn and $^{126}$Sn but in the lighter isotopes, neither the Aldermaston nor the Los Alamos group has seen them.

The classic reaction which excites primarily states of high excitation energy is the $(\gamma, n)$ reaction. A broad peak, the giant dipole resonance,

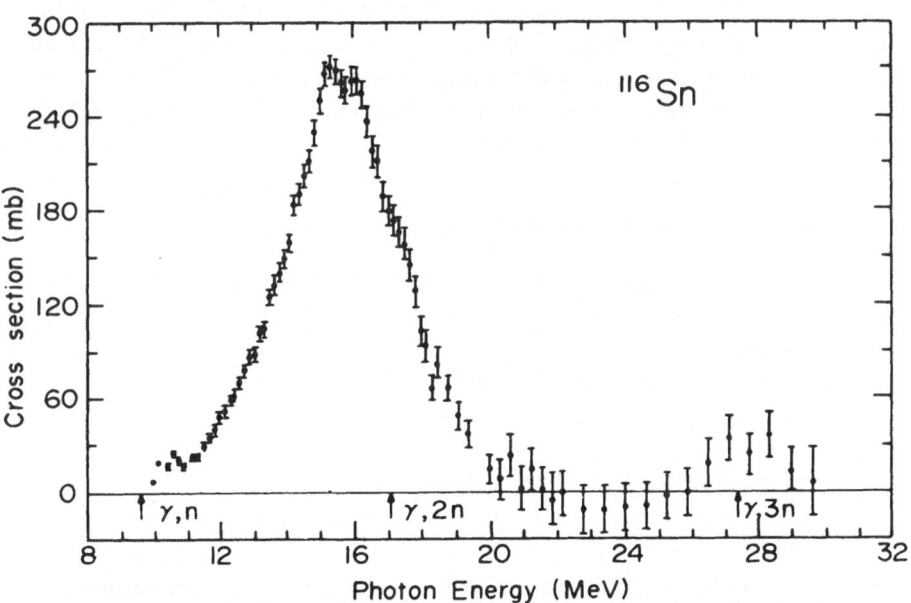

**Fig. 5.** Single-photoneutron cross section $\sigma[(\gamma, n) + (\gamma, pn)]$ for $\gamma$-rays incident on $^{120}$Sn. The resolution is 300–400 keV [from (Ful+ 69)].

occurs in this cross section for all nuclei at an excitation energy which varies smoothly with $A$. The cross section for $^{116}$Sn is shown in Fig. 5: the resolution is 300–400 keV. The variation of the energy of the peak of the total photoneutron cross section and its width in going from $^{116}$Sn to $^{124}$Sn is small; for instance, in $^{116}$Sn these are 15.67 $\pm$ 0.04 MeV and 4.19 $\pm$ 0.06 MeV while in $^{24}$Sn they are 15.18 $\pm$ 0.04 MeV and 4.80 $\pm$ 0.06 MeV (Ful+ 69). The isotope $^{116}$Sn has the narrowest peak for all the isotopes measured. From a microscopic point of view, states which occur so widely and vary with A so smoothly must be core-excited states. They are considered to be combinations of particle-hole states of spin 1⁻, which are collectively excited by electric dipole radiation.

The large number of stable tin isotopes means that it is possible to investigate changes in the nuclear charge and matter distributions with neutron number. The best determination of the matter distribution (GMP 70) has been made by analyzing proton elastic scattering and polarization data with the optical model. It is found that the neutron rms radius is only slightly larger than the proton radius; i.e., by 0.15 fm. The small changes in the proton charge distribution as neutrons are added have been detected by measuring the isotope shifts for optical, $K$ x-ray, and $\mu$-mesic atom transitions (WW 69, DD 69). The size changes less rapidly than $A^{1/3}$, in agreement with results in other regions of the periodic table. Accurate differences in the mean square radii of neighboring isotopes have been extracted (BBL 69). The results decrease gradually from the value $\delta\langle r^2 \rangle = 0.165 \pm 0.011$ fm² for the isotopes $^{112}$Sn to $^{114}$Sn to a value of $\delta\langle r^2 \rangle = 0.05\epsilon \pm 0.011$ fm² for $^{122}$Sn to $^{124}$Sn.

## 2.2. Odd-Mass Tin Isotopes

The neutron separation energies for the tin isotopes, $S_n(A)$, are listec in Table II. It takes more energy to remove a neutron from an even nucleus, where all neutrons are paired, than from an odd nucleus, where one neutror is unpaired. The odd–even mass difference, $P_n(A)$, which is a measure o the pairing energy, is defined as

$$P_n(A) = \tfrac{1}{2}[S_n(A + 1) - S_n(A)]$$
$$= \tfrac{1}{2}[BE(A + 1) + BE(A - 1)] - BE(A), \qquad A \text{ odd}$$

where $BE(A)$ is the binding energy of the tin isotope of atomic mass $A$. The mass difference $P_n(A)$ is plotted below in Fig. 13. It has a fairly constan value of about 1.2 MeV except for a marked dip at $A = 115$.

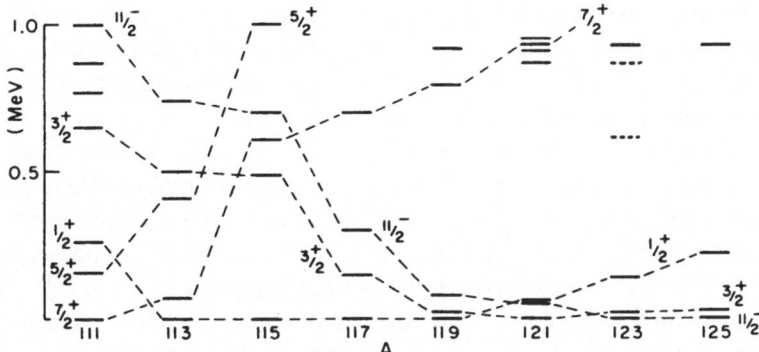

**Fig. 6. Levels below 1 MeV in the odd tin isotopes.** Where possible, the data were taken from the spectroscopic information listed in "Table of Isotopes" (LHP 68); these were supplemented for the higher lying states with the reaction data given in (SPC 67, Cav 69, Bor+ 69, FHM 70, and private communication).

**TABLE II. Neutron Separation Energies and the Odd Even Mass Difference[a]**

| Nucleus | $S_n(A)$, keV | $P_n(A)$, keV |
|---------|---------------|---------------|
| $^{111}$Sn | $8,177 \pm 40^b$ | $1,307 \pm 35$ |
| $^{112}$Sn | $10,790 \pm 25^c$ | |
| $^{113}$Sn | $7,444 \pm 17$ | $1,288 \pm 16$ |
| $^{114}$Sn | $10,320 \pm 16$ | |
| $^{115}$Sn | $7,537 \pm 9$ | $1,013 \pm 8$ |
| $^{116}$Sn | $9,563 \pm 7$ | |
| $^{117}$Sn | $6,941 \pm 5$ | $1,195 \pm 5$ |
| $^{118}$Sn | $9,331 \pm 4.6$ | |
| $^{119}$Sn | $6,481 \pm 4.6$ | $1,315 \pm 4$ |
| $^{120}$Sn | $9,110 \pm 4.3$ | |
| $^{121}$Sn | $6,181 \pm 6$ | $1,312 \pm 6$ |
| $^{122}$Sn | $8,804 \pm 7$ | |
| $^{123}$Sn | $5,932 \pm 11$ | $1,287 \pm 11$ |
| $^{124}$Sn | $8,506 \pm 11$ | |
| $^{125}$Sn | $5,767 \pm 12$ | $1,197 \pm 20$ |
| $^{126}$Sn | $8,160 \pm 27^d$ | |

[a] All numbers are taken from (MTW 65) except those labelled ([b]), ([c]), and ([d]).
[b] See (Cav + 70, Fle + 70).
[c] See (Cav + 70).
[d] See (Bje + 69).

The lowest levels in the odd tin isotopes are plotted in Fig. 6. The fact that each nucleus has several levels below 1 MeV and that the spectra differ drastically among each other is in marked contrast to the spectra of the even isotopes, shown in Fig. 2. The BCS theory describes these low levels as neutron 1QP states—one for each single-particle state—and predicts that their energies change in a regular way with neutron number. There are no low levels which do not fit into this 1QP description and the excitation energies do vary qualitatively as expected. The $11/2^-$ and $3/2^+$ levels, which can be described as nearly single-particle $h_{11/2}$ and $d_{3/2}$ states in $^{111}$Sn, start to fill as neutrons are added and become the ground states in the heavier isotopes. The $7/2^+$ and $5/2^+$, which are approximately single-hole $g_{7/2}$ and $d_{5/2}$ states, rise, and the $1/2^+$ falls, turns, and rises as the $s_{1/2}$ level becomes filled.

If these states were pure 1QP states, then the $(d, p)$ and $(p, d)$ strength to levels of a definite spin and parity should be concentrated in one level. Furthermore, the $(d, p)$ spectroscopic factor to this level would yield $u_J{}^2$, which is the probability that the $J$th single-particle level is empty and the $(p, d)$ spectroscopic factor would yield $v_J{}^2$, the probability that it is filled. Indeed, little fractionization is observed except for the $5/2^+$ levels for $A > 117$, in which cases the 1QP level is no longer at low excitation energy. For the other spins, neither experiment reveals any level with a strength more than 10% of the main transition for $A \geq 115$; for $A = 111$ and 113 there is some evidence of fractionization of the $1/2^+$ strength (SPC 67, Bor+ 69, Cav+ 69). The $(p, d)$ and $(d, p)$ spectroscopic factors to these low levels are plotted in Fig. 14 below. If the states were pure and if DWBA theory were perfect, the experimental curves would coincide. Indeed the agreement between them is quite good. All the spectroscopic factors show the correct trend with $A$, i.e., $u_J{}^2$ decreases and $v_J{}^2$ increases as neutrons are added. They agree quite well with theoretical predictions, as will be discussed later. Thus these states act like 1QP states, but the inaccuracy of DWBA analysis makes more quantitative statements difficult.

There is evidence on this question from experiments which are more sensitive to deviations from the 1QP picture. First, all the known magnetic moments, which are listed in Table III, deviate from the Schmidt values, which are the values expected for 1QP states. Secondly, $M1$ transitions have been observed between the $3/2^+$ and $1/2^+$ levels in $^{119}$Sn and $^{121}$Sn. These would be forbidden transitions ("$l$-forbidden") if the levels were $d_{3/2}$-, $s_{1/2}$-1QP states. Thirdly, the experimental log $ft$ values for transitions involving a $d_{5/2}$ proton particle and a $d_{3/2}$ neutron quasi-particle or a $g_{9/2}$ proton hole and a $g_{7/2}$ neutron quasi-particle are all larger than the 1QP

TABLE III. Magnetic Moments of States in Odd-Mass Isotopes of Indium, Tin, and Antimony

|  | $J^\pi$ | $\mu$ (exp) | $\mu$ (sp) |
|---|---|---|---|
| *Indium Isotopes* | | | |
| $^{111}_{49}\text{In}_{62}$ | 9/2+ | 5.53 | 6.79 |
| $^{113}_{49}\text{In}_{64}$ | 9/2+ | 5.523 | 6.79 |
|  | 1/2− | −0.210 | −0.26 |
| $^{115}_{49}\text{In}_{66}$ | 9/2+ | 5.534 | 7.69 |
|  | 1/2− | −0.244 | −0.26 |
| $^{117}_{49}\text{In}_{68}$ | 1/2− | −0.2515 | −0.26 |
| *Tin Isotopes* | | | |
| $^{113}_{50}\text{Sn}_{63}$ | 1/2+ | ±0.88 | −1.91 |
| $^{115}_{50}\text{Sn}_{65}$ | 1/2+ | −0.918 | −1.91 |
| $^{117}_{50}\text{Sn}_{67}$ | 1/2+ | −1.000 | −1.91 |
| $^{119}\text{Sn}_{69}$ | 1/2+ | −1.046 | −1.91 |
|  | 3/2+ | 0.68 | 1.15 |
| $^{121}_{50}\text{Sn}_{71}$ | 3/2+ | ±0.70 | 1.15 |
| *Antimony Isotopes* | | | |
| $^{115}_{50}\text{Sb}_{64}$ | 5/2+ | +3.46 | 4.79 |
| $^{117}_{51}\text{Sb}_{66}$ | 5/2+ | +2.67 | 4.79 |
| $^{119}_{51}\text{Sb}_{68}$ | 5/2+ | +3.45 | 4.79 |
| $^{121}_{51}\text{Sb}_{70}$ | 5/2+ | +3.359 | 4.79 |
|  | 7/2+ | +2.51 | 1.72 |
| $^{123}_{51}\text{Sb}_{72}$ | 7/2+ | +2.547 | 1.72 |
| $^{125}_{51}\text{Sb}_{74}$ | 7/2+ | ±2.6 | 1.72 |

All values are taken from "Table of Rounded-off Values and Index" which appears in the article "Nuclear Spins and Moments," G. H. Fuller and V. W. Cohen, in *Nuclear Data* A5:433 (1969).
(exp) — experimental.
(sp) — single particle.

value. This is shown in Table IV, where the experimental log $ft$ values are compared with the single-particle and the 1QP values. The 1QP results, (in which the proton is treated as a pure particle or hole), are slightly inhibited compared to single-particle values, but they are not as large as the experimental values. These facts can probably all be explained by reasonably small admixtures of certain 3QP states into the 1QP state. Calculations using a pairing-plus-quadrupole residual interaction fail to

TABLE IV. Log $ft$ values for $2d$ and $1g$ decays[a]

| Decay | $j_i$ | $j_f$ | $(\log ft)_{exp}$ | $(\log ft)_{sp}$ | $(\log ft)_{QP}$ |
|---|---|---|---|---|---|
| $_{51}Sb_{64}^{115} \rightarrow {}_{50}Sn_{65}^{115}$ | $5/2^+ \rightarrow 3/2^+$ | | 4.7 | 3.52 | 3.62 |
| $Sb^{117} \rightarrow Sn^{117}$ | $5/2^+ \rightarrow 3/2^+$ | | 4.9 | 3.52 | 3.68 |
| $Sb^{119} \rightarrow Sb^{119}$ | $5/2^+ \rightarrow 3/2^+$ | | 5.0 | 3.52 | 3.76 |
| $Sn^{121} \rightarrow Sb^{121}$ | $3/2^+ \rightarrow 5/2^+$ | | 5.0 | 3.34 | 3.68 |
| $Sn^{123} \rightarrow Sb^{123}$ | $3/2^+ \rightarrow 5/2^+$ | | 5.2 | 3.34 | 3.77 |
| $Sn^{125} \rightarrow Sb^{125}$ | $3/2^+ \rightarrow 5/2^+$ | | 5.4 | 3.34 | 3.91 |
| $_{50}Sn_{61}^{111} \rightarrow {}_{49}In_{62}^{111}$ | $7/2^+ \rightarrow 9/2^+$ | | 4.7 | 3.38 | 3.57 |
| $Sn^{113} \rightarrow In^{113}$ | $7/2^+ \rightarrow 9/2^+$ | | 4.6 | 3.38 | 3.51 |
| $In^{115} \rightarrow Sn^{115}$ | — | — | — | 3.38 | 3.47 |
| $In^{117} \rightarrow Sn^{117}$ | $9/2^+ \rightarrow 7/2^+$ | | 4.5 | 3.47 | 3.53 |
| $In^{119} \rightarrow Sn^{119}$ | $9/2^+ \rightarrow 7/2^+$ | | 4.4 | 3.47 | 3.52 |
| $In^{121} \rightarrow Sn^{121}$ | $9/2^+ \rightarrow 7/2^+$ | | 4.7 | 3.47 | 3.51 |
| $In^{123} \rightarrow Sn^{123}$ | $9/2^+ \rightarrow 7/2^+$ | | $\sim$4.5 | 3.47 | 3.50 |

[a] The experimental values are taken from *Table of Isotopes* (LHP 68) and *Nuclear Data Sheets* (NDS 60); the 1QP values were computed using (CB 68).

achieve mixing of the proper type (BS 69), but calculations made with more realistic and thus more complicated interactions do bring better agreement between experiment and theory (Ham 65, HB 67, FK 61, KBB 66a).

The fact that these low levels can be described as predominantly 1QP states does not mean that we have proved that the wave functions are literally a neutron quasi-particle in addition to a BCS-neutron state and a closed-shell proton state. The experiments we have used to show the 1QP nature of the levels involve relations between two states and tell us little about the real wave functions but only the relationship between them. (This point has been emphasized by Halbert and Krieger (HK 70) in a discussion of the single-particle states in the lead region.) More realistically we can say that these states have been shown to be predominantly 1QP states with respect to the ground state of the even isotopes, but this ground state itself may be very far from the simple state described so far.

The level density increases above 1 MeV and new types of states appear. The weak-coupling model is useful in discussing this particular energy region. According to this model the levels in the odd tin isotopes will be grouped into multiplets whose wave functions consist of one of the eigenfunctions of the neighboring even tin isotopes, usually referred to as a "phonon," coupled to a quasi-particle. The energy of the multiplet above the 1QP level will be, to first approximation, just the excitation energy of the phonon. If the particle–phonon coupling is strong, then the degeneracy between states of different angular momentum is removed; if several phonon-plus-quasi-particle states lie close enough together, then there will be mixing of different multiplets. This description is completely destroyed if the interaction between the extra quasi-particle and one of the particles in the phonon is strong, so that the phonon is broken up. Corrections due to neglect of the Pauli principle can also be important in changing the results. The model has not yet been justified in this region through a reliable microscopic structure calculation. One attempt showed large disagreement between the energies of the microscopic and quasi-particle-plus-phonon results (KBB 66a). As will be discussed below, the experimental data are not extensive; agreement with a limited amount of data cannot prove a model. The multiplets which are most likely to have some validity are those based on the lowest $2^+$ "phonon," which lies at an excitation energy of only 1.2 MeV and is rather separated from other levels.

Experimental information about the energy region between 1 MeV and 2 MeV is now emerging; the nucleus $^{117}$Sn will be treated in detail as an example. Somewhat above 1 MeV, the weak-coupling model predicts states which consist of the lowest $2^+$ state in the neighboring even tin nucleus coupled to an $s_{1/2}$ or $d_{3/2}$ quasi-particle. From Fig. 6 it can be seen that the $d_{5/2}$ single-quasi-particle state is expected to be in this same energy region. Thus this model predicts three $5/2^+$, two $3/2^+$, one $1/2^+$, and one $7/2^+$ levels at about 1.0 to 1.5 MeV. In Fig. 7, all the known levels in $^{117}$Sn below 1.5 MeV are shown, together with the results of a variety of experiments. We use these to make simple statements about the levels using this weak-coupling model and to examine the validity of the model.

The $3/2^+$ level at 1005 keV, which has a large $B(E2)$ to the ground state, also has a strongly inhibited $M1$ rate to the ground state (Ste 67). Both these facts are consistent with it being the collective $2^+$ level coupled to an $s_{1/2}$ quasi-particle. The $5/2^+$ levels at 1020, 1180, and 1445 keV are more complicated. The $(p, d)$ reaction picks out the $d_{5/2}$ single-quasi-particle strength and it is split between the 1020- and 1180-keV levels. On the other hand both of these levels show strong collective properties in their large $B(E2)$ values. As Stelson has pointed out, the large $B(E2)$ to the $3/2^+$ state for the level at 1180 keV indicates it has a large component consisting of a collective state built on the $d_{3/2}$-1QP state. These experiments show that there is strong mixing between the 1QP state and the collective states, but that the collective states do not mix with each other.

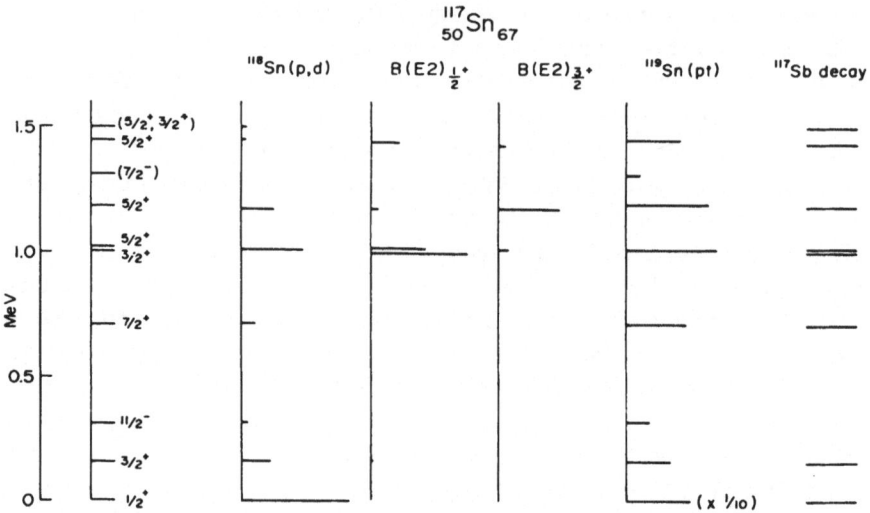

Fig. 7. Cross sections of reactions leading to levels in $^{117}$Sn. The first column gives all known levels below 1.5 MeV (Bee+ 69b, Rob+ 69, SPC 67); the second column, $(p, d)$ peak cross sections with 30 MeV protons and an energy resolution of 55–70 keV (Cav 70); the third and fourth columns, $B(E2)$ values to the ground state and to the $3/2^+$ first excited state (Ste 67 and private communication); the fifth column, $(p, t)$ integrated cross sections with 20 MeV protons and an energy resolution of 60 keV (Hol+ 69); the sixth column, the levels seen in the decay of $^{117}$Sb (Bee+ 69).

The $(p, t)$ results seem to contradict this. The simplest interpretation of $(p, t)$ experiments with the target in a 1QP state is that the multiplets consisting of the $2^+$ phonon coupled to this 1QP state would be enhanced since the $2^+$ level in the even tin isotopes is enhanced in such reactions. The extra quasi-particle would act as a spectator. But as seen in Fig. 7, the $(p, t)$ reaction on $^{119}$Sn, which has a $1/2^+$ ground state, does not show a preference for states which have large values of $B(E2)_{1/2^+}$. This leads to the suspicion that the $(p, t)$ reaction mechanism is not so simple. In addition, a different picture might appear if the experiment were done with better resolution.

In this nucleus there is a need for reaction experiments with better resolution. Some such experiments have been done in other odd-mass tin isotopes, i.e., $^{123}$Sn excited by $^{122}$Sn$(d, p)$ with 10 keV resolution (Bor+ 69) and $^{121}$Sn excited by $^{119}$Sn$(t, p)$ and $^{120}$Sn$(t, d)$ with 15 and 10 keV resolution, respectively (FHM 70 and private communication). This is the kind of resolution needed if the spectrum from 1–2 MeV is to become well established.

As the excitation energy increases so does the level density, and the structure becomes more complicated. Little is known about this region above 2 MeV. At high excitation energy ($\sim$15.6 MeV) the giant dipole resonance has been observed in $^{117}$Sn and $^{119}$Sn (Ful+ 69). A level of very high spin (23/2$^+$) has been tentatively identified at 3.683 MeV in $^{117}$Sn (YE 69).

## 2.3. Odd-Mass Antimony Isotopes

The low states of these isotopes are described according to the simplest shell model as one proton outside the $Z = 50$ closed shell with neutrons in the BCS ground state. There are four low levels in the odd isotopes, $^{113}$Sb to $^{129}$Sb, with spins 5/2$^+$, 7/2$^+$, 1/2$^+$, 3/2$^+$. (In most cases these are the lowest known levels.) These spins are consistent with those of the single-particle proton levels in this region (see Fig. 1), but analyses of various experiments, discussed below, show this to be an inaccurate description of the levels.

Spectroscopic factors for the ($^3$He, *d*) reaction on the even tin isotopes leading to the 5/2$^+$ and 7/2$^+$ states are close to the single-particle value: Conjeaud *et al.* (CHC 68) find values of 0.7–0.9 for both states of the odd isotopes $^{113}$Sb to $^{125}$Sb; Ishimatsu *et al.* (Ish+ 67) and Auble *et al.* (ABF 68a) find similar values for the 7/2$^+$ level and higher values, closer to unity, for the 5/2$^+$. On the other hand, the 1/2$^+$ and 3/2$^+$ levels have only 0.3–0.6 of the single-particle value; a major part of the $s_{1/2}$ and $d_{3/2}$ single-particle amplitude lies at higher excitation energy. These levels are excited by Coulomb excitation (Bar+ 64) which indicates that their wave functions have collective components. An 11/2$^-$ level is strongly excited in the stripping experiments; it is possibly the $h_{11/2}$ single-particle level.

The magnetic moments for the 5/2$^+$ and 7/2$^+$ levels, listed in Table III, show deviations from the Schmidt values, which corroborates that they are not pure $d_{5/2}$ and $g_{7/2}$ proton single-particle states. Jackson *et al.* (JRG 68) calculated by perturbation theory the effect on the magnetic moments of admixtures of more complicated states, using Kuo–Brown matrix elements to calculate the admixtures, and found good agreement except for $^{117}$Sb. The experimental value for this isotope differs by 0.8 nuclear magnetons from the value for the others and such a variation for one isotope is very hard to explain theoretically. However, except for this puzzle, all the data indicate that the 5/2$^+$ and 7/2$^+$ are predominantly single-particle proton levels; the 1/2$^+$ and 3/2$^+$ are not. Again, as in Section 2.1, we caution against a literal interpretation of this statement. The data all involve re-

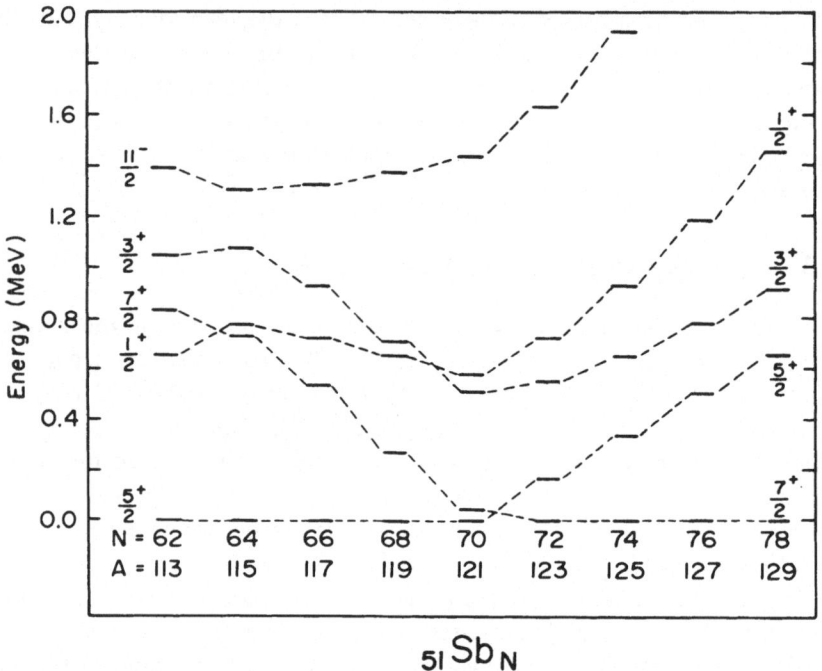

Fig. 8. Energies of four low levels in the odd antimony isotopes with spins 1/2⁺, 3/2⁺, 5/2⁺, 7/2⁺. The high 11/2⁻ level is plotted in addition. The data for $A \leq 125$ are taken from (³He, $d$) reaction results (CHC 68); for $A \geq 125$ from ($d$, ³He) results (ABF 68b). These assignments are confirmed by numerous $\beta$-decay studies. (See references at the end of article.)

lationships between two states. The 5/2⁺ and 7/2⁺ levels may be single-proton states with respect to the tin ground state but this itself may be far from a single-closed-shell-BCS state.

The energies of the four lowest levels, as well as that of the higher 11/2⁻ level, are plotted in Fig. 8. The crossing of the 5/2⁺ and 7/2⁺ levels is particularly dramatic. There are two obvious physical effects which can contribute to this. First, the single-particle proton energies can change due to interactions with the additional neutrons; secondly, the admixtures of more complicated states can change thus producing a change in the energy eigenvalues. The crossings of 1/2⁺ and 3/2⁺ levels are less spectacular and less surprising considering that they are presumably more complicated states.

The proton separation energies for ¹¹³Sb to ¹²⁵Sb are given in Table V and Fig. 15. Since the low states are predominantly single-particle proton

TABLE V. Proton Separation Energies[a]

| Nucleus | $S_p(A)$, keV | Nucleus | $S_p(A)$, keV |
|---------|---------------|---------|---------------|
| $^{112}$Sn | $7,540 \pm 50$[b] | $^{113}$Sb | $\begin{cases} 3,090 \pm 40^c \\ 2,491 \pm 44 \end{cases}$ |
| $^{114}$Sn | $8,515 \pm 12$ | $^{115}$Sb | $3,725 \pm 22$ |
| $^{116}$Sn | $9,270 \pm 9$ | $^{117}$Sb | $4,339 \pm 30$ |
| $^{118}$Sn | $10,016 \pm 10$ | $^{119}$Sb | $5,120 \pm 21$ |
| $^{120}$Sn | $10,654 \pm 50$[b] | $^{121}$Sb | $5,782 \pm 4$ |
| $^{122}$Sn | $11,404 \pm 50$[b] | $^{123}$Sb | $6,570 \pm 5$ |
| $^{124}$Sn | $12,100 \pm 50$[b] | $^{125}$Sb | $7,327 \pm 9$ |

[a] All numbers taken from (MTW 65) except those labeled ([b]) and ([c]). There are two contradictory values for $^{113}$Sb.
[b] See (CHT 69a, Har 68).
[c] See (CHC 68, Har 68).

states these show the variation of the single-particle proton energies as neutrons are added to the nucleus. The separation energies increase by more than 4 MeV over this range of isotopes. The last proton becomes more strongly bound as the neutron number increases. The effect makes the dramatic crossing of the $5/2^+$ and $7/2^+$ levels of Fig. 8 appear to be a fine detail.

As in the odd tin isotopes there is now some information about the states above 1 MeV, particularly in $^{121}$Sb and $^{123}$Sb, and again the weak-coupling model described in Section 2.2 is useful in interpreting them. As an example, we treat $^{121}$Sb in detail drawing heavily on the arguments of Barnes et al. (Bar+ 64, Bar+ 67).

Some experimental results are shown in Fig. 9. The levels just below 1.5 MeV and the $1/2^+$ and $3/2^+$ levels discussed above are excited by both ($^3$He, $d$) and ($d, d'$) and can be interpreted as mixtures of single-particle and collective states. This is in contrast to the levels between 1.0 and 1.2 MeV which are collective in character. The two levels at 1.024 and 1.143 MeV, strongly excited in the ($d, d'$) reaction and in Coulomb excitation (Bar+ 64), can be considered as the high spin components of the multiplet of states formed when the collective $2^+$ of an even tin isotope is coupled to a $d_{5/2}$ proton. The $^{123}$Sb($p, t$) reaction excites levels at 1.031 and 1.135 MeV, which are new levels and not at the same energies as those excited in ($d, d'$) (Bar+ 67). The occurrence of four distinct levels has recently been confirmed. Booth and Goldman (BG 70) use bremsstrahlung resonance fluorescence to excite the 1.025 and 1.143 MeV levels while two different levels, at

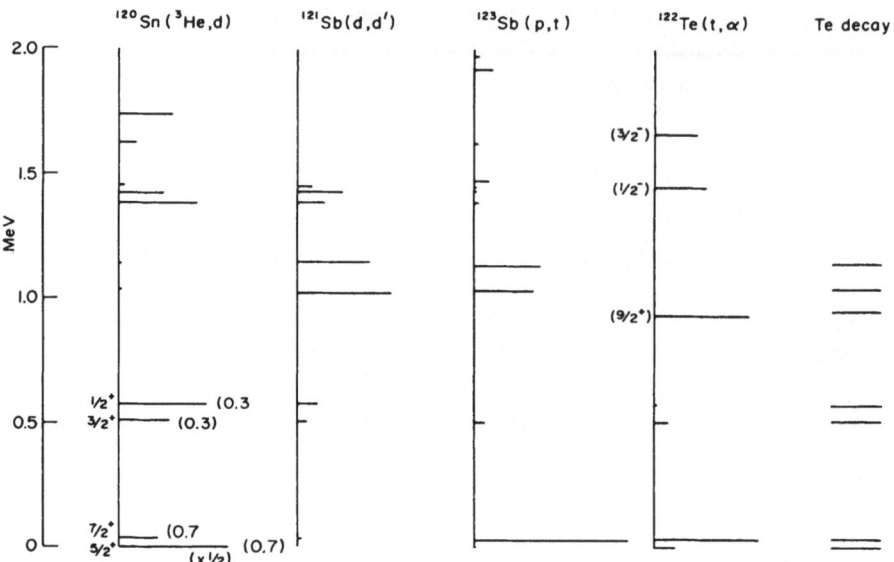

**Fig. 9. Cross sections of reactions leading to levels in** $^{121}$Sb. The first column gives $d\sigma/d\Omega$ at $\theta = 40°$ in arbitrary units for the ($^3$He, $d$) reaction with 19 MeV $^3$He particles and 30 keV resolution (Bar+ 64). The numbers in parentheses are spectroscopic factors (CHC 68). The second column gives $d\sigma/d\Omega$ at $\theta = 125°$ for ($d$, $d'$) with 12 MeV deuterons and 7 to 10 keV resolution (Bar+ 64); the third column $d\sigma/d\Omega$ at $\theta = 20°$ for ($p$, $t$) with 22-MeV protons and 10-keV resolution (Bar+ 67, private communication); the fourth column gives the relative spectroscopic factors from ($t$, $\alpha$) with 11.8-MeV tritons and 25-keV resolution (Har+ 70).

1035.6 $\pm$ 0.5 keV and 1139.3 $\pm$ 0.5 keV, are observed in the tellurium $\beta$-decay (M. Walters and K. Apt, private communication). The two ($p$, $t$) levels are interpreted as being part of the multiplet consisting of the collective $2^+$ of an even tin isotope coupled to the $g_{7/2}$ proton for the following reasons. The ground state of $^{123}$Sb with spin $7/2^+$ is predominantly a single-proton state with the neutrons in the BCS ground state. Since the ($p$, $t$) reaction is strongly enhanced when two neutrons are transferred between two BCS ground states, the lowest $7/2^+$ level in $^{121}$Sb should be strongly excited, in agreement with the experimental results. In addition, since the $2^+$ level is somewhat enhanced in ($p$, $t$) reactions on the even tin isotopes, we expect states to be excited which are collective $2^+$ states of the even isotopes coupled to a $g_{7/2}$ proton. This argument about the reaction mechanism of the ($p$, $t$) reaction is more valid here than in the odd tin isotopes because the spectator particle is a proton, not a neutron, and cannot be picked up by the incident proton to form a triton. The levels at 1.031 and 1.135 MeV are of this type.

Thus because of the level crossing of the two low states it is possible to excite collective states based on the first excited as well as on the ground state. It appears there are four very pure states lying close together in energy, and that the two types of collective states are not mixed. This interpretation of these levels is somewhat speculative since no spin assignments have been made and since all members of the multiplet have not been identified.

As shown in Fig. 9, column 4 the proton pickup experiment, $^{122}\text{Te}(t, \alpha)$ $^{121}\text{Sb}$, excites the four lowest levels and, in addition, three levels at higher energy. The even tellurium isotopes have $Z = 52$ and therefore there are two protons outside the closed $Z = 50$ shell, presumably in the $d_{5/2}$, $g_{7/2}$, $s_{1/2}$, and $d_{3/2}$ single-particle levels. Thus the four lowest states should indeed be excited in this reaction and the reaction can be used to analyze the tellurium ground state. The higher levels can be interpreted as states with substantial proton two-particle–one-hole $2p1h$ configurations in their wave function. The measured $l$-transfer value are $l = 4$ for the level at 0.935 keV and $l = 1$ for those at 1.448 and 1.659 MeV, which is consistent with their being the expected $g_{9/2}$, $p_{1/2}$, and $p_{3/2}$ hole states. The spectroscopic factors are only 0.12, 0.36, and 0.14 of their single-hole value if the normalization is chosen so the sum of the spectroscopic factors to the four lowest levels is two (because there are two particles outside the core). This normalization is, however, quite arbitrary and the values could be even smaller (Har+ 70). No strong states are seen at higher energy, so either the remaining strength is very fragmented, or the extraction of spectroscopic factors is wrong by a sizeable factor.

Similar levels are seen in other odd antimony isotopes. Two types of experiments have been done: Auble et al. (ABF 68b) studied $^{126,128,130}\text{Te}(d,$ $^3\text{He})$; Conjeaud et al., Caballero et al., and Harar et al. (Con+ 69, Cab+ 69, Har+ 70) studied $^A\text{Te}(t, \alpha)^{A-1}\text{Sb}$ for $A = 122$ to 130. There is always a group of a few states excited strongly, but at an excitation energy which varies rapidly with $A$. For instance, the lowest level occurs at 2.71 MeV in $^{129}\text{Sb}$ (Con+ 69), 1.82 in $^{125}\text{Sb}$ (ABF 68b), and has decreased to 0.94 MeV (Har+ 70) in $^{121}\text{Sb}$.

Experimental binding energies can be used to estimate the position of $2p1h$ states. If the energy to create a hole in the even tin nucleus is known, as well as the energy relative to tin of the ground state of the even tellurium nucleus with the same neutron number, then if the particle–hole interaction is negligible, the position of the lowest $2p1h$ state in antimony relative to the $1p$ state can be predicted. Values lying between 2.7 and 3.0 MeV for the isotopes $^{119}\text{Sb}$ to $^{125}\text{Sb}$ are obtained. This is smaller than the 5 MeV excitation energy of the $1p1h$ states in tin because of the pairing energy of the two protons. We see that while the experimental values for the heavy isotopes are reasonable, the energy of the level in $^{121}$Sb is surprisingly low. In addition, the rapid change with neutron number is hard to understand, even qualitatively. Similar states have been observed in the odd indium isotopes and this problem is discussed again the next section.

## 2.4. Odd-Mass Indium Isotopes

The separation energies of the tin isotopes are compared to those of antimony in Table V. Since the lowest states of the odd indium isotopes are mainly proton single-hole states, the separation energies are approximately the negative of the single-hole energies. These are plotted in Fig. 15. More energy ($\sim$5 MeV) is needed to remove the fiftieth proton than to remove the fifty-first and the gap between major shells is large compared to the separation of levels in major shells, all of which is consistent with $Z = 50$ being a good closed shell. The gap of about 5 MeV remains surprisingly constant with changing neutron number. As in antimony, the separation energies increase rapidly as the number of neutrons increases and it becomes more difficult to remove the last proton.

The excitation energies of the lowest levels of the odd indium isotopes are plotted in Fig. 10 as a function of $A$. The spins $9/2^+$, $1/2^-$ and $3/2^-$ suggest that they are the $g_{9/2}$, $p_{1/2}$, and $p_{3/2}$ proton single-hole states, but this is deceptive. The $(d, {}^3He)$ reaction on the even tin isotopes enables a rough measurement of the magnitude of the single-hole components. The measured spectroscopic factors (CHT 69a, WT 68) are only about 0.6 of

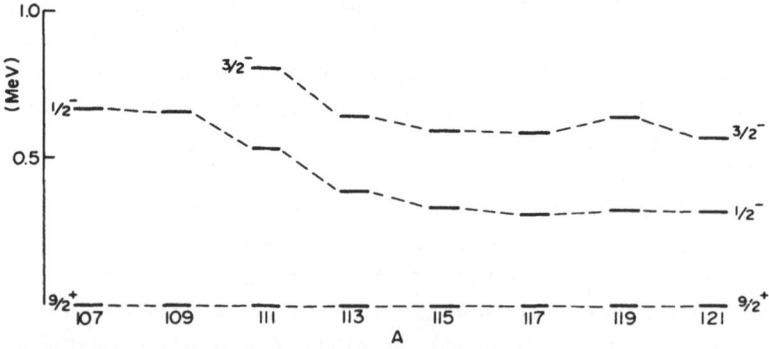

**Fig. 10. Energies of low levels in the odd-mass indium isotopes.** The three lowest levels are plotted for $A = 111$ to 121 and the two lowest in 107, 109. The data are taken from (CHT 69a, GG 68, Thu 70).

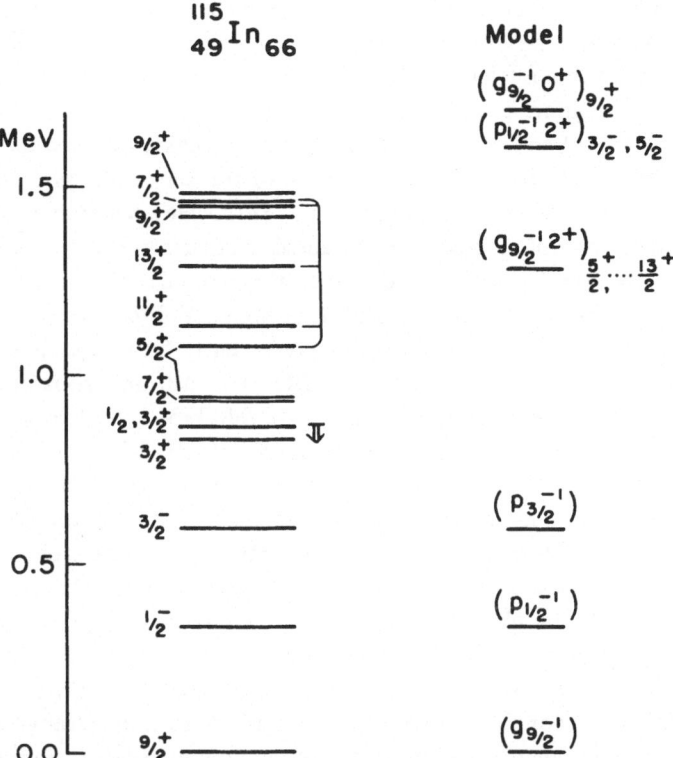

**Fig. 11. Energy levels of** [115]In. Experimental data are taken from (Die+ 70, BFM 67, Gra+ 66). The levels according to the simple weak-coupling model are shown on the right.

the single-hole value for the $9/2^+$ and $1/2^+$ levels for the odd isotopes, [111]In to [121]In. Fractionization of the $l = 4$ strength to higher levels has been observed (WT 68). In addition, the magnetic moments, listed in Table III, show deviations from the pure single-hole predictions, and the ground state quadrupole moment is three times the single-hole value. The low $3/2^-$ level has an even smaller single-hole component; the spectroscopic factor for this level is approximately 0.4–0.5 of the single-hole value. It has been excited by the $(d, d')$ reaction, which indicates it has collective components (HA 67, Die+ 69). It is surprising that a $5/2^+$ level, corresponding to a $f_{5/2}$ hole state, has not been observed in any isotope. The excitation energies of the $1/2^-$ and $3/2^-$ levels vary only slightly as the neutron number is changed and thus, in contrast to those in antimony, do not behave anomalously.

As in the tin and antimony isotopes, more complicated states appear around 1 MeV. (In fact, in [117]In these levels have been observed as low as 659 keV.) The isotope [115]In will be chosen as an example. In Fig. 11 the known levels of this nucleus up to 1.5 MeV are plotted in the first column. In the second column a naive level scheme is given, where the states in the odd nucleus are taken to be a single proton hole weakly coupled to the even tin states; the excitation energies are those of the states of [116]Sn. The three lowest levels are excited strongly in the [116]Sn($d$, [3]He) reaction (CHT 69a, WT 68); spectroscopic factors are extracted, as discussed above. There is a splitting of the $l = 4$ peak; in addition to the ground state transition there is an $l = 4$ transition to the state at 1.45 MeV (WT 68).

Coulomb excitation experiments have recently been repeated by Dietrich et al. (Die+ 70), who measured $B(E2)$ transition probabilities and made spin assignments for the levels above 940 keV. The five levels connected by the bracket in Fig. 11 have large $B(E2)$ values. (The $5/2^+$ at 942 keV and the $9/2^+$ at 1486 keV are weakly excited.) Thus they can be identified as the five members of the multiplet consisting of the $2^+$ phonon coupled to a $g_{9/2}$ proton hole. This is strengthened by the results of a phenomenological calculation made by (Die+ 70) which involved strong coupling of the $g_{9/2}$ hole to the phonon state. The results are in very good agreement with the level positions, $B(E2)$ values, the ground-state quadrupole moment, and with the relative spectroscopic factors extracted from the ($d$, [3]He) reaction, and in reasonable agreement with $M1$ transition strengths. Since there is only one *positive* parity hole state at low energies in indium, it is unnecessary to include mixing with phonon state coupled to a different hole. Thus, as (Die+ 70) point out, this model is particularly simple for the odd indium isotopes, and has more chance of accuracy than in more complicated cases.

However, in addition to the levels excited by Coulomb excitation, there are lower levels. In 1966, Bäcklin et al. (BFM 67) studied the levels at 828 and 864 keV by the $\beta$-decay of [115]Cd and were able to make spin assignments and measure some transition rates. The spins $1/2^+$ and $3/2^+$ do not fit into the weak-coupling model. Furthermore the $B(E2)$ transition probability between the two levels is 100 single-particle units. Similar states are found in [117]In; the transition probability between the levels at 749 and 659 keV is also 100 single-particle units. It was suggested that they represent the start of a rotational band. This is not necessarily in contradiction with the core-particle coupling model except that it means the coupling is very strong and not weak.

The reaction [114]Cd([3]He, $d$) has been used to study these same levels (CHT 69b, Thu 70); the levels at 864 and 829 keV were excited, although not resolved. Since cadmium has two protons less than fifty, the results indicate that these states have sizeable components of the proton $2h1p$ state, with the two holes coupled to $J = 0$. Similarly, in $^{117}$In the highly collective states at 659 and 749 keV have been excited by both $^{116}$Cd($^{3}$He, *d*) and $^{116}$Cd($\alpha$, *t*) (HH 70). As in antimony the position of the lowest $2h1p$ state relative to the $1h$ state can be estimated from experimental binding energies. The result is 2.48 MeV in $^{115}$In; the levels are observed much lower than expected. While the lowest $1p1h$ states in tin start only 1 MeV below the estimated position, the $2p1h$ states in $^{121}$Sb and the $2h1p$ in $^{115}$In start 1.5–2.0 MeV too low. Also unlike the even tin isotopes, it is these lowest levels which seem to have the most strength. However, they are not pure $2p1h$ or $2h1p$ states since the strength is far below the sum rule limit.

In summary, three types of states coexist in $^{115}$In. There are the predominantly $1h$ states, such as the lowest $9/2^+$ and $1/2^-$ levels; the predominantly vibrational states, where a proton hole is coupled to the vibration of the neutrons, which lie between 1.0 and 1.5 MeV; and the "rotational" states at 829 and 864 keV, which have sizeable $2h1p$ components.

## 3. THEORETICAL CALCULATIONS

### 3.1. General Discussion

In this section, I sketch the state of structure calculations in the $Z = 50$ region. The discussion is restricted to microscopic calculations which use a realistic residual interaction, which means an interaction which fits the nucleon–nucleon scattering data. Two groups have been involved in such calculations for several years. One was located at Pittsburgh, where the research was done by T. T. S. Kuo, David Clement, R. Mercier, Michel Baranger, and myself. The other was located in Trieste, where the main instigator was Jerzy Sawicki, a dynamic physicist who was tragically killed in an airplane accident in the summer of 1968. A group at Chalk River has very recently made calculations of this type. At the end of the present article, references to the papers of each group are listed. All use the BCS method, which has been explained in many places. The reader is referred to the references listed at the end or to (Bar 60, Bar 63, Jea 67) if he is unfamiliar with this approximation or if he wishes more details.

There are several reasons why realistic interactions are used in shell-model calculations. The tedium of searching for the parameters of an assumed type of force is eliminated. The interactions are rich in form and have the possibility of describing different properties of a variety of states. However, one appeal of this approach is less practical and more aesthetic.

It has always been an aspiration of nuclear theorists to derive the properties of finite nuclei starting from the free nucleon–nucleon interaction. Methods are known for doing this, i.e., Hartree–Fock (HF) theory if the interaction is smooth and nonsingular and Brueckner–Hartree–Fock (BHF) theory if there is a hard core. Then in a shell-model calculation one should use as single-particle energies and wave functions those derived by HF or BHF and as the residual interaction the original nucleon–nucleon interaction. If the forces are such that the theories converge well, then the shell-model vector space should be of reasonable size. If not, then methods exist for formulating the effective interaction suitable for a given space. Although this majestic outline exists, our practical techniques are insufficient to carry it out. Shell-model programs can not deal with a large enough vector space, and the calculation of effective interactions for use in a small vector space is exceedingly complicated. In view of these limitations, calculations are done, first, to examine the problems of convergence, and, second, to generate results which can be compared with experiment to determine empirically how successful the whole approach will be.

It was early realized (KBB 66) that the $J = 0^+$ matrix elements of realistic interactions are too weak to reproduce the experimental odd–even mass difference if the model space is restricted to the usual five levels. This statement can be made without specifying the particular nucleon–nucleon interaction or the method used to calculate the Brueckner $G$-matrix because the $T = 1$ matrix elements of various realistic forces calculated by various methods agree with each other fairly well (KBB 66, CB 68, Sau 70). It is clearly necessary to include more configurations.

One way to do this in a BCS calculation is to increase the number of levels in the gap equation. When this is done, the odd–even mass difference (which is approximately the smallest 1QP energy) increases, and seems to level off at twelve levels, as is shown in Fig. 12. In doing this, one includes the possibility that pairs of particles coupled to angular momentum zero are excited out of the previously completely occupied levels and into the unoccupied ones. The excited states in the even tin isotopes can then be found by diagonalizing in a 2QP space, which includes all 2QP neutron states constructed from the twelve levels shown in Fig. 12, column 3 as well as proton hole–particle states constructed from these same levels. In the Pittsburgh work this procedure was followed; the Tabakin interaction was used, which is a nonlocal potential which fits the nucleon–nucleon phase shifts and has no hard core (Tab 64).

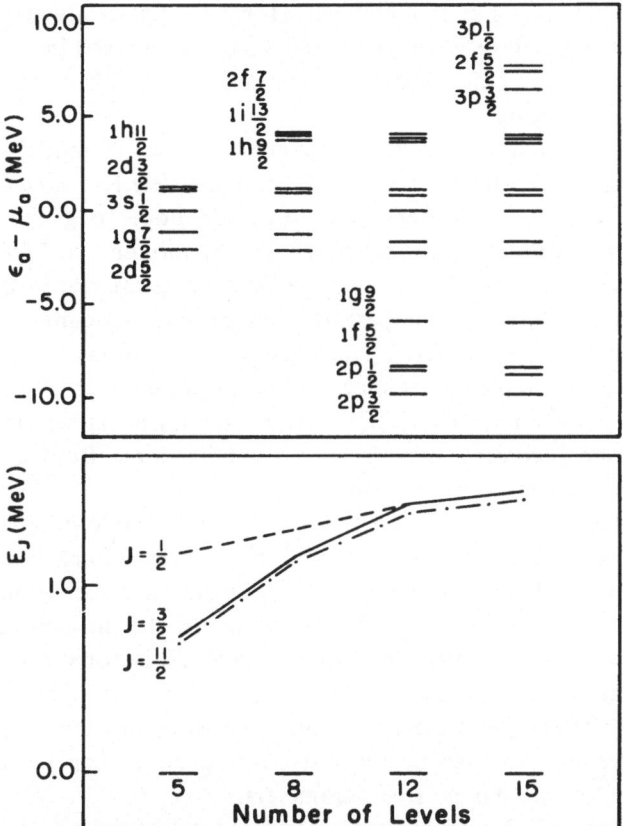

**Fig. 12. The 1QP energies $E_J$ and the relative values of $\varepsilon_a - \mu_a$.**
In the lower part of the figure, the 1QP energies $E_J$ for $J^\pi =$
$1/2^+$, $3/2^+$ and $11/2^-$ for $A = 119$ as the number of single-
particle levels included in the calculation is increased. In the
upper part of the figure the relative values of $\varepsilon_a - \mu_a$ obtained
in each case. The quantum numbers a associated with each
$\varepsilon_a - \mu_a$ are listed to the left in the same order as the levels
themselves [from (KBB 66)].

The Trieste group did some calculations along this line but have
mainly proceeded in a different direction. They included the required extra
configuration mixing by adding the core polarization diagram of Bertsch–
Kuo–Brown (Ber 65, KB 66) to the bare two-particle matrix elements. These
"effective" matrix elements were calculated in order to restrict the model
space to five single-particle levels. The main effect of the core-polarization
corrections is to strengthen the $J = 0$ elements, the pairing elements, and

thus the size of the gap is increased. However, physically it is different from the first procedure since the states which are admixed into the ground state are not simply pairs of particles coupled to zero, but are of a more complicated structure.

The two calculations make a different choice as to model space. The Trieste group, since they have only five single-particle levels, can do a 4QP calculation in the even isotopes and thus can hope to generate the $0^+$, $2^+$, and $4^+$ states at 2–3 MeV, which should be mixtures of 2QP and 4QP states. On the other hand, since particle–hole states are not in their model space, but are included onlt in perturbation theory, they cannot study such states as the $3^-$ collective state. The Pittsburgh group has the core-excited states explicitly in their model space but since the number of levels is now 12 instead of 5 the number of 4QP states becomes prohibitively large and the calculation including such states cannot be done. Both groups have concentrated on the even tin isotopes.

The interaction is fixed and there are no adjustable parameters in it. In principle one could use the potential in a Hartree–Fock calculation to determine the single-particle energies; however, in most shell-model calculations so far, these are taken from experiment. In the tin isotopes this can be done by assuming that the low states in the odd isotopes are 1QP states and then varying the single-particle energies until a good fit to the odd–even mass difference and the excitation energies of low levels is obtained. This was the method used by the Pittsburgh group. Then the calculation of the even isotopes has no free parameters.

In Section 3.2, I discuss those calculated quantities which agree well with the experimental data and those which do not and show some comparisons between theory and experiment. The theoretical calculations are those of the Pittsburgh group. First, the odd tin isotopes are discussed where only the simple single quasi-particle approximation was used in the calculations, with a few remarks about the odd antimony and indium isotopes. This is followed by a discussion of the even tin isotopes, in which, as discussed above, the calculations used the 2QP approximation with interactions between the quasi-particles and were more extensive.

## 3.2. Comparison with Experimental Results

### 3.2.1. *Odd Isotopes*

The order of magnitude of the odd–even mass difference and the general behavior of the energies of the low levels of the odd tin isotopes fit

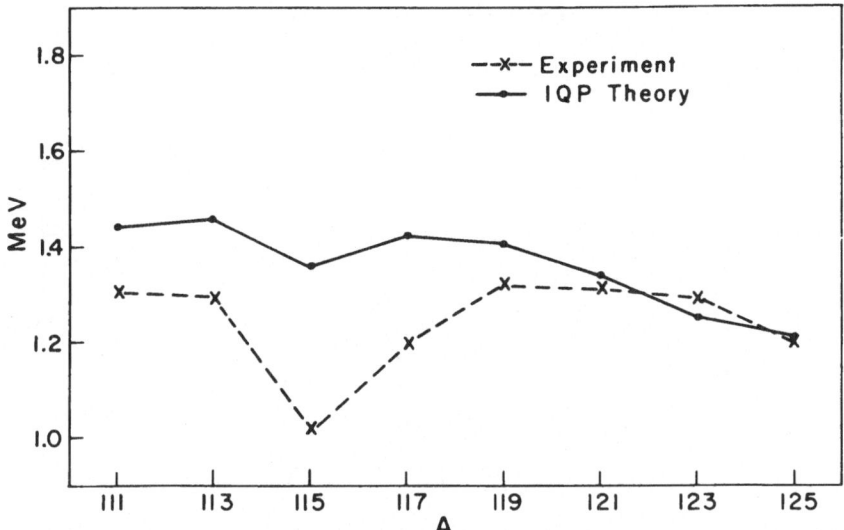

**Fig. 13. Odd–even mass difference, $P_n(A)$.** The solid line is calculated with the Tabakin interaction (CB 68). The data are taken from Table II.

the experimental values with a completely reasonable set of single-particle energies and come easily out of the calculation. However, it seems impossible to get the shape of the odd–even mass difference curve and to get quantitative agreement with the energies. For instance the $1/2^+$ level in the heavy isotopes is 500 keV too high. A typical fit to the odd–even mass difference is shown in Fig. 13. There are several possible explanations for the lack of agreement. First, the BCS states do not have a unique particle number, and the states in the odd isotopes may be sensitive to this. Second, 3QP admixtures into the 1QP state may have important effects. Third, it may be necessary to change the single-particle energies with mass number. These energies change due to interaction with the added neutrons, and this effect is included in the calculation. But the interaction with the core is so far taken as constant for the various isotopes. On the other hand, the spectroscopic factors for both $(d, p)$ and $(p, d)$ on the even isotopes are reproduced very well, within the error of extraction of spectroscopic factors from the data, and are insensitive to the single-particle energies. Figure 14 shows a comparison of calculated spectroscopic factors with those obtained from both $(d, p)$ and $(p, d)$ experiments.

   The low states in the odd antimony and indium isotopes can be approximated roughly as a proton single-particle or hole state plus the neutrons in the BCS ground state. The experimental proton separation energy of an

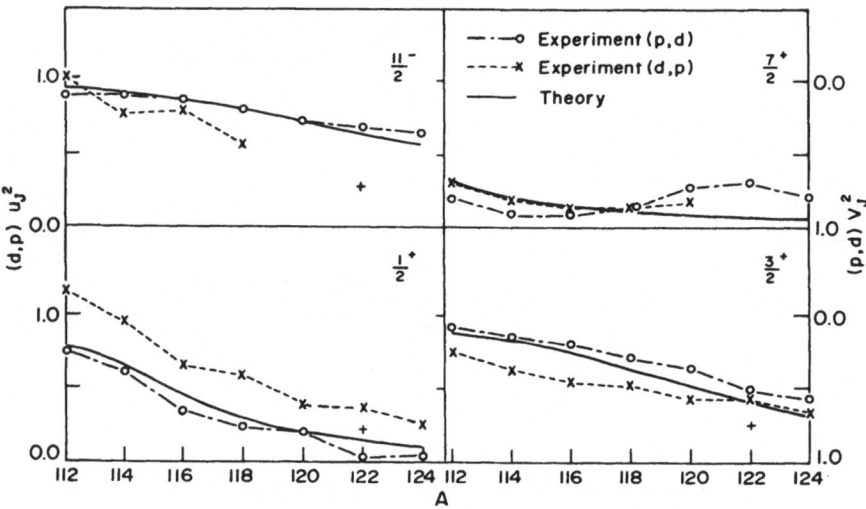

**Fig. 14.** Some $(p, d)$ and $(d, p)$ spectroscopic factors for transitions leading to states of a given $J^\pi$ in an odd-mass tin nucleus as a function of the atomic weight of the even nucleus involved in the reaction. For the $(p, d)$ reaction the sum of the spectroscopic factor divided by $(2J + 1)$ is plotted (Cav 68a, Cav 68b); the scale to the right is to be used. For the $(d, p)$ reaction, the spectroscopic factor itself is plotted (SPC 67); the scale to the left is to be used. The new values for $^{123}$Sn (Bor+ 69) are marked $+$. The 1QP theoretical results are $v_J^2$ and $u_J^2$ respectively and are taken from (CB 68).

odd antimony isotope gives the energy required to remove the proton particle; the separation energy of an even tin leading to an odd indium isotope gives the energy required to create the hole state. These energies change as we go through the isotopes because of the interaction of the proton with the added neutrons and this change can be calculated. The results are shown in Fig. 15. The calculated numbers are fit to experiment at $A = 118$. The overall slope of the curves agrees well with experiment; but the details are not reproduced. For example, the crossing of the $d_{5/2}$ and $g_{7/2}$ levels, which is discussed in Section 2.3, is not evident. The slopes of these curves are much larger than the slopes for the neutron single-particle levels (KBB 66) because the neutron–proton force is much stronger than the neutron–neutron force.

### 3.2.2. Collective Levels of the Even Tin Isotopes

In the even isotopes the general increase in level density occurs at the correct place. Of course, this is guaranteed once there is qualitative agreement with the odd–even mass difference, which is the reason it is so important to calculate properties in the odd isotopes before proceeding to the even ones. The collective 2+ and 3− levels are brought to their correct positions when the quasi-particle interaction is included. This is a stringent test of the interaction since these are shifted down by 1 to 2 MeV from their unperturbed 2QP position. In Fig. 16 the results calculated by D. Clement are shown (CB 68). In the upper curve, the bare Tabakin interaction is used; in the lower one, the so-called second-order Born corrections are included since the Tabakin interaction is not a completely smooth potential. In calculating these curves no parameters are adjusted to get a fit. The 3− is mainly a core-excited state. The sum of the squares of amplitudes which belong to configurations of two quasi-particles both in one of the usual five neutron levels is only 0.35. The 2+, on the other hand, is mainly a

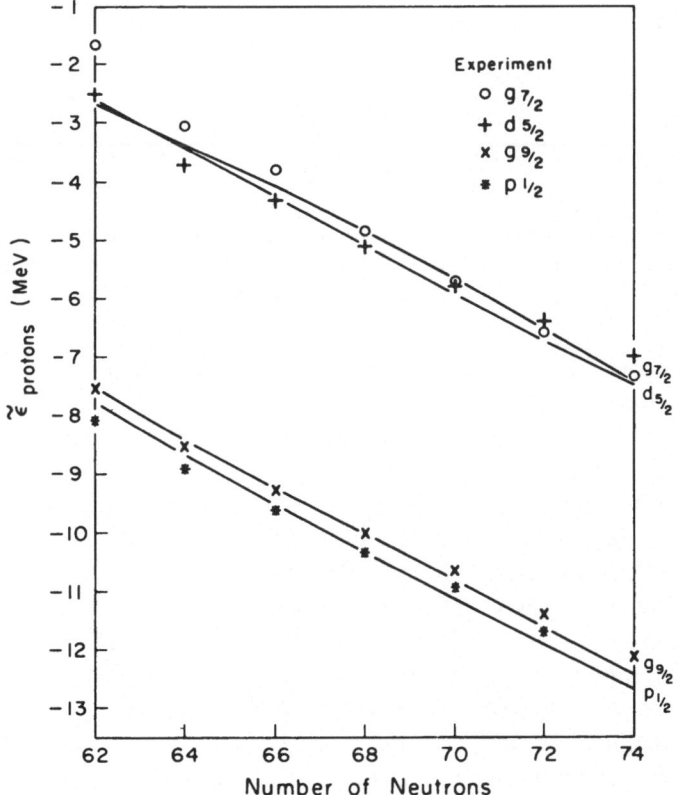

Fig. 15. **Single-particle proton energies.** The experimental values are obtained using the separation energies listed in Table V and excitation energies plotted in Figs. 8 and 10. (Figure from CB 68).

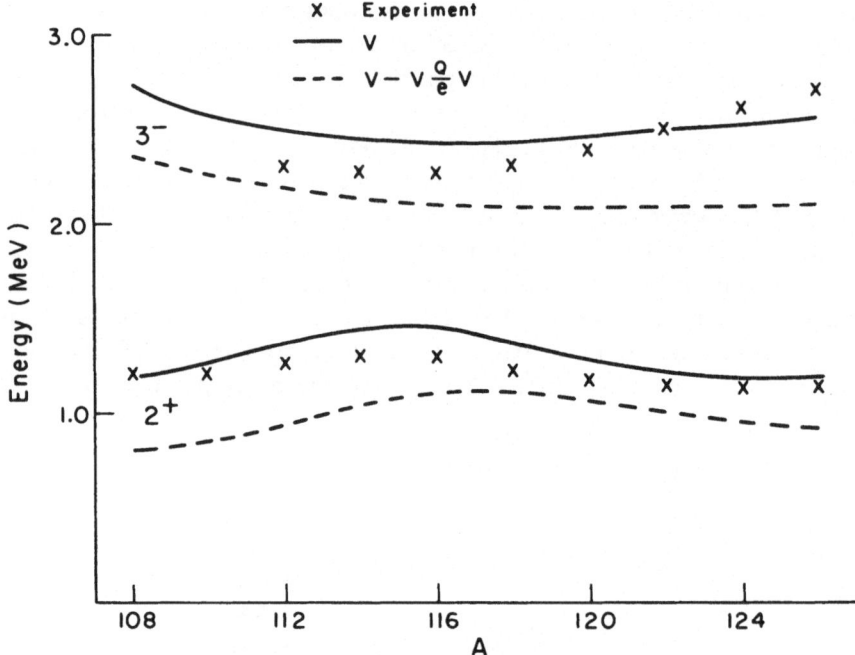

Fig. 16. Excitation energies of the lowest $2^+$ and $3^-$ states. The theoretical values (CB 68) are calculated both with (– – –) and without (——) second-order correction, and with a model space of 12 single-particle proton and neutron levels.

valence state; the sum of squares of amplitudes which belong to these configurations is 0.72.

Is this agreement a happy accident? If the model space is enlarged would it disappear? One way to enlarge the model space is to do a random phase (RPA) calculation. In the results presented here, the residual inter-action was diagonalized in a 2QP space; i.e., a Tamm–Dancoff approxi-mation (TDA) was used. With residual interactions of Gaussian form there is little difference between TDA and RPA either for $3^-$ states in closed shell nuclei (GGS 66) or for $2^+$ states in single-closed shell BCS nuclei (Arv 63, Arv+ 63). But in $^{208}$Pb (KBB 70), and in $^{40}$Ca (DLB 68, BK 69) large effects were found on the lowest $3^-$ state when a RPA calculation is done with real-istic interactions. In fact, the eigenvalue becomes imaginary; inclusion of some higher-order terms in addition to the bare interaction helps. Gmitro et al. (Gmi+ 70) have calculated the energy of the lowest $2^+$ level in the tin isotopes using a five level model space with an effective interaction derived from realistic interactions and find little difference between TDA and RPA.

On the other hand, an RPA calculation of these same levels by (HL 70) using twelve single-particle levels and a "bare" uncorrected realistic interaction suggested by S. Kahana find an imaginary energy. It seems we do not know how to include higher-order effects in a convergent way, and that merely doing the RPA is not correct. The problem of convergence plagues all shell-model calculations which use realistic interactions. The common Kuo–Brown (KB 66) prescription for treating non-closed-shell nuclei, which is to use as an effective interaction the bare interaction plus the core-polarization correction, also has problems of convergence; i.e., higher order diagrams do not seem to be small (BK 70, KZ 70).

Another way to enlarge the model space is to include 4QP states in the TDA calculation. The Trieste group has shown that 4QP admixtures into the lowest $2^+$ level are small. In these calculations they include the core polarization contribution to the interaction and have only five single-particle states. The energies are probably unaffected by 4QP admixtures, but certain matrix elements can be quite sensitive to them. Only a limited investigation of 4QP admixtures has been made when the larger twelve-level space is used.

The degree of collectivity in these wave functions can be examined by calculating physical processes which are particularly enhanced; all components in the wave function should add in phase. By comparing with experimental values we can test if our model space is indeed large enough. This is a stringent test since with a truly collective process even small admixtures into the wave function can affect the answer. Examples of processes which demonstrate collectivity are radiation to the ground state, $(p, p')$ cross-sections, and to a much lesser degree $(p, t)$ cross sections. These have been calculated and compared with experiment to some extent for the $2^+$ and $3^-$ levels.

The most straightforward process to compute is the radiative transition probability. The $B(E2)_{0 \to 2}$ for the lowest $2^+$ level calculated with no effective charge with (CB 68) wave functions is $0.41 \times 10^{-49}$ cm$^4$ and $0.34 \times 10^{-49}$ cm$^4$ for $^{116}$Sn and $^{124}$Sn respectively. The experimental values are $2.16 \times 10^{-49}$ cm$^4$ and $1.61 \times 10^{-49}$ cm$^4$. The calculated values can be increased by including more proton particle–hole states; only a small number (four) were included in the calculation. Also certain 4QP admixtures would change the radiative matrix elements but not the relative energy. This effect, plus that of increasing the number of hole states, has been estimated (Mer 70) and the calculated $B(E2)$'s can be brought into qualitative agreement with experiment. The quadrupole moments of the $2^+$ levels calculated by (CB 68) are small, i.e., $-0.02$ barn, for $^{116}$Sn. Small 4QP admixtures could increase them substantially. There is, however, no disagreement with experiment; the measured value is 0.07 ± 0.16 barn.

The inelastic scattering cross sections for protons of energy 156 MeV have been measured and calculated (Com+ 69, Com 70). The calculations use the distorted wave impulse approximation together with the wave functions of (CB 68). The calculated values are about five times less than the experimental ones. The collectivity is high in that components add coherently. A similar analysis has been carried out for inelastic α-particle scattering with the calculated values 3.5 times smaller than the experimental ones (TB 70). These cross sections are more sensitive to the neutron rather than the proton part of the wave function, and show that the neutron model space is also too small. However, as with the $B(E2)$'s, small admixtures of additional particle–hole states and of 4QP states can probably affect the results considerably.

The results of $(p, t)$ experiments on the even tin isotopes $^{112}$Sn to $^{124}$Sn, have been analyzed by Fleming *et al.* (Fle+ 70) using the wave functions of (CB 68). The ground-state transitions are greatly enhanced and the variation with mass number is in rough agreement with BCS calculations. (The same is true of $(t, p)$ ground-state transitions (BHR 70).) The lowest 2+ level is always the second most strongly excited level; the 3− is the third in most of the nuclei. The calculations bear this out. For instance, all of the components of the lowest 2+ wave function contribute in phase, which leads to an enhancement. In Fig. 17 the ratio of the cross section to the ground-state cross section is shown for the lowest levels of spin 2+ and 3−. The agreement between theory and experiment as to magnitude and variation with $A$ is reasonably good. The calculations are extremely sensitive to the details of the wave functions; (Fle+ 70) claim that the cross section to the 2+ level in the heavier isotopes would be enhanced if the single-particle levels $2f_{5/2}$, $3p_{1/2}$, and $3p_{3/2}$ had been included in the calculation.

### 3.2.3. *Other Levels of the Even Tin Isotopes*

In Fig. 18 the calculated spectrum for $^{116}$Sn is compared with the experimental one. The 0+, 2+, and 4+ energies are plotted in one column, the energies belonging to other spins and parities in the second. The dashed lines give the lowest energy of a particular spin if the QP interaction is neglected, so that the effect of the residual interaction is evident.

The agreement between theory and experiment is reasonably good for levels in the second column for $E < 3$ MeV. Above that, the level density gets high, and the experimental resolution is not good enough to see all of

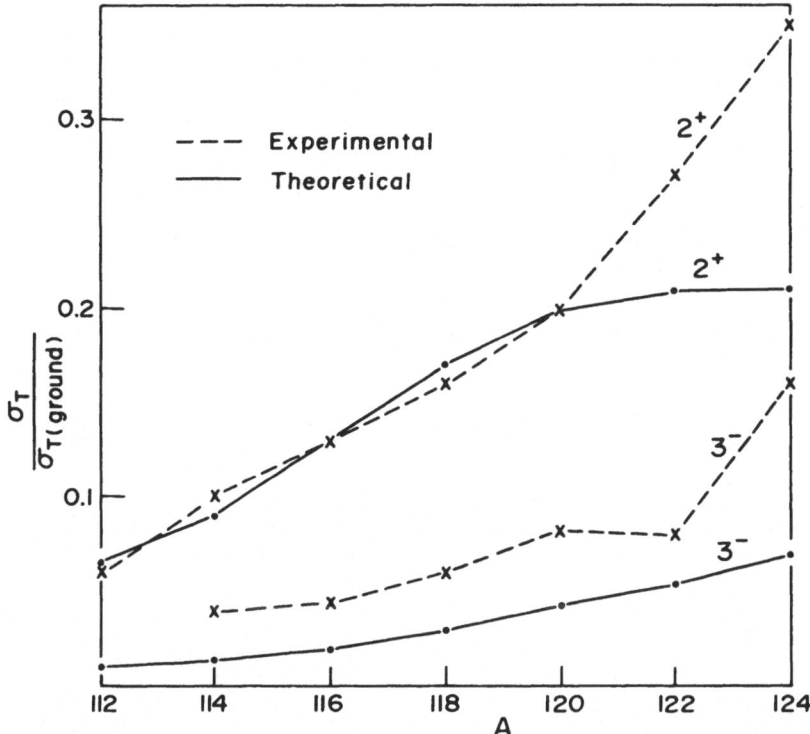

Fig. 17. The ratio of $(p, t)$ cross sections $\sigma_T$ to the ground-state cross section $\sigma_{T(\text{ground})}$ for various tin isotopes. Here $\sigma_T$ is the cross section to the lowest $2^+(3^-)$ level, integrated from 10 to 50 degrees in the center-of-mass system. The experimental (– – –) and theoretical (——) results are from (Fle+ 70), who use the wave functions of (CB 68) in their analysis.

the levels. On the whole, the energy is close to the value without the QP interaction so the overall agreement with experiment is confirmation of the quasi-particle approach, not a test of the QP residual interaction. The major exception is the lowest $3^-$ level, discussed above. In addition, the lowest $5^-$, $7^-$, and $6^+$ are shifted down slightly from their unperturbed position and they agree better with experiment when the interaction is included.

On the other hand, the lowest levels with spin $0^+$, $2^+$, and $4^+$ are shifted down substantially from the position without the residual QP interaction, and agreement with experiment is vastly improved when it is included. But there are fewer theoretical than experimental levels with spins $0^+$, $2^+$, and $4^+$. In particular the lowest $0^+$ is not in the correct position, and there are not two low $0^+$ states as the data show. We would like to think that we are

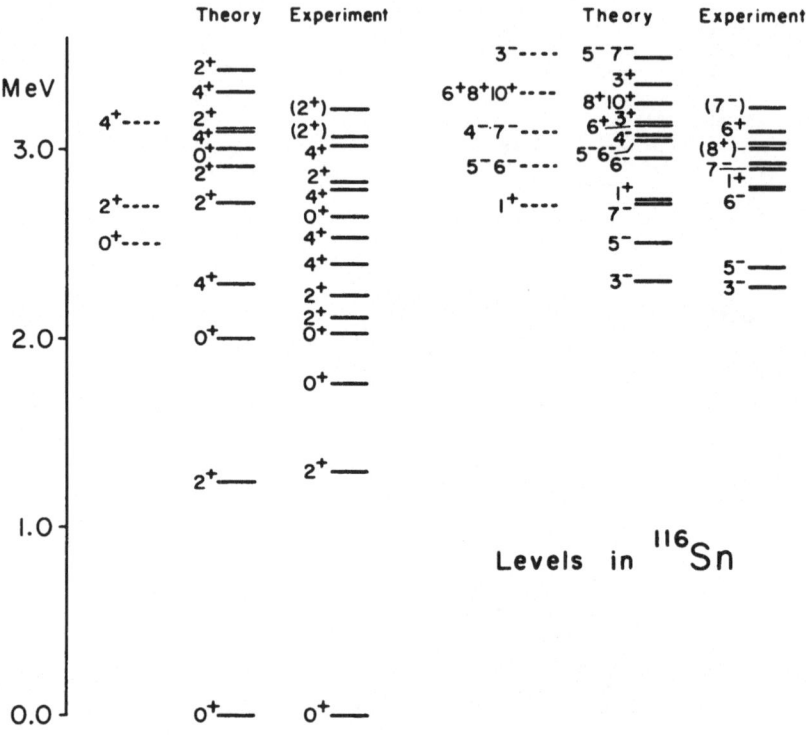

**Fig. 18. Excitation energies in** $^{116}$Sn. The levels with spin 0+, 2+, and 4+ are listed on the left; all others are on the right. The dashed lines are calculated positions if the QP residual interaction is ignored. The theoretical results are from (CB 68), the experimental from (Bee+ 70).

missing 2 phonon or 4QP states in this region, which would be expected to have spins 0+, 2+, and 4+, and which have not been included in the calculation. The Trieste group, with their modified matrix elements and only five single-particle levels, have made several 4QP calculations with different single-particle energies and with different realistic interactions (in particular the Yale–Shakin interaction and the Tabakin). They find that all the states in this energy region are well represented as 2QP states; the lowest 4QP state seems to lie at 3 MeV. However, the energy difference between the 2QP and 4QP states has not been consistently calculated because, while the effect of the 4QP states on the 2QP states has been included, the comparable effect of 6QP on 4QP has not. This would probably depress the 4QP states. Thus, neither theoretical calculation yields satisfactory agreement for these states. One aspect of the BCS approximation is that the states do not have a definite number of particles. Studies of how much error this causes have been made in the nickel region (Mac 66) and more recently in the tin region (OS 70), but the interactions used are of Gaussian form and it is difficult to draw conclusions from them about the present calculations. It is probably true that shifts of a few hundred keV can occur.

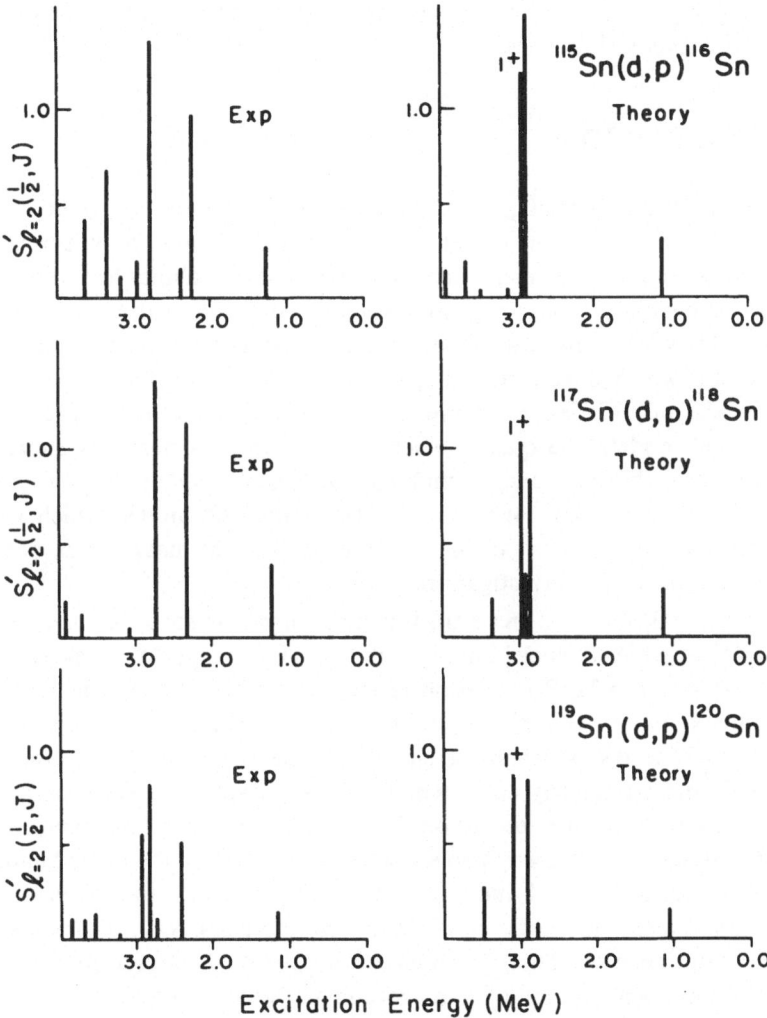

Fig. 19. Spectroscopic factors for states excited by $l = 2$ transfer. Calculated states with spins 1+ are labeled explicitly; all others have spin 2+. The theoretical results are from (CB 68); the experimental from (SPC 67) [from (CB 68)].

One can still hope that the wave functions for states in this region can be useful to interpret experiments which pick out 2QP amplitudes as long as one is interested in a general picture rather than a level by level fit. In Fig. 19 the stripping results on an odd tin target are shown; these are sensitive to the component in the wave functions of the 2QP $s_{1/2}d_{3/2}$ amplitude. The strength is predicted in the correct position, the lowest $2^+$ is accurately reproduced, but there is more spread in the experimental strengths than in the theoretical ones.

## 4. CONCLUSION

In the past five years, there has been an enormous influx of experimental information about nuclei in this region. We now know something about the structure of nuclei much further from the stability line and about spectra in energy regions of greater level density. The variety of reaction experiments which can now be performed yields a wealth of data. Some of the features which have emerged are expected, lie within the framework of our models and theories, and serve to strengthen our faith in the predictions of the shell model. The quasi-particle framework is remarkably successful in correlating the data and in making our description of the low energy levels simple in concept. Whether the weak-coupling model, which would yield such a simple picture of higher states in the odd mass nuclei, has any validity is still under investigation.

In general, the even mass tin isotopes show few surprises and behave as single-closed-shell nuclei should behave. The sharpness of levels which appear at 4 and 5 MeV excitation energy is surprising but their existence does not contradict our picture of these nuclei. The quadrupole moments of the first $2^+$ levels, which are so shockingly large in the cadmium isotopes, are small and completely consistent with both a spherical shell model and a quasi-particle picture based on it. The odd-mass tin isotopes, whose spectra are less well-known, also show no surprises. On the other hand, the odd-mass antimony and indium isotopes have some anomalous levels. There are states with large core-excited components lying at surprisingly low excitation energy. There is evidence for deformed states at low energies. These coexist with states of a more conventional description. The extra proton outside the closed shell seems to make the core more deformable and results in a spectrum of greater variety.

As for our theoretical understanding, it has been shown that a shell-model description using a realistic nucleon–nucleon interaction is capable of describing many of the properties of nuclei in this region. Exactly how accurate is this description is difficult to assess. We are still plagued with unknown inaccuracies due to not using a fixed number of particles in our basis states. So far our calculations have contained adjustable parameters, in particular the single-particle energies, which in a unified theory should be derived. Moreover, the problems of higher-order effects and convergence have hardly been touched.

## ACKNOWLEDGMENTS

The author wishes to acknowledge those experimentalists who discussed their results with her and generously made unpublished data available to her. They are P. Barnes, G. Berzins, E. C. Booth, M. E. Cage, P. E. Cavanaugh, C. F. Coleman, M. Conjeaud, F. Dietrich, C. Ellegaard, D. G. Fleming, E. Flynn, O. Hansen, S. Harar, G. E. Holland, F. Metzger, K. K. Seth, P. H. Stelson, G. Vourvopoulos, W. B. Walters, and C. Weiffenbach.

She wishes to thank M. Baranger for many helpful suggestions, H. MacManus and W. Turchinetz for some useful discussions, and W. C. Parkinson for requesting a talk which stimulated this article.

## REFERENCES

The references are divided into three groups. The first consists of general references explicitly mentioned in the text which do not contain new experimental information about nuclei with $Z = 49$, 50, and 51 nor results of calculations in this region with realistic forces. Next are listed references to recent papers yielding experimental information on nuclei with $Z = 49$, 50, and 51. Only a fraction of these is referred to in the text of the paper. Last are listed references to theoretical papers which use "realistic" interactions in microscopic structure calculations. Again, most of these are not referred to individually in the text.

The period covered in the list of experimental papers extends from 1965 to April, 1970. In the literature search heavy use was made of *Recent References, Nuclear Data B1–3* (1966); *B2–2* (1967); *B3–1* (1969); *B4–1,2* (1970). Earlier references can be found in the *Table of Isotopes* by C. M. Lederer, J. M. Hollander, and I. Perlman [John Wiley and Sons, Inc., New York (1968)] and *Nuclear Data Sheets* [U. S. Government Printing Office, Washington, D. C. (1961)]. Selection of papers is such as to include only those dealing with the *structure* of $Z = 49$, 50, and 51 nuclei. For instance, decays resulting in a final nucleus outside this range of $Z$ are not included.

There have been several recent reviews of certain types of data. References which are listed in the following reviews or compilations are not given in the list below.

"Nuclear Spins and Moments," G. M. Fuller and C. W. Cohen, in *Nuclear Data Tables*, **A5**:433 (1969).

"Muonic Atoms and Nuclear Structure," C. S. Wu and Lawrence Wilets, in *Annual Rev. of Nucl. Sci.* (E. Segre, ed.) Annual Reviews, Inc., Palo Alto, Calif. (1969).

"Muonic Atoms," S. Devons and I. Duerdoth, in *Advances in Nuclear Physics*, Vol. 2 (M. Baranger and E. Vogt, eds.) Plenum Press, New York (1969).

"Compendium of Thermal-Neutron-Capture $\gamma$-Ray Measurements, Part II: $Z = 47$ to $Z = 67$ (Ag to Ho)," L. V. Groshev *et al.*, in *Nuclear Data Tables* **5**:1 (1968).

"Table of Nuclear Lifetimes," A. Merelius, P. Sparrman, and T. Sundström, in *Hyperfine Structure and Nuclear Radiations* (E. Matthias and D. A. Shirley, eds.) North-Holland, Amsterdam (1968).

"1964 Atomic Mass Table and Consistent Set of $Q$-Values," J. H. E. Mattauch, W. Thiele, and A. H. Wapstra, in *Nucl. Phys.*, **67**:1 (1965).

## GENERAL REFERENCES

(Arv 63)　　R. Arvieu, *Ann. Phys.*, **8**:407 (1963).

(Arv+ 63)　R. Arvieu, E. Baranger, M. Veneroni, M. Baranger, and V. Gillet, *Phys. Letters*, **4**:119 (1963).

(ASS 68)　　W. P. Alford, J. P. Schiffer, and J. J. Swartz, *Phys. Rev. Letters*, **21**:156 (1968).

(Bar 60)　　M. Baranger, *Phys. Rev.*, **120**:957 (1960).

(Bar 63)　　M. Baranger, in *Cargese Lectures in Theoretical Physics* (M. Levy, ed.), W. A. Benjamin Co., New York (1963).

(Ber 65)　　G. F. Bertsch, *Nucl. Phys.*, **74**:234 (1965).

(Bje+ 68b)　J. H. Bjerregaard, Ole Hansen, O. Nathan, R. Chapman, and S. Hinds, *Nucl. Phys.*, **A107**:241 (1968).

(BK 69)　　J. Blomquist and T. T. S. Kuo, *Phys. Letters*, **28B**:544 (1969).

(BK 70)　　B. Barrett and M. Kirson, *Nucl. Phys.*, **A148**:145 (1970).

(Boh 68)　　A. Bohr, *Proceedings of the International Symposium on Nuclear Structure, Dubna, 1968* (IAEA, eds.), Vienna (1968).

(BRS 68)　　R. Broglia, C. Riedel, and B. Sorensen, *Nucl. Phys.*, **A115**:273 (1968).

(BS 69)　　D. R. Bes and R. A. Sørensen, in *Advances in Nuclear Physics*, Vol. 2 (M. Baranger and E. Vogt, eds.) Plenum Press, New York (1969).

(BT 67)　　J. Bardwick and R. Tickle, *Phys. Rev.*, **161**:1217 (1967).

(CB 68)       D. M. Clement and E. U. Baranger, *Nucl. Phys.*, **A120**:25 (1968).

(CB 68a)      D. M. Clement and E. U. Baranger, *Nucl. Phys.*, **A108**:27 (1968).

(Coh+ 69)     S. Cohen, R. D. Lawson, M. H. Macfarlane, and M. Soga, in *Methods in Computational Physics* (B. Alder, S. Fernbach, and M. Rosenberg, eds.) Academic Press, New York and London (1969).

(DD 69)       S. Devons and I. Duerdoth, in *Advances in Nuclear Physics*, Vol. 2 (M. Baranger and E. Vogt, eds.) Plenum Press, New York (1969).

(DLB 68)      A. E. L. Dieperinck, H. P. Leenhouk, and P. J. Brussaard, *Nucl. Phys.*, **A116**:556 (1968).

(FK 61)       N. Freed and L. S. Kisslinger, *Nucl. Phys.*, **A116**:401 (1968).

(Fre+ 69)     J. B. French, E. C. Halbert, J. B. McGrory, and S. S. M. Wong, in *Advances in Nuclear Physics*, Vol. 3 (M. Baranger and E. Vogt, eds.) Plenum Press, N. Y. (1969).

(GGS 66)      V. Gillet, A. M. Green, and E. A. Sanderson, *Nucl. Phys.*, **88**:321 (1966).

(Ham 65)      I. Hamamoto, *Nucl. Phys.*, **62**:49 (1965).

(HB 67)       J. A. Halbleib and R. A. Sorensen, *Nucl. Phys.*, **A98**:542 (1967).

(HK 70)       E. C. Halbert and S. J. Krieger, *Comments on Nuclear and Particle Physics* IV: 33 (1970).

(Jea 67)      M. Jean, in *International School of Physics Enrico Fermi*, Varenna 1967, Academic Press, New York (1967).

(KB 66)       T. T. S. Kuo and G. E. Brown, *Nucl. Phys.*, **85**:40 (1966).

(KBB 66a)     T. T. S. Kuo, E. U. Baranger, and Michel Baranger, *Nucl. Phys.* **79**:513 (1966).

(KBB 70) ·    T. T. S. Kuo, J. Blomquist, and G. E. Brown, *Phys. Letters*, **31B**:93 (1970).

(KS 59)       L. S. Kisslinger and R. A. Sorensen, *Mat. Fys. Medd. Dan. Vid. Selsk.*, **32**, No. 9 (1959).

(KS 63)       L. S. Kisslinger and R. A. Sorensen, *Revs. Mod. Phys.*, **35**:853 (1963).

(KZ 70)       M. Kirson and L. Zamick, *Ann. Phys.* **60**:188 (1970).

(Lan 64)      A. M. Lane, *Nuclear Theory*, W. A. Benjamin, New York (1964).

(LHP 68)      C. M. Lederer, J. M. Hollander, and I. Perlman, *Table of Isotopes*, John Wiley and Sons, New York (1968).

(LP 60)       A. M. Lane and E. D. Pendlebury, *Nucl. Phys.*, **15**:39 (1960).

(Mac 66)      M. H. Macfarlane, in *Lectures in Theoretical Physics*, Vol. VIIIc, The University of Colorado Press, Boulder (1966).

(MTW 65)      J. H. E. Mattauch, W. Thiele, and A. H. Wapstra, *Nucl. Phys.*, **67**:1 (1965).

(NDS 60)      *Nuclear Data Sheets*, U. S. Government Printing Office, Washington, D. C. (1960).

(OS 70)       P. L. Ottaviani and M. Savoia, *Phys. Rev.*, **187**:1306 (1969).

(Sau 70)      P. Sauer, *Nucl. Phys.* **A150**:467 (1970).

(Tab 64)      Frank Tabakin, *Ann. Phys.*, **30**:51 (1964).

(WW 69)       C. S. Wu and Lawrence Wilets, in *Annual Review of Nuclear Science*, (E. Segre, ed.) Annual Reviews Inc., Palo Alto, California (1969).

## Even-Mass Tin Isotopes

(AAD 65)      D. L. Allan, B. H. Armitage, and B. A. Doran, *Nucl. Phys.*, **66**:481 (1965). Energy Levels in Even Tin Isotopes from 11 MeV Proton Inelastic Scattering.

(All 68)    D. L. Allan, *Nucl. Phys.*, **A114**:211 (1968).
            An Experimental Study of the Configurational Connections Between the
            States of an Odd- and an Even-Mass Nucleus

(AM 70)     H. Arenhövel and J. M. Maison, *Nucl. Phys.*, **A147**:305 (1970).
            Study of the Giant Resonance by Photon Scattering in Medium Spherical
            Nuclei

(Bar+ 66)   N. Baron, R. F. Leonard, J. L. Need, W. M. Stewart, and V. A. Madsen,
            *Phys. Rev.*, **146**:861 (1966).
            Inelastic Alpha-Particle Excitation in the Even Tin Isotopes

(Bas+ 65)   G. Bassani, N. M. Heintz, C. D. Kavaloski, J. R. Maxwell, and G. M.
            Reynolds, *Phys. Rev.* **139**:B830 (1965).
            $(p, t)$ Ground-State $L = 0$ Transitions in the Even Isotopes of Sn and Cd
            at 40 MeV, $N = 62$ to 74

(BB 67)     P. Barreau and J. B. Bellicard, *Phys. Rev. Letters*, **19**:1444 (1967); *Phys.
            Letters*, **25B**:470 (1967).
            Study of Collective Levels in Tin-116, Tin-120, and Tin-124 by Inelastic
            Electron Scattering

(BBL 69)    S. K. Bhattacherjee, F. Boehm, and P. L. Lee, *Phys. Rev.*, **188**:1919 (1969).
            Nuclear Charge Radii from Atomic K x-rays

(BCH 67)    L. Bianchi, M. Conjeaud, and S. Harar, CEN Saclay CEA-N 844 Compte
            Rendu d'activite 1966–1967.
            Reactions $^{113}$In($^3$He, $d$)$^{114}$Sn et $^{115}$In($^8$He, $d$)$^{116}$Sn

(Bee+ 70)   O. Beer, A. El Bahay, P. Lopato, Y. Terrien, G. Vallois, and Kamal K.
            Seth, *Nucl. Phys.*, **A147**:326 (1970).
            Spectroscopy of Even Tin Isotopes by Inelastic Scattering of 24.5 MeV
            Protons

(BG 68)     Richard N. Boyd and G. W. Greenlees, *Phys. Rev.*, **176**:1394 (1968).
            Nuclear-Matter Sizes in the Tin Isotopic Sequence

(BHQ 69)    C. R. Bingham, M. L. Halbert, and A. R. Quinton, *Phys. Rev.*, **180**:1197
            (1969).
            Scattering of 65-MeV Alpha Particles from $^{89}$Y, $^{92}$Zr, $^{94}$Zr, $^{96}$Zr, and $^{116}$Sn

(Big+ 67)   J. A. Biggerstaff, C. Bingham, P. D. Miller, J. Solomon, and K. K. Seth,
            *Phys. Letters*, **25B**:273 (1967).
            Proton Particle–Hole States in $^{116}$Sn

(Bje+ 68a)  J. H. Bjerregaard, O. Hansen, O. Nathan, L. Vistisen, R. Chapman, and
            S. Hinds, *Nucl. Phys.*, **A110**:1 (1968).
            The Reactions $^{116,118}$Sn($t, p$)$^{118,120}$Sn

(Bje+ 69)   J. H. Bjerregaard, O. Hansen, O. Nathan, R. Chapman, and S. Hinds,
            *Nucl. Phys.*, **A131**:481 (1969).
            $(t, p)$ Reactions on $^{112,124}$Sn and the Pairing Model

(BLL 69)    N. Baron, R. F. Leonard, and D. A. Lind, *Phys. Rev.*, **180**:978 (1969).
            Elastic Scattering of 21-MeV Protons from Nitrogen-14, Oxygen-16, Argon-
            40, Nickel-58, and Tin-116

(Ces+ 69)   R. Cesareo, M. Giannini, P. Oliva, D. Prosperi, and M. C. Ramorino,
            *Nucl. Phys.*, **A132**:512 (1969).
            Nuclear Resonant Scattering of Gamma Rays in Ni, Cd, Sn, and Bi

(CHP 66)    M. Conjeaud, S. Harar, and J. Picard, *Phys. Letters*, **23**:104 (1966).
            A Possible Process for Core Excitation in Even Single-Closed-Shell Nuclei

(CHY 69)     C. H. Chang, G. B. Hagemann, and T. Yamazaki, *Nucl. Phys.*, **134**:110 (1969).
Excited States of $^{116}$Sn Populated by the $(\alpha, 2n)$ Reaction

(Com+ 69)   V. Comparat, A. Desnoyers, N. Marty, and A. Willis, *International Conference on Properties of Nuclear States, Montreal*, 1969, Les Presses de l'Université de Montreal (1969).
Analyse de la Diffusion $pp'$ a 155 MeV Utilisant des Descriptions Microscopic des Isotopes $^{116}$Sn, $^{118}$Sn, $^{120}$Sn

(Com 70)    V. Comparat, Thesis, University of Paris, 1970.
Etude de la Diffusion Elastique et Inelastique des Protons de 156 MeV par les Isotopes Pairs de l'Etain

(Con+ 69)   M. Conjeaud, B. Fernandez, S. Harar and E. Thurière, M. L. Chatterjee, N. Cindro, and E. Turk, *International Conference on Properties of Nuclear States, Montreal*, 1969, Les Presses de l'Université de Montreal (1969).
Etats Particule-Trou de Proton dans les Isotopes 120 et 122 de l'Etain

(Coo 66)    Cooperation of the Angular Correlation Groups of Bonn and Hamburg, *Nucl. Phys.*, **89**:305 (1966).
The Magnetic Moment of the $5^-$ State of $^{116}$Sn and Other Spectroscopic Investigations in the Decay of $^{116}$Sn and $^{116}$In

(CR 68)     N. Cue and Patrick Richard, *Phys. Rev.*, **173**:1108 (1968).
Proton Decay of Analog States Formed in the $(d, n)$ Reaction

(Cur+ 69)   T. H. Curtis, R. A. Eisenstein, D. W. Madsen, and C. K. Bockelman, *Phys. Rev.*, **184**:1162 (1969).
Quadrupole and Octupole Excitations of the Even Tin Isotopes by Electron Scattering

(DHS 67)    O. Dietzsch, E. W. Hamburger, and K. Schechet, *Bull. Am. Phys. Soc.*, **12**:19 (1967).
Isobaric Analog Resonances in the Scattering of Protons from Tin

(FB 70)     E. R. Flynn and J. G. Beery, *Phys. Rev. Letters*, **24**:143 (1970).
Evidence for Particle Quasi-Hole States in $^{124}$Sn

(FBB 70)    E. R. Flynn, J. G. Beery, and A. G. Blair, *Nucl. Phys.* **A154**:225 (1970).
The $^{122,124}$Sn$(t, p)^{124,126}$Sn Reaction at 20 MeV

(FBF 69)    Donald G. Fleming, Marshall Blann, and H. W. Fulbright, *Bull. Am. Phys. Soc.*, **14**:490 (1969).

(Fle+ 70)   Donald G. Fleming, Marshall Blann, H. W. Fulbright, and John A. Robbins, *Nucl. Phys.* **A157**:1 (1970).
$(p, t)$ Reactions on the Even Tin Isotopes

(Ful+ 69)   S. C. Fultz, B. L. Berman, J. T. Caldwell, R. L. Bramblett, and M. A. Kelly, *Phys. Rev.*, **186**:1255 (1969).
Photoneutron Cross Sections for Sn$^{116}$, Sn$^{117}$, Sn$^{118}$, Sn$^{119}$, Sn$^{120}$, Sn$^{124}$, and Indium

(GMP 70)    G. W. Greenlees, W. Makofske, and G. J. Pyle *Phys. Rev.*, **C1**:1145 (1970).
Analysis of Proton Elastic Scattering Using Potentials Derived from Nucleon-Density Distributions and Two-Body Potentials

(GPT 68)    G. W. Greenlees, G. J. Pyle, and Y. C. Tang, *Phys. Rev.*, **171**:1115 (1968).
Nuclear Matter Radii from a Reformulated Optical Model

(Har 68)    S. Harar, Thesis, University of Paris, 1968 and CEA-R3530 Saclay Report.
Spectroscopic Study of Near Closed-Shell Nuclei $Z = 50$ by Means of

the ($^3$He, $d$) and $d$, $^3$He) Reactions on All Even Tin Isotopes, and the ($^3$He, $d$) Reaction on the $A = 113$ and $A = 115$ Indium Isotopes

(HKV 66)   B. Hrastnik, V. Knapp, and M. Vlatkovic, *Nucl. Phys.*, **89**:412 (1966). Lifetimes of the First Excited States in $^{118}$Sn and $^{120}$Sn by the Nuclear Gamma-Ray Resonance Method

(HLJ 68)   S. A. Hjorth, E. K. Lin, and A. Johnson, *Nucl. Phys.*, **A116**:1 (1968). Analysis of the Elastic Scattering of 14.5 MeV Deuterons on Isotopes between $^{54}$Fe and $^{124}$Sn

(Hol+ 67)  G. E. Holland, J. Maher, C. A. Whitten, and D. A. Bromley, *Bull. Am. Phys. Soc.*, **12**:19 (1967). $^{120}$Sn($p, t$)$^{118}$Sn Reaction at 20.00 MeV

(Jam+ 69)  A. N. James, P. T. Andrews, P. Kirkby, and B. G. Lowe, *Nucl. Phys.*, **A138**:145 (1969). Measurement of ($p, 2p$) Cross Sections in Medium Weight and Heavy Nuclei

(Jar+ 67)  O. N. Jarvis, B. G. Harvey, D. L. Hendrie, and Jeanette Mahoney, *Nucl. Phys.*, **A102**:625 (1967). The Inelastic Scattering of 17.8 MeV Protons from $^{58}$Ni, $^{60}$Ni, and $^{120}$Sn and the Determination of Spin and Parity Assignments for $^{58}$Ni from $\alpha$-particle Scattering

(JK 69)    Cleland H. Johnson and Robert L. Kernell, *Bull. Am. Phys. Soc.*, **14**:490 (1969). Proton Strength Functions from ($p, n$) Reactions on In and Five Sn Isotopes

(Jol 65)   R. K. Jolly, *Phys. Rev.*, **139B**:318 (1965). Analysis of Single Excitations in Inelastic Deuteron Scattering from $^{60}$Ni, $^{92}$Zr, and $^{120}$Sn Nuclei

(Kar+ 70)  O. Karban, P. D. Greaves, V. Hnizdo, J. Lowe, and G. W. Greenlees, *Nucl. Phys.*, **A147**:461 (1970). Inelastic Scattering of Polarized Protons at 30.3 MeV by $^{54,56}$Fe, $^{58}$Ni, $^{120}$Sn, and $^{208}$Pb

(Kas+ 69)  J. W. Kast, S. Bernow, S. C. Cheng, D. Hitlin, W. Y. Lee, E. R. Macagno, A. M. Rushton, and C. W. Wu, *Bull. Am. Phys. Soc.*, **14**:1241 (1969). Isotone Shift in Muonic X-rays in the Tin Region

(KB 69)    B. G. Kiselev and V. R. Burmistrov, *Soviet Journal of Nuclear Physics*, **8**:613 (1969). Investigation of the $^{118}$Te $\rightarrow$ $^{118}$Sb $\rightarrow$ $^{118}$Sn Decay Chain

(KC 66)    Y. S. Kim and B. L. Cohen, *Phys. Rev.*, **142**:788 (1966). Studies of ($d, d'$) Reactions on Sn, Cd, Te, and Mo Isotopes

(Khv+ 70)  V. M. Khvastunov *et al.*, *Nucl. Phys.*, **A146**:15 (1970). Elastic Electron Scattering on $^{58,60,64}$Ni and $^{112,118}$Sn Isotopes

(Kle+ 70)  A. M. Kleinfeld, R. Covello-Moro, H. Ogata, G. G. Seaman, S. G. Steadman, and J. de Boer, *Nucl. Phys.* **A154**:499 (1970). Reorientation Measurements in $^{116}$Sn and $^{124}$Sn

(Koi+ 68)  M. Koike, I. Nonaka, H. Nakamura, H. Taketani, Y. Awaya, T. Wada, and K. Matsuda, *J. Phys. Soc. Japan*, **25**:626 (1968). Collective States of $^{116}$Sn Excited by Inelastic Scattering of 14.695 MeV Protons

(Koi+ 69)    M. Koike, R. Bigler, J. Horton, and S. Ali, *Bull. Am. Phys. Soc.*, **14**:1219 (1969).
Nuclear Structure Studies of $^{120}$Sn with $(p, p')$ and $(d, p)$ Reactions

(Kum+ 68)    I. Kumabe, H. Ogata, T. H. Kim, M. Inoue, Y. Okuma, and M. Matoba, *J. Phys. Soc. Japan*, **25**:14 (1968).
Scattering of 34.4 MeV Alpha Particles by the Tin Isotopes

(Kol+ 69)    J. J. Kolata, A. Galonsky, R. Howell, and R. Sager, *Bull. Am. Phys. Soc.*, **14**:490 (1969).
Proton Spin Flip at 30 MeV in the Reaction $^{120}$Sn$(p, p'\gamma)^{120}$Sn.

(Lin+ 69)    D. A. Lind, M. E. Cage, P. D. Kunz, and R. R. Johnson, *Bull. Am. Phys. Soc.*, **14**:56 (1969).
$^{113}$In$(^3$He, $d)^{114}$Sn Reaction at 37.7 MeV

(Mac+ 70)    E. R. Macagno, S. Bernow, S. C. Cheng, S. Devons, I. Duerdoth, D. Hitlin, J. W. Kast, W. Y. Lee, J. Rainwater, C. S. Wu, and R. C. Barrett, *Phys. Rev.*, **C1**:1202 (1970).
Muonic Atoms: II. Isotope Shifts

(Mak+ 67)    W. Makofske, M. Slagowitz, W. Savin, H. Ogata, T. H. Kruse, and T. Tamura, *Phys. Letters*, **25B**:322 (1967).
Scattering of 16-MeV Protons from Even Tin Isotopes

(Mak+ 68)    W. Makofske, W. Savin, H. Ogata, and T. H. Kruse, *Phys. Rev.*, **174**:1429 (1968).
Elastic and Inelastic Proton Scattering from Even Isotopes of Cd, Sn, and Tl

(MM 69)     V. E. Michalk and J. A. McIntyre, *Nucl. Phys.*, **A137**:115 (1969).
Population of Nuclear Energy Levels via Photon-Excited Levels

(Sam+ 68)    C. Samour, J. Julien, J. M. Kuchly, R. N. Alves, and J. Morganstern, *Nucl. Phys.*, **A122**:512 (1968).
Capture Radiative Partielle des Neutions de Resonance dans Quelques Isotopes de l'Etain

(SFV 69)     R. Shoup, J. D. Fox, and G. Vourvopoulos, *Nucl. Phys.*, **A135**:689 (1969).
Proton Particle–Hole States in $^{116}$Sn from the $^{115}$In$(^3$He, $d)^{116}$Sn Reaction

(SHC 67)     E. J. Schneid, E. W. Hamburger, and B. L. Cohen, *Phys. Rev.* **161**:1208 (1967).
Study of Inelastic Scattering of Protons from $^{116}$Sn, $^{122}$Sn and $^{124}$Sn at Isobaric Analog Resonances

(SPC 67)     E. Schneid, A. Prakash, and B. L. Cohen, *Phys. Rev.*, **156**:1316 (1967).
$(d, p)$ and $(d, t)$ Reactions on the Isotopes of Tin

(Ste+ 65)    P. H. Stelson, R. L. Robinson, M. J. Kim, J. Rapaport, and G. R. Satchler, *Nucl. Phys.*, **68**:97 (1965).
Excitation of Collective States by the Inelastic Scattering of 14–MeV Neutrons

(Ste+ 68)    P. H. Stelson, F. K. McGowan, R. L. Robinson, W. T. Milner, and R. O. Sayer, *Phys. Rev.*, **170**:1172 (1968).
Coulomb Excitation of the Even Tin Nuclei

(Ste+ 70)    P. H. Stelson, F. K. McGowan, R. L. Robinson, and W. T. Milner, *Phys. Rev.*, **C2**:2015 (1970).
Static Quadrupole Moment of the First 2$^+$ States of the Even Tin Nuclei

(TB 70)      B. Tatischeff and I. Brissaud, *Nucl. Phys.*, **A155**:89 (1970).
166-MeV Elastic and Inelastic Alpha Particle Scattering; Macroscopic and Microscopic Analysis

(Yag+ 68a)   K. Yagi, Y. Saji, T. Ishimatsu, Y. Ishizaki, M. Matoba, Y. Nakajima, and
             C. Y. Huang, *Nucl. Phys.*, **A111**:129 (1968).
             Level Structure of $^{116}$Sn and $^{117}$Sn from $(p, d)$ $(p, t)$, and $(p, p')$ Reactions
(Yag+ 68b)   K. Yagi, Y. Saji, T. Ishimatsu, Y. Ishizaki, M. Matoba, Y. Nakajima, and
             C. Y. Huang, *J. Phys. Soc. Japan*, **24**:1167 (1968).
             Level Structure of $^{118}$Sn from a $^{119}$Sn$(p, d)$ Reaction
(YE 69)      T. Yamazaki and G. T. Ewan, *Nucl. Phys.*, **A134**:81 (1969).
             Level and Isomer Systematics in Even Tin Isotopes from $^{108}$Sn to $^{118}$Sn
             Observed in Cd$(\alpha, xn)$ Reactions
(YEP 68)     T. Yamazaki, G. T. Ewan, and S. G. Prussin, *Phys. Rev. Letters*, **20**:1376
             (1968).
             Level and Isomer Systematics in Even Sn Isotopes

## Odd-Mass Tin Isotopes

(All 68)     See reference under **Even Mass Tin Isotopes.**
(Ark+ 69)    R. Arking, R. N. Boyd, B. Gonsior, J. C. Lombardi, and A. B. Robbins,
             *Bull. Am. Phys. Soc.*, **14**:491 (1969).
             A Study of $f_{7/2}$ Isobaric Analogue Resonances in Odd Isotopes of Sb
(Bee+ 69)    D. B. Beery, G. Berzins, W. B. Chaffee, W. H. Kelly, and W. C. McHarris,
             *Nucl. Phys.*, **A123**:659 (1969).
             The States of $^{117}$Sn
(BH 70)      C. R. Bingham and M. L. Halbert, *Phys. Rev.*, **C1**:244 (1970).
             Reaction $^{116}$Sn$(\alpha, {}^3$He$)$ at 65.7 MeV
(Bha+ 68)    M. R. Bhat, R. E. Chrien, O. A. Wasson, M. Beer, and M. A. Lone,
             *Phys. Rev.*, **166**:111 (1968).
             Investigation of $\gamma$-Rays following $S$- and $P$-wave Neutron Capture in Tin
             Isotopes
(Boc+ 68)    J. P. Bocquet, Y. Y. Chu, G. T. Emergy, and M. L. Perlman, *Phys. Rev.*,
             **167**:1117 (1968).
             Internal Conversion Studies with $^{119m}$Sn and $^{117m}$Sn
(Bor+ 69)    T. Borello, O. Dietzsch, E. Frota-Pessoa, E. W. Hamburger, and C. Q.
             Orsini, *International Conference on Properties of Nuclear States, Montreal*,
             1969, Les Presses de l'Université de Montreal (1969).
             Energy Levels of $^{113}$Sn and $^{123}$Sn
(BPW 70)     P. A. Baedecker, A. Pakkanen, and W. B. Walters, to be published.
             Decay of $^{117}$In Isomers and Search for a Low-Lying 9/2$^-$ Level in $^{117}$Sn
(Cav 68a)    P. E. Cavanagh, AERE-R 5801 United Kingdom Atomic Energy Authority,
             Research Group Report (1968).
             $j$-dependence and Spectroscopic Factors in the $2d$ shell for the Tin Isotopes
(Cav 68b)    P. E. Cavanagh, AERE-R 5901 United Kingdom Atomic Energy Authority
             Research Group Report (1968).
             Spectroscopic Factors, Symmetry Effects, and Nucleon Density Distribu-
             tions for the Tin Isotopes
(Cav+ 70)    P. E. Cavanagh, C. F. Coleman, A. G. Hardacre, G. A. Gard, and J. F.
             Turner, *Nucl. Phys.*, **A141**:97 (1970).
             A study of the Nuclear Structure of the Odd Tin Isotopes by Means of the
             $(p, d)$ Reaction

(DHS 67)    See reference under **Even-Mass Tin Isotopes.**

(FBF 69)    See reference under **Even-Mass Tin Isotopes.**

(FHM 70)    E. R. Flynn, O. Hansen, and T. Mulligan, *Bull. Am. Phys. Soc.*, **15**:622 (1970).
            One and Three Quasiparticle States in $^{121}$Sn

(Fle+ 68)   D. G. Fleming, M. Blann, H. W. Fulbright, J. A. Robbins, and Yu-Wen Yu, *Bull. Am. Phys. Soc.*, **13**:1429 (1968).
            $(p, t)$ Reactions on the Tin Isotopes

(Ful+ 69)   See reference under **Even-Mass Tin Isotopes.**

(Gon+ 68)   B. Gensior, R. Arking, J. Lombardi and A. B. Robbins, *Bull. Am. Phys. Soc.*, **13**:1562 (1968).
            Spin and Parity Measurements on Isobaric Analogue States in $^{125}$Sb

(Hol+ 69)   G. E. Holland, C. A. Whitten, Jr., J. Maher, and D. A. Bromley, *International Conference on Properties of Nuclear States, Montreal*, 1969. Les Presses de l'Université de Montreal (1969).
            The $(p, t)$ Reaction on Sn$^{119}$ and Sn$^{117}$

(Kum+ 68)   See reference under **Even-Mass Tin Isotopes.**

(Rah 68)    O. Rahmouni, *Nuovo Cim.*, **52B**:289 (1968).
            De croissance de $^{116}$Sb

(Rob+ 69)   R. L. Robinson, F. K. McGowan, P. H. Stelson, W. T. Milner, and R. A. Sayer, *Nucl. Phys.*, **A123**:193 (1969).
            Gamma–gamma Angular Correlations Following Coulomb Excitation

(Sam+ 68)   See reference under **Even-Mass Tin Isotopes.**

(SB 68)     R. E. Snyder and G. B. Beard, *Nucl. Phys.*, **A113**:581 (1968).
            Decay of $^{121m}$Sn and $^{121}$Sn

(SHC 67)    See reference under **Even-Mass Tin Isotopes.**

(SPC 67)    See reference under **Even-Mass Tin Isotopes.**

(Ste+ 67)   P. H. Stelson, W. T. Milner, F. K. McGowan, and R. L. Robinson, *Bull. Am. Phys. Soc.*, **12**:19 (1967).
            Coulomb Excitation of $^{117,119}$Sn

(VE 68)     L. Veeser and J. Ellis, *Nucl. Phys.*, **A115**:185 (1968).
            Polarization Measurements near Isobaric Analogue Resonances in $^{117}$Sb, $^{119}$Sb, and $^{121}$Sb

(Yag+ 68a)  See reference under **Even-Mass Tin Isotopes.**

(YE 69)     See reference under **Even-Mass Tin Isotopes.**

## Odd-Mass Antimony Isotopes

(AW 70)     K. E. Apt and W. B. Walters, *Bull. Am. Phys. Soc.*, **15**:622 (1970).
            Decay of 2.2 h $^{127}$Sn to Levels of $^{127}$Sb

(ABF 68a)   R. L. Auble, J. B. Ball, and C. B. Fulmer, *Phys. Rev.*, **169**:955 (1968).
            Levels in Odd-Mass Sb and I Isotopes Studied with the $(^3$He, $d)$ Reaction

(ABF 68b)   R. L. Auble, J. B. Ball, and C. B. Fulmer, *Nucl. Phys.*, **A116**:14 (1968).
            The $(d, {}^3$He$)$ Reaction on $^{126}$Te, $^{128}$Te, and $^{130}$Te Nuclei

(AK 66a)    R. L. Auble and W. H. Kelly, *Nucl. Phys.*, **79**:577 (1966).
            $\beta$ and $\gamma$ Spectroscopic Studies of 9.7 d $^{125}$Sn

(AK 66b)    R. L. Auble and W. H. Kelly, *Nucl. Phys.*, **81**:442 (1966).

A Study of $^{123}$Sb Levels Populated in the Beta Decay of the High Spin Isomer of $^{123}$Sn

(AKB 64)    R. L. Auble, W. H. Kelly, and H. H. Bolotin, *Nucl. Phys.*, **58**:337 (1964).
The Decay of $^{121}$Te and $^{121m}$Te

(Ark+ 69)   See reference under **Odd-Mass Tin Isotopes**.

(BA 68)     G. Berzins and R. L. Auble, *Nucl. Phys.*, **A109**:316 (1968).
The Decay of 9.7 min $^{125}$Sn

(Bar+ 64)   P. D. Barnes, C. Ellegaard, B. Herskind, and M. C. Joshi, *Phys. Letters*, **23**:266 (1964).
Properties of Levels in $^{121}$Sb and $^{123}$Sb Excited with the Reactions ($^3$He, $d$) ($d$, $d'$), and ($^{16}$O, $^{16}$O'$\gamma$)

(Bar+ 67)   P. D. Barnes, Edward R. Flynn, George J. Igo, and Richard Woods, *Bull. Am. Phys. Soc.*, **12**:19 (1967).
Particle-Core Coupling in the Sb Isotopes and the Reactions: $^{121}$Sb($t$, $p$)$^{123}$Sb and $^{123}$Sb($p$, $t$)$^{121}$Sb

(Bas+ 66)   G. Bassani, M. Conjeaud, J. Gastebois, S. Harar, J. M. Laget, J. Picard, and Y. Cassagnou, *Phys. Letters*, **22**:189 (1966).
($^3$He, $d$) Reactions on the Even Sn Isotopes

(BBS 69)    L. M. Beyer, G. Berzins, and J. W. Starner, *Bull. Am. Phys. Soc.*, **14**:568 (1969).
Decay of $^{115}$Te

(Ber+ 67)   G. Berzins, W. H. Kelly, G. Graeffe, and W. B. Walters, *Nucl. Phys.*, **A104**:241 (1967).
Ge(Li)–NaI(Tl) Studies of the Decay of $^{117}$Te

(BG 70)     E. C. Booth and Mark Goldman, *Bull. Am. Phys. Soc.*, **15**:100 (1970).
Resonance Fluorescence Measurements of Partial Widths in Sb$^{121}$ and $^{123}$Sb

(BK 67a)    G. Berzins and W. H. Kelly, *Nucl. Phys.*, **A92**:65 (1967).
High Resolution Gamma-Ray Spectroscopic Studies of $^{119}$Te Isomers

(BK 67b)    G. Berzins and W. H. Kelly, *Nucl. Phys.*, **A104**:263 (1967).
Ge(Li)–NaI(Tl) Angular Correlation Studies of Gamma Cascades in the Decay of $^{119m}$Te

(BS 67)     G. B. Beard and R. E. Snyder, *Phys. Letters*, **25B**:18 (1967).
First Excited State of $^{121}$Sb

(BW 68)     P. A. Baedecker and W. B. Walters, *Nucl. Phys.*, **A107**:449 (1968).
Decay Schemes of 42 min $^{123}$Sn and 10 min $^{125m}$Sn

(Cab+ 69)   M. Caballero, M. Conjeaud, B. Fernandez, S. Harar, and E. Thuriere, *Compte Rendu d'Activite*, CEN Saclay, CEA-N 1232 (1968–9).
Reactions $^{A}_{52}$Te($t$, $\alpha$)$^{A-1}_{51}$Sb, ($A$ = 122, 124, 126, 128, 130)

(CHC 68)    M. Conjeaud, S. Harar, and Y. Cassagnou, *Nucl. Phys.*, **A117**:449 (1968).
($^3$He, $d$) Reactions on the Even Tin Isotopes

(Con+ 69)   M. Conjeaud, B. Fernandez, S. Harar, E. Thuriere, M. L. Chatterjee, B. Kostalac, and M. Turk, in: *International Conference on Properties of Nuclear States, Montreal*, 1969, Les Presses de l'Université de Montreal (1969).
Reaction $^{130}_{52}$Te($t$, $\alpha$)$^{129}_{51}$Sb

(GHS 67)    G. Graeffe, E. J. Hoffman, and D. G. Sarantites, *Phys. Rev.*, **158**:1183 (1967).
Decay Schemes of the $^{119}$Te Isomers

(Har 68) See reference under **Even-Tin Mass Isotopes.**

(Har+ 70) S. Harar, M. Conjeaud, B. Fernandez, E. Thuriere, and M. Caballero, to be published in *Phys. Letters.*
The $(t, \alpha)$ Reaction on $^{122}$Te

(Hei+ 70) C. Heiser, H. F. Brinckmann, W. D. Fromm, and U. Hagemann, *Nucl. Phys.*, **A145**:81 (1970). Ein hochangeregter Isomer Kernzustand in $^{117}$Sb

(Hjo 67) S. A. Hjorth, *Arkiv Fysik*, **33**:183 (1967).
Analysis of Some Deuteron-Induced Reactions in the Sb Isotopes

(HMM 67) S. Hinds, H. Merchant, and R. Middleton, *Phys. Rev.*, **24B**:89 (1967).
The Strongest Transitions from the $(t, p)$ Reactions on Cd and Sb

(Ish+ 67) T. Ishimatsu, K. Yagi, H. Ohmura, Y. Nakajima, T. Nakagawa, and H. Orihara, *Nucl. Phys.*, **A104**:481 (1967).

(Jer+ 68) J. M. F. Jeronymo, N. Leal da Costa, A. G. de Pinho, I. D. Goldman, and J. A. Guillaumon, *Nuovo Cim.*, **55B**:49 (1968).
Level Scheme of $^{123}$Sb

(JRG 68) A. D. Jackson, Jr., E. H. Rogers, Jr., and G. J. Garrett, *Phys. Rev.*, **175**:65 (1968).
Spins and Nuclear Moments of $^{115}$Sb, $^{117}$Sb, $^{118}$Sb, $^{119}$Sb, and $^{120}$Sb

(Kan+ 67) J. Kantele, J. Hattula, T. Hattula, H. Kalm, and O. J. Marttila, *Physica*, **259**:20 (1967).
Decay of $^{119}$Te Isomers and Structure of $^{119}$Sb

(Kim+ 68) H. J. Kim, R. L. Kernell, R. L. Robinson, and C. H. Johnson, *Bull. Am. Phys. Soc.*, **13**:657 (1968).
Spin-Parity Assignments via $^{117}$Sn$(p, n)^{117}$Sb Reaction

(PAW 69) G. M. Palmer, A. P. Arya, and C. L. Walls, *Bull. Am. Phys. Soc.*, **14**:1569 (1969).
The 437-keV $\beta$ Transition in the Decay of $^{125}$Sb

(SBL 68) W. M. Stewart, N. Baron, and R. F. Leonard, *Phys. Rev.*, **171**:1316 (1968).
Core Excitations in $^{107}$Ag, $^{109}$Ag, $^{113}$In, $^{115}$In, $^{121}$Sb, and $^{123}$Sb Resulting from Inelastic Scattering of 42-MeV Alpha Particles

(SHL 69) E. N. Shipley, R. E. Holland, and F. J. Lynch, *Phys. Rev.*, **182**:1165 (1969).
Lifetimes of Excited States of $^{51}$V, $^{61}$Ni, $^{69}$Ga, $^{75}$As, $^{79}$Br, $^{85}$Rb, and $^{123}$Sb

(VE 68) See reference under **Odd-Tin Mass Isotopes.**

(WW 67) J. F. Wild and W. B. Walters, *Nucl. Phys.*, **A103**:601 (1967).
Decay of 9.6 d $^{125}$Sn to Levels of $^{125}$Sb

## Odd-Mass Indium Isotopes

(Als 69) William J. Alston III, *Phys. Rev.*, **188**:1837 (1969).
Resonance Fluorescence Measurements of $^{115}$In Transition Strengths below 3 MeV

(BB 67b) Edward C. Booth and John Brownson, *Nucl. Phys.*, **A98**:529 (1967).
Electron and Photon Excitation of Nuclear Isomers

(BCH 69) M. Bovin, Y. Cauchots, and Y. Heno, *Nucl. Phys.*, **A137**:520 (1969).
Photoactivation Nucleaire du $^{77}$Se, $^{107,109}$Ar, $^{111}$Cd, $^{115}$In, et $^{199}$Hg

(BFM 67) A. Bäcklin, B. Fogelberg, and S. G. Malmskog, *Nucl. Phys.*, **A96**:539 (1967).
Possible Deformed States in $^{115}$In and $^{117}$In

(BKS 69)    R. B. Begzhanov, M. Kh. Khodzhaev, and Sh. Sh. Sharipov, *Sov. J. of Nucl. Phys.*, **8**:142 (1969).
            Spectroscopy of the Nucleus In$^{117}$

(Bos+ 67)   H. E. Bosch, M. C. Simon, E. Szichman, L. Gatto, and S. M. Abecasis, *Phys. Rev.*, **159**:1029 (1967). Disintegration of $^{113}$Sn

(BSP 70)    E. M. Bernstein, G. G. Seaman, and J. M. Palms, *Nucl. Phys.*, **A141**:67 (1970).
            Coulomb Excitation of $^{113}$In and $^{115}$In with Oxygen Ions

(CB 65)     Benson T. Chertok and Edward C. Booth, *Nucl. Phys.*, **66**:230 (1965).
            Nuclear Excitation by 1 to 3 MeV Electrons

(CHT 69a)   M. Conjeaud, S. Harar, and E. Thuriere, *Nucl. Phys.*, **A129**:10 (1969).
            Reaction ($d$, $^3$He) sur les Isotopes Doublement Pair de l'Etain

(CHT 69b)   M. Conjeaud, S. Harar, and E. Thuriere, *International Conference on Properties of Nuclear States, Montreal, 1969*, Les Presses de l'Université de Montreal (1969), and *Compte rendu d'activite* CEN Saclay CEA-N 1232 (1968–69).
            Reactions $^{106,114}_{48}$Cd(He$^3$, $d$)$^{107,115}_{49}$In

(CJ 68)     B. T. Chertok and W. T. K. Johnson, *Phys. Rev.*, **174**:1525 (1968).
            Comparison of Experiment and Theory for Nuclear Excitation in $^{115}$In by Threshold-Energy Electrons

(CVT 68)    G. Chilosi, J. R. Van Hise, C. W. Tang, *Phys. Rev.*, **168**:1409 (1968).
            Lifetimes of Three Low-Lying Excited States in $^{117}$In

(Die+ 70)   F. Dietrich, B. Herskind, R. Naumann, R. Stokstad, and G. Walker, *Nucl. Phys.*, **A155**:209 (1970).
            Hole–Vibration Coupling in $^{115}$In

(Ful+ 69)   See reference under **Even-Mass Tin Isotopes.**

(GG 68)     G. Graeffe and G. E. Gordon, *Nucl. Phys.*, **A114**:321 (1968).
            Decay of 35 min $^{111}$Sn to Levels of $^{111}$In

(Gra+ 66)   G. Graeffe, C. W. Tang, C. D. Coryell, and G. E. Gordon, *Phys. Rev.*, **149**:884 (1966).
            Decay Schemes of 43 d $^{115m}$Cd and 2.3 d $^{115g}$Cd

(HA 67)     S. A. Hjorth and L. H. Allen, *Arkiv Fysik*, **33**:121 (1967).
            Analysis of Some Deuteron-Induced Reactions in the Indium Isotopes

(Har 68)    See reference under **Even-Mass Tin Isotopes.**

(HH 70)     S. Harar and L. N. Horoshko, *Bull. Am. Phys. Soc.*, **15**:552 (1970).
            Study of $^{117}$In Levels by the ($^3$He, $d$) Reactions

(HLJ 68)    See reference under **Even-Mass Tin Isotopes.**

(KRJ 69)    H. J. Kim, R. L. Robinson, and C. H. Johnson, *Phys. Rev.*, **180**:1175 (1969).
            Low-Lying States of $^{111}$In via the Reaction $^{111}$Cd($p$, $n$)$^{111}$In

(Lee+ 69)   W. Y. Lee, S. Bernow, M. Y. Chen, S. C. Cheng, D. Hitlin, J. W. Kost, E. R. Macagno, A. M. Rushton, C. S. Wu, and B. Budick, *Phys. Rev. Letters*, **23**:648 (1969).
            Finite Distribution of Nuclear $M1$ and $E2$ Moments in Muonic $^{115}$In, $^{133}$Cs, and $^{141}$Pr

(MG 69)     E. der Mateosian and M. Goldhaber, *Phys. Rev.*, **186**:1285 (1969). Study of the Isobars $^{113m}$Cd and $^{113m}$In

(Mor 69)    Roger Moret, *Le Journal de Phys.*, **30**:501 (1969).
            Etude du Schema de Desintegration du Cadmium 117
(MPS 67)    J. McDonald, D. Porter, and D. T. Stewart, *Nucl. Phys.*, **A104**:177 (1967).
            Excited States in $^{115}$In
(Pan+ 66)   V. R. Pandharipande, K. G. Prasad, R. M. Singer, and R. P. Sharma,
            *Phys. Rev.*, **143**:741 (1966).
            Low-Lying Excited States in $^{115}$In and $^{117}$In
(Pan+ 68)   V. R. Pandharipande, K. G. Prasad, R. P. Sharma, and B. V. Thosar,
            *Nucl. Phys.*, **A109**:81 (1968).
            Level Structure of $^{117}$In from the Decay of $^{117}$Cd
(PPS 67)    V. R. Pandharipande, K. G. Prasad, and R. P. Sharma, *Nucl. Phys.*, **A104**:
            525 (1967).
            Magnetic Moment of the 660-keV State in $^{117}$In
(SBL 68)    See reference under **Odd-Mass Antimony Isotopes.**
(SBL 70)    S. Shastry, H. Bakhru, and I. M. Ladenbauer-Bellis, *Phys. Rev.*, **C1**:1835
            (1970).
            Decay of Sn$^{109}$ and the level Structure of In$^{109}$
(SPS 69)    C. L. Starke, E. A. Phillips, and E. J. Spejewski, *Nucl. Phys.*, **A139**:33
            (1969).
            Radioactivity of $^{105}$Cd and $^{105}$In
(Thu 70)    Evelyne Thuriere, Thesis, Université de Paris (1970).
            Etude de la Structure Nucleaire des Isotopes 107 et 115 de l'Indium et
            des Correlations dans l'Etat Fondamental des Noyaux Cibles de Cadmium,
            a l'Aide des Reactions Cd($^3$He, *d*)In
(WT 68)     C. Weiffenbach and Robert Tickle, *Bull. Am. Phys. Soc.*, **13**:1429 (1968).
            The (*d*, $^3$He) Reaction on Even Tin Isotopes

## Pittsburgh Calculations

(CB 68)     D. M. Clement and E. U. Baranger, *Nucl. Phys.*, **A120**:25 (1968).
            A Shell-Model Calculation of the Even-Mass Tin Isotopes Including Core
            Excitation
(Cle 67)    D. M. Clement, Thesis, University of Pittsburgh (1967).
            The Use of Realistic Interactions in Nuclear Shell-Modal Calculations
            Including Core Excitation
(KBB 66)    T. T. S. Kuo, E. Baranger, and M. Baranger, *Nucl. Phys.*, **89**:145 (1966).
            Calculation of the Tin Isotopes Using Realistic Nucleon–Nucleon Inter-
            actions
(Kuo 64)    T. T. S. Kuo, Thesis, University of Pittsburgh (1964).
            Shell-Model Study of the Tin Isotopes
(Mer 70)    R. E. Mercier, Thesis, University of Pittsburgh (1970).
            The Shell Model with Reaction Matrices and Pairing Correlations

## Trieste Calculations

(AS 68)     R. Alzetta and J. Sawicki, *Phys. Rev.*, **173**:1185 (1968).
            One-Neutron Transfer Reactions in Even and Odd Tin Isotopes and a
            Realistic Nucleon–Nucleon Potential

(GHS 68)    M. Gmitro, J. Hendekovic, and J. Sawicki, *Phys. Rev.*, **169**:983 (1968).
Core Polarization and Quasi-Particle Theories of Even Tin Isotopes with
a Realistic Nucleon–Nucleon Force

(Gmi+ 68a)  M. Gmitro, A. Rimini, J. Sawicki, and T. Weber, *Phys. Rev.*, **173**:964
(1968).
Success and Limitations of Two and Four Quasi-Particle Tamm–Dancoff
Theories of Vibrational States: Applications to the Even Tin Isotopes with
a Realistic Nucleon–Nucleon Potential

(Gmi+ 68b)  M. Gmitro, A. Rimini, J. Sawicki, and T. Weber, *Phys. Rev.*, **175**:1243
(1968).
Microscopic Theory of Effective Operators of Electromagnetic Interactions in Nuclei

(Gmi+ 70)   M. Gmitro, A. Rimini, P. Rossi, and T. Weber, *Phys. Rev.*, C1:1801 (1970).
Comparison between the Quasi-Particle Tamm–Dancoff and Random-
Phase Approximation in Even Tin Isotopes using Realistic Interaction

(GS 68)     B. Gyarmati and J. Sawicki, *Phys. Rev.*, **169**:966 (1969).
Spectroscopic Factors for Two-Neutron Transfer Reactions in Vibrational
Nuclei

## Chalk River Calculations

(HL 70)     K. Hara, and H. C. Lee, *Bull. Am. Phys. Soc.*, **15**:622 (1970); and to be
published.
Quasi-Particle Calculation of the Low Energy Structure of Odd and Even
$_{50}$Sn, Odd $_{51}$Sb, and Odd $_{49}$In Isotopes

*Chapter 6*

# AN *s–d*-SHELL-MODEL STUDY
# FOR *A* = 18–22*

## E. C. Halbert, J. B. McGrory,
## B. H. Wildenthal,[†] and S. P. Pandya[‡]

*Oak Ridge National Laboratory*
*Oak Ridge, Tennessee*

## ABSTRACT

Using the full basis of Pauli-allowed $(0s)^4(0p)^{12}(1s, 0d)^{A-16}$ states, we have performed conventional shell-model calculations for $A = 18$–22. Seven alternative $(1 + 2)$-body Hamiltonians were used. Two of these were the "realistic" Hamiltonians proposed by Kuo and Brown, and by Kuo, for $A = 18$. Five others were obtained by choosing five different adjustable forms for the $(1 + 2)$-body Hamiltonian operator, and then optimizing the free parameters in these forms so as to get least-square fits to measured energy-level data in $A = 17$–22. For levels which seem to be members of ground-state bands, all seven Hamiltonians yield fair-to-good reproduction of experimental data for the following kinds of observables: excitation energies, spectroscopic factors for single-nucleon transfer, electric quadrupole moments, intraband $B(E2)$ strengths, and magnetic moments. Within ground-state bands there is generally very striking similarity between our shell-model $E2$ results, and the $E2$ results predicted by a strong-coupling rotational model. This general similarity, and the exceptions to it, can be explained qualitatively by arguments involving $SU_3$ shell models. For ob-

* Research sponsored by the U. S. Atomic Energy Commission under contract with the Union Carbide Corporation.
† Now at Michigan State University, East Lansing, Michigan.
‡ Permanent address: Physical Research Laboratory, Navrangpura, Ahmedabad, India.

servables involving levels outside ground-state bands, the agreement with experiment is spotty and there is considerable sensitivity to the choice among our seven Hamiltonians. Again, $SU_3$ considerations are helpful in explaining these features. Some very simple shell models can account for the pattern of weakness and strength seen among our shell-model $B(M1)$ values.

# 1. INTRODUCTION

We present here the results of a series of shell-model calculations ($1$) for the structure of low-energy even-parity states in nuclei of masses $A = 18$–$22$. This article reports a specific piece of original research, but because of the article's scope and style of presentation, we feel it can serve to introduce and illustrate "modern shell-model calculations." For instance, this article illustrates three frequent characteristics of modern shell-model studies:

(a) the use of "realistic" effective interactions, based on data from nucleon–nucleon scattering experiments;

(b) the interest in relating microscopic and macroscopic descriptions of many-body nuclei; and

(c) the use of large fast computers. (Such computers make it feasible to perform unified calculations involving many nuclei, many observables, large shell-model vector spaces, and a variety of alternative model Hamiltonians.)

The nuclei $A = 18$–$22$ are light "$s$–$d$-shell" nuclei, and in our calculation the three shells $0d_{5/2}$, $1s_{1/2}$, and $0d_{3/2}$ are all treated as active while the $(0s)^4(0p)^{12}$ core remains closed. No further restrictions are imposed; thus the shell-model basis includes all Pauli-allowed states formed from $A - 16$ active nucleons distributed among the three active shells. Our general procedure is standard. For each $A, J, T$ combination we diagonalize a model Hamiltonian in the vector space spanned by our basis. The eigenvalues are then interpreted as energy levels, and the eigenvectors are used to calculate nuclear moments and transition rates. We have calculated $A = 18$–$22$ results with a variety of model Hamiltonians—some "realistic" ones based on the Hamada–Johnston ($2$) scattering potential, and some "phenomenological" ones obtained via least-square fits to experimentally observed energy levels. This variety of Hamiltonians has been used in an attempt to determine which characteristics of the computed results are sensitive to details of the model Hamiltonian, and which characteristics seem to be more closely associated with the shell-model vector space assumed.

There are several reasons for making these extensive calculations. First, there are now available a large number of experimental data on energy levels, nucleon-transfer spectroscopic factors, and electromagnetic transition rates and static moments. We would like to know whether (or which of) these data can be understood in terms of a comprehensive, many-body model. Some of the lighter nuclei which we shall discuss have already been studied many times in shell-model investigations, but we know of no earlier calculation which has been successful in correlating so many of the $A = 17$–$22$ data with one comprehensive shell model.

A second interest is stimulated by the current availability of realistic effective interactions. These are interactions derived from nucleon–nucleon scattering data, and designed expressly for use in shell-model calculations. Several of our alternative model Hamiltonians incorporate effective nucleon–nucleon interactions that were derived from the Hamada–Johnston scattering potential by Kuo and Brown (*3*), and by Kuo (*4*). These realistic interactions were intended for use in the same unrestricted *s–d*-shell vector space that we use; in particular, they were designed and used for calculations of $^{18}$O and $^{18}$F, the *s–d*-shell nuclei which have two extra-core nucleons. By using the Kuo–Brown and Kuo two-body interactions for systems of three, four, five, and six active nucleons, we study whether these realistic interactions are useful in conventional many-body shell models.

A third reason for doing these large calculations is that the theoretical results can be used as data for comparison with results from other nuclear-structure models. Of special interest are alternative models which make the calculations simpler to perform, and yet allow active particles in $0d_{5/2}$, $1s_{1/2}$, and $0d_{3/2}$ orbits (so that, for example, nuclear reactions transferring a $d_{5/2}$, $s_{1/2}$, or $d_{3/2}$ nucleon can all be described). The particular set of many-particle basis states which we use makes it quick and easy for us to calculate Hamiltonian matrix elements between any two basis states (*5*). However, even in a low-lying energy eigenstate, the intensity is usually spread out over a large number of components. If we were to try reducing the labor of computation simply by omitting many of our basis states from the representation (while keeping the Hamiltonian fixed) we would introduce significant errors into some of the calculated results. But truncation is of great practical interest, for in a *full* *s–d*-shell model the fixed-$(A, J, T)$ energy matrices increase rapidly in size as the number of active particles (or holes) is increased beyond one or two. For $23 \le A \le 33$ the matrix dimensions are prohibitively large (6); it is for this reason that we have cut off the present calculations at $A = 22$. There have been many investigations aimed at finding a good basis-state representation for even-parity states in $s$–$d$-shell nuclei. By "good" here we mean a representation in which a small number of basis states would allow a satisfactory description of the observed even-parity states. Elliott and others (7–11) have shown that the $SU_3$ group and its subgroups are useful, in this respect, for many nuclear states in light $s$–$d$-shell nuclei. Other workers (12–16) have used restricted basis sets defined in terms of Nilsson orbits, or have made deformed Hartree–Fock calculations to determine a good representation.[1] If one is willing to forego the possibility of calculating $d_{3/2}$ transfer rates, then an obvious simplification (17, 18) is to allow active nucleons in the $0d_{5/2}$ and $1s_{1/2}$ orbits only.[2] In many investigations involving restricted $s$–$d$-shell vector spaces, the idea is to approximate the results that would emerge from calculation in a complete $s$–$d$-shell vector space. A comparison of the "full $s$–$d$-shell" energies, spectroscopic factors, and transition rates reported here, against those calculated from a more restricted model, should help to determine whether the more restricted model can be applied with some confidence, either to nuclei which cannot be treated within the complete $s$–$d$-shell space, or to the nuclei $A = 18$–22 (so as to speed up calculation).

Some comments on our computational techniques are given in Appendix A. In Section 2 of the main text, we describe the several different effective Hamiltonians we have used. In Section 3 we discuss the shell-model results for excitation energies, and for spectroscopic factors in one-nucleon transfer reactions. In Section 4 we present the results for ground-state

[1] We imply, here, the use of Nilsson or Hartree–Fock orbits in constructing states quantized in $A$, $J$, and (sometimes) $T$. The most usual procedure of this kind is to obtain a model for the lowest state of $(A, J, T)$ by projecting the appropriate good-$J$ good-$T$ part from the lowest-lying many-particle determinant-state involving deformed Hartree–Fock (or deformed Nilsson) orbits. Thus, the shell-model basis is reduced to *one* state for each $(A, J, T)$ combination. For examples, see (12–14). An extension of this procedure is to diagonalize the $(1 + 2)$-body model Hamiltonian in a small $(A, J, T)$ vector space, a space spanned by the good-$J$ good-$T$ parts projected from *several* of the lowest-lying Hartree–Fock (or Nilsson) determinants. For examples, see (15) and (16).

[2] This constitutes a severe reduction of the full $s$–$d$-shell vector space. Nevertheless, for many physical observables, calculation within a $0d_{5/2}$–$1s_{1/2}$ model can yield results very similar to those obtained with a $0d_{5/2}$–$1s_{1/2}$–$0d_{3/2}$ model. (This similarity can exist if the model Hamiltonian, and other model operators, are adjusted to compensate for neglect of $d_{3/2}$ active nucleons in the explicit basis.) For examples of $0d_{5/2}1s_{1/2}$ calculations, see (17 and 18).

binding energies. In Section 5 we consider electric quadrupole moments and $B(E2)$ strengths; and in Section 6, magnetic moments and $B(M1)$ strengths. Section 7 contains a summary.

*Throughout this article*, terms such as "s–d model," "s–d-shell model," and "s–d-shell vector space" imply a restriction to $A$–16 active $1s$–$0d$ nucleons, outside a closed $(0s)^4(0p)^{12}$ core.

## 2. MODEL HAMILTONIANS

We make the usual assumption, that the effective Hamiltonian is a sum of one-body and two-body terms. Then, in first-quantized notation, we have for the energy with respect to $^{16}$O,

$$H_{\text{eff}} = \sum_{i=1}^{A-16} \varepsilon(j_i) + \sum_{i<j}^{A-16} V_{ij} \tag{1}$$

Here $H_{\text{eff}}$ acts only within the s–d-shell vector space of our many-particle states, and $A - 16$ is the total number of active nucleons in the model nucleus. The most general $(1 + 2)$-body Hamiltonian in our $d_{5/2}$–$s_{1/2}$–$d_{3/2}$ vector space is specified by three "single-particle energies"

$$\langle j_a \mid \varepsilon(j) \mid j_b \rangle = \delta_{ab}\varepsilon_a$$

and by 63 two-body matrix elements between normalized antisymmetric states,

$$\langle j_a j_b \, J T \mid V \mid j_c j_d \, J T \rangle$$

## 2.1. List of the Hamiltonian Used

In this article we shall report on results from seven alternative Hamiltonians. Five of these were adjusted via least-squares search so as to fit observed ground-state binding energies and level excitations in the nuclei $A = 17$–22. Since it was impractical to vary 66 parameters independently, we assumed various restricted forms for these Hamiltonians, and then searched to optimize the parameters of these restricted forms. We now list the seven kinds of Hamiltonians used. For each Hamiltonian we give first its shorthand label, then the number of parameters which were adjusted via least-squares search, and then a capsule description (to be enlarged upon later):

$K+{}^{17}O$ (*no free parameters*): This is the realistic effective Hamiltonian derived from the Hamada–Johnston potential by Kuo (*4*), and used by him for ${}^{18}O$ and ${}^{18}F$. The single-particle energies in this Hamiltonian are the negatives of the measured binding energies of the lowest $5/2^+$, $1/2^+$, and $3/2^+$ states in ${}^{17}O$, with respect to the ground state of ${}^{16}O$. These single-particle energies are $-4.15$, $-3.28$, and $0.93$ MeV, respectively.[3]

$KB+{}^{17}O$ (*no free parameters*): This is the realistic effective Hamiltonian derived by Kuo and Brown (*3*) for ${}^{18}O$ and ${}^{18}F$. This Hamiltonian, too, incorporates the ${}^{17}O$ single-particle energies.

$K+SPE$ (*3 free parameters*): The two-body part of this Hamiltonian is the Kuo interaction. The values of the three single-particle energies are adjusted to give a least-square fit to experimentally observed energies of levels in the nuclei $A = 17$–$22$.

$KB+SPE$ (*3 free parameters*): The two-body part is the Kuo–Brown interaction, and the three single-particle energies are adjusted to give a least-square fit to experimentally observed energies in $A = 17$–$22$.

$K+12FP$ (*12 free parameters*): The adjusted parameters are the three single-particle energies, plus nine two-body matrix elements involving only the $d_{5/2}$ and $s_{1/2}$ orbits. All other two-body matrix elements are held fixed at the Kuo values.

$RIP$ (*14 free parameters*): The letters $RIP$ stand for "radial integral parameterization." The 14 adjusted parameters are the three single-particle energies, plus 11 linearly independent combinations of radial Talmi integrals describing an effective central interaction.

$MSDI$ (*7 free parameters*): The letters $MSDI$ stand for "modified surface-delta interaction." The seven free parameters are the three single-particle energies, plus two surface-delta (*20*) strengths (one for $T = 0$ interactions and one for $T = 1$ interactions), plus two modifying "monopole" strengths (one for $T = 0$ and one for $T = 1$). The two monopole strengths specify an additional Hamiltonian term which depends only on the isospin $T$ of the interacting nucleon-pair.

In Table I we collect for comparison all the one- and two-body matrix elements of the seven Hamiltonians we have used. There are striking differ-

---

[3] More accurately: The measured binding energy of ${}^{17}O$ with respect to ${}^{16}O$ is 4.142 MeV (*19*). The energy $-4.15$ MeV, which we use, is an average of $-4.142$ MeV with $-4.155$ MeV, the latter being an estimate of the nuclear binding energy of ${}^{17}F$ relative to ${}^{16}O$ (see Section 4 ahead). The single-particle energies $-3.28$ and $0.93$ MeV were obtained by adding $-4.15$ MeV to the pertinent observed excitation energies in ${}^{17}O$.

**TABLE I. The Two-Body and One-Body Matrix Elements Defining the Seven Hamiltonians Used.** The units are MeV, and the phase conventions are explained in Section A.2 of Appendix A. The upper part of column 1 shows labels $2j_a$, $2j_b$, $2j_c$, $2j_d$, $JT$ which identify the 63 two-body matrix elements $\langle j_a j_b JT \mid V \mid j_c j_d JT \rangle$ shown in columns 2–6. The lower part of column 1 shows labels $j$ for the three single-particle energies $\varepsilon(j)$; and it also shows the $^{17}O$ single-particle energies which were used in the two unadjusted Hamiltonians $K+^{17}O$ and $KB+^{17}O$. (The two-body matrix elements of $K+^{17}O$ and $KB+^{17}O$ are the same as those shown in columns 2 and 3, respectively.) The starred numbers are matrix elements affected by parameters which were adjusted to give least-square fits.

|   | 1 | 2<br>$K+SPE^a$ | 3<br>$KB+SPE^b$ | 4<br>$K+12FP$ | 5<br>$RIP$ | 6<br>$MSDI$ |
|---|---|---|---|---|---|---|
| 1 | 5555,01 | −2.4381 | −2.5264 | −1.6502* | −2.7173* | −2.4920* |
| 2 | 5555,10 | −1.0284 | −0.7848 | −1.0284 | −1.1072* | −3.7633* |
| 3 | 5555,21 | −1.0358 | −0.9426 | −1.3492* | −0.5862* | −0.2839* |
| 4 | 5555,30 | −0.8589 | −0.8355 | −0.8589 | −0.7550* | −3.2107* |
| 5 | 5555,41 | −0.0502 | 0.1366 | 0.2788* | 0.3424* | 0.0977* |
| 6 | 5555,50 | −3.6640 | −3.6872 | −3.7282* | −3.6526* | −3.6086* |
| 7 | 5551,21 | −0.8542 | −0.8093 | −0.0602* | −0.8911* | −0.8655* |
| 8 | 5551,30 | −1.5654 | −1.5220 | −1.5654 | −1.6566* | −1.0274* |
| 9 | 5553,10 | 3.1651 | 3.1676 | 3.1651 | 1.1804* | 1.4036* |
| 10 | 5553,21 | −0.3969 | −0.3719 | −0.3969 | 0.1425* | −0.4626* |
| 11 | 5553,30 | 1.8746 | 1.9338 | 1.8746 | 0.3000* | 0.4594* |
| 12 | 5553,41 | −1.3626 | −1.2946 | −1.3626 | −1.1333* | −0.5452* |
| 13 | 5511,01 | −0.9677 | −1.0818 | −1.5051* | −1.2611* | −1.6526* |
| 14 | 5511,10 | −0.5959 | −0.5960 | −0.5959 | −2.0642* | −0.9154* |
| 15 | 5513,10 | −0.2368 | −0.3602 | −0.2368 | −0.8434* | −0.5231* |
| 16 | 5513,21 | −0.8364 | −0.7803 | −0.8364 | −0.7274* | −0.7067* |
| 17 | 5533,01 | −3.7882 | −4.1090 | −3.7882 | −1.7235* | −2.3371* |
| 18 | 5533,10 | 1.6209 | 1.6502 | 1.6209 | 0.8243* | 0.9098* |
| 19 | 5533,21 | −0.9034 | −0.9390 | −0.9034 | −0.4019* | −0.4997* |
| 20 | 5533,30 | 0.4996 | 0.5442 | 0.4996 | 0.6128* | 0.2653* |
| 21 | 5151,20 | −0.6222 | −0.4218 | −0.6222 | −1.5860* | −3.1223* |
| 22 | 5151,21 | −1.2879 | −1.0917 | −0.6325* | −0.7951* | −0.7746* |
| 23 | 5151,30 | −3.6919 | −3.6317 | −3.6919 | −3.9650* | −4.0507* |
| 24 | 5151,31 | 0.1723 | 0.2426 | 0.6312* | 0.4544* | 0.3703* |
| 25 | 5153,20 | −1.4488 | −1.1559 | −1.4488 | −1.3337* | −0.8271* |
| 26 | 5153,21 | −0.2181 | −0.1456 | −0.2181 | −0.6300* | −0.6120* |
| 27 | 5153,30 | 1.1561 | 1.0324 | 1.1561 | 1.2901* | 0.7909* |
| 28 | 5153,31 | −0.0892 | −0.1037 | −0.0892 | 0 | 0 |
| 29 | 5113,20 | −2.5788 | −2.3922 | −2.5788 | −1.9424* | −0.7581* |
| 30 | 5113,21 | −1.5511 | −1.3613 | −1.5511 | −1.0204* | −0.9348* |
| 31 | 5133,21 | −0.7436 | −0.6296 | −0.7436 | −0.6804* | −0.6610* |

**TABLE I** (*Continued*)

|  | 1 | 2<br>$K+SPE^a$ | 3<br>$KB+SPE^b$ | 4<br>$K+12FP$ | 5<br>RIP | 6<br>MSDI |
|---|---|---|---|---|---|---|
| 32 | 5133,30 | 0.0269 | 0.0364 | 0.0269 | 0.2761* | 0.1712* |
| 33 | 5353,10 | −5.8276 | −5.2826 | −5.8276 | −2.4374* | −5.2886* |
| 34 | 5353,11 | −0.1257 | −0.0532 | −0.1257 | −0.6063* | 0.3703* |
| 35 | 5353,20 | −4.5271 | −4.2293 | −4.5271 | −1.0880* | −3.6086* |
| 36 | 5353,21 | −0.2037 | −0.0469 | −0.2037 | −0.2631* | 0.0431* |
| 37 | 5353,30 | −1.1313 | −1.0046 | −1.1313 | −0.9615* | −3.0781* |
| 38 | 5353,31 | 0.1316 | 0.1910 | 0.1316 | 0.9090* | 0.3703* |
| 39 | 5353,40 | −4.3137 | −3.9699 | −4.3137 | −3.6526* | −3.6086* |
| 40 | 5353,41 | −1.6603 | −1.2292 | −1.6603 | −1.3576 | −0.7201* |
| 41 | 5311,10 | 1.7125 | 1.7247 | 1.7125 | 3.1209* | 1.3841* |
| 42 | 5313,10 | −1.9132 | −1.6663 | −1.9132 | −1.1160* | −0.6920* |
| 43 | 5313,11 | −0.0976 | −0.0513 | −0.0976 | 0 | 0 |
| 44 | 5313,20 | −1.5404 | −1.1414 | −1.5404 | −1.6335* | −1.0130* |
| 45 | 5313,21 | −0.7697 | −0.7660 | −0.7697 | −0.5143* | −0.4997* |
| 46 | 5333,10 | 0.0383 | 0.1738 | 0.0383 | −0.3231* | −0.4377* |
| 47 | 5333,21 | −1.0101 | −1.0367 | −1.0101 | −0.6767* | −0.3533* |
| 48 | 5333,30 | 2.1579 | 2.1500 | 2.1579 | 1.2152* | 0.3063* |
| 49 | 1111,01 | −1.9493 | −2.2130 | −2.4564* | −1.7253* | −0.5838* |
| 50 | 1111,10 | −3.1839 | −3.6241 | −3.1839 | −4.3222* | −3.2770* |
| 51 | 1113,10 | 0.3085 | 0.5560 | 0.3085 | 0 | 0 |
| 52 | 1133,01 | −0.7448 | −0.8387 | −0.7448 | −1.0303* | −1.3493* |
| 53 | 1133,10 | −0.2127 | −0.2533 | −0.2127 | 1.1033* | 0.4894* |
| 54 | 1313,10 | −3.2771 | −3.1107 | −3.2771 | −3.9650* | −4.0507* |
| 55 | 1313,11 | 0.2167 | 0.3456 | 0.2167 | 0.4544* | 0.3703* |
| 56 | 1313,20 | −1.6099 | −1.2296 | −1.6099 | −2.3790* | −3.4317* |
| 57 | 1313,21 | −0.3267 | −0.0617 | −0.3267 | −0.3792* | −0.3930* |
| 58 | 1333,10 | 0.7995 | 0.7323 | 0.7995 | 1.5782* | 0.9787* |
| 59 | 1333,21 | −0.2071 | −0.1484 | −0.2071 | −0.5557* | −0.5397* |
| 60 | 3333,01 | −0.8076 | −0.5396 | −0.8076 | −2.0141* | −1.5379* |
| 61 | 3333,10 | −0.4695 | −0.0662 | −0.4695 | −0.8759* | −3.4317* |
| 62 | 3333,21 | 0.0770 | 0.2575 | 0.0770 | 0.3604* | −0.0113* |
| 63 | 3333,30 | −2.5872 | −2.3485 | −2.5872 | −3.0242* | −3.4317* |
|  | $^{17}O$ |  |  |  |  |  |
| 5/2 | −4.15 | −4.03* | −4.24* | −4.24* | −4.48* | −4.49* |
| 1/2 | −3.28 | −2.45* | −2.84* | −3.34* | −2.81* | −3.16* |
| 3/2 | 0.93 | 1.58* | 1.03* | 1.51* | 1.07* | 1.04* |

[a] The two-body matrix elements are from (4).
[b] The two-body matrix elements are from (3).

ences in some important two-body matrix elements $\langle j_a j_b \, JT \,|\, V \,|\, j_c j_d \, JT \rangle$, e.g., in $\langle \frac{5}{2} \frac{5}{2} 10 \,|\, V \,|\, \frac{5}{2} \frac{5}{2} 10 \rangle$, $\langle \frac{5}{2} \frac{5}{2} 10 \,|\, V \,|\, \frac{5}{2} \frac{3}{2} 10 \rangle$, and $\langle \frac{5}{2} \frac{3}{2} 10 \,|\, V \,|\, \frac{5}{2} \frac{3}{2} 10 \rangle$. (See entries 2, 9, and 33 in Table I.)

## 2.2. Further Comments on the Realistic Two-Body Interactions *K* and *KB*

The Kuo–Brown realistic interaction (*3*) *KB* was derived from the Hamada–Johnston scattering potential via reaction matrix methods. Each of its 63 two-body matrix elements included two different kinds of contributions: a "bare *G*-matrix" term, symbolized by Fig. 1(a), and a "three-particle one-hole" renormalization term, symbolized by Fig. 1(b). The bare *G*-matrix term is designed to make up for the fact that, in the shell-model calculations to be made with $H_{\text{eff}}$, the explicit vector space omits high-lying states (i.e., high-lying on a harmonic-oscillator model). The three-particle one-hole term is a second-order-perturbation correction. It is designed to make up for the fact that the explicit shell-model basis omits states in which only 15 nucleons occupy 0*s*–0*p* orbits, while the sixteenth one is excited by an energy $2\hbar\omega$ either to the 0*d*–1*s* shell or to the 0*f*–1*p* shell. There are several other second-order $2\hbar\omega$ corrections, as shown by (c), (d), and (e)

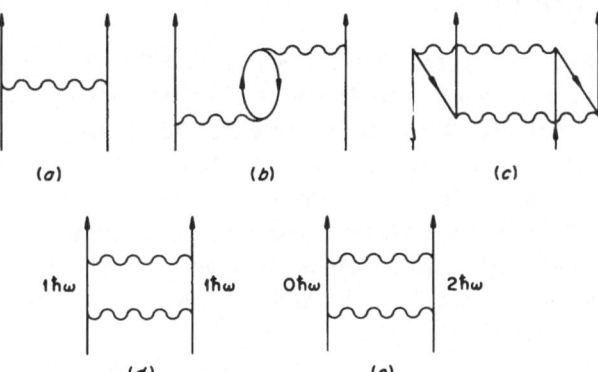

Fig. 1. Diagrams symbolizing the first-order reaction-matrix contribution (a), and the second-order $2\hbar\omega$-renormalization contributions (b), (c), (d), and (e), to a realistic two-nucleon interaction suitable for use in shell-model calculations. In the last two diagrams, the intermediate-state lines are labeled so as to distinguish between two related cases. In case (d), each of the two intermediate-state particles lies $1\hbar\omega$ above the active shells. In case (e), one of the two intermediate-state particles occupies an active shell while the other lies $2\hbar\omega$ above.

in Fig. 1, but these were omitted from the Kuo–Brown interaction. These omissions were defended on various theoretical grounds (3). In addition, there were pragmatic grounds, for the calculated mass-18 spectra looked very similar to the experimentally observed spectra when these further renormalization terms were omitted, but looked unlike the observed spectra when these terms were included.

The Kuo (4) interaction $K$, which may be considered a later version of the Kuo–Brown interaction, was calculated with several refinements in the bare-$G$-matrix techniques; furthermore, the four terms represented by Fig. 1(a), (b), (c), and (d) are all included in its two-body part. The resulting interaction, together with $^{17}O$ single-particle energies, forms a Hamiltonian which leads to $A = 18$ spectra clearly resembling the experimentally observed spectra.

The two total interactions, Kuo and Kuo–Brown, are quite similar even though they differ in their bare-$G$ terms and in their renormalization contributions. This circumstance illustrates the point that a successful shell-model reproduction of measured data does not in itself prove that the procedure for obtaining the effective Hamiltonian is correct in all its aspects.[4,5]

## 2.3. Further Description of the Least-Square Hamiltonians

In our least-square searches for $K+SPE$, $KB+SPE$, and $MSDI$ we minimized the rms deviation to 41 pieces of experimentally observed data. These data comprised 29 excitation energies, and 12 ground-state binding energies with respect to $^{16}O$. (The ground-state binding energies were all corrected to subtract Coulomb contributions as described in Section 4 later.) All but three of the 41 fitted levels are shown within the first columns of Figs. 2–9, and marked at the left with a black dot. The other three fitted levels are just the lowest $5/2^+$, $1/2^+$, and $3/2^+$ levels in $^{17}O$ (not illustrated).

---

[4] In fact, an arithmetic error common to the bare-$G$ parts of the interactions $K$ and $KB$ has recently been discovered. In the second-order tensor term $V_t (Q/e) V_t \approx (V_t{}^2/e_{eff})$, the effective-energy denominator $e_{eff}$ was given only half the value it should have had for a correct mathematical treatment of the physical approximations intended. However, it is now thought that a more accurate treatment of the Pauli operator $Q$ will tend to counterbalance the changes in $K$ and $KB$ that would result from correction of this factor-of-two error in $e_{eff}$. A new calculation, improved in several ways, is in progress (21).

[5] There is no clear justification for limiting the renormalization corrections to second-order effects [see (22)].

The ground state of each nucleus $(A, T)$ was fitted as a binding energy with respect to $^{16}O$, while other states of $(A, T)$ were fitted as excitations with respect to the ground state of $(A, T)$. For most of those levels for which measured excitation energies in "mirror" nuclei were available, the measured mirror energies were averaged before being fitted. But for $A = 17$, we fitted to $-4.15$, $-3.28$, and $0.93$ MeV, the same $^{17}O$ energies as we used in the unadjusted Hamiltonians $KB+^{17}O$ and $K+^{17}O$. In our *RIP* search, we used these same criteria except that we added two more $^{20}Ne$ levels ($6^+$ and $8^+$) to the data set. In our $K+12FP$ search, we included these two $^{20}Ne$ levels, added the first $5^+$ level in $^{22}Na$, and then fitted all 44 levels as binding energies with respect to $^{16}O$.

Keeping the two-body-interaction matrix elements fixed while varying only the single-particle energies is an easy time-honored way of adjusting the theory to fit data. The optimized single-particle energies for $KB+SPE$ and $K+SPE$ are included in Table I.

The Hamiltonian form $K+12FP$ is a modification of an 18-parameter version which we tried first. In this 18-parameter version, the free parameters were the $d_{5/2}$ and $s_{1/2}$ single-particle energies, plus all of the 16 two-body-interaction matrix elements involving $d_{5/2}$ or $s_{1/2}$ nucleons but not $d_{3/2}$ nucleons. The idea was that occupation of the $d_{3/2}$ shell might be considered a perturbation, and that in this spirit we would satisfy ourselves with the Kuo matrix elements for the perturbative influence, but adjust the more important matrix elements (i.e., all non-$d_{3/2}$ matrix elements) to fit experimentally observed level data. (But note that even this 18-parameter form keeps constant the very strong $d_{5/2}$–$d_{3/2}$ interaction that is a characteristic of the Kuo Hamiltonian.) Preliminary searches with this 18-parameter form indicated that several of the adjustable two-body matrix elements were not well determined by our least-square search criteria. Therefore we reduced the number of adjustable two-body matrix elements to only nine, the ones indicated by the starred numbers in Table I. When all three single-particle energies are also made adjustable, we get the 12 free parameters implied by the abbreviation $K+12FP$.

As indicated in our descriptions above, the adjustable forms $K+SPE$, $KB+SPE$, and $K+12FP$ are constrained to be similar to the two realistic Hamiltonians we have used. However, the two forms *RIP* and *MSDI* have no such formal constraints.

In the *RIP* form $(23, 24)$, the adjustable parameters of the two-body interaction are linear combinations of the diagonal radial integrals

$$(nl)_T = \int_0^\infty R_{nl}^2(r) \, V_{nlT}(r) r^2 \, dr \tag{2}$$

which define a central (but generally nonlocal) interaction between nucleons. Here $R_{nl}$ is the radial part of a harmonic-oscillator wave function describing the relative motion of the two interacting nucleons, and $V_{nlT}$ is an effective central potential which depends on their isospin $T$ as well as on the quantum numbers $nl$ describing their relative motion. For each value of $T$, all the two-body matrix elements for $sd$-shell nucleons are specified by the following seven linear combinations of the radial integrals $(nl)_T$ of (2):

$$(0s + 0g)_T, \ (0s + 1d)_T, \ (0s + 2s)_T, \ (1s)_T, \ (0d)_T,$$

$$(0p + 0f)_T, \ (0p + 1p)_T \tag{3}$$

Since $T$ is 0 or 1, there are 14 $RIP$ parameters needed to completely determine 63 two-body matrix elements. Adding three variable single-particle energies would make 17 parameters in all. In the course of searching with these 17 adjustable parameters, we found that values for the $T = 0$ parameters $(0d)_0$, $(0p + 0f)_0$, and $(0p + 1p)_0$ were not well-determined by our search criteria. That is, the calculated energy spectra seemed to be quite insensitive to the values of these parameters. Therefore the parameters $(0d)_0$, $(0p + 0f)_0$, and $(0p + 1p)_0$ were held fixed at zero, while the other 11 radial–integral combinations were allowed to vary. The optimized values of these 11 radial–integral parameters are listed in Table II; and the resulting 63 two-body matrix elements $\langle j_a j_b JT \,|\, V \,|\, j_c j_d JT \rangle$ are shown in Table I. The three optimized single-particle energies are also shown in Table I.

TABLE II. Optimized Values of the Two-Body Parameters for $RIP$ and $MSDI$. All entries are in MeV.

| | RIP | | MSDI |
| --- | --- | --- | --- |
| | $T = 0$ | $T = 1$ | |
| $0s + 0g$ | $-9.74$ | $-5.81$ | $A_0 = \phantom{-}0.774$ |
| $0s + 1d$ | $-13.61$ | $-6.77$ | $A_1 = \phantom{-}0.954$ |
| $0s + 2s$ | $-20.78$ | $-10.39$ | $B_0 = -2.503$ |
| $1s$ | $-0.24$ | $-4.52$ | $B_1 = \phantom{-}0.370$ |
| $0d$ | $0^a$ | $1.02$ | |
| $0p + 0f$ | $0^a$ | $1.82$ | |
| $0p + 1p$ | $0^a$ | $-1.21$ | |

$^a$ Held fixed at zero during the least-squares search.

The _MSDI_ Hamiltonian form (25, 26) has a two-body part which is central but not translationally invariant. It is

$$V_T(ij) = -4\pi A_T \,\delta(r_i - r_j)f_{ij} + B_T \qquad (4)$$

Here $T$ indicates the isospin $T$ (0 or 1) of the interacting nucleon-pair; $A_T$ and $B_T$ are strengths depending only on $T$; and $f_{ij}$ is an operator which has the following effect on the radial part of $\langle j_a j_b JT \mid V \mid j_c j_d JT \rangle$:

$$\int_0^\infty \int_0^\infty r_i dr_i \, r_j \, dr_j \, R_a(r_i)R_b(r_j) \, \delta(r_i - r_j)f_{ij}R_c(r_i)R_d(r_j) = (-)^{n_a+n_b+n_c+n_d} \quad (5)$$

Here $R_a$, $R_b$, $R_c$, $R_d$ are the single-particle radial wave functions involved in $\langle j_a j_b \, J T \mid V \mid j_c j_d \, J T \rangle$; and $n_a$ refers to the principal quantum number of the orbit associated with $j_a$, with our convention, that the lowest orbit of given $l$ has principal quantum number zero.[6] Thus all the _s–d_-shell two-body matrix elements are specified by the four parameters $A_0$, $A_1$, $B_0$, and $B_1$ in (4). Their optimized values are included in Table II, and the resulting 63 two-body matrix elements are shown in Table I. The three optimized single-particle energies are also listed in Table I.

## 3. EXCITATION ENERGIES AND SPECTROSCOPIC FACTORS

Although our basic shell model involves no collective-rotational coordinates (nor is it limited to favored $SU_3$ representations), our Hamiltonians lead to low-lying eigenvalues and eigenvectors which exhibit rotational-band features. Except for ¹⁸F, all the nuclei in Figs. 2–9 show, in their experimentally observed energy spectra and in their shell-model spectra, at least a few members of what appears to be a ground-state rotational band. In some nuclei, higher bands can also be identified. As we shall see, within each ground-state band our shell models yield generally good agreement with experimentally determined excitation energies and single-nucleon spectroscopic factors. The calculated energies of other band-heads, with respect to the ground state, agree less well with measured results. These

---

[6] Equation (5) differs in phase from the analogous statement made in (26). The convention implicit in (26) was that all single-particle wave functions were chosen to have the same sign at the nuclear radius. (This was the convention assumed in the original development of the surface-delta interaction [see (20)].) But now, in the present article, we use a different convention, that all $R_{nl}(r)$ are positive at small values of $r$.

experimental–theoretical energy discrepancies, coupled with experimental uncertainties for the higher states, cause difficulties in correlating observed and shell-model levels outside ground-state bands. Nevertheless, we can point out some modest successes outside ground-state bands, and a few recognizable disappointments.

The experimental information used in this Section 3 comes from References (27) to (84). Our detailed nucleus-by-nucleus discussion in Section 3.3 will make it clear why we treat excitation energies and spectroscopic factors together. The reason is that it is extremely useful to have at least one more characteristic, besides excitation energy, spin, isospin, and parity, on which to base the correlation between a calculated level and an experimentally observed level.

## 3.1. Excitation Energies: Introductory Remarks

Figures 2–9 show the experimentally observed excitation spectra for $A = 18$–22, and calculated spectra from six different model Hamiltonians. (To avoid an eighth column, we have omitted results from the $KB+SPE$ model. They are, as expected, most similar to those from the $KB+{}^{17}O$ model.)

### 3.1.1. Notation in Figs. 2–9

Bold lines indicate levels which seem to be members of a ground-state rotational band. (We shall refer to them, hereafter without apology, as ground-state-band members.) In each figure, all the shell-model spectra are lined up at the observed ground state of minimum $T$. The vertical scale represents *nuclear* energy. Hence in each experimentally observed spectrum, the plotted energy difference between two ground states of different $T$ is the measured energy difference, minus an estimated Coulomb-energy difference.[7] The lowest known state of established odd parity is included as a dashed line. All other known odd-parity states are omitted, and levels of uncertain existence are omitted too. Thin solid lines indicate levels of established even parity. Dotted lines indicate levels of uncertain parity. When a spin-label appears, an added parity-label means that the parity is *not* established as even. For example, the label 6 or (6) implies established even parity; $6^{(+)}$ or (6+) implies probably-even parity; and $6^{\pm}$ or (6)$^{\pm}$ implies unknown parity. Since all the shell-model levels have even parity, they are all drawn as solid

---

[7] See Section 4.

lines with no parity-labels. For levels of half-integral spin, the spin-labels are twice $J$ instead of $J$ itself. A black dot at the left of an experimentally observed level means that this level was included in the least-square fits leading to the adjusted Hamiltonians $K+SPE$, $KB+SPE$, $K+12FP$, $RIP$, and $MSDI$.

For each isospin $T$, a given spectrum includes all known levels of even parity, or possibly-even parity, up to the highest level shown. (Exceptions are noted in the figure captions.) In cases where experimentally observed states in mirror nuclei seem obviously analogous, we have plotted the average of their observed excitations. Close-lying levels have been drawn far enough apart to make them distinguishable.

### 3.1.2. *"Intruder States"*

Our explicit shell-model basis is limited to configurations of the type $(0s)^4(0p)^{12}(1s, 0d)^{4-16}$. Hence our models yield no theoretical partners for any of the odd-parity levels observed in $A = 18$–22. Furthermore, we find that there are certain observed states of *even* parity for which our *s–d*-shell models yield no theoretical partners. For some of these states we believe that *no* simple $(1 + 2)$-body Hamiltonian, acting within our *s–d*-shell space, could yield satisfactory partners. Thus we consider that all odd-parity observed states, and some even-parity observed states, are "intruder states" with respect to any $(0s)^4(0p)^{12}(1s, 0d)^{4-16}$ model.

In this article we shall sometimes refer to intruder states as "dominated by configurations outside our *s–d*-shell model," or "core-excited." Also, we shall use the term "predominantly *s–d*-shell" in referring to observed states whose properties *can* be satisfactorily reproduced within a $(0s)^4(0p)^{12}(1s, 0d)^{4-16}$ model. *These descriptions are statements about the kind of model vector-spaces needed to satisfactorily reproduce the observed data of interest.* They are not to be interpreted as literal descriptions of the true nuclear wave functions. Indeed, many writers have pointed out that shell-model wave functions should be considered as transformations of actual nuclear wave functions, not as approximations to actual nuclear wave functions (*85*). We shall now expand on these points.

For our brief discussion here let us consider, as "true nuclear wave functions," the many-body eigenfunctions of a Hamiltonian incorporating any modern phenomenological scattering potential, e.g., the Hamada–Johnston potential. (Thus we ignore questions about meson coordinates, etc.). It is customary to distinguish between two ways in which true nuclear wave functions may differ substantially from shell-model nuclear wave functions, even when the shell model successfully reproduces many measured data:

(a) Brueckner-theory studies (86) have emphasized that successful model wave functions may differ substantially from the true wave functions in their behavior *at very close nucleon–nucleon approaches* (say, within distances three times the Hamada–Johnston hard-core radius).

(b) "Pseudonium studies" (87), and some other shell-model studies, suggest that successful model wave functions may differ substantially from the true wave functions *even in long-range nucleon–nucleon correlations.*

As an illustration of point (b), we cite the work of Zuker (88). He discusses a model (88, 89) in which nuclear states are represented by eigenstates of a Hamiltonian acting within the vector space defined by configurations $(0s)^4(0p_{3/2})^8(0p_{1/2}, 1s_{1/2}, 0d_{5/2})^{4-12}$. In this model the ground state of $^{16}$O turns out to be only 65% $(0s)^4(0p)^{12}$. Zuker finds that most of the lowest-lying $A = 18$ eigenstates in this model can be closely approximated by putting two nucleons in the $1s_{1/2}$, $0d_{5/2}$ orbits, and coupling these two nucleons to the $^{16}$O model ground state. Thus, these low-lying $A = 18$ states are only $\approx 65\%$ $(0s)^4(0p)^{12}(1s_{1/2}, 0d_{5/2})^2$. Nevertheless, for many "observables" involving these states, e.g., for $d_{5/2}$ spectroscopic factors, the calculated values are essentially the same whether the 65%-closed $^{16}$O core is used, or whether that core is replaced by a fully-closed $(0s)^4(0p)^{12}$ core. Thus, for the purpose of calculating these observables, a $(0s)^4(0p)^{12}(1s_{1/2}, 0d_{5/2})^{4-16}$ model would have been adequate; and for this reason we call these $A = 18$ states "predominantly s–d-shell." Suppose that the model were altered to allow additional kinds of core excitation, and suppose that the 65% figure then dropped below 50% but the calculated observables remained almost invariant. We would still call these states predominantly sd-shell. In a completely analogous way: we classify some observed even-parity states as predominantly s–d-shell, and we would still call them predominantly sd-shell even if we found out that the long-range correlations in their true wave functions were better described by wave functions having <50% contribution from $(0s)^4(0p)^{12}(1s, 0d)^{4-16}$.

We hope that the above illustration conveys an idea of the sense in which we shall use terms like "core-excited" and "predominantly s–d-shell" when we discuss the characteristics of observed nuclear states.

Our remarks above, and the general course of our shell-model work, have been influenced by this simplifying assumption: that most low-lying nuclear states are either "strongly dominated by $(0s)^4(0p)^{12}(1s, 0d)^{A-16}$ configurations," or else "strongly dominated by other (non-s–d-shell) configurations." This assumption implies that, for the most part, observed states can be neatly divided into *two classes*, one class involving energies that should be matched by s–d-shell-model energies, and another class that should be entirely missing in s–d-shell-model spectra. In constructing our least-squares Hamiltonians, we have not searched on observed states which are thought to be intruder states. (For example, we have not tried to fit the second observed 1⁺ state in ¹⁸F.) Sometimes, for a given pair of observed states, we suspect that there is significant mixing between "s–d-shell character" and "intruder character." Then there are questions: Should the model eigenvalue be expected to coincide with the observed energy of the nuclear state which is more predominantly $(0s)^4(0p)^{12}(1s, 0d)^{A-16}$? Or should the model eigenvalue coincide with some kind of *centroid* for the two states? Or should we just give up trying to describe the situation with a $(0s)^4(0p)^{12}$-$(1s, 0d)^{A-16}$ model? In constructing our least-squares Hamiltonians, we have generally not searched on the measured energies of states which are suspected to involve significant mixing of s–d-shell character and intruder character. (Examples are the second and third 2⁺ states in ²⁰Ne.)

## 3.2. Spectroscopic Factors: Introductory Remarks

The importance of spectroscopic factors to an understanding of nuclear structure has been discussed by Macfarlane and French (*90*) [see also (*91, 92*)]. We define S to be the spectroscopic factor for single-nucleon transfer, *in an isospin formalism.* For example, our calculated S is exactly the same for ¹⁷O(³He, d)¹⁸F$_{T=1}$ and ¹⁷O(d, p)¹⁸O$_{T=1}$. (But the calculated *cross sections* are different because they are proportional to $(C)^2 S$, where C is a Clebsch–Gordan coefficient depending on $T_z$.) The spectroscopic factor for $0d_{5/2}$ transfer from an sd-shell state $\Psi^{J\,T}$ involving $n = A - 16$ active nucleons, to an sd-shell state $\psi^{J_0 T_0}$ involving $n - 1$ active nucleons, is

$$S(0d_{5/2}) = n \langle \Psi^{J\;T}_{J_z T_z}(1, \ldots, n) \,|\, \{\psi^{J_0 T_0}(1, \ldots, n-1) \times \varphi_{0d}^{j=5/2,\,t=1/2}(n)\}^{J\;T}_{J_z T_z} \rangle^2 \quad (6a)$$

where $\varphi_{0d}^{j=5/2,\,t=1/2}(n)$ is the $0d_{5/2}$ single-particle wave function for the nth nucleon. Analogous definitions hold for $S(0d_{3/2})$ and $S(1s_{1/2})$. Hereafter we shall write $S(0d_{5/2})$ simply as $S(d_{5/2})$, etc. When comparing shell-model spectroscopic factors with experimental spectroscopic factors, we shall sometimes be concerned with

$$S(l = 2) \equiv S(d_{3/2}) + S(d_{5/2}) \quad (6b)$$

rather than with the separate $j$-contributions. That is because present analyses of experiments do not usually distinguish between $d_{5/2}$ and $d_{3/2}$, in cases where the initial and final nuclear spins allow both kinds of transfer.

### 3.2.1. Reliability of Numerical S-Values

In choosing experimentally determined $S$-values to compare with our shell-model values, we have preferred experimental values that were determined via DWBA (distorted wave Born approximation) over $S$-values determined via plane-wave (Butler–Born) analysis. In the best of circumstances, when appropriate elastic scattering data are available to determine the optical-model parameters, DWBA spectroscopic factors are uncertain by at least 20% in relative value, and 40% in absolute value. These percentages are associated with ambiguities in the optical-model parameters, and estimated uncertainties in finite-range adjustments, nonlocality corrections, etc. Under less favorable circumstances (or with more pessimistic estimates of uncertainties in the finite-range adjustments, etc.), DWBA "errors" of course run larger. For weak transitions, especially, there are further errors because of physical processes other than the one-step direct-reaction process that is assumed in DWBA (and assumed also in plane wave treatments).

Plane-wave results are recognized as generally less reliable than DWBA results. For several of the nuclei $A = 18$–22, we were tempted to compare our shell-model $S$-values with $S$-values determined from plane-wave analyses of experimental data; that was because the available plane-wave results extended to a larger number of final states than the available DWBA results. However, we found that plane-wave and DWBA results often deviate markedly from each other, even in relative values of $S$ for a fixed $l$-value. An example is shown in Table III. This, and other examples, persuaded us to limit our comparisons to experimental $S$-values based on DWBA analyses.

But it is also true that, for light $s$–$d$-shell nuclei, even the DWBA results reported by different experimenters are seldom in close agreement. Some examples of the differences are included in our detailed nucleus-by-nucleus discussion (see especially Tables IX and X). We hope that these examples will serve as practical indications of the general degree of reliability to be associated with recent experimental determinations of $S$-values for light $s$–$d$-shell nuclei.

As to shell-model spectroscopic factors, we have calculated these straightforwardly from our effective-Hamiltonian eigenvectors, without any

**TABLE III. Comparison of the Spectroscopic Factors $S$ as Determined from a PW (Plane-Wave) Analysis of $^{17}O(d, p)^{18}O$ Data, and as Determined from a DWBA (Distorted-Wave) Analysis of $^{17}O(^3He, d)^{18}F_{T-1}$ Data.** The final $^{18}F_{T-1}$ states are isobaric analogs of those in $^{18}O$; and the energies in this table refer to excitation energies in $^{18}O$. A dotted entry implies a spectroscopic strength too weak to determine from the data. A dash means that the angular distribution was not measured. The table shows all final levels for which (28) reported spectroscopic strengths.

| $J^\pi$ | Excitation energy in $^{18}O$, MeV | $l$ | PW $^{17}O(d, p)^{18}O$ $S^a$ | DWBA $^{17}O(^3He, d)^{18}F_{T-1}$ $S^b$ | $S^a/S^b$ |
|---|---|---|---|---|---|
| $0^+$ | 0 | 2 | 0.81 | 2.32 | 0.35 |
| $2^+$ | 1.98 | 0 | 0.10 | 0.32 | 0.31 |
|  |  | 2 | 0.72 | 1.48 | 0.49 |
| $4^+$ | 3.55 | 2 | 1.23 | 1.96 | 0.63 |
| $0^+$ | 3.63 | 2 | (0.22) | — |  |
| $2^+$ | 3.92 | 0 | 0.32 | 0.44 | 0.73 |
|  |  | 2 | ...$^c$ | 0.98 |  |
| $2^+$ | 5.25 | 0 | 0.30 | — |  |
|  |  | 2 | ... |  |  |
| $0^+$ | 5.33 | 2 | <0.2 | — |  |
| $(3)^+$ | 5.37 | 0 | 1 | — |  |
| $4^+$ | 7.10 | 2 | (0.19) | — |  |

$^a$ From values $(2J + 1)\theta^2$ in (28), normalized as suggested in (28).
$^b$ From (27).
$^c$ All $l = 2$ strength was attributed to $^{16}O(d, p)^{17}O$.

correction for contamination of the real wave functions by configurations other than $(0s)^4(0p)^{12}(1s, 0d)^{4-16}$. Recall that these contaminations warranted the use of a renormalized $G$-matrix interaction in our shell-model calculations, rather than either the nucleon–nucleon scattering potential or an unrenormalized $G$-matrix interaction. An optimistic viewpoint is this: since $S$ involves an overlap between target-nucleus and final-nucleus wave functions, and since core-excitation is presumably similar in the target nucleus and final nucleus, $S$ will be insensitive to core excitation. Similarly, $S$ will be insensitive to contamination from other excited configurations. This optimistic viewpoint is encouraged by spectroscopic-factor results from "pseudonium" studies (87), and also, by comparison of results from other shell-model studies that have treated the same nuclear levels with various (smaller and larger) shell-model vector spaces. [See Section 3.1.2 and (*88*).] Furthermore, this optimistic viewpoint is encouraged by the fact that shell-model $S$-values often agree with experimentally determined $S$-values. (But perhaps the errors in DWBA-determined $S$-factors are fortuitously similar to the errors in shell-model $S$-values!)

*In any case, we believe that qualitative agreement with experimentally determined S-factors is an extremely useful indication as to whether a shell-model level should be associated with an experimentally observed level.* We shall consider agreement within 30% as good agreement.

### 3.2.2. Spectroscopic Factors from "Strong-Coupling" Rotational Models

Since we shall be discussing the energy spectra in terms of rotational bands, it is appropriate to add here a comment on the estimation of $S$-factors from simple "strong-coupling" rotational models. We should like to point out that for $s$–$d$-shell nuclei, such estimates are often poor. (And that is why we have not made routine comparisons of our shell-model spectroscopic factors with strong-coupling rotational results.)

Consider, as a target, the ground state of an even–even nucleus, with $J = T = 0$. Using a strong-coupling model, we describe this ground state as the lowest member of the lowest $A$-nucleon rotational band; and we take the microscopic part of its wave function $\psi^{00}$ to be an axially symmetric deformed intrinsic state $\chi$. In $\chi$, all single-particle deformed orbits up through orbit $N$ are occupied. Now consider an $(A + 1)$-nucleon final state $\Psi^{JT}$, described as one member of the rotational band that is based on the intrinsic state $X$ formed by adding a nucleon in orbit $N'$ to $\chi$. The spectroscopic factor connecting $\Psi^{JT}$ to $\psi^{00}$ is, according to a simple and often used strong-coupling rotational-model prescription (*93*),

$$S_j = \frac{2}{2j + 1} \langle N' \mid j \rangle^2 \, \delta_{jJ} \tag{7}$$

Here $\langle N' \mid j \rangle$ represents the overlap of the single-particle deformed state $|N'\rangle$ with the appropriate single-particle spherical-shell-model state $|j\rangle$. (We have taken the overlap of final and initial *vibrational* states to be unity.) The maximum of $S_j$ according to (7) is $2/(2j + 1)$; but for $j > 1/2$, a spherical-shell-model calculation quite often exceeds $2/(2j + 1)$. As an example, for the reaction $^{20}Ne(d, p)^{21}Ne$ to the lowest $5/2^+$ state in $^{21}Ne$, the strong-coupling rotational prescription (7) yields a maximum of only $S = 1/3$, but a "reasonable" spherical-shell-model calculation yields

$S \approx 2/3$.[8] The fault is in the strong-coupling model, but the fault is _not_ in the assumption of an axially symmetric deformed intrinsic state. Rather, the fault arises from dealing with simple product wave functions, each formed from an intrinsic state times a macroscopic rotational wave function. If the "rotational" nuclear wave functions are formed instead by _projecting_ states of good $J$ out of the microscopic intrinsic states $\chi$ and $X$, then the calculated $S$ values become reasonable. For the $^{20}$Ne$(d, p)^{21}$Ne case just discussed a recent _projection_ calculation (_15_), using Hartree–Fock intrinsic states derived from the $K+^{17}O$ Hamiltonian, yields $S = 0.62$, in excellent agreement with our $K+^{17}O$ shell-model calculation.

## 3.3. Nucleus-by-Nucleus Comparison Between Experimental Results and $K+^{17}O$ Results

The $K+^{17}O$ Hamiltonian is of special interest because it is a "realistic" Hamiltonian which has not been adjusted specifically to fit energy-level data throughout the $A = 18–22$ region. For simplicity, we restrict our nucleus-by-nucleus discussion here to this $K+^{17}O$ model. (Results from the other Hamiltonians will be considered in Section 3.4.) The agreement between $K+^{17}O$ results and experimental results is good enough so that we can correlate many $K+^{17}O$ states with experimentally observed states.

### 3.3.1. The A = 18, T = 0 System

Figure 2 displays energy-level spectra for the $T = 0$ states in $^{18}$F, and Table IV lists spectroscopic factors $S$ for $^{17}O(^3He, d)^{18}F_{T=0}$. The experimental $S$-values (_27_) are normalized so that $S = 2$ for the $J = 5$ level in $^{18}$F. (As we shall discuss later, there are hints that this normalization makes the experimental $S(l = 2)$ values larger than they should be.) Above the dashed line in Table IV we list levels for which there is clear-cut correlation between observed results and $K+^{17}O$ shell-model results. The $K+^{17}O$ model yields satisfactory agreement with experimentally determined excitation energies and $S$-values for the lowest three levels: $1^+$, $3^+$, and $5^+$. Also, we consider that there is a clear-cut correlation between the second $K+^{17}O$

---

[8] This value $S = 2/3$ holds exactly for the connection between a pure $d_{5/2}^4$ $J = T = 0$ state and a pure $d_{5/2}^5$ $J = 5/2$, $T = 1/2$ state. But even in the full _sd_-shell vector space, a reasonable Hamiltonian such as $K+^{17}O$ yields $S \approx 2/3$ also, connecting Hamiltonian eigenfunctions which are not at all pure $d_{5/2}^n$. (The $K+^{17}O$ ground state for $^{20}$Ne is only 21% pure $d_{5/2}^4$. The lowest $5/2^+$ state of $^{21}$Ne is 25% pure $d_{5/2}^5$.) The preservation of the pure $d_{5/2}^n$ $S$-value is a pseudonium phenomenon [see Section 3.1.2 and (_87_)].

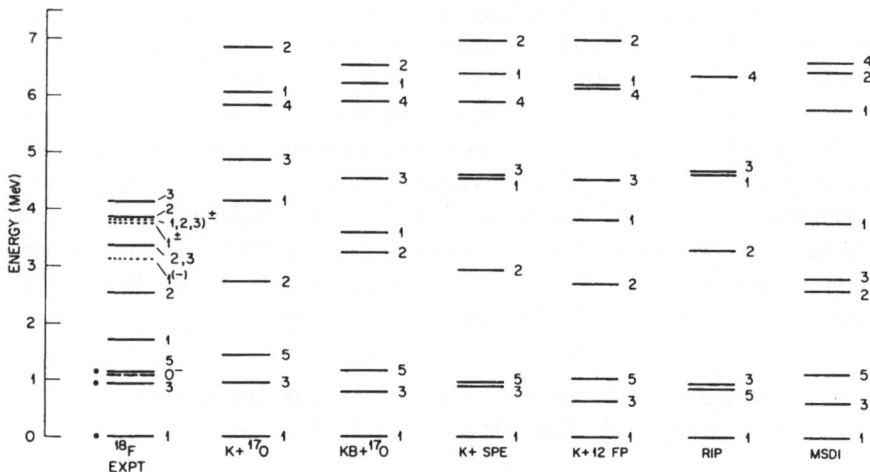

**Fig. 2.** $T = 0$ **levels of** $^{18}F$. For notation conventions, see Section 3.1.1. The experimentally observed spectrum is taken from (27).

$3^+$ level and the observed 4.12-MeV $3^+$ level, because there is no other observed state having possible spin 3 and also large $S(l = 2)$.

Between the low-lying $1^+$, $3^+$, $5^+$ triad and the observed 4.12-MeV $3^+$ level, there are several observed levels not matched by any shell-model levels. Two of these are of special interest, the second observed $1^+$ and the first observed $2^+$. We shall consider these in turn.

The second observed $1^+$ state appears at 1.7 MeV and has a very small spectroscopic factor, but the second $K+^{17}O$ $1^+$ state appears near 4 MeV and has a substantial spectroscopic factor. This discrepancy suggests that the observed 1.7-MeV $1^+$ state is dominated by configurations outside our $s$–$d$-shell model; it would then be "missing" in our calculated spectra, just as the odd-parity states are missing. The experimental partner to the second $K+^{17}O$ $1^+$ state could be either the $1^+$ state observed at 3.72 MeV, or the $1^\pm$, $2^\pm$, $3^\pm$ state observed at 3.79 MeV. Though spectroscopic strengths for these two states were not experimentally determined, each of them was produced with strength consistent with $S(l = 2) \approx 0.5$, at angles where impurities in the target did not obscure the $^{17}O(^3He, d)$ cross section (27).

The lowest observed $2^+$ state in $^{18}F$ appears at 2.52 MeV, and the lowest $K+^{17}O$ $2^+$ state appears nearby, at 2.72 MeV. But the observed state shows very small $l = 0$ and $l = 2$ strengths, while the calculated state has substantial $l = 0$ and $l = 2$ strengths. The second $K+^{17}O$ $2^+$ level lies much higher, at 6.9 MeV. Thus the observed $2^+$ state at 2.52 MeV, like the ob-

**TABLE IV. Spectroscopic Factors $S$ for $^{17}O(^3He, d)^{18}F_{T=0}$.** Above the long dashed line there is clear-cut correlation between experimental and shell-model levels. In the experimental columns, a dash implies that the angular distribution was not measured, while a dotted entry denotes an $S$ value too small to measure. In the shell-model columns, a dotted entry denotes an $S$-value that is exactly zero because of angular-momentum selection rules. The experimental columns include all the $T = 0$ $^{18}F$ levels observed below 4.12 MeV (except those known to have negative parity). The shell-model columns include all the $T = 0$ $^{18}F$ levels calculated from $K + ^{17}O$ below 4.9 MeV, plus the second $2^+$ level calculated from $K + ^{17}O$.

| | Experiment[a] | | | | $K + ^{17}O$ shell model | | |
|---|---|---|---|---|---|---|---|
| $J^\pi$ | Excitation energy in $^{18}F$, MeV | $S(l=0)$ | $S(l=2)$ | $J^\pi$ | Excitation energy in $^{18}F$, MeV | $S(l=0)$ | $S(l=2)$ |
| $1^+$ | 0 | . . . | 1.50 | $1^+$ | 0 | . . . | 1.15 |
| $3^+$ | 0.94 | 0.64 | <0.72 | $3^+$ | 0.92 | 0.67 | 0.60 |
| $5^+$ | 1.13 | . . . | 2.00 | $5^+$ | 1.42 | . . . | 2.00 |
| $3^+$ | 4.12 | 0.26 | 1.50 | $3^+$ | 4.87 | 0.32 | 1.31 |
| $1^+$ | 1.70 | — | — | | | | |
| $1^{(-)}$ | 3.13 | — | — | | | | |
| $1^\pm$ | 3.72 | —[b] | —[b] | $1^+$ | 4.13 | . . . | 0.53 |
| 1, 2, 3)$\pm$ | 3.79[c] | —[b] | —[b] | | | | |
| $2^+$ | 2.52 | 0.05 | <0.01 | | | | |
| $2^+, 3^+,$ | 3.36[d] | — | — | | | | |
| $2^+$ | 3.84 | 1.0 | <1.1 | $2^+$ | 2.72 | 0.51 | 0.32 |
| | | | | $2^+$ | 6.88 | 0.34 | 0.65 |

[a] From (27). The experimental $S$-values were deduced via DWBA analysis, then normalized so that $S = 2$ for the $J = 5$ level.
[b] Complete angular distributions were not taken for the 3.72-MeV and 3.79-MeV states. But the measured results are consistent with either of these states being a partner to the 4.13-MeV $K + ^{17}O$ state [see (27)].
[c] Arguments involving a weak-coupling model suggest $3^-$ [see (97)].
[d] Arguments involving a weak-coupling model suggest $3^+$ [see (97)].

served $1^+$ at 1.7 MeV, is "extra"; it is probably dominated by configurations outside our *s–d*-shell model. A possible empirical partner for the 2.72-MeV $K + ^{17}O$ state is the observed $2^+$ state at 3.84 MeV; for as Table IV shows, there is fair agreement in their $S$-values.

For discussion of "two-hole four-particle" configurations in the observed 1.7-MeV and 2.52-MeV states, see (88, 94–97).

## 3.3.2. *The A = 18, T = 1 System*

For mass 18 the difference in nuclear binding energy between the lowest $T = 1$ and $T = 0$ states is, empirically, 1.09 MeV (*84*). The $K+^{17}O$ shell model yields a good match, 1.19 MeV.

Figure 3 shows the low-energy levels of $^{18}O$, and Table V lists spectroscopic factors $S$ for $^{17}O(^3He, d)^{18}F_{T=1}$. For every level except $J = 5$, the experimental $S(l = 2)$ values in Tables IV and V exceed the corresponding $K+^{17}O$ values. We note that the experimental $S$-values (*27*) in Tables IV and V are deduced from the same set of $^{18}F$ data, and that all these experimental values are normalized so that $S = 2$ for the observed state having $J = 5$, $T = 0$. This normalization choice gives exact agreement with the $s$–$d$-shell-model $S$ for the $J = 5$ state. But when this same normalization is applied to data for the lowest state of $J = 0$ and $T = 1$, the resulting empirical value is $S = 2.32$, a number which exceeds the $s$–$d$-shell-model limit of 2 for this state. Furthermore, for the $2^+$ states in Table IV, the resulting empirical $S(l = 2)$ values sum to 2.46; and this also exceeds the $s$–$d$-shell-model limit of 2. These excesses hint that in Tables IV and V, the empirically determined $S(l = 2)$ values may be high by $\gtrsim 20\%$.

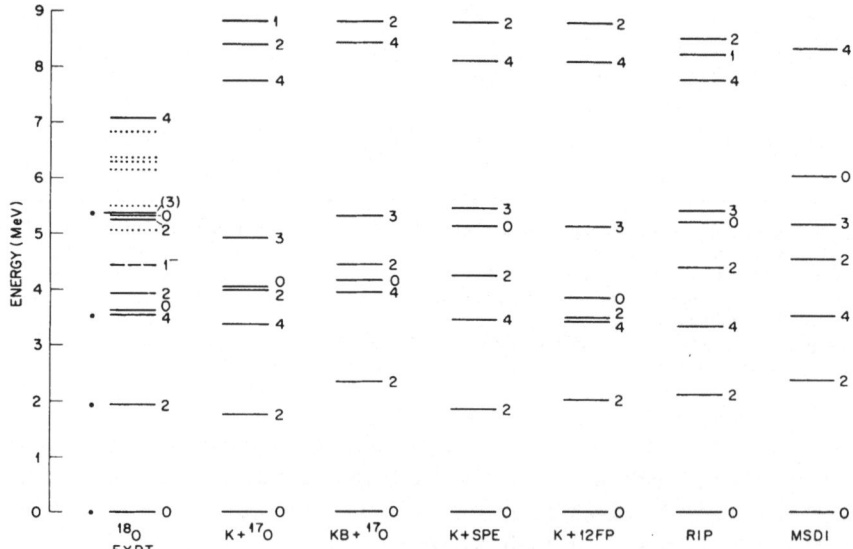

**Fig. 3.** $T = 1$ **levels of** $^{18}O$. For notation conventions, see Section 3.1.1. The experimentally observed spectrum is based on information in (*28–30*).

**TABLE V. Spectroscopic Factors $S$ for $^{17}O(^3He, d)^{18}F_{T=1}$.** All the final states are isobaric analogs of those in $^{18}O$, and the energies in this table refer to excitation energies in $^{18}O$. In the experimental columns, a dash implies that the angular distribution was not measured, while a dotted entry denotes an $S$-value too small to measure. In the shell-model columns, a dotted entry denotes an $S$-value that is exactly zero because of angular-momentum selection rules. The experimental and shell-model columns include all $^{18}O$ levels below 4 MeV. Above the dashed line there is clear-cut correlation between experimental and shell-model levels.

| | Experiment[a] | | | $K+^{17}O$ shell model | | | |
|---|---|---|---|---|---|---|---|
| $J^\pi$ | Excitation energy in $^{18}O$, MeV | $S(l=0)$ | $S(l=2)$ | $J^\pi$ | Excitation energy in $^{18}O$, MeV | $S(l=0)$ | $S(l=2)$ |
| $0^+$ | 0 | ... | 2.32 | $0^+$ | 0 | ... | 1.67 |
| $2^+$ | 1.98 | 0.32 | 1.48 | $2^+$ | 1.74 | 0.38 | 1.12 |
| $4^+$ | 3.55 | ... | 1.96 | $4^+$ | 3.38 | ... | 1.89 |
| $0^+$ | 3.63 | ... | —[b] | $0^+$ | 4.03 | ... | 0.17 |
| $2^+$ | 3.92 | 0.44 | 0.98 | $2^+$ | 3.99 | 0.56 | 0.84 |

[a] From (27). The experimental $S$-values were deduced via DWBA analysis, then normalized so that $S = 2$ for the $J = 5$ state of $^{18}F_{T=0}$ (see Table IV).
[b] Known to be small, from $^{17}O(d, p)^{18}O$ measurements [see (28)].

All the $^{18}F_{T=1}$ states are isobaric analogs of states in $^{18}O$, and in our subsequent discussion we shall refer to these $T = 1$ states by their excitation energies in $^{18}O$. (That is the way they are listed in Table V.)

As Fig. 3 and Table V show, the $K+^{17}O$ Hamiltonian yields satisfactory agreement with experimental excitation energies and $S$-values for the lowest five $T = 1$ levels in mass 18. Above these five levels, Fig. 3 shows four experimentally observed even-parity levels—$2^+$, $0^+$, $3^+$, and $4^+$. There are no DWBA $S$-values available for these levels. The $K+^{17}O$ spectrum has obvious partners for the observed $3^+$ and $4^+$ levels in this upper group (though the energy match leaves something to be desired). However, the $K+^{17}O$ spectrum has no partners for the observed $2^+$ and $0^+$ states near 5.2 MeV; or, to put it more accurately, the $K+^{17}O$ model yields only one $0^+$, $2^+$ pair between 3 and 6 MeV, instead of the two pairs observed experimentally. The agreement shown in the next-to-last line of Table V indicates that the observed 3.63-MeV $0^+$ state and the $K+^{17}O$ 4.03-MeV $0^+$ state are both produced weakly via $^{17}F(^3He, d)$. Though this agreement is consistent with correlation between the second shell-model and observed $0^+$ levels, it would also be consistent with any other division of $s$–$d$-shell character between the two $0^+$ states observed at 3.63 MeV and 5.33 MeV. The agreement shown in the last line of Table V is more positive. It supports correlation of the $K+^{17}O$ 3.99-MeV $2^+$ state with the observed 3.92-MeV $2^+$ state and this correlation suggests that the observed 5.25-MeV $2^+$ state is dominated by configurations outside our $s$–$d$-shell model.

Our $E2$ comparisons in Section 5 will cast further doubt on correlation between the second observed and second $K+^{17}O$ $0^+$ states.

It seems fair to mention also some empirical results which suggest that there is significant $s$–$d$-shell character in *both* the second and third observed $2^+$ states. In particular, a plane-wave analysis of $^{17}O(d, p)^{18}O$ data yields *equal* $S(l = 0)$ strengths for these 3.92-MeV and 5.25-MeV $2^+$ states (*28*). However, this result is questionable because for the lowest five $T = 1$ states this same plane-wave analysis (*28*) leads to $S(l = 0)$ values which are not proportional to the DWBA $S(l = 0)$ values, and to $S(l = 2)$ values which are not proportional to the DWBA $S(l = 2)$ values. (See Table III.)

For discussion of "two-hole four-particle" configurations in the $0^+$ and $2^+$ states of $^{18}O$, see (*88, 95, 96, 98*).

### 3.3.3. The A = 19, T = 1/2 System

Energy-level spectra for the $T = 1/2$ states of $^{19}F$ are displayed in the lower part of Fig. 4. The experimentally observed spectrum shows the pattern of a rotational band based on the $J = 1/2$ ground state, and the shell-model spectra show this pattern too. But in each spectrum the situation is complicated by the presence of *two* candidates for the $7/2^+$ member of the ground-state band. In drawing the experimental spectrum we have indicated, by a thick line, that the 5.47-MeV $7/2^+$ state behaves like the ground-state-band member. This upper observed $7/2^+$ state decays strongly to the $3/2^+$ ground-state-band member (*35*), whereas the lower observed $7/2^+$ state shows no branching to that $3/2^+$ state (*32, 33*). In drawing the $K+^{17}O$ spectrum we have indicated, by a thick line, that the 5.12-MeV $7/2^+$ state has almost all of the ground-state-band character. Its calculated $B(E2)$ to the $3/2^+$ ground-state-band member is nine times that from the next $K+^{17}O$ $7/2^+$ state (at 5.68 MeV).

As Fig. 4 shows, the $K+^{17}O$ model yields satisfactory reproduction of observed (*32–37*) excitation energies within the ground-state band. The lowest three members of this band can be reached by $l = 0$ or $l = 2$ proton

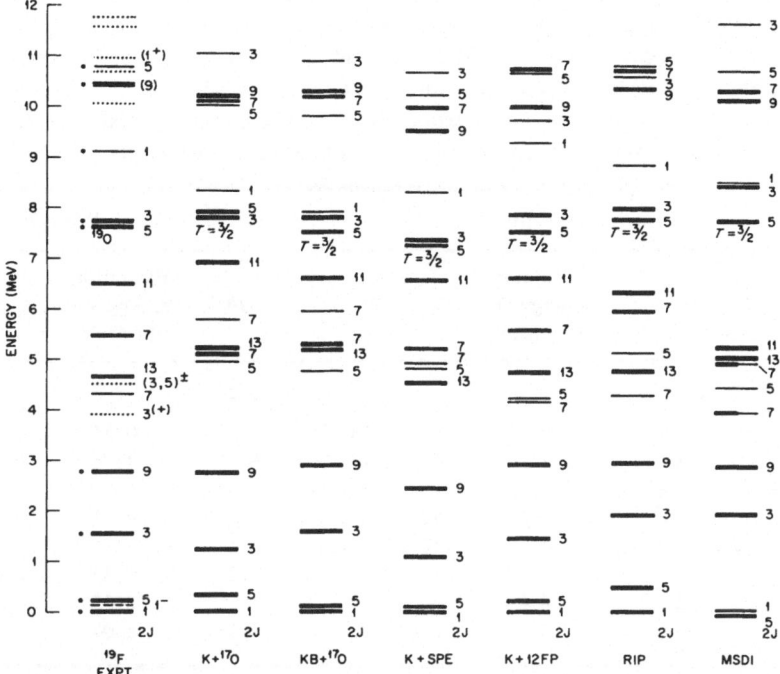

**Fig. 4. Levels of $^{19}$F and $^{19}$O.** For notation conventions, see Section 3.3.1. The spectrum marked EXPT is based on $^{19}$F information in (32–37) and (45–48); on $^{19}$O information in (49–53); and on a semiempirical estimate of the nuclear-energy difference between isobars (see Section 4). But this spectrum EXPT omits the following $^{19}$F levels of possible or established even parity: (7/2⁻) and (9/2⁻) at 4.00 MeV and 4.04 MeV [see (38–40)], and fifteen levels between 4.68 and 6.50 MeV [see (40–47)]. The column $K+^{17}O$ omits five levels between 6.45 and 6.93 MeV. The other shell-model columns include only the analogs of those levels shown for $K+^{17}O$, i.e. the ground-state band, the second 5/2⁺ level, and the second 7/2⁺ level.

transfer in the reaction $^{18}O(^{3}He, d)^{19}F$. Table VI shows that for these three states, the $K+^{17}O$ model yields $S$-values in good agreement with experimentally determined values (45).

Outside the ground-state band, correlations between the $K+^{17}O$ and observed levels are unclear, partly because of uncertainties in the experimental spectrum. Besides ground-state-band members, the $K+^{17}O$ model yields seven other levels below its 11/2⁺ member of the ground-state band. These seven consist of the second 5/2⁺ and second 7/2⁺ states, drawn with light lines in Fig. 4; and states 3/2⁺, 5/2⁺, 1/2⁺, 7/2⁺, and 3/2⁺, all lying

**TABLE VI.** Spectroscopic Factors $S$ for $^{18}O(^{3}He, d)^{19}F_{T-1/2}$. The experimental columns include all the levels below 6.25 MeV which have been assigned an unparenthesized even value of $l$. The shell-model columns include all the $1/2^+$, $3/2^+$, and $5/2^+$ levels calculated from $K+^{17}O$ below 6.45 MeV, plus the second and third $1/2^+$ states calculated from $K+^{17}O$. Above the dashed line there is clear-cut correlation between experimental and shell-model levels.

| | Experiment[a] | | | $K+^{17}O$ shell model | | |
|---|---|---|---|---|---|---|
| $l$ | $J^\pi$ | Excitation energy in $^{19}F$, MeV | $S$ | $J^\pi$ | Excitation energy in $^{19}F$, MeV | $S$ |
| 0 | $1/2^+$ | 0 | 0.45 | $1/2^+$ | 0 | 0.49 |
| 2 | $5/2^+$ | 0.20 | 0.63 | $5/2^+$ | 0.36 | 0.65 |
| 2 | $3/2^+$ | 1.56 | 0.44 | $3/2^+$ | 1.25 | 0.34 |
| 2 | $(3/2^+, 5/2^+)^b$ | $4.56^b$ | (0.36, 0.24) | $5/2^+$ | 4.97 | 0.14 |
| 3 | $(5/2^-, 7/2^-)^b$ | $4.56^b$ | (0.17, 0.13) | | | |
| | | | | $3/2^+$ | 6.45 | 0.03 |
| 0 | $1/2^+$ | 6.25 | 0.31 | $1/2^+$ | 6.75 | 0.002 |
| | | | | $1/2^+$ | 8.04 | 0.30 |

[a] The $l$ and $S$ values are from (45); they were deduced via DWBA analysis. The excitation energies and spin-parity assignments are based on (32, 41, 45–47).

[b] In (45) the 5.46-MeV level is reported as a possible doublet with $l = 2, 3$. Also, the presence of a doublet at 5.46 MeV is suggested by the fact that in the $^{19}F(p, p'\gamma)^{19}F$ reaction, the gamma-ray branching pattern varies with bombarding energy [see (46, 47)]. *Some* gamma-ray correlation results indicate $J = 3/2$ if only one state contributes [see (32, 46, 47)]; but other results indicate $J = 5/2$ if only one state contributes [see (48)].

between 6.45 MeV and 6.93 MeV but not shown in Fig. 4. Besides ground-state-band members the experimentally observed spectrum includes, below its first $11/2^+$ level, 21 states of possible or established even parity.[9] These consist of a $3/2^{(+)}$ state at 3.91 MeV (32, 33, 45–48), states $(7/2^-)$ and $(9/2^-)$ at 4.00 and 4.04 MeV (38–40, 45–48), the *non*ground-state-band $7/2^+$ state at 4.39 MeV (32–34), a probable doublet $J = (3/2, 5/2)$ at 4.56 MeV (32, 45–48), and 15 more states (40–47) all lying between 4.68 MeV and 6.50 MeV but not shown in Fig. 4. Thus, a comparison between model and experiment poses several obvious questions. Is either of the observed states at 3.91 and 4.56 MeV a $3/2^+$ state with predominantly $s$–$d$-shell configuration? (If

[9] We are counting the $^{19}F$ states at 4.00, 4.04, 4.68, and 5.42 MeV as having $(7/2^-)$, $(9/2^-)$, $5/2^{(-)}$, and $7/2^{(-)}$, respectively, though in some $^{19}F$ references these assignments are given without any parentheses.

so, then the $K+^{17}O$ model yields its second 3/2$^+$ state about 2.5 MeV too high.) Should the observed 7/2$^+$ state at 4.39 MeV be correlated with the second $K+^{17}O$ 7/2$^+$ state, at 5.68 MeV? (If so, then this $K+^{17}O$ state is 1.3 MeV too high.) Where is the experimental counterpart to the 4.97-MeV $K+^{17}O$ 5/2$^+$ state? (At 4.6 MeV?) Our present calculations, and the present experimental data, are not sufficient to provide definitive answers. Future answers may come from beta- and gamma-ray data, and from models including non-_s–d_-shell states. In the meantime, we turn our attention to spectroscopic factors.

Below its dashed line, Table VI lists _S_-factors for $^{18}O(^{3}He, d)^{19}F$ transitions to $^{19}F$ states outside the ground-state band. The $K+^{17}O$ shell-model columns include the lowest three states of spin 1/2, and the lowest two states for each of spins 3/2 and 5/2. The experimental columns include all those $^{19}F$ levels up to 6.25 MeV which in (_45_) were assigned an _un_parenthesized even value of $l_p$. (All other possibly-even levels below 6.25 MeV were very weakly excited.) The clearest discrepancy shown in Table VI is that the experimental data indicate a 6.25-MeV 1/2$^+$ state with $S = 0.31$, but the $K+^{17}O$ model has no counterpart until 8.04 MeV.

### 3.3.4. _The A = 19, T = 3/2 System_

Figure 4 displays energy-level spectra for $^{19}O$ as well as $^{19}F$. As this figure shows, the empirically determined splitting between the $T = 1/2$ and $T = 3/2$ levels is well reproduced by the $K+^{17}O$ model.

In the observed spectrum of $^{19}O$, the lowest 3/2$^+$, 5/2$^+$, and (9/2)$^+$ levels are viewed as members of a $K = 3/2$ band in which the 7/2$^+$ member has not yet been identified. The ordering and spacing of the observed 5/2$^+$ and 3/2$^+$ levels imply that the rotational-band character is not very "pure." Therefore one is _not_ justified in assuming 7/2$^+$ for the fourth observed $^{19}O$ level, even though it is the sole candidate lying between the observed 5/2$^+$ and (9/2)$^+$ states. (Indeed, several of the shell-model spectra shown in Fig. 4 indicate that the lowest 7/2$^+$ in $^{19}O$ may lie above the first 9/2$^+$.) The $K+^{17}O$ shell-model spectrum shows the standard ordering, though not the standard spacing, of a rotational $K = 3/2$ band. Its 3/2$^+$, 5/2$^+$, and 9/2$^+$ members are listed in the first three lines of Table VII. For these three levels, Fig. 4 and Table VII show that the $K+^{17}O$ model yields fair agreement with measured excitation energies, and with experimental _S_-factors for $^{18}O(d, p)^{19}O$.

Besides the lowest 3/2$^+$, 5/2$^+$, and 9/2$^+$ states, there are good reasons to make experimental–theoretical correlations for three other $^{19}O$ states. These are listed just above the dashed line in Table VII. For the lowest

**TABLE VII. Spectroscopic Factors** $S$ **for** $^{18}O(d, p)^{19}O_{T=3/2}$. The experimental columns include all levels below 5.45 MeV (re-ordered to facilitate comparison with theory). The $K+^{17}O$ columns include all levels below 6.23 MeV (also re-ordered). Above the dashed line there is clear-cut correlation between experimental and shell-model levels. Below the dash–dot line, no correlation at all is implied by the listing of a theoretical state on the same line with an experimentally observed state.

| | | Experiment[a] | | | $K+^{17}O$ shell model | |
|---|---|---|---|---|---|---|
| Footnote | $J^{\pi}$ | Excitation energy in $^{19}O$, MeV | $S$ | $J^{\pi}$ | Excitation energy in $^{19}O$, MeV | $S$ |
| b | $5/2^+$ | 0 | 0.41 | $5/2^+$ | 0 | 0.67 |
| b | $3/2^+$ | 0.10 | (0.03) | $3/2^+$ | −0.12 | 0.01 |
| b | $(9/2)^+$ | 2.78 | weak | $9/2^+$ | 2.29 | zero |
| b | $1/2^+$ | 1.47 | 0.48 | $1/2^+$ | 0.44 | 0.89 |
| b | $5/2^+$ | 3.15 | weak | $5/2^+$ | 2.15 | 0.02 |
| | | 5.45 | strong | $3/2^+$ | 4.53 | 0.79 |
| - - - | - - - | - - - | - - - | - - - | - - - | - - - |
| c | $5/2^+$ | 4.71 | weak | $5/2^+$ | 4.11 | 0.00 |
| c | $5/2^+$ | 5.16 | weak | $5/2^+$ | 6.23 | 0.03 |
| -·-·- | -·-·- | -·-·- | -·-·- | -·-·- | -·-·- | -·-·- |
| d | | (0.35) | weak | $7/2^+$ | 2.17 | zero |
| d | | 2.37 | weak | $3/2^{\cap}$ | 3.14 | 0.08 |
| d | | (2.62) | weak | $9/2^+$ | 4.25 | zero |
| d | | 3.06 | weak | $7/2^+$ | 5.39 | zero |
| d | $(1/2^+)$ | 3.23 | weak | $1/2^+$ | 5.54 | 0.01 |
| d | | 3.95 | weak | | | |
| d | | 4.11 | weak | | | |
| d | | (4.33) | weak | | | |
| d | | 4.41 | weak | | | |
| d | | 4.59 | weak | | | |
| d | | 4.98 | weak | | | |
| d | | (5.11) | weak | | | |

[a] From (49–54). The numerical $S$-factors are values determined via DWBA analysis (54). The qualitative descriptions "weak" or "strong" are based on plane-wave Butler–Born analyses (49–52).
[b] Clear-cut correlation between observed and calculated level.
[c] Possible correlation between observed and calculated level.
[d] No correlation is implied by the listing of a shell-model level on the same line with an observed level.

$1/2^+$ state, the experimental–theoretical correlation is suggested quite obviously by Fig. 4, even though the $K+^{17}O$ level lies significantly lower than the observed level. (Also, the theoretical $S$ is considerably larger than the experimental $S$.) The next-listed state in Table VII is the second $5/2^+$ state. We identify it as the second member of a rotational band based on the lowest $1/2^+$ level. In the $K+^{17}O$ spectrum this identification is obvious because the third $5/2^+$ state lies almost 2 MeV higher. In the experimental spectrum, the $K = 1/2$ rotational-band character is surmised from the experimental observation that both of its presumed members, the lowest $1/2^+$ and the second $5/2^+$, are strongly excited in $^{17}O(t, p)^{19}O$ (*49, 53*). Further support for the $K = 1/2$ band structure comes from the fact that the energy spacing of this $1/2^+$, $5/2^+$ pair is nearly the same for the experimental spectrum and all our shell-model spectra. The fifth and last of our "firm" experimental–theoretical correlations for $^{19}O$ connects the third $K+^{17}O$ $3/2^+$ state with a level observed at 5.45 MeV. The experimental partner is the only remaining level seen to be strongly excited in $^{18}O(d, p)^{19}O$ measurements (*49*), while the theoretical partner is the only remaining $K+^{17}O$ level below 5.5 MeV that has a large $S$-value for this reaction.

There are many other observed $^{19}O$ states which have possibly-even parity, and Table VII lists all such states which are seen in $^{18}O(d, p)^{19}O$ measurements up to 5.45-MeV excitation in $^{19}O$. The shell-model columns of Table VII include all the $K+^{17}O$ states below 6.23 MeV in $^{19}O$. We see that there are no serious inconsistencies between experiment and model. The worst disagreement is that the fourth $K+^{17}O$ $5/2^+$ level lies 1 MeV higher within $^{19}O$ than the fourth experimentally identified $5/2^+$ level.

### 3.3.5. *The A = 20, T = 0 System*

Energy-level spectra for the $T = 0$ states of $^{20}Ne$ are displayed in the lower part of Fig. 5. (In each column the cross-hatched line shows the lowest $T = 1$ level, which corresponds to the ground state of $^{20}F$.) The observed $T = 0$ spectrum of $^{20}Ne$ includes a ground-state rotational band $0^+$, $2^+$, $4^+$, $6^+$, $8^+$; and the $K+^{17}O$ model reproduces the appearance of this band quite well. The $0^+$ and $2^+$ states in this band can be reached by $l = 0$ and $l = 2$ proton transfer, respectively, in the reaction $^{19}F(^3He, d)^{20}Ne$; and Table VIII shows that for these two states, the $K+^{17}O$ model yields $S$-values in good agreement with experiment.

Next we consider $T = 0$ states outside the ground-state band but below 10 MeV in excitation. The $K+^{17}O$ model gives only two such states: $0^+$ at 6.8 MeV, and $2^+$ at 8.4 MeV. But the experimental spectrum drawn in

Fig. 5 shows *six* such states, including two assigned 0⁺, and three assigned 2⁺. Furthermore, there are other observed states below 10 MeV, not drawn in Fig. 5 but described and referenced in its caption. The appearance of "extra" observed 0⁺ and 2⁺ states, like the appearance of the extra 0⁺ and 2⁺ states in ¹⁸O, is attributed (at least in part) to core excitation. That is, each of the two 0⁺ states observed near 7 MeV is thought to be some mixture of

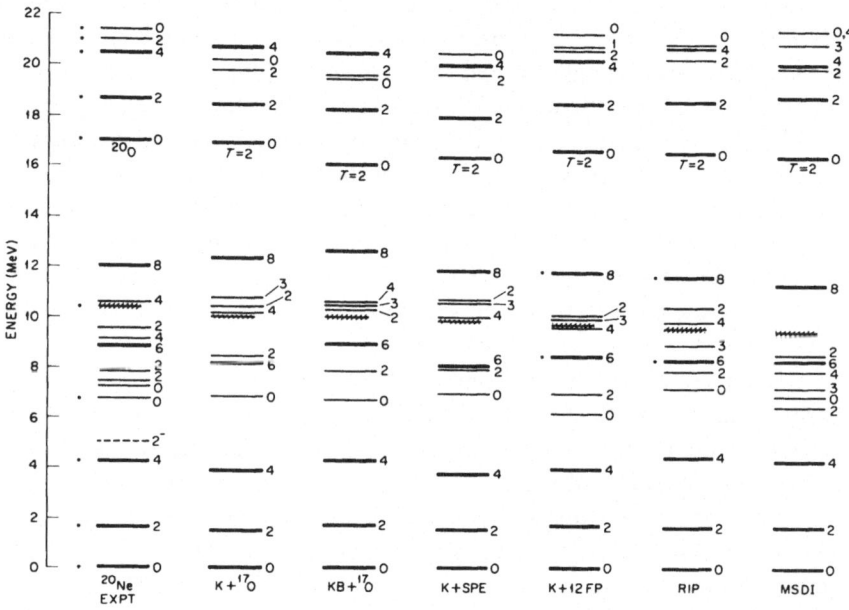

**Fig. 5.** $T = 0$ **levels of** ²⁰Ne, **and** $T = 2$ **levels of** ²⁰O. The cross-hatched line shows the lowest $T = 1$ level, which is the ground state of ²⁰F, but all other levels plotted below 14 MeV are $T = 0$ levels. For other notation conventions see Section 3.1.1. In the spectrum marked EXPT the ²⁰Ne levels are taken from (*55–58*); the ²⁰O levels are taken from (*68, 69*); and the plotted separations between ground states of ²⁰Ne, ²⁰F, and ²⁰O are nuclear-energy differences estimated as described in Section 4. The experimentally observed 10.27-MeV level in ²⁰Ne has $J^\pi = 2^+$ and $T = (1)$; we have assumed it to be the lowest $T = 1$ state, and not drawn it separately. To simplify the figure, we have omitted the following other observed ²⁰Ne levels of possible or established even parity: broad levels 0⁺, 2⁺, and 4⁺ at 8.6, 8.8, and 10.0 MeV, respectively [see (*58, 59*)]; states of uncertain existence at 7.93 and 9.31 MeV [see (*57*)]; the (5⁻) state at 10.3 MeV [see (*57*)]; and all states above the third known 4⁺ but below the first known 8⁺ [see (*58, 59*)]. In each shell-model column we show all $T = 0$ levels up through the third 2⁺, first 3⁺, and second 4⁺; then we omit all higher $T = 0$ levels except for the first 8⁺. In columns $K+12FP$ and $RIP$, black dots mark two levels which were included in the least-square fits leading to $K+12FP$ and $RIP$, though they were not included in fits leading to the other three least-square Hamiltonians.

**TABLE VIII. Spectroscopic Factors $S$ for $^{19}F(^{3}He, d)^{20}Ne_{T=0}$.** The table includes all of the experimentally known levels below 7.5 MeV which can be reached by $l = 0$ or $l = 2$ transfer to $^{19}F_{J=1/2}$, and it includes all of the $K+^{17}O$ levels below 10 MeV which can be reached by such transfer. For $J^{\pi} = 0^{+}$ the transfer is $l = 0$, while for $J^{\pi} = 2^{+}$ the transfer is $l = 2$. Above the dashed line there is clear-cut correlation between observed and shell-model levels.

| | Experiment[a] | | | $K+^{17}O$ shell model | |
|---|---|---|---|---|---|
| $J^{\pi}$ | Excitation energy in $^{20}$Ne, MeV | $S$ | $J^{\pi}$ | Excitation energy in $^{20}$Ne, MeV | $S$ |
| $0^{+}$ | 0 | $0.62^{b}$ | $0^{+}$ | 0 | 0.88 |
| $2^{+}$ | 1.63 | 1.26 | $2^{+}$ | 1.46 | 1.00 |
| $0^{+}$ | 6.72 | 0.94 | $0^{+}$ | 6.81 | 1.02 |
| $0^{+}$ | 7.20 | <0.06 | | | |
| $2^{+}$ | 7.43 | 0.32 | $2^{+}$ | 8.43 | 0.19 |

[a] From (55).
[b] The ground-state $S$-value is particularly sensitive to the optical-model parameters, so that $0.3 \lesssim S \lesssim 1.0$.

one s-d-shell state, and one state lying outside our s-d-shell model; and similarly, the two $2^{+}$ states observed near 7.6 MeV are presumed to be mixtures of s-d-shell and core-excited parts. To get some hints about the relative importance of s-d-shell and core-excited configurations in these states, we consider $S$-values for $^{19}F(^{3}He, d)^{20}Ne$. (See Table VIII.) Comparison between the experimental and the $K+^{17}O$ $S$-values suggests that, of the two $0^{+}$ states observed near 7 MeV, the *lower* one has essentially all of the s-d-shell character. And the same comment holds for the *lower* state in the $2^{+}$ pair observed near 7.6 MeV, though the evidence is somewhat weaker in this case.

### 3.3.6. *The A = 20, T = 1 System*

As Fig. 5 shows, for $A = 20$ the $K+^{17}O$ model satisfactorily reproduces the empirically determined splitting between the lowest $T = 1$ level (drawn as a cross-hatched line) and the lowest $T = 0$ level.

Spectra for the $T = 1$ states of $^{20}$F are shown on an expanded scale in Fig. 6. As the figure indicates, there are many uncertain spin-parity assignments for the experimentally observed levels.

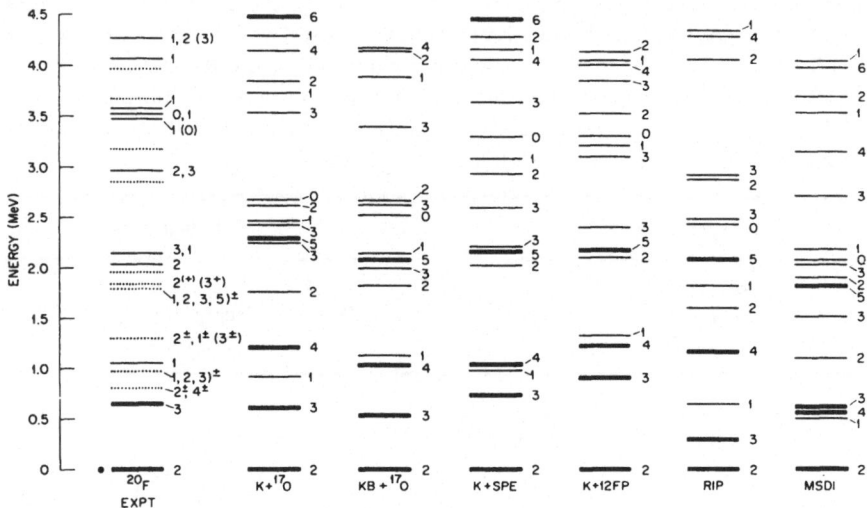

**Fig. 6. Levels of** $^{20}$**F.** For notation conventions see Section 3.1.1. The experimentally observed spectrum is based on information in (*59–66*). The assignment 1, 2, 3, 5)$^{\pm}$ means spin 1 *or* 2 *or* 3 *or* 5, all with unknown parity.

Spectroscopic strengths for the reaction $^{19}$F$(d, p)^{20}$F are listed in Table IX. In the experimental columns of this table, we include two sets of experimentally determined strengths $(2J + 1)S$. Both sets are "absolute" values determined by DWBA analysis (*60*); but one set is based on data obtained with 8.9-MeV deuterons (*67*), while the other set is based on more recent data obtained with 16-MeV deuterons (*60*). We shall briefly compare these experimental sets. For the two strongest $l = 2$ transitions (to levels at 0.66 MeV and 2.04 MeV), the relative strengths are in excellent agreement from one experimental set to the other, but the absolute values are in poor agreement. For the two strongest $l = 0$ transitions (unresolved in the 8.9-MeV experiment), the summed strength relative to the dominant $l = 2$ strength is roughly the same from one experimental set to the other. The discrepancies in absolute values may possibly stem from defects in the DWBA treatment, or perhaps they stem from some experimental difficulties in normalization. For the aforementioned strongly produced states, there is a pattern of greater strength for the 16-MeV data. However, this pattern is broken by the equality of the two experimentally determined strengths for the 2.20-MeV state. For some other levels, numerical comparison is not possible because the pertinent angular distribution was not measured in both experiments.

As Table IX shows, the $K+^{17}O$ spectrum exhibits the same major features as the experimental results: two very strong $l = 2$ states, and one dominant $l = 0$ state. Though the agreement for other states is not clear, we shall discuss the probable and possible correlations beginning, as usual, with the lowest band.

In the ground-state band, only the first two members, $2^+$ and $3^+$, can be clearly identified in the experimentally observed spectrum. As Table IX shows, the $K+^{17}O$ results agree qualitatively with experiment in that the ground-state-band $2^+$ member is produced very weakly in $^{19}F(d, p)$, while the ground-state-band $3^+$ member is produced very strongly. The $K+^{17}O$ model yields $4^+$ and $5^+$ members at 1.21 and 2.29 MeV, respectively. If we look for experimentally observed partners below 2.8 MeV, we see that the spin assignments shown in Fig. 6 allow, as experimentally observed candidates for the $4^+$ and $5^+$ members, only the states seen at 0.82, 1.82, and 1.97 MeV. Since the 1.97-MeV state has a 20% branch to the $2^+$ ground state (*62, 64*), it is unlikely to be a $5^+$ state. Hence the most plausible candidates for the $4^+$ and $5^+$ members are the state observed at 0.82 and 1.82 MeV. The 1.82-MeV state decays with a branching ratio $>80\%$ to the 0.82-MeV state. Its other branches, if any, are uncertain (*62, 63*). The 0.82-MeV state decays both to the first $3^+$ state and to the $2^+$ ground state (*62–64*). All these gamma-ray data are consistent with our suggested $4^+$ and $5^+$ assignments.

Next we consider states outside the ground-state band. Aside from the $3^+$ ground-state-band member, there are two other states which measurements show to be very strongly produced in $^{19}F(d, p)$. These are (1) the observed 2.04-MeV $2^+$ state, which is therefore clearly correlated with the second $K+^{17}O$ $2^+$ state, at 1.76 MeV, and (2) the 3.49-MeV $l = 0$ state, which is therefore clearly correlated with the second $K+^{17}O$ $1^+$ state, even though the latter appears quite low, at 2.47 MeV. The correlation for the $2^+$ states is further supported by the experimental observation of a strong gamma decay from the 2.04-MeV level to the ground-state-band $3^+$ level (*62–65*); for this observation matches the $K+^{17}O$ shell-model calculation of a particularly strong $M1$ decay from the second $2^+$ to the first $3^+$. (See Table XXIV ahead).

Besides the aforementioned strong $M1$ transition from the second $2^+$ to the first $3^+$, the $K+^{17}O$ model yields two other exceptionally strong $M1$ transitions between low-lying states. One of these is from the first $1^+$ to the ground-state $2^+$. Among the observed candidates for this first $1^+$ level, only the $1^+$ state at 1.06 MeV has the predicted rapid decay to ground (*62–65*). (It was the most likely candidate, even without this argument).

TABLE IX. Spectroscopic Strengths $(2J + 1)S$ for $^{19}F(d, p)^{20}F$. A dotted entry implies a weak or vanishing spectroscopic factor. The experimental columns show all known levels below 4.28 MeV, and list all $l$-values for which at least one of the two experiments indicated a definite $l$-value and strength. The shell-model columns show all $K+^{17}O$ levels calculated below 4.31 MeV (but only the lowest four levels of each spin were calculated). Above the dashed line, there is clear-cut correlation between each observed level and the shell-model level listed on that line. Below the dash-dot line, no correlation at all is implied by the listing of a shell-model level on the same line with an observed level.

| | | Experiment[a] | | | | | $K+^{17}O$ shell model | | |
| | | | | $(2J + 1)S$ | | | | | |
| Note | $J^\pi$ | Excitation in $^{20}F$, MeV | $l^c$ | $E_d = 8.9$ MeV[b] | $E_d = 16$ MeV[c] | $J$ | Excitation in $^{20}F$, MeV | $l$ | $(2J + 1)S$ |
|---|---|---|---|---|---|---|---|---|---|
| $i$ | $2^+$ | 0 | ... | ... | ... | $2_1$ | 0 | 2 | 0.06 |
| $i$ | $3^+$ | 0.66 | 2 | 1.71 | 3.19 | $3_1$ | 0.61 | 2 | 4.72 |
| $i$ | $1^+$ | 1.06 | (0) | 0.03 | (0.03) | $1_1$ | 0.92 | 0 | 0.00 |
| | | | | | | | | 2 | 0.03 |
| $i$ | $2^+$ | 2.04 | 2 | 1.39 | 2.49 | $2_2$ | 1.76 | 2 | 3.28 |
| $i$ | $1^+(0^+)$ | 3.49 | 0 | $0.74^d$ | 1.25 | $1_2$ | 2.47 | 0 | 1.67 |
| | | | (2) | ... | (0.18) | | | 2 | 0.02 |
| $j$ | $2\pm, 4\pm$ | 0.82 | (2) | 0.11 | (0.03) | $4_1$ | 1.21 | ... | ... |
| $j$ | $1, 2, 3, 5)\pm$ | 1.82 | ... | ... | ... | $5_1$ | 2.29 | ... | ... |
| $j$ | $0^+, 1^+$ | 3.53 | 0 | $—^d$ | 0.50 | $0_1$ | 2.68 | 0 | 0.55 |
| $k$ | $1, 2, 3)\pm$ | 0.98 | (2) | 0.09 | (0.02) | $3_2$ | 2.26 | 2 | 0.11 |

| k | $J^\pi$ (obs) | $E_x$ (obs) | $l$ | $S$ | $S$ | $J_n$ | $E_x$ (calc) | $l$ | $S$ |
|---|---|---|---|---|---|---|---|---|---|
| k | $2^\pm, 1^\pm(3^\pm)$ | 1.31 | ... | ... | ... | | | | |
| k | $2^{(+)}(3^+)$ | 1.84 | $[2]^b$ | ... | $[0.10]^b$ | $3_3$ | 2.43 | 2 | 0.12 |
| k | | 1.97 | ... | ... | ... | $2_3$ | 2.63 | 2 | 0.11 |
| k | $3^+, 1^+$ | 2.20 | 2 | 0.41 | 0.36 | $3_4$ | 3.55 | 2 | 0.35 |
| k | | 2.87 | | $—^e$ | | $1_3$ | 3.75 | 0 | 0.01 |
| | | | | | | | | 2 | 0.18 |
| k | $2^+, 3^+$ | 2.97 | 2 | $—^e$ | 0.40 | $2_4$ | 3.86 | 2 | 0.21 |
| k | | 3.18 | | | | $4_2$ | 4.17 | ... | ... |
| k | | 3.68 | $[0]^{g,h}$ | | $[0.04]^{g,h}$ | $1_4$ | 4.31 | 0 | 0.20 |
| | | | | | | | | 2 | 0.72 |
| k | | 3.98 | 2 | | 0.22 | | | | |
| k | $1^+$ | 4.09 | 0 | 0.10 | 0.18 | | | | |
| k | $1^+, 2^+(3^+)$ | 4.28 | 2 | | 0.13 | | | | |

a. Excitation energies and spin-parity assignments based on (59–66).

b. From the $(d, p)$ data of (67) (deuteron energy 8.9 MeV). The S-values came from a DWBA analysis performed by H. T. Fortune [and reported in (60)].

c. From the $(d, p)$ data and DWBA analysis of (60) (deuteron energy 16 MeV).

d. The 3.49- and 3.53-MeV levels were unresolved, but the DWBA analysis of their summed angular distribution yielded 0.74 for $(2J + 1)S(l = 0)$.

e. The 2.87- and 2.97-MeV levels were unresolved, and the combined angular distribution was not analyzed.

f. No angular distribution was measured.

g. The square brackets are ours; they indicate an uncertainty inferred from the smallness of the experimentally determined strength.

h. The 3.68-MeV state is observed to decay with 67% probability to the 0.66-MeV $3^+$ state, and 33% to the $2^+$ ground state [see (64)]. These decays suggest a $2^+, 3^+$ assignment, rather than the $0^+, 1^+$ assignment that would be implied by an $l = 0$ stripping strength.

i. Clear-cut correlation between observed and calculated level.

j. Probable correlation between observed and calculated level (see text).

k. No correlation is implied by the listing of a shell-model level on the same line with the observed level.

Another plausible correlation is between the observed and shell-model states having the second-largest $l = 0$ strength. In this correlation, the $K + {}^{17}O$ partner is the lowest $0^+$ state, and the experimental partner, at 3.53 MeV, is assigned $0^+$ or $1^+$. (Since this 3.53-MeV state decays 100% to the 1.06-MeV $1^+$ state, and not at all to the $2^+$ ground state (64), an assignment of $0^+$ is preferred.)

The above suggestions complete our list of proposed one-to-one correlations. Hence in the lowest section of Table IX (below the dash–dot line), we simply list the levels in order of increasing excitation. The third $K + {}^{17}O$ $1^+$ state, at 3.75 MeV, may seem an obvious candidate for partner to the $1^+$ state observed at 3.59 MeV. However, a comparison of observed and shell-model spectroscopic strengths does not support that candidacy. In fact, the strengths measured for this 3.59-MeV state resemble those of the 4.31-MeV $K + {}^{17}O$ $1^+$ state more closely than they resemble those of the 3.75-MeV $K + {}^{17}O$ $1^+$ state. Next, consider $3^+$ states. The $K + {}^{17}O$ model predicts a very strongly enhanced $M1$ transition from the second $3^+$ to the first $4^+$. (See Table XXIV below.) However, none of the observed states at 0.99, 1.31, or 1.84 MeV exhibit a rapid decay to the presumed $4^+$ at 0.82 MeV (62–65). It seems probable that the second observed $3^+$ is among the group of states at 1.97, 2.20, 2.87, and 2.97 MeV. Further correlations are even more speculative; and again, they are not supported by agreement between $K + {}^{17}O$ and measured stripping strengths.

Among the proposed correlations listed in the upper two sections of Table IX, we note two large energy discrepancies. For the observed 3.49- and 3.53-MeV states, the proposed $K + {}^{17}O$ partners are too low by 1.02 and 0.85 MeV, respectively.

### 3.3.7. The A = 20, T = 2 System

As Fig. 5 shows, for mass 20 the $K + {}^{17}O$ model reproduces the empirically determined nuclear energy separation between the lowest states of $T = 2$, $T = 1$, and $T = 0$. In ${}^{20}O$ the $K + {}^{17}O$ model reproduces the observed energies in the $0^+$, $2^+$, $4^+$ ground-state band. However, the $K + {}^{17}O$ energies of the second $2^+$ and second $0^+$ states are depressed more than an MeV below their observed excitations.

### 3.3.8. The A = 21, T = 1/2 System

Observed and calculated energy levels of the $A = 21$ nuclei are displayed in Fig. 7. Spectroscopic strengths $(2J + 1)S$ for ${}^{20}Ne(d, p){}^{21}Ne$ are shown in Table X, and S-factors for ${}^{22}Ne(p, d){}^{21}Ne$ are shown in Table XI.

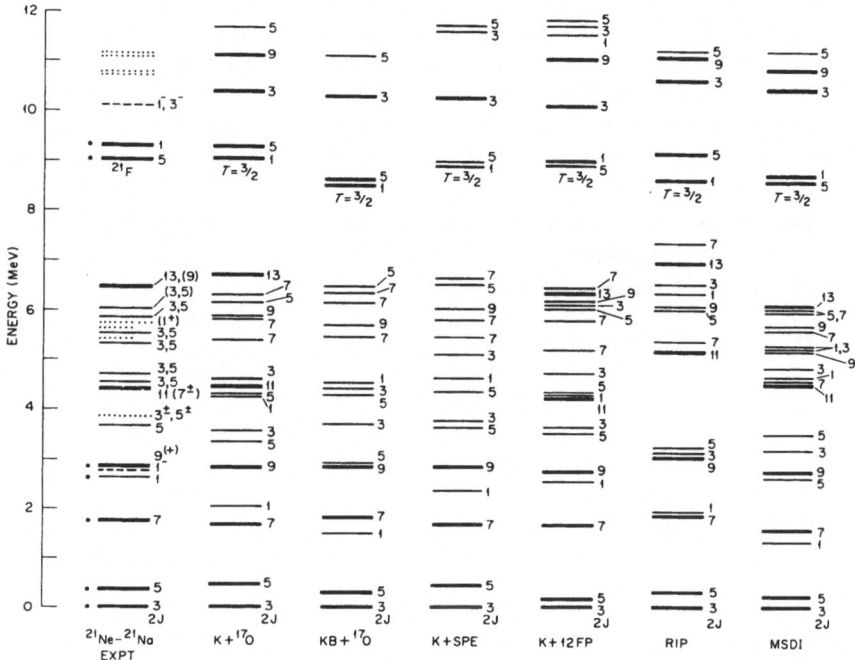

**Fig. 7. Levels of ²¹Ne–²¹Na, and levels of ²¹F.** In the spectrum marked EXPT, a label 3,5 describes a single level and implies 3 *or* 5. But in the shell-model spectra, a label 5, 3, 9 describes three separate levels and implies 5 *and* 3 *and* 9 (in order of increasing energy). For other notation conventions see Section 3.1.1. The spectrum marked EXPT is based on ²¹Ne and ²¹Na information in (*70–75*); on *T* = 3/2 information in (*76, 77*); and on a semiempirical estimate of the nuclear energy difference between isobars (see Section 4). In some of the shell-model spectra, the following states are missing because they were not calculated for that Hamiltonian: *J* = 11/2 for *T* = 1/2, *J* = 13/2 for *T* = 1/2, and *J* = 9/2 for *T* = 3/2. In other cases, only the lowest four states of each *J*, *T* combination were calculated. Hence in some of the *T* = 1/2 shell-model spectra (those in which four 5/2 or four 7/2 levels are shown), there may exist more 5/2 or 7/2 states which would lie above 6 MeV but below the highest *T* = 1/2 level plotted.

All the experimental strengths and *S*-values listed in these tables are "absolute" values, deduced via DWBA analysis. In each of these tables we have included two sets of experimental results. In Table X, for the 20 → 21 reaction, the empirical strengths from (*71*) extend to higher-energy ²¹Ne levels than those from (*72*); but the results from (*72*) are derived with a more modern optical-model potential. Omitted from Table X are the results from two studies in which absolute strengths were determined via DWBA analysis of ²⁰Ne(*d, n*)²¹Na data (*73, 74*). For the first 5/2⁺ state in *A* = 21,

**TABLE X. Spectroscopic Strengths** $(2J + 1)S$ for $^{20}$Ne$(d, p)^{21}$Ne$_{T=1/2}$. In the experimental columns, a dotted entry implies a strength too small to be determined from the data. The experimental columns include all $^{21}$Ne levels below 6.00 MeV which are observed to be produced by $l = 0$ or $l = 2$ transfer. The shell-model columns show the lowest four $K+^{17}O$ levels for each of $J = 1/2$, $3/2$, and $5/2$. Above the long dashed line there is clear-cut correlation between experimental and shell-model levels.

| | Experiment[a] | | | | $K+^{17}O$ shell model | |
| | Excitation | $(2J + 1)S$ | | | Excitation | |
| $J^\pi$ | energy[c] in | | | $J^\pi$ | energy[c] in | $(2J + 1)S$ |
| | $^{21}$Ne | Ref. 71 | Ref. 72 | | $^{22}$Ne–$^{21}$Na | |
|---|---|---|---|---|---|---|
| $3/2^+$ | 0 | . . . | . . . | $3/2^+$ | 0 | 0.07 |
| $5/2^+$ | 0.35 | 3.0 | 3.7 | $5/2^+$ | 0.46 | 3.69 |
| $1/2^+$ | 2.80 | 0.8 | 1.6 | $1/2^+$ | 2.04 | 1.29 |
| $5/2^+$ | 3.74 | . . . | 0.2[b] | $5/2^+$ | 3.37 | 0.01 |
| $3/2^\pm, 5/2^\pm$ | 3.89 | . . . | . . . | | | |
| $3/2^+, 5/2^+$ | 4.53 | 0.99 | 0.8[b] | $3/2^+$ | 3.58 | 1.14 |
| $3/2^+, 5/2^+$ | 4.69 | 0.93 | | $5/2^+$ | 4.33 | 0.95 |
| $3/2^+, 5/2^+$ | 5.34 | 0.44 | | $3/2^+$ | 4.62 | 0.96 |
| $3/2^+, 5/2^+$ | 5.55 | 0.56 | | | | |
| $(1/2^+)$ | 5.78 | (0.05) | | $1/2^+$ | 4.27 | 0.18 |
| $3/2^+, 5/2^+$ | 5.82 | 0.17 | | | | |
| $(3/2^+, 5/2^+)$ | 6.00 | (0.10) | | | | |
| | | | | $5/2^+$ | 6.22 | 0.04 |
| | | | | $3/2^+$ | 7.32 | 0.14 |
| | | | | $1/2^+$ | 7.56 | 0.05 |
| | | | | $1/2^+$ | 8.42 | 0.01 |

[a] From (70–72).
[b] Preliminary result.
[c] In MeV.

the $(2J + 1)S$ values reported in (73) and (74) are 2.3 and 2.1, compared to 3.0 and 3.7 from the $(d, p)$ references quoted in Table X. For the first $1/2^+$ state, the $(2J + 1)S$ values reported in (73) and (74) are 0.4 and 0.9,[10] compared to 0.8 and 1.6 from the $(d, p)$ references quoted in Table X. In subsequent discussion we shall, for simplicity, limit our experimental-theoretical comparisons to the $(d, p)$ results (71, 72).

[10] For the first $1/2^+$ state in $^{21}$Na, there is a disagreement in the *measured data* from the two $(d, n)$ experiments (73 and 74). (6-MeV deuterons were used in both experiments.)

In the observed $^{21}$Ne spectrum, six members of the $K = 3/2$ ground-state band have been identified (some tentatively, see Fig. 7). For these six states, the $K+{}^{17}O$ excitations agree satisfactorily with observed excitations. The $3/2^+$ and $5/2^+$ ground-state-band members can be reached by $l = 2$ nucleon transfer in $^{20}$Ne$(d, p)$ or in $^{22}$Ne$(p, d)$. Table X and XI show that for these two states there is satisfactory agreement between observed and $K+{}^{17}O$ spectroscopic strengths.

The lowest observed state which is not a member of the ground-state band is the first $1/2^+$ state. This lies at 2.4 MeV in $^{21}$Na, and at 2.80 MeV in $^{21}$Ne. The first $K+{}^{17}O$ $1/2^+$ state lies 0.6 MeV lower than the average of the observed $^{21}$Na and $^{21}$Ne states. We consider this first $1/2^+$ state to be the band-head of a $K = 1/2$ band. The next observed member of this $K = 1/2$ band is probably the 3.74-MeV state, recently (70) assigned $5/2^+$; and we correlate it with the 3.37-MeV $K+{}^{17}O$ $5/2^+$ state. Tables X and XI show that for these $1/2^+$ and $5/2^+$ states, the $K+{}^{17}O$ model yields satisfactory agreement with observed spectroscopic strengths.

At higher energies some discrepancies are apparent. Table X shows a strong $l = 2$ $K+{}^{17}O$ state at 3.58 MeV. The first candidate for its experimental partner occurs 1 MeV higher. In the observed spectrum the fourth and fifth strong $l = 2$ states occur at 5.34 MeV and 5.55 MeV; but in the $K+{}^{17}O$ spectrum the fourth strong $l = 2$ state occurs at 4.62 MeV, while the fifth either does not exist or else must occur above 6.22 MeV (where we stopped computation of $5/2^+$ states). The second $1/2^+$ state in the $K+{}^{17}O$ spectrum occurs at 4.27 MeV, but the lowest candidate for its experimental partner is the $(1/2^+)$ state observed at 5.78 MeV.

**TABLE XI. Spectroscopic Factors for $^{22}$Ne$(p, d)^{21}$Ne.** A dotted entry implies an $S$ factor too small to be determined from the data.

| | Experiment | | | | $K+{}^{17}O$ shell model | |
|---|---|---|---|---|---|---|
| $J^\pi$ | Excitation energy in $^{21}$Ne, MeV | $S$ Ref. 75 | Ref. 72 | $J^\pi$ | Excitation energy in $^{21}$Ne, MeV | $S$ |
| $3/2^+$ | 0 | . . . | 0.25 | $3/2^+$ | 0 | 0.11 |
| $5/2^+$ | 0.35 | 1.86 | 2.5 | $5/2^+$ | 0.46 | 2.16 |
| $1/2^+$ | 2.80 | . . . | ≤0.11 | $1/2^+$ | 2.04 | 0.15 |

### 3.3.9. The A = 21, T = 3/2 System

As Fig. 7 indicates, the $K+^{17}O$ model inverts the ordering of the two lowest states in the observed $^{21}F$ spectrum. The observed ground state of $^{21}F$ has $J = 5/2$, $T = 3/2$ (76). For the nuclear energy difference between the lowest $J = 5/2$, $T = 3/2$ state and the lowest $J = 3/2$, $T = 1/2$ state, we find that (a) the $K+^{17}O$ model yields 9.35 MeV, (b) the experimental spectrum of Fig. 7 shows 9.05 MeV, and (c) the measured excitation energy of the first $T = 3/2$ state in $^{21}Ne$ is 8.86 MeV (77). The splitting 9.05 MeV (drawn in Fig. 7) equals the measured energy difference between ground states of $^{21}F$ and $^{21}Ne$, minus a Coulomb-energy difference estimated as described in Section 4.

Figure 7 shows only the seven experimentally observed $^{21}F$ levels (76) lying below 2.1 MeV. [The negative parity marked for the third $^{21}F$ state is based (77) on the assignment $1/2^-$, $3/2^-$ given to the third $T = 3/2$ state in $^{21}Ne$.] Because spins and parities are not known for several of these displayed states, we cannot determine whether they include the expected $3/2^+$ and $9/2^+$ ground-state-band members.

The $^{21}F$ spectrum up through 5.9 MeV excitation has been investigated by $^{19}F(t, p)^{21}F$ measurements (76). The data from these measurements indicate the existence of seven levels below 2.1 MeV, no levels in the region 2.1 to 3.4 MeV, then seven or eight levels within the next 1-MeV interval (3.45 to 4.45 MeV). The $K+^{17}O$ model yields a total of ten even-parity levels below 3.45 MeV. From these facts we conclude that outside the ground-state band, the $K+^{17}O$ model yields some $^{21}F$ levels which are too low by $\gtrsim 1$ MeV.

### 3.3.10. The A = 22, T = 0 System

Figure 8 shows the low-lying $T = 0$ levels of $^{22}Na$. In the observed spectrum there is a ground-state rotational band with members $3^+$, $4^+$, and $5^+$ at 0, 0.89, and 1.53 MeV. In addition, the observed 3.71-MeV state, assigned $J \geq 2$, exhibits branching ratios which suggest that it is the expected $6^+$ member (78). The measured energy spacings within this band are satisfactorily reproduced by the $K+^{17}O$ model.

Other excited states in the $K+^{17}O$ spectrum appear too low with respect to this $3^+$, $4^+$, $5^+$ band. In particular, the first $1^+$ (instead of the first $3^+$) is the lowest state of the $K+^{17}O$ spectrum. Also, the second $K+^{17}O$ $1^+$ level is too low by 1.7 MeV with respect to the lowest $3^+$ state.

There seem to be no experimentally determined spectroscopic factors

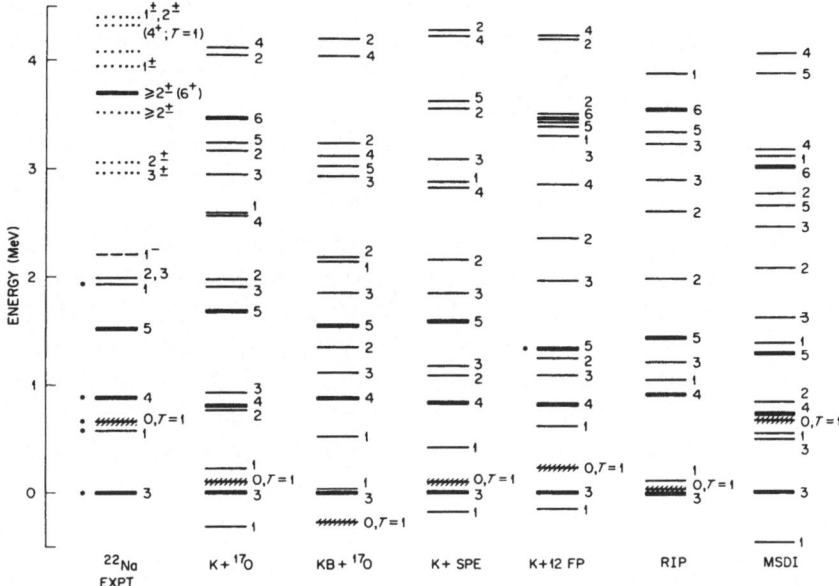

**Fig. 8.** $T = 0$ **levels of** $^{22}$**Na.** The lowest $T = 1$ level is also shown; it corresponds to the ground state of $^{22}$Ne, and it is drawn as a cross-hatched line. For other notation conventions see Section 3.1.1. The experimentally observed $T = 0$ levels are taken from (78, 79). In the columns $KB+^{17}O$ and $K+SPE$, the 6$^+$ level is missing because it was not calculated. For $J = 0$ to $J = 5$, only the lowest four levels of each spin were calculated. Hence in some of the shell-model spectra (those in which four $J = 3$ levels are shown), there may exist more $J = 3$ states below the highest level plotted. In the $K+12FP$ spectrum, the black dot marks a level which was included in the least-square fit leading to $K+12FP$, though it was not included in the fits leading to the other four least-square Hamiltonians. *Note*: this figure corrects a few small plotting errors in the $^{22}$Na figures of (1).

available for single-nucleon transfer connecting the excited $T = 0$ states of mass 22 with states of mass 21. One possible experiment of this type is $^{21}$Ne($^3$He, $d$)$^{22}$Na.

### 3.3.11. *The A = 22, T = 1 System*

The cross-hatched line in Fig. 8 shows the position of the lowest $T = 1$ state in mass 22, with respect to the lowest $T = 0$ state. The $K+^{17}O$ Hamiltonian gives this $T = 1$ position as 0.7 MeV lower than the empirical result.

Figure 9 shows that within the $K = 0$ ground-state band of $^{22}$Ne, the $K+^{17}O$ shell model yields excitations in good agreement with measured excitations. Above the 4+ member of this band, the $K+^{17}O$ spectrum shows a higher density of states than the observed spectrum. However, if we are willing to rest content with $K+^{17}O$ levels which lie about 1 MeV lower than their experimental counterpart, then there are no inconsistencies between the $K+^{17}O$ spectrum and the observed spectrum. One possible set of correlations is shown in Table XII, and we shall discuss these correlations briefly.

Consider, first, the state appearing next above the ground-state-band 4+ level. The $K+^{17}O$ spectrum suggests that an excited $K = 2$ band is built on this state, with members 2+, 3+, and 4+ at 3.53, 4.51, and 5.54 MeV. The experimental spectrum shows the following possible partners for these 2+, 3+, and 4+ levels: $J^\pi = 2^+$ at 4.46 MeV; $J^\pi = 3^+$, 2+ at 5.63 MeV; and $2 \leq J \leq 6$ at 6.68 MeV. Another plausible correlation is between the $K+^{17}O$ 0+ state at 4.95 MeV, and the (0+) state at 6.24 MeV. Then aside

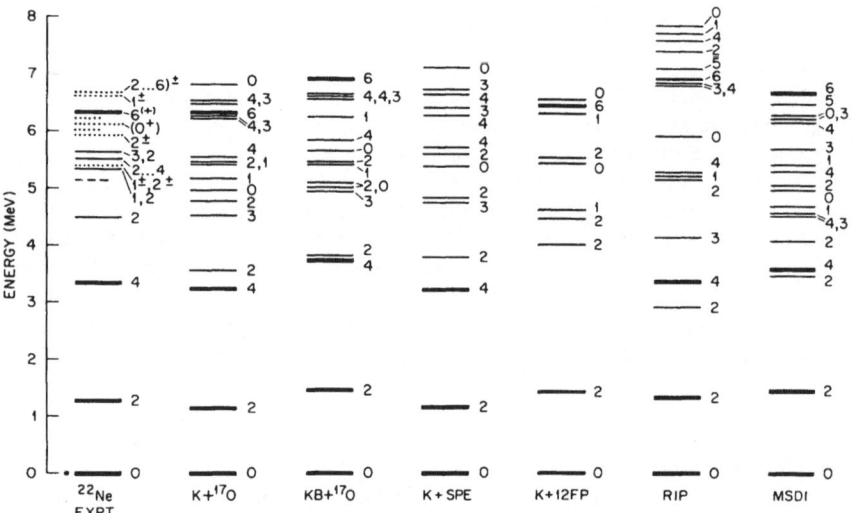

**Fig. 9.** Levels of $^{22}$Ne. In the experimentally observed spectrum a label 1, 2 describes a single level and implies 1 *or* 2; but in the shell-model spectra a label 4, 3 describes two separate levels and implies 4 *and* 3 (4 being lower). The assignment $2...6)^\pm$ means $2 \leq J \leq 6$ with parity unknown. For other notation conventions see Section 3.1.1. The experimentally observed spectrum is based on (*80–83*). In the $K+SPE$ spectrum, the 6+, 1+, and 5+ levels are missing because we did not calculate them. Also, in the $K+12FP$ spectrum, the 3+ and 4+ levels are missing because we did not calculate them. Some other levels are missing because, in general, for each Hamiltonian we calculated only the lowest four levels for 0+ and 2+, the lowest three levels for 0+, and the lowest two levels for 6+, 1+, and 5+.

**TABLE XII. Levels of $^{22}$Ne, and Spectroscopic Strengths for $^{21}$N$(d, p)^{22}$Ne.** The experimental columns include all observed levels below 6.68 MeV (except a known odd-parity level). The shell-model columns include these members of the $K+^{17}O$ spectrum: the lowest two $1^+$ states, the lowest four $2^+$ states, the lowest $6^+$ state, and all states below 6 MeV having $J = 0$ or $J \geq 3$. In the experimental columns a dotted entry indicates a spectroscopic strength too weak to measure, and a dash indicates that the level was not investigated. In the $K+^{17}O$ columns a dotted entry indicates a strength that vanishes because of angular-momentum selection rules, and a dash indicates that the $S$-value was not calculated. Above the dashed line there is clear-cut correlation between experimental and shell-model levels. Below the dashed line, the indicated correlations are conjectures.

| Note | Experiment[a] | | | | $K+^{17}O$ shell model | | | |
|------|-------|--------------------------------|-------|-----------|-------|--------------------------------|-------|-----------|
|      | $J^\pi$ | Excitation in $^{22}$Ne, MeV | $l$[b] | $(2J+1)S$ | $J^\pi$ | Excitation in $^{22}$Ne, MeV | $l$ | $(2J+1)S$ |
| $d$ | $0^+$ | 0 | 2 | $\leq 0.22$ | $0_1$ | 0 | 2 | 0.11 |
| $d$ | $2^+$ | 1.28 | 0 | $\leq 0.15$ | $2_1$ | 1.14 | 0 | 0.04 |
|     |       |      | 2 | 7.0 |       |      | 2 | 5.47 |
| $d$ | $4^+$ | 3.36 | 2 | $\leq 0.50$ | $4_1$ | 3.21 | 2 | 0.49 |
| $e$ | $2^+$ | 4.46 | 0 | 0.32 | $2_2$ | 3.53 | 0 | 1.28 |
|     |       |      | 2 | 1.20 |       |      | 2 | 1.26 |
| $d$ | $6^{(+)}$ | 6.35 | — | — | $6_1$ | 6.30 | ... | ... |
|     | $1^\pm, 2^\pm$ | 5.33 }[c] | 0 | 2.2 | $2_3$ | 4.77 | 0 | 0.02 |
|     |                |          | 2 | $\leq 2.5$ |     |      | 2 | — |
|     | $1^\pm, 2^\pm$ | 5.36 |   |     | $1_1$ | 5.18 | 0 | 1.53 |
|     |                |      |   |     |       |      | 2 | — |
| $e$ | $2, 3, 4)^+$ | 5.52 | 2 | 3.0 | $3_1$ | 4.51 | 2 | 2.30 |
|     | $2^+, 3^+$ | 5.63 | 2 | 0.85 | $2_4$ | 5.43 | 0 | — |
|     |            |      |   |      |       |      | 2 | — |
|     | $2^\pm$ | 5.92 | ... | ... |       |      |   |   |
|     |         | 6.10 | — | — |       |      |   |   |
| $f$ | $(0^+)$ | 6.24 | ... | ... | $0_2$ | 4.95 | 2 | 0.04 |
|     |         | 6.30 | ... | ... |       |      |   |   |
|     | $1^\pm$ | 6.64 } |   |     | $1_2$ | 5.45 | 0 | 0.13 |
|     |         |      | 2 | 0.74 |       |      | 2 | — |
| $e$ | $2, \ldots, 6)^\pm$ | 6.68 } |   |     | $4_2$ | 5.54 | 2 | 0.03 |

[a] From ($80$–$83$).

[b] From ($82$).

[c] At least one of the 5.33,5.36-MeV doublet has $J = 2$ [see ($80$)]. In the $(d, p)$ experiment of ($82$) this doublet was unresolved, but the measurements indicated that the observed $l = 0$ strength is most likely associated with the 5.36-MeV state.

[d] Ground-state-band levels.

[e] Possible members of a $K = 2$ band.

[f] Excited $0^+$ and $(0^+)$ levels.

from the correlations we have already mentioned, Table XII shows five other possible correlations between $K+^{17}O$ and observed levels. For all of these eight correlations outside the ground-state band, our $K+^{17}O$ levels lie about 1 MeV below their proposed partners in the observed spectrum.

Spectroscopic factors for $^{21}Ne(d, p)^{22}Ne$ would probably help to determine the proper correlations in $^{22}Ne$. Some very recent experimental results (82) are shown in Table XII, along with the $K+^{17}O$ spectroscopic strengths that we have calculated.

## 3.4. Comparison of Results from Different Hamiltonians

Figures 2 through 9 allow comparisons of excitation energies from six different Hamiltonians; Tables XIII and XIV allow comparisons of spectroscopic factors from four different Hamiltonians. Rather than discuss these comparisons nucleus-by-nucleus, we shall point out some general features, with examples.

We start with *excitation energies*, first noting those features common to all our shell models. As Figs. 2–9 show, the common "successes" are concentrated in the ground-state rotational bands. That is, excitation energies from the different Hamiltonians are generally most similar to each other, and to experimental data, for levels within these ground-state bands. This result is partly explained by the criteria for determining our least-square Hamiltonians, for as Figs. 2–9 show, most of the fitted levels belong to ground-state bands. But some of the fitted levels do fall outside the ground-state bands, and such levels often give larger-than-average contributions to the rms deviation. We shall return to this point later.

How about low-lying observed levels which are "missing" in the $K+^{17}O$ spectrum (presumably because they are dominated by configuration outside our $s$–$d$-shell model)? Such levels are also missing in the shell-model spectra computed from our other model Hamiltonians. In all seven models we notice:

    (a) the lack of a theoretical partner for the observed 1.7-MeV $1^+$ state in $^{18}F$;

    (b) the lack of a theoretical partner for the observed 2.52-MeV $2^+$ in $^{18}F$ [a state weakly produced via $^{17}O(^3He, d)$];

    (c) the appearance of only one pair of shell-model states $0^+$, $2^+$ in the region of 3 to 6 MeV excitation for $^{18}O$, rather than the two pairs experimentally observed;

(d) the appearance of only one $0^+$, $2^+$ shell-model pair in the region of 6 to 9 MeV for $^{20}$Ne, rather than the two pairs experimentally observed; and

(e) the lack of a theoretical partner for the $J = 3/2^{(+)}$ level that is observed at 3.91 MeV in $^{19}$F.

Another common feature is that levels outside the ground-state bands are often predicted too low. Thus, all our models yield excitations lower than those experimentally observed for the first two $1^+$ levels in $^{22}$Na, and for the first $2^+$ level in $^{22}$Ne. For another example, consider the second-lowest level to be produced with moderate strength in $^{20}$Ne$(d, p)^{21}$Ne. All our models yield an excitation energy that is lower than the 4.53 MeV experimentally observed for this state. (See Table X.)

Next we consider the *differences* among the energy spectra given by our alternative models. To establish a criterion for this discussion we shall say that the excitation of a given level is "sensitive to the choice among our model Hamiltonians" if all its excitations, as given by our seven Hamiltonians, scatter over an energy range greater than 1 MeV. Levels whose excitations are sensitive in this way generally fall outside the ground-state band. Some examples illustrated in Figs. 2-9 are the second $0^+$ in $^{18}$O, the second $4^+$ in $^{20}$Ne, the second $0^+$ in $^{20}$O, the second $2^+$ in $^{20}$Ne, the third $5/2^+$ in $^{21}$Ne, and the second $1^+$ in $^{22}$Ne. Even for fitted levels, there is sometimes sensitivity. For example, the first $1/2^+$ in $^{21}$Ne is calculated with excitations ranging from 1.31 MeV to 2.56 MeV. Within ground-state rotational bands, the excitations of a few high-spin states are sensitive, e.g., the $11/2^+$ excitation in $^{19}$F. (This state was not included in our least-square fits.)

Much of the sensitivity in energies has to do with the models *RIP* and *MSDI*. Recall that, among the five least-square Hamiltonians, only the two central-force Hamiltonians *RIP* and *MSDI* are completely independent of the realistic interactions. (That is, only *RIP* and *MSDI* have no two-body matrix elements in common with either $K+^{17}O$ or $KB+^{17}O$.) For levels whose energies are sensitive to the choice among our model Hamiltonians, *RIP* and *MSDI* are the models which most often yield the extreme values of excitation. In $^{21}$Ne, $^{21}$Na, and $^{22}$Ne, the *RIP* model has a general tendency to yield higher excitations than the other models do, at least beyond the first eight levels or so. When $T \leq 1/2$, the *MSDI* model tends to yield lower-than-usual excitations for levels outside the ground-state band. (See the level diagrams for $^{18}$F, $^{20}$Ne, and $^{21}$Ne.) But when $T$ is a maximum, the *MSDI* model tends to yield higher-than-usual excitations for levels outside the ground-state band. (See the level diagrams for $^{18}$O and $^{19}$O.)

In fitting observed excitations for $A = 18$–$22$, none of our seven Hamiltonians is dramatically superior to all the others. However, the best all-around record belongs to $K+12FP$.

Next we discuss the sensitivity of single-nucleon spectroscopic factors. As Tables XIII and XIV show, many of these spectroscopic factors are insensitive to the choice among our Hamiltonians. For example, Table XIII shows that the $S$-values for $^{18}O(d, p)^{19}O$ are insensitive, and Table XIV shows that the $S$-values for $^{21}Ne(p, d)^{20}Ne$ are insensitive. But these tables show some very sensitive cases too; an example is in the first line in Table XIV. In fact, we see from Tables XIII and XIV that the general insensitivity of excitations within ground-state bands does *not* extend to spectroscopic factors. (Remember that $S$-values were not included in the least-square searches!) Also, we see that discrepancies in excitation do not necessarily go hand in hand with discrepancies in spectroscopic factors.

The criteria for calling $S$-factors "sensitive" are of course moot. We now comment on two cases which illustrate this—the second and third $3^+$ states produced in the reaction $^{19}F(d, p)^{20}F$. These cases are shown in Table XIV. [See the entries labeled 19, 1/2, 1/2 and 20, 3, 1 in columns 1 and 2, for energies (2.20) and (2.97) in column 3]. Assuming the experimental–theoretical correlations which Table XIV suggests for these states, the pertinent $S(l = 2)$ values are 0.05 and 0.07 from experiment, 0.02 and 0.02 from $K+17O$, and 0.08 and 0.05 from $K+12FP$. Since all these numbers are small, we are inclined to say that there is no significant discrepancy between the three pairs of $S$-values. But suppose we consider, instead of $S$-values, the strengths $(2J_f + 1)S$. Then the three pertinent pairs of values are 0.36 and 0.40 from experiment, 0.11 and 0.12 from $K+17O$, and 0.54 and 0.36 from $K+12FP$. And when we consider these strength values, we are swayed to say that the $K+12FP$ results resemble the experimental values while the $K+17O$ results look quite different.[11]

---

[11] In fact, it was the resemblance between these experimental and $K+12FP$ strengths which prompted us to make the tentative theoretical–experimental correlations shown in Table XIII for the observed 2.20-MeV and 2.97-MeV states in $^{20}F$. These particular correlations are not clearly suggested by the $K+17O$ results, and that is why these correlations were not indicated in Table IX. Indeed, for $^{20}F$ there are other ways in which the $K+12FP$ results seem to resemble experimental results more closely than the $K+17O$ results do. Table XIV indicates proposed experimental–theoretical correlations for the observed states at 3.53, 3.49, and 4.09 MeV. For these three states, the $K+17O$ and $K+12FP$ both give fair agreement with the experimentally determined $S$-values, but the $K+12FP$ model yields much better agreement with the experimentally determined energies.

TABLE XIII. Single Nucleon Spectroscopic Factors $S$, for Reactions Involving $J = 0$ Targets. Where the experimentally observed $E$-value is parenthesized, there is considerable uncertainty us to whether the observed level should be associated with the shell-model levels listed on the same line of the table; and in these cases we have parenthesized the experimental $S$-value also. A dotted entry means that the $S$-value was too weak to measure. An entry nc means that the shell-model value was not calculated. All excitation energies are given with respect to the ground state of a nucleus having $T = T_z$.

| Initial state, $A_i J_i T_i$ | Final state, $A_f J_f T_f$ | Excitation energy $E_f$, MeV | | | | | $S(J_f)$ | | | | |
|---|---|---|---|---|---|---|---|---|---|---|---|
| | | Expt.[a] | $K+^{17}O$ | $K+12FP$ | $RIP$ | $MSDI$ | Expt.[a] | $K+^{17}O$ | $K+12FP$ | $RIP$ | $MSDI$ |
| 18, 0, 1 | 19, 1/2, 1/2 | 0 | 0 | 0 | 0 | 0 | .45 | .49 | .63 | .60 | .58 |
| | | (6.25) | 6.75 | 5.67 | 5.75 | 4.93 | (.31) | .00 | .09 | .12 | .41 |
| | | | 8.04 | 7.79 | 8.05 | 6.18 | | .30 | .21 | .15 | .00 |
| | | | 9.99 | 8.79 | 10.50 | 8.64 | | .23 | .02 | .00 | .04 |
| | 19, 3/2, 1/2 | 1.56 | 1.25 | 1.45 | 1.95 | 1.91 | .44 | .34 | .35 | .17 | .18 |
| | | | 6.45 | 5.90 | 6.87 | 5.40 | | .03 | .00 | .000 | .009 |
| | | | 6.93 | nc | 8.72 | 5.70 | | nc | nc | nc | nc |
| | | | 7.94 | nc | 9.08 | 7.11 | | nc | nc | nc | nc |
| | 19, 5/2, 1/2 | 0.20 | 0.36 | 0.22 | 0.48 | −0.03 | .63 | .65 | .71 | .65 | .88 |
| | | (4.56)[b] | 4.97 | 4.22 | 5.15 | 4.43 | (.24)[b] | .14 | .03 | .12 | .04 |
| | | | 6.59 | nc | 5.92 | 4.69 | | .01 | nc | nc | nc |
| | | | 7.72 | nc | 7.26 | 6.47 | | nc | nc | nc | nc |
| 20, 0, 0 | 19, 1/2, 1/2 | 0 | 0 | 0 | 0 | 0 | | .88 | 1.01 | 1.70 | .88 |
| | | (6.25) | 6.75 | 5.67 | 5.75 | 4.93 | | .00 | .00 | .06 | .00 |
| | | | 8.04 | 7.79 | 8.05 | 6.18 | | .00 | .00 | .00 | .00 |
| | | | 9.99 | 8.79 | 10.50 | 8.64 | | .00 | .00 | .00 | .00 |
| | 19, 3/2, 1/2 | 1.56 | 1.25 | 1.45 | 1.95 | 1.91 | | .71 | .67 | .33 | .30 |
| | | | 6.45 | 5.90 | 6.87 | 5.40 | | .001 | .00 | .00 | .00 |
| | | | 6.93 | nc | 8.72 | 5.70 | | .002 | nc | nc | nc |
| | | | 7.94 | nc | 9.08 | 7.11 | | .002 | nc | nc | nc |

TABLE XIII (Continued)

| Initial state, $A_iJ_iT_i$ | Final state, $A_fJ_fT_f$ | Excitation energy $E_f$, MeV | | | | | $S(J_f)$ | | | | |
|---|---|---|---|---|---|---|---|---|---|---|---|
| | | Expt.[a] | K+[17]O | L+12FP | RIP | MSDI | Expt.[a] | K+[17]O | K+12FP | RIP | MSDI |
| 18, 0, 1 | 19, 5/2, 1/2 | 0.20 | 0.36 | 0.22 | 0.48 | −0.03 | | 2.33 | 2.24 | 1.82 | 2.72 |
| | | (4.56)[b] | 4.97 | 4.22 | 5.15 | 4.43 | | .00 | .00 | .01 | .00 |
| | | | 6.59 | nc | 5.92 | 4.69 | | .01 | nc | nc | nc |
| | 19, 1/2, 3/2 | 1.47 | 0.44 | 1.75 | 1.07 | 0.75 | .48 | .89 | .75 | .89 | .88 |
| | | | 5.54 | 6.45 | 6.12 | 6.46 | | .01 | .01 | .04 | .02 |
| | | | 9.68 | nc | 9.43 | 9.93 | | .01 | nc | nc | nc |
| | | | 13.85 | nc | 13.38 | 13.79 | | .00 | nc | nc | nc |
| | 19, 3/2, 3/2 | 0.10 | −0.12 | 0.33 | 0.21 | 0.71 | (.03) | .01 | .01 | .03 | .00 |
| | | | 3.14 | 2.22 | 2.96 | 3.95 | | .08 | .00 | .40 | .03 |
| | | 5.45[b] | 4.53 | nc | 4.71 | 4.69 | strong | .76 | nc | nc | nc |
| | | | 6.74 | nc | 7.53 | 7.37 | | .10 | nc | nc | nc |
| | 19, 5/2, 3/2 | 0 | 0 | 0 | 0 | 0 | .41 | .67 | .71 | .65 | .70 |
| | | 3.15 | 2.15 | 3.16 | 3.06 | 3.00 | weak | .02 | .00 | .01 | .01 |
| | | (4.71) | 4.11 | 3.64 | 4.95 | 5.46 | (weak) | .00 | .02 | .01 | .00 |
| | | (5.16) | 6.23 | 6.20 | 6.39 | 7.08 | (weak) | .03 | .01 | .03 | .00 |
| 20, 0, 0 | 21, 1/2, 1/2 | 2.6 | 2.04 | 2.56 | 1.94 | 1.31 | ≈0.6 | .64 | .63 | .25 | .62 |
| | | (5.78) | 4.27 | 4.38 | 6.41 | 4.69 | (.03) | .09 | .08 | .21 | .02 |
| | | | 7.56 | 7.16 | 8.34 | 5.26 | | .02 | .02 | .04 | .11 |
| | | | 8.42 | 8.10 | 8.98 | 6.79 | | .00 | .00 | .05 | .00 |
| | 21, 3/2, 1/2 | 0 | 3.58 | 3.68 | 3.11 | 3.20 | ... | .29 | .34 | .41 | .30 |
| | | (4.53)[b] | 4.62 | 4.76 | 6.60 | 4.89 | (.25)[b] | .24 | .18 | .21 | .25 |
| | | (5.34)[b] | 7.32 | 6.14 | 7.90 | 5.28 | (.11)[b] | .04 | .03 | .01 | .14 |

| | | | | | | $S_{exp}$ | | | | |
|---|---|---|---|---|---|---|---|---|---|---|
| 21, 5/2, 1/2 | 0.35 | 0.46 | 0.18 | 0.31 | 0.22 | ≈0.5 | .62 | .63 | .66 | .70 |
| | 3.74 | 3.37 | 3.55 | 3.21 | 2.62 | 0.2 | .00 | .03 | .01 | .00 |
| | (4.69)[b] | 4.33 | 4.33 | 6.09 | 3.51 | (.16)[b] | .16 | .15 | .15 | .06 |
| | | 6.22 | 6.11 | 7.84 | 6.12 | | .01 | .00 | .00 | .00 |
| 22, 0, 1 / 21, 1/2, 1/2 | 2.6 | 2.04 | 2.56 | 1.94 | 1.31 | ≤0.11 | .15 | .16 | .25 | .32 |
| | (5.78) | 4.27 | 4.38 | 6.41 | 4.69 | | .01 | .01 | .14 | .08 |
| | | 7.56 | 7.16 | 8.34 | 5.26 | | .01 | .01 | .01 | .02 |
| | | 8.42 | 8.10 | 8.98 | 6.79 | | .00 | .00 | .08 | .02 |
| 21, 3/2, 1/2 | 0 | 0 | 0 | 0 | 0 | 0.25 | .11 | .13 | .10 | .02 |
| | (4.53)[b] | 3.58 | 3.68 | 3.11 | 3.20 | | .23 | .25 | .21 | .14 |
| | (5.34)[b] | 4.62 | 4.76 | 6.60 | 4.89 | | .03 | .01 | .02 | .04 |
| | | 7.32 | 6.14 | 7.90 | 5.28 | | .00 | .00 | .00 | .00 |
| 21, 5/2, 1/2 | 0.35 | 0.46 | 0.18 | 0.31 | 0.22 | 2.5 | 2.16 | 2.19 | 2.08 | 2.44 |
| | 3.74 | 3.37 | 3.55 | 3.21 | 2.62 | | .02 | .00 | .13 | .00 |
| | (4.69)[b] | 4.33 | 4.33 | 6.09 | 3.51 | | .01 | .02 | .00 | .04 |
| | | 6.22 | 6.11 | 7.84 | 6.12 | | .04 | .03 | .13 | .12 |
| 22, 0, 1 / 21, 1/2, 3/2 | 0.28 | -0.25 | 0.06 | -0.57 | 0.14 | | .66 | .60 | 1.18 | .60 |
| | | 2.78 | 2.64 | 3.94 | 3.62 | | .00 | .02 | .01 | .00 |
| | | 4.14 | 4.33 | 4.67 | 4.62 | | .00 | .00 | .00 | .00 |
| | | 4.96 | 5.69 | 6.35 | 5.77 | | .00 | .00 | .00 | .00 |
| 21, 3/2, 3/2 | | 1.11 | 1.19 | 1.46 | 1.90 | | .32 | .27 | .18 | .07 |
| | | 2.57 | 2.85 | 2.64 | 3.00 | | .01 | .07 | .00 | .00 |
| | | 3.32 | 3.14 | 3.94 | 3.42 | | .03 | .01 | .00 | .07 |
| | | 3.62 | 3.72 | 4.86 | 3.57 | | .00 | .01 | .00 | .00 |
| 21, 5/2, 3/2 | 0 | 0 | 0 | 0 | 0 | | 1.58 | 1.62 | 1.23 | 1.86 |
| | | 2.37 | 2.90 | 2.07 | 2.66 | | .00 | .00 | .01 | .00 |
| | | 3.38 | 3.56 | 3.94 | 3.11 | | .00 | .00 | .00 | .00 |
| | | 3.84 | 3.96 | 4.40 | 4.26 | | .00 | .00 | .00 | .00 |

[a] For more information about the experimental spins and S values, see Section 3.3 and Tables VI-XI.
[b] The spin (or parity) of the observed state is uncertain. Hence if we list an experimentally determined S value, there is uncertainty in our conversion from $(2J + 1)S$ to $S$.

**TABLE XIV. Single-Nucleon Spectroscopic Factors $S$, for Reactions Involving $J \neq 0$** than the separate contributions $S(j = 3/2)$ and $S(j = 5/2)$. Where the experimentally level should be associated with the shell-model levels listed on the same line of the table; mental $S$ columns, a dotted entry means that the spectroscopic strength was too weak to because of Pauli or angular-momentum selection rules. An entry nc means that the shell-state of a nucleus having $T = T_z$.

| Initial state, $A_iJ_iT_i$ | Final state, $A_fJ_fT_f$ | Excitation energy $E_f$, MeV | | | | | $S(l=2)$ Expt.[a] | $S(d_{5/2})$ | |
|---|---|---|---|---|---|---|---|---|---|
| | | Expt.[a] | K+17O | K+12FP | RIP | MSDI | | K+17O | K+12FP |
| 17, 5/2, 1/2 | 18, 1, 0 | 0 | 0 | 0 | 0 | 0 | 1.50 | .70 | .76 |
| | | (3.72)[b] | 4.13 | 3.88 | 4.76 | 3.84 | | .52 | .54 |
| | | | 6.08 | 6.25 | 7.35 | 5.85 | | .47 | .45 |
| | 18, 2, 0 | (3.84) | 2.72 | 2.73 | 3.33 | 2.62 | (<1.1) | ... | ... |
| | | | 6.88 | 7.03 | 8.75 | 6.50 | | ... | ... |
| | | | 10.80 | 11.09 | 10.89 | 8.71 | | ... | ... |
| | 18, 3, 0 | 0.94 | 0.92 | .67 | 0.97 | 0.62 | <0.72 | .54 | .55 |
| | | 4.12 | 4.87 | 4.59 | 4.76 | 2.82 | 1.50 | 1.27 | 1.30 |
| | | | 8.55 | 8.99 | 8.94 | 7.32 | | .15 | .13 |
| | 18, 4, 0 | | 5.85 | 6.21 | 6.42 | 6.68 | | ... | ... |
| | 18, 5, 0 | 1.13 | 1.42 | 1.05 | 0.88 | 1.14 | 2.00 | 2.00 | 2.00 |
| 19, 1/2, 1/2 | 18, 1, 0 | 0 | 0 | 0 | 0 | 0 | | ... | ... |
| | | (3.72)[b] | 4.13 | 3.88 | 4.76 | 3.84 | | ... | ... |
| | | | 6.08 | 6.25 | 7.35 | 5.85 | | ... | ... |
| | 18, 2, 0 | (3.84) | 2.72 | 2.73 | 3.33 | 2.62 | | .17 | .15 |
| | | | 6.88 | 7.03 | 8.75 | 6.50 | | .00 | .00 |
| | | | 10.80 | 11.09 | 10.89 | 8.71 | | .00 | .00 |
| | 18, 3, 0 | 0.94 | 0.92 | 0.67 | 0.97 | 0.62 | | .73 | .65 |
| | | 4.12 | 4.87 | 4.59 | 4.76 | 2.82 | | .00 | .00 |
| | | | 8.55 | 8.99 | 8.94 | 7.32 | | .00 | .00 |
| 17, 5/2, 1/2 | 18, 0, 1 | 0 | 0 | 0 | 0 | 0 | 2.32 | 1.67 | 1.41 |
| | | (3.63) | 4.03 | 3.89 | 5.28 | 6.14 | (...) | .17 | .45 |
| | | | 14.39 | 15.40 | 12.78 | 13.66 | | .15 | .14 |
| | 18, 1, 1 | | 8.85 | 9.25 | 8.28 | 9.47 | | ... | ... |
| | | | 10.08 | 10.51 | 11.03 | 10.80 | | ... | ... |
| | 18, 2, 1 | 1.98 | 1.74 | 2.03 | 2.14 | 2.41 | 1.48 | 1.11 | 1.84 |
| | | (3.92) | 3.99 | 3.53 | 4.46 | 4.61 | (.98) | .85 | .11 |
| | | | 8.42 | 8.85 | 8.60 | 9.11 | | .03 | .03 |
| | 18, 3, 1 | 5.37 | 4.94 | 5.17 | 5.47 | 5.26 | | ... | ... |
| | | | 9.12 | 9.52 | 9.78 | 9.47 | | ... | ... |
| | 18, 4, 1 | 3.55 | 3.38 | 3.48 | 3.40 | 3.60 | 1.96 | 1.79 | 1.81 |
| | | (7.10) | 7.79 | 8.16 | 7.87 | 8.44 | | .21 | .19 |

**Targets.** For the experimentally determined spectroscopic factors we list $S(l = 2)$ rather observed $E$ value is parenthesized, there is some uncertainty as to whether the observed and in these cases we have parenthesized the experimental $S$-value also. In the experi- measure. In the shell-model $S$ columns, a dotted entry means that the $S$-value vanishes model value was not computed. All excitation energies are given with respect to the ground

| $S(d_{5/2})$ | | $S(d_{3/2})$ | | | | $S(s_{1/2})$ | | | | |
|---|---|---|---|---|---|---|---|---|---|---|
| RIP | MSDI | K+$^{17}$O | K+12FP | RIP | MSDI | Expt.$^{a}$ | K+$^{17}$O | K+12FP | RIP | MSDI |
| .69 | 1.55 | .44 | .40 | .14 | .11 | ... | ... | ... | ... | ... |
| 1.10 | .39 | .008 | .007 | .11 | .21 | ... | ... | ... | ... | ... |
| .19 | .01 | .24 | .29 | .09 | .56 | ... | ... | ... | ... | ... |
| ... | ... | .32 | .27 | .10 | .06 | (1.0) | .51 | .58 | .77 | .91 |
| ... | ... | .65 | .70 | .54 | .76 |  | .34 | .30 | .22 | .08 |
| ... | ... | .03 | .03 | .36 | .18 |  | .15 | .12 | .01 | .00 |
| .54 | 1.18 | .06 | .05 | .02 | .02 | .64 | .67 | .67 | .71 | 40 |
| 1.45 | .82 | .04 | .03 | .01 | .00 | .26 | .32 | .32 | .26 | 59 |
| .003 | .01 | .69 | .75 | .87 | .97 |  | .01 | .01 | .03 | 02 |
| ... | ... | 1.00 | 1.00 | 1.00 | 1.00 |  | ... | ... | ... | .. |
| 2.00 | 2.00 | ... | ... | ... | ... |  | ... | ... | ... | .. |
| ... | ... | .00 | .00 | .00 | .00 |  | .44 | .49 | .61 | .31 |
| ... | ... | .00 | .00 | .01 | .01 |  | .01 | .06 | .01 | .05 |
| ... | ... | .03 | .03 | .02 | .01 |  | .01 | .01 | .00 | .00 |
| .12 | .12 | .09 | .09 | .04 | .03 |  | ... | ... | ... | ... |
| .001 | .01 | .000 | .00 | .002 | .01 |  | ... | ... | ... | ... |
| .001 | .00 | .003 | .00 | .001 | .01 |  | ... | ... | ... | .. |
| .68 | .84 | ... | ... | ... | ... |  | ... | ... | ... | ... |
| .01 | .09 | ... | ... | ... | ... |  | ... | ... | ... | ... |
| .00 | .00 | ... | ... | ... | ... |  | ... | ... | ... | ... |
| 1.80 | 1.72 | ... | ... | ... | ... |  | ... | ... | ... | ... |
| .16 | .23 | ... | ... | ... | ... |  | ... | ... | ... | ... |
| .03 | .05 | ... | ... | ... | ... |  | ... | ... | ... | ... |
| ... | ... | .99 | .99 | 1.00 | 1.00 |  | ... | ... | ... | ... |
| ... | ... | .01 | .01 | .00 | .00 |  | ... | ... | ... | ... |
| 1.50 | 1.28 | .01 | .01 | .00 | .01 | .32 | .38 | .04 | .22 | .32 |
| .49 | .70 | .000 | .00 | .03 | .00 | (.44) | .56 | .89 | .71 | .64 |
| .001 | .02 | .82 | .83 | .90 | .87 |  | .02 | .03 | .04 | .03 |
| ... | ... | .00 | .00 | .00 | .00 | 1.00 | 1.00 | 1.00 | 1.00 |  |
| ... | ... | 1.00 | 1.00 | 1.00 | 1.00 | .00 | .00 | .00 | .00 |  |
| 1.86 | 1.97 | .11 | .09 | .07 | .01 |  | ... | ... | ... | ... |
| .14 | .03 | .89 | .91 | .93 | .99 |  | ... | ... | ... | ... |

**TABLE XIV** (*Continued*)

| Initial state, $A_iJ_iT_i$ | Final state, $A_fJ_fT_f$ | Excitation energy $E_f$, MeV | | | | | $S(l=2)$ Expt.[a] | $S(d_{5/2})$ | |
|---|---|---|---|---|---|---|---|---|---|
| | | Expt.[a] | $K+^{17}O$ | $K+12FP$ | RIP | MSDI | | $K+^{17}O$ | $K+12FP$ |
| 19, 1/2, 1/2 | 18, 0, 1 | 0 | 0 | 0 | 0 | 0 | | ... | ... |
| | | (3.63) | 4.03 | 3.89 | 5.28 | 6.14 | | ... | ... |
| | | | 14.39 | 15.40 | 12.78 | 13.66 | | ... | ... |
| | 18, 1, 1 | | 8.85 | 9.25 | 8.28 | 9.47 | | ... | ... |
| | | | 10.08 | 10.51 | 11.03 | 10.80 | | ... | ... |
| | 18, 2, 1 | 1.98 | 1.74 | 2.03 | 2.14 | 2.41 | $\approx 0.5^c$ | .64 | .30 |
| | | (3.92) | 3.99 | 3.53 | 4.46 | 4.61 | | .05 | .29 |
| | | | 8.42 | 8.85 | 8.60 | 9.11 | | .03 | .02 |
| | 18, 3, 1 | | 4.94 | 5.17 | 5.47 | 5.26 | | .00 | .00 |
| 19, 1/2, 1/2 | 20, 0, 0 | 0 | 0 | 0 | 0 | 0 | | ... | ... |
| | | (6.72) | 6.81 | 6.21 | 7.19 | 6.89 | | ... | ... |
| | | | 12.81 | 11.79 | 11.87 | 20.23 | | ... | ... |
| | 20, 2, 0 | 1.63 | 1.46 | 1.74 | 1.62 | 1.61 | 1.26 | .76 | .71 |
| | | (7.43) | 8.43 | 7.04 | 7.90 | 6.44 | (0.32) | .18 | .26 |
| | | | 10.37 | 9.93 | 10.39 | 8.54 | | .12 | .04 |
| | | | 11.10 | 10.97 | 11.61 | 9.75 | | .09 | .08 |
| | 20, 3, 0 | | 10.69 | 9.91 | 8.99 | 7.30 | | .57 | .41 |
| | | | 11.90 | 11.24 | 11.64 | 9.53 | | | |
| 21, 3/2, 1/2 | 20, 0, 0 | 0 | 0 | 0 | 0 | 0 | | ... | ... |
| | | (6.72) | 6.81 | 6.21 | 7.19 | 6.89 | | ... | ... |
| | 20, 1, 0 | | 11.78 | 11.79 | 11.03 | 9.42 | | .00 | .00 |
| | | | 13.38 | 13.09 | 12.86 | 11.18 | | .00 | .00 |
| | 20, 2, 0 | 1.63 | 1.46 | 1.74 | 1.62 | 1.61 | | 1.06 | 1.05 |
| | | (7.43) | 8.43 | 7.04 | 7.90 | 6.44 | | .01 | .03 |
| | | | 10.37 | 9.93 | 10.39 | 8.54 | | .03 | .01 |
| | | | 11.10 | 10.97 | 11.61 | 9.75 | | .04 | .04 |
| | 20, 3, 0 | | 10.69 | 9.91 | 8.99 | 7.30 | | .15 | .13 |
| | | | 11.90 | 11.24 | 11.64 | 9.53 | | .00 | .00 |
| | 20, 4, 0 | 4.25 | 3.86 | 4.04 | 4.41 | 4.26 | | .16 | .17 |
| | | | 10.18 | 9.64 | 9.85 | 7.86 | | .01 | .04 |
| 19, 1/2, 1/2 | 20, 0, 1 | (3.53)[b] | 2.68 | 3.33 | 2.48 | 2.07 | | ... | ... |
| | | | 5.91 | 6.06 | 6.63 | 6.80 | | ... | ... |
| | 20, 1, 1 | 1.06 | 0.92 | 1.34 | 0.65 | 0.52 | | ... | ... |
| | | 3.49 | 2.47 | 3.29 | 1.84 | 2.20 | (.06) | ... | ... |
| | | (4.09) | 3.75 | 4.09 | 4.38 | 3.57 | (.06) | ... | ... |
| | | | 4.31 | 4.69 | 5.65 | 4.08 | | ... | ... |

| $S(d_{5/2})$ | | $S(d_{3/2})$ | | | | $S(s_{1/2})$ | | | | |
| RIP | MSDI | $K+^{17}O$ | $K+12FP$ | RIP | MSDI | Expt.[a] | $K+^{17}O$ | $K+12FP$ | RIP | MSDI |
|---|---|---|---|---|---|---|---|---|---|---|
| ... | ... | ... | ... | ... | ... | ≈0.4[c] | .49 | .63 | .60 | .58 |
| ... | ... | ... | ... | ... | ... | | .03 | .03 | .19 | .02 |
| ... | ... | ... | ... | ... | ... | | .00 | .00 | .00 | .00 |
| ... | ... | .00 | .00 | .00 | .00 | | .00 | .00 | .00 | .00 |
| ... | ... | .00 | .00 | .00 | .00 | | .00 | .00 | .00 | .00 |
| .42 | .72 | .22 | .10 | .08 | .09 | | ... | ... | ... | ... |
| .16 | .07 | .02 | .10 | .03 | .01 | | ... | ... | ... | ... |
| .00 | .00 | .00 | .00 | .00 | .00 | | ... | ... | ... | ... |
| .00 | .00 | ... | ... | ... | ... | | ... | ... | ... | ... |
| ... | ... | ... | ... | ... | ... | .62 | .88 | 1.01 | 1.70 | .88 |
| ... | ... | ... | ... | ... | ... | (.94) | 1.02 | 1.16 | .50 | 0.86 |
| ... | ... | ... | ... | ... | ... | | .03 | nc | nc | nc |
| .80 | .84 | .24 | .23 | .15 | .12 | | .02 | nc | nc | nc |
| .11 | .25 | .01 | .01 | .00 | .00 | | ... | ... | ... | ... |
| .13 | .07 | .00 | .00 | .13 | .00 | | ... | ... | ... | ... |
| .15 | .17 | .00 | .00 | .06 | .16 | | ... | ... | ... | ... |
| .70 | .77 | ... | ... | ... | ... | | ... | ... | ... | ... |
| ... | ... | .02 | .02 | .02 | .00 | | ... | ... | ... | ... |
| ... | ... | .00 | .00 | .00 | .00 | | ... | ... | ... | ... |
| .00 | .00 | .01 | .01 | .01 | .00 | | .02 | .02 | .03 | .01 |
| .00 | .00 | .00 | .00 | .00 | .00 | | .00 | .00 | .01 | .00 |
| 1.08 | 1.17 | .07 | .06 | .08 | .03 | | .02 | .02 | .03 | .05 |
| .04 | .04 | .01 | .02 | .02 | .02 | | .05 | .05 | .17 | .13 |
| .01 | .02 | .01 | .01 | .01 | .00 | | .04 | .03 | .02 | .03 |
| .01 | .00 | .00 | .00 | .00 | .00 | | nc | .04 | .02 | .00 |
| .16 | .13 | .02 | .02 | .01 | .01 | | ... | ... | ... | ... |
| .00 | .00 | .01 | .00 | .01 | .01 | | ... | ... | ... | ... |
| .14 | .19 | ... | ... | ... | ... | | ... | ... | ... | ... |
| .06 | .03 | ... | ... | ... | ... | | ... | ... | ... | ... |
| ... | ... | ... | ... | ... | ... | (.50)[b] | .55 | .53 | .47 | .75 |
| ... | ... | ... | ... | ... | ... | | .14 | .01 | .17 | .08 |
| ... | ... | .01 | .01 | .01 | .00 | (.01) | .00 | .08 | .00 | .03 |
| ... | ... | .01 | .03 | .01 | .00 | .42 | .56 | .32 | .28 | .42 |
| ... | ... | .06 | .10 | .50 | .00 | (.07) | .00 | .22 | .01 | .04 |
| ... | ... | .24 | .09 | .02 | .00 | | .06 | .01 | .02 | .09 |

**TABLE XIV** (*Continued*)

| Initial state, $A_iJ_iT_i$ | Final state, $A_fJ_fT_f$ | Excitation energy $E_f$, MeV | | | | | $S(l=2)$ Expt.[a] | $S(d_{5/2})$ | |
|---|---|---|---|---|---|---|---|---|---|
| | | Expt.[a] | $K+^{17}O$ | $K+12FP$ | RIP | MSDI | | $K+^{17}O$ | $K+12FP$ |
| | 20, 2, 1 | 0 | 0.00 | 0.00 | 0.00 | 0.00 | ... | .00 | .02 |
| | | 2.04 | 1.76 | 2.12 | 1.61 | 1.11 | .50 | .65 | .68 |
| | | | 2.63 | 3.56 | 2.90 | 1.91 | | .01 | .00 |
| | | | 3.86 | 4.17 | 4.10 | 3.72 | | .00 | .00 |
| | 20, 3, 1 | 0.66 | 0.61 | 0.92 | 0.30 | 0.60 | .46 | .67 | .64 |
| | | (2.20)[b] | 2.26 | 2.41 | 2.49 | 1.52 | (.05)[b] | .02 | .08 |
| | | (2.97)[b] | 2.43 | 3.12 | 2.92 | 2.07 | (.07)[b] | .02 | .05 |
| | | | 3.55 | 3.88 | 4.51 | 2.73 | | .05 | .01 |
| 21, 3/2, 1/2 | 22, 0, 0 | | 5.41 | 5.44 | 9.21 | 6.81 | | ... | ... |
| | | | 8.25 | 8.52 | 11.51 | 7.74 | | ... | ... |
| | | | 10.30 | 8.94 | 12.66 | 8.15 | | ... | ... |
| | | | 12.92 | 11.96 | 14.48 | 10.47 | | ... | ... |
| | 22, 1, 0 | (0.58) | −0.31 | −0.15 | 0.12 | −0.48 | | .50 | .48 |
| | | (1.94) | 0.24 | 0.64 | 1.05 | 0̄.55 | | .12 | .08 |
| | | | 2.60 | 3.42 | 3.92 | 1.39 | | .10 | .27 |
| | | | 4.48 | 4.50 | 5.62 | 3.15 | | .01 | .01 |
| | 22, 2, 0 | | 0.77 | 1.25 | 2.00 | 1.35 | | .08 | nc |
| | | | 1.99 | 2.39 | 2.64 | 2.10 | | .02 | nc |
| | | | 3.17 | 3.54 | 5.29 | 2.80 | | .00 | nc |
| | | | 4.06 | 4.24 | 5.99 | 4.65 | | .02 | nc |
| | 22, 3, 0 | 0 | 0 | 0 | 0 | 0 | | .45 | nc |
| | | | 0.93 | 1.11 | 1.23 | 0.50 | | .49 | nc |
| | | | 1.86 | 1.99 | 2.93 | 1.64 | | .01 | nc |
| | | | 2.96 | 3.35 | 3.26 | 2.48 | | .02 | nc |
| | 22, 4, 0 | 0.89 | 0.81 | 0.83 | 0.94 | 0.74 | | 1.00 | nc |
| | | | 2.59 | 2.88 | 4.29 | 3.21 | | .01 | nc |
| | | | 4.13 | 4.27 | 4.49 | 4.13 | | .00 | nc |
| | | | 5.28 | 4.48 | 7.05 | 4.98 | | nc | nc |
| 21, 3/2, 1/2 | 22, 0, 1 | 0 | 0 | 0 | 0 | 0 | | ... | ... |
| | | (6.24)[b] | 4.95 | 5.49 | 5.97 | 5.02 | | ... | ... |
| | | | 6.81 | 6.61 | 7.87 | 6.29 | | ... | ... |

[a] For more information about the experimental spins and $S$-values, see Section 3.3 and Tables IV–XII.
[b] The spin (or parity) of the observed state is uncertain. Hence if we list an experimentally determined
[c] From (*31*).

| S($d_{5/2}$) | | S($d_{3/2}$) | | | | S($s_{1/2}$) | | | | |
| RIP | MSDI | K+$^{17}$O | K+12FP | RIP | MSDI | Expt.[a] | K+$^{17}$O | K+12FP | RIP | MSDI |
|---|---|---|---|---|---|---|---|---|---|---|
| .00 | .00 | .01 | .01 | .01 | .01 | ... | ... | ... | ... | |
| .71 | .60 | .01 | .01 | .01 | .00 | ... | ... | ... | ... | |
| .01 | .09 | .01 | .01 | .15 | .01 | ... | ... | ... | ... | |
| .00 | .01 | .04 | .01 | .08 | .00 | ... | ... | ... | ... | |
| .69 | .74 | ... | ... | ... | ... | ... | ... | ... | ... | |
| .02 | .01 | ... | ... | ... | ... | ... | ... | ... | ... | |
| .00 | .00 | ... | ... | ... | ... | ... | ... | ... | ... | |
| .12 | .05 | ... | ... | ... | ... | ... | ... | ... | ... | |
| ... | ... | .52 | .54 | .13 | .72 | ... | ... | ... | ... | |
| ... | ... | .37 | .24 | .62 | .05 | ... | ... | ... | ... | |
| ... | ... | .00 | .10 | .13 | .18 | ... | ... | ... | ... | |
| ... | ... | .00 | .01 | .01 | .01 | ... | ... | ... | ... | |
| .49 | .69 | .01 | .02 | .01 | .03 | .06 | nc | nc | nc | |
| .24 | .20 | .35 | .34 | .27 | .11 | .02 | nc | nc | nc | |
| .02 | .19 | .02 | .01 | .04 | .05 | .08 | nc | nc | nc | |
| .01 | .14 | .18 | .21 | .05 | .02 | nc | nc | nc | nc | |
| nc | .05 | .26 | nc | nc | .15 | .07 | nc | nc | .15 | |
| nc | .03 | .20 | nc | nc | .13 | .07 | nc | nc | .21 | |
| nc | .16 | .03 | nc | nc | .00 | .32 | nc | nc | .09 | |
| nc | .02 | .21 | nc | nc | .01 | nc | nc | nc | .07 | |
| nc | .46 | .04 | nc | nc | .01 | ... | ... | ... | ... | |
| nc | .54 | .02 | nc | nc | .00 | ... | ... | ... | ... | |
| nc | .02 | .02 | nc | nc | .00 | ... | ... | ... | ... | |
| nc | .00 | nc | nc | nc | .09 | ... | ... | ... | ... | |
| nc | 1.05 | ... | ... | ... | ... | ... | ... | ... | ... | |
| nc | .02 | ... | ... | ... | ... | ... | ... | ... | ... | |
| nc | .00 | ... | ... | ... | ... | ... | ... | ... | ... | |
| nc | .02 | ... | ... | ... | ... | ... | ... | ... | ... | |
| ... | ... | .11 | nc | nc | .02 | ... | ... | ... | ... | |
| ... | ... | .04 | nc | nc | .07 | ... | ... | ... | ... | |
| ... | ... | .00 | nc | nc | .01 | ... | ... | ... | ... | |

S-value, there is uncertainty in our conversion form $(2J + 1)S$ to S.

The $^{20}$F cases considered on p. 362 are not typical; i.e., the $K+12FP$ model is not generally superior for fitting experimental $S$-values. We are especially interested in superiority for cases in which the $S$-values are large as well as sensitive. Therefore we have selected, from Tables XIII and XIV, some particular cases to examine, *viz.*, those which satisfy the following three criteria: (a) an $S$-value is available from experiment, (b) at least one of the experimental or shell-model $S$-values is $\geq 0.30$, and (c) the extremes in $S$ differ by a factor $\geq 2$. These selected cases of extreme sensitivity are collected in Table XV. In this table, the shell-model $S$-values which differ substantially from the empirical results can be spotted immediately, for we have emphasized them with stars. Also (as in Tables XIII and XIV), we have parenthesized those experimentally determined $S$-values for which there is considerable doubt as to whether the observed final state should be correlated with all the shell-model states on the same line of the table. Consider first the second $2^+$ state in $^{18}$O. It may be that because of core excitation in the real nucleus $^{18}$O, the second and third observed $2^+$ states should be considered as *sharing* the character of the second $2^+$ state in the $s$–$d$-shell-model spectrum. In that event, the shell-model spectroscopic factors $S(^{17}$O $\rightarrow ^{18}$O$)$ should equal or exceed the experimentally determined values for this second $2^+$ state. Thus for this second $2^+$ state in $^{18}$O, the only "significant" experimental–theoretical deviation in Table XV is for the model $K+12FP$; it gives too small an $S(l = 2)$ value. Similar core-excitation considerations apply to $S(^{19}$F $\rightarrow ^{20}$Ne$)$ for the second $0^+$ state in $^{20}$Ne. In this case, the experimental–theoretical deviation shown for $RIP$ is "significant," because it involves a shell-model $S$-factor *smaller* than the experimentally determined result.

Our previous discussions have made it clear that one should be *cautious* about using agreement with experimental $S$-factors as a criterion for deciding whether one model Hamiltonian is superior to another—cautious because of uncertainties in experimentally determined $S$-values, and cautious because of uncertainties in the interpretation of $S$-values computed from our simple $s$–$d$-shell model. (See Sections 3.2 and 3.3.) However, we feel it is worth pointing out that the information shown in Table XV does favor $K+^{17}O$ over the three least-square Hamiltonians $K+12FP$, $RIP$, and $MSDI$. In Table XV each of the models $K+12FP$, $RIP$, and $MSDI$ shows deviations from experiment which are larger than those shown by $K+^{17}O$. The significant deviations for $K+12FP$ are confined to the $A = 18$ cases. This result, together with the relative success of $K+12FP$ in fitting energies, indicates that our least-square criteria have led to a $K+12FP$ Hamiltonian which is generally more "effective" for $A = 19$–22 than it is for $A = 18$.

**TABLE XV. Spectroscopic Factors S in Cases for Which Shell-Model Results Are Particularly Sensitive to the Model Hamiltonian and Experimental S-values Are Available.** A spin assignment without a prime indicates the lowest-energy state of that $(A, J, T)$. A spin assignment with a prime indicates the second state of that $(A, J, T)$. Stars (*) mark those shell-model results which differ by more than 30% from the experimental results.

| Initial state, $A_iJ_iT_i$ | Final state, $A_fJ_fT_f$ | $l$ | Spectroscopic factor $S$ | | | | |
|---|---|---|---|---|---|---|---|
| | | | expt.[a] | $K+^{17}O$ | $K+12FP$ | $RIP$ | $MSDI$ |
| 17, 5/2, 1/2 | 18, 1, 0 | 2 | 1.50 | 1.10 | 1.16 | 0.83 | 1.66 |
| | 18, 3, 0 | 0 | 0.64 | 0.67 | 0.67 | 0.71 | 0.40* |
| | | 2 | <0.72 | 0.60 | 0.60 | 0.56 | 1.20* |
| | 18, 3′, 0 | 0 | 0.26 | 0.32 | 0.32 | 0.26 | 0.59* |
| | | 2 | 1.50 | 1.31 | 1.33 | 1.46 | 0.82* |
| | 18, 2, 1 | 0 | 0.32 | 0.38 | 0.04* | 0.22 | 0.32 |
| | | 2 | 1.48 | 1.12 | 1.85 | 1.50 | 1.29 |
| | 18, 2′, 1 | 0 | (0.44) | 0.56 | 0.89* | 0.71* | 0.64* |
| | | 2 | (0.98) | 0.85 | 0.11* | 0.76 | 0.70 |
| 18, 0, 1 | 19, 3/2, 1/2 | 2 | 0.44 | 0.34 | 0.35 | 0.17* | 0.18* |
| 19, 1/2, 1/2 | 20, 0, 0 | 2 | 0.62[b] | 0.88 | 1.01 | 1.70* | 0.88 |
| | 20, 0′, 0 | 2 | (0.94) | 1.02 | 1.16 | 0.50* | 0.86 |
| 20, 0, 0 | 21, 1/2, 1/2 | 0 | ≈0.6 | 0.64 | 0.63 | 0.25* | 0.62 |
| 22, 0, 1 | 21, 1/2, 1/2 | 1 | ≤0.11 | 0.15* | 0.16* | 0.25* | 0.32* |

[a] For more information about the experimentally determined S-values, see Tables III–X.
[b] This result is especially sensitive to the optical-model parameters, so that $0.3 \lesssim S \lesssim 1.0$.

## 4. GROUND-STATE BINDING ENERGIES

The quantity we are concerned with here is the *negative of the ground-state nuclear binding energy with respect to* $^{16}O$:

$$\mathscr{E} \equiv E(A, Z) - E(^{16}O) - [E_{\text{Coul}}(A, Z) - E_{\text{Coul}}(^{16}O)] \qquad (8)$$

On the right, $E$ denotes the kinetic energy plus potential interaction energy of all the nucleons in the ground state, and $E_{\text{Coul}}$ denotes the Coulomb interaction energy of all these nucleons. Our shell-model Hamiltonians include no Coulomb terms; therefore our shell-model answers for $\mathscr{E}$ are simply the ground-state eigenvalues of the model Hamiltonian. But measured energies do include Coulomb contributions; hence to convert measured energies $E$ into empirical energies $\mathscr{E}$, we need some estimate of the square-bracketed Coulomb term in (8). We are grateful to P. W. M. Glaudemans for providing us with these estimates. We shall sketch his procedure (99). The square-bracketed term in (8) was assumed to have the form (100)

$$[E_{\text{Coul}}(A, Z) - E_{\text{Coul}}(^{16}\text{O})] = CZ' + \tfrac{1}{2}aZ'(Z' - 1) + b[Z'/2] \qquad (9)$$

where $Z' = (Z - 8)$ is the number of active protons, and $[Z'/2]$ is the largest integer $\leq Z'/2$. The parameters $C$, $a$, and $b$ in (9) were then determined by requiring a least-square fit to the empirically determined values (84) of 12 energy differences $[E(A, Z) - E(A, Z')]$. These differences were for the ground states of the 12 pairs of mirror nuclei having $17 \leq A \leq 28$ and either $T_z = \pm T = \pm 1/2$ or $T_z = \pm T = \pm 1$. The resulting least-square solution turned out to fit the experimentally determined energy differences within 0.13 MeV for $A = 17$–21, and within 0.27 MeV for $^{22}\text{Mg}$–$^{22}\text{Ne}$. The optimized Coulomb parameters were $C = 3.557$ MeV, $a = 0.376$ MeV, and $b = 0.194$ MeV. These parameters, when combined with measured energies $E$ according to (8) and (9), yield empirical values of nuclear energy $\mathscr{E}$. But *at this stage*, these empirical values of $\mathscr{E}$ depend on $A$ and $Z$. In contrast, our shell-model values of $\mathscr{E}$ depend on $A$ and $T$. Hence for nuclei with $T = 1/2$ or $T = 1$, we average the empirical $\mathscr{E}$ values for $T_z = \pm T$. This gives us $Z$-independent empirical energies $\mathscr{E}$ with which to compare our shell-model energies. The results for $A = 17$–22 are listed in Table XVI. Also listed are empirical values of a "reduced" nuclear energy,

$$\mathscr{E}' = \mathscr{E} - (-4.15n) \qquad (10)$$

Here $n = A - 16$, and $-4.15$ MeV is the empirical $\mathscr{E}$ value obtained for $A = 17$. Thus, $\mathscr{E}'$ is a rough measure of the nuclear energy associated with interactions among the active nucleons.

Before comparing empirical and shell-model results, we remark that shell-model values of $\mathscr{E}$ are sensitive to characteristics of the Hamiltonian quite different from those which affect shell-model excitation energies. Any change in the shell-model Hamiltonian which adds an *n-dependent term* to all eigenvalues will of course influence the $\mathscr{E}$ values but not the excitation energies. Furthermore, *small* terms of this kind, terms which affect the

TABLE XVI. Comparison between Empirical and Shell-Model Ground-State Energies. Here $\mathscr{E}$ is the nuclear (nonCoulomb) energy with respect to $^{16}$O, and $\mathscr{E}'$ is the reduced energy obtained by subtracting $(-4.15) \times (A - 16)$ from $\mathscr{E}$. By shell-model ground state, we mean the lowest state having $J$ equal to the observed ground-state spin.

| A | T | Empirical nuclear energy[a] | | Deviation of shell-model from empirical nuclear energies, MeV | | | | | | |
|---|---|---|---|---|---|---|---|---|---|---|
| | | $\mathscr{E}$, MeV | $\mathscr{E}'$, MeV | $K+^{17}O$ | $KB+^{17}O$ | $K+SPE$ | $KB+SPE$ | $K+12FP$ | $RIP$ | $MSDI$ |
| 17 | 1/2 | − 4.15 | − 0 | 0 | 0 | 0.12 | −0.09 | −0.09 | −0.33 | −0.34 |
| 18 | 0 | −13.30 | − 5.00 | −0.09 | 0.17 | 0.63 | 0.31 | 0.05 | −0.19 | −0.44 |
| 18 | 1 | −12.21 | − 3.91 | 0.01 | −0.31 | 0.43 | −0.36 | 0.09 | −0.12 | −0.34 |
| 19 | 1/2 | −23.79 | −11.34 | −0.68 | −0.16 | 0.55 | 0.20 | −0.02 | −0.19 | −0.15 |
| 19 | 3/2 | −16.15 | − 3.7 | −0.36 | −0.21 | 0.18 | −0.34 | −0.09 | −0.01 | −0.03 |
| 20 | 0 | −40.71 | −24.11 | −0.90 | 0.09 | 0.55 | 0.30 | 0.25 | 0.61 | 0.82 |
| 20 | 1 | −30.34 | −13.74 | −1.14 | −0.21 | 0.01 | −0.17 | −0.34 | −0.17 | −0.05 |
| 20 | 2 | −23.75 | − 7.15 | −0.92 | −0.72 | −0.03 | −0.76 | 0.01 | 0.28 | 0.34 |
| 21 | 1/2 | −47.47 | −26.72 | −1.78 | 0.04 | −0.13 | 0.25 | 0.10 | −0.17 | 0.25 |
| 21 | 3/2 | −38.42 | −17.67 | −1.48 | −0.34 | −0.15 | −0.34 | 0.09 | 0.06 | −0.12 |
| 22 | 0 | −58.52 | −33.62 | −1.93 | 0.53 | −0.09 | 0.68 | 0.31 | 0.26 | −0.18 |
| 22 | 1 | −57.75 | −32.85 | −2.60 | −0.51 | −0.76 | −0.37 | −0.22 | −0.50 | −0.27 |
| | | | rms deviation, MeV | 1.27 | 0.34 | 0.39 | 0.35 | 0.17 | 0.29 | 0.34 |

[a] For $T = 3/2$, the listed values are Coulomb-corrected energies of the nuclei having maximum neutron excess. For $T = 1/2$ and $T = 1$, the listed values are averages for the two mirror nuclei (as described in the text).

Hamiltonian matrix elements by only a few tenths of an MeV, can affect the $\mathscr{E}$ values by several MeV. For example, a uniform shift of 0.2 MeV in all diagonal two-body matrix elements would alter the $\mathscr{E}$ value for $A = 22$ by 3 MeV; a uniform shift of only 0.2 MeV in all three single-particle energies would alter the $\mathscr{E}$ value for $A = 22$ by 1.2 MeV. As another kind of small change, consider an overall strengthening of the Hamiltonian by 5%. This would increase the excitation energies by no more than 0.5 MeV, for levels up to 10 MeV in the spectrum, but it would increase $\mathscr{E}$ by as much as 3 MeV, for nuclei up to mass 22. In fact, our calculated excitation energies are insensitive to any change in the model Hamiltonian which adds an essentially $n$-dependent energy to the low-lying states without depressing highly excited levels down into the low-energy region. In contrast, our calculated $\mathscr{E}$ values are sensitive to such changes.

Columns 3 through 9 of Table XVI list deviations between shell-model and empirical values of $\mathscr{E}$. (These deviations are, of course, the same as those between the shell-model and empirical values of $\mathscr{E}'$.) Except for $K+^{17}O$, all the models yield rms deviations of only a few tenths of an MeV. Even the unadjusted Hamiltonian $KB+^{17}O$ yields generally good agreement with the empirical values. The overbinding yielded by $K+^{17}O$ is very mild for $^{18}F$, but becomes noticeable at $A = 19$ and increases with $A$. An adjustment of the $K+^{17}O$ single-particle energies, determined by least-square-fitting 12 empirical $\mathscr{E}$ values and 29 excitation energies, results in the Hamiltonian $K+SPE$. This adjustment substantially lowers the rms deviation from empirical $\mathscr{E}$ values, at the cost of increasing the previously small deviation for mass 18.

## 5. ELECTRIC QUADRUPOLE MOMENTS AND $E2$ TRANSITIONS

We have used the shell-model wave functions associated with our several model Hamiltonians to compute electric quadrupole moments and $E2$ transition strengths. The results depend of course not only on the shell-model wave functions, but also on the effective $E2$ operator assumed. In this section we shall discuss this effective operator, compare our computed moments and strengths with experimentally determined values, and display the variation among results from different Hamiltonians. Then, as a way of investigating the rotational-band characteristics of our shell-model wave functions, we shall make some comparisons to $E2$ results obtained from a strong-coupling rotational model, and from an $SU_3$ model.

## 5.1. The Effective Operator $Q^2$

### 5.1.1. *Basic Assumptions*

In calculating electric quadrupole observables, we use the effective operator

$$\mathbf{Q}^2 = \sum_{k=1}^{A} \tilde{e}_k r_k^2 \, \mathbf{Y}^2(\Omega_k) \tag{10}$$

Here $\tilde{e}_k$ is the total effective charge of the $k$th nucleon, $r_k^2$ is the square of its radial coordinate, and $\mathbf{Y}^2$ is the usual (*101*) spherical harmonic operator of rank 2. Effective charges are introduced because past experience shows that in computing $E2$ matrix elements, it is important to compensate for a restriction of the model wave functions to one active major shell. The symbol $\tilde{e}_k$ in (10) implies a charge that depends only on the state of particle $k$. Thus, the significant assumption implicit in (10) is that the effective operator $\mathbf{Q}^2$ has one-body terms only. We make the further assumption that $\tilde{e}_k$ depends only on whether the particle $k$ is a neutron or a proton; then we have

$$\mathbf{Q}^2 = (\tilde{e}_p + \tilde{e}_n) \sum_{k=1}^{A} \frac{1}{2} r_k^2 \mathbf{Y}^2(\Omega_k) + (\tilde{e}_p - \tilde{e}_n) \sum_{k=1}^{A} t_z(k) r_k^2 \, \mathbf{Y}^2(\Omega_k) \tag{11}$$

Here $\tilde{e}_p$ and $\tilde{e}_n$ are the total effective charges of proton and neutron, respectively; and we use the convention $t_z = +\frac{1}{2}$ for a proton, $t_z = -\frac{1}{2}$ for a neutron. In our numerical work we have assumed

$$\tilde{e}_p = 1.5e, \qquad \tilde{e}_n = 0.5e \tag{12}$$

(These values are rather commonly used, and we shall discuss them further, below.)

For the electric quadrupole moment $Q$ of a many-particle state $\psi_{T_z}^{JT}(\alpha)$, we take the usual definition,

$$Q = \left( \frac{16\pi}{5} \right)^{1/2} \langle \alpha J T \quad J_z = J \quad T_z | Q_0^2 | \alpha J T \quad J_z = J \quad T_z \rangle \tag{13}$$

Here $Q_0^2$ is the $z$-component of $\mathbf{Q}^2$. To describe the strength of an $E2$ transition from $\psi_{T_{z_i}}^{J_i T_i}(\alpha_i)$ to $\psi_{T_{z_f}}^{J_f T_f}(\alpha_f)$ we use

$$B(E2)_{i \to f} = (2J_i + 1)^{-1} \langle \alpha_f J_f T_f, T_{z_f} \| \mathbf{Q}^2 \| \alpha_i J_i T_i, T_{z_i} \rangle^2 \tag{14}$$

where the double-barred matrix element (*5*) is reduced with respect to $J$ but not $T$. In order to calculate the many-body matrix elements on the right of

(13) and (14), we need numerical values for the single-nucleon matrix elements of $r^2$. To estimate these we assume that the single-nucleon functions in each $\psi^{JT}(\alpha)$ are harmonic-oscillator eigenfunctions with (102)

$$\hbar\omega = 41A^{-1/3} \text{ MeV} \tag{15}$$

This choice gives $\hbar\omega = 15.9$ MeV at $A = 17$, and $\hbar\omega = 14.6$ MeV at $A = 22$. For harmonic-oscillator functions, the matrix elements $\langle r^2 \rangle$ are proportional to $b^2 = \hbar/m\omega$. Our assumption (15) corresponds to $b^2 = 1.03A^{1/3}$ fm$^2$.

## 5.1.2. Further Discussion of Effective Charges

Here we comment further on our assumptions $\tilde{e}_p = 1.5e$ and $\tilde{e}_n = 0.5e$. There are two main points:

(a) Our simple choices for $\hbar\omega$, $\tilde{e}_p$, and $\tilde{e}_n$ represent a rather crude attempt to construct an effective $\mathbf{Q}^2$-operator for $s$-$d$-shell calculations. However, this crudeness is consistent with the current state of knowledge about this effective operator.

(b) In many cases we can easily estimate how our computed $Q$ and $B(E2)$ values would change if we were to substitute in $\mathbf{Q}^2$ a set of state-independent effective charges different from $\tilde{e}_p = 1.5e$ and $\tilde{e}_n = 0.5e$.

We start with point (b). Two useful exact relations are evident from equation (11). For oxygen isotopes, all of the calculated moments $Q$ are exactly proportional to $\tilde{e}_n$, while the $B(E2)$ values are exactly proportional to $\tilde{e}_n{}^2$. For $T = 0$ nuclei, all quadrupole moments $Q$ are exactly proportional to the scalar effective charge

$$\tilde{e}_S = (\tilde{e}_p + \tilde{e}_n)/2 \tag{16}$$

while the $B(E2)$ values are exactly proportional to $\tilde{e}_S{}^2$. Some approximate relations are evident from our numerical work. In particular, for $\Delta T = 0$ transitions in light $s$-$d$-shell nuclei having one or two active protons *and* one or more active neutrons, many $B(E2)$ strengths are approximately proportional to $\tilde{e}_S{}^2$. This last result holds especially well within ground-state bands. (It is connected with the validity of a simple strong-coupling-rotational model, to be described in Section 5.4.)

The simple $\tilde{e}_n$ and $\tilde{e}_p$ dependencies mentioned above are very helpful when we try to find out whether there is a set of state-independent charges which will lead to agreement between calculated and measured $E2$ results.

We shall now describe some empirical determinations of $\tilde{e}_p$ and $\tilde{e}_n$. (As we shall indicate, the determined values depend on the choice of experimental data to be fit, and on the details of the *s–d*-shell model assumed.)

First, suppose we require that our *s–d*-shell models fit the measured quadrupole moment of $^{17}$O, and the measured $B(E2)$ values of the first-to-ground transitions in $^{17}$F and $^{17}$O. These requirements lead to empirical values for the one-body matrix elements of $\tilde{e}_n r^2 Y^2$ in the $\langle d_{5/2} | \ldots | d_{5/2} \rangle$ case, and of the one-body matrix elements of $\tilde{e}_p r^2 Y^2$ and $\tilde{e}_n r^2 Y^2$ in the $\langle d_{5/2} | \ldots | s_{1/2} \rangle$ case. With our choice of $b^2$, the empirical results can be reproduced by

$$\tilde{e}_n = (0.51 \ \pm 0.06)e \quad \text{to fit the measured } {}^{17}\text{O quadrupole moment} \quad (103)$$

$$\tilde{e}_n = (0.611 \ \pm 0.004)e \quad \text{to fit the measured } {}^{17}\text{O first-to-ground } B(E2) \quad (104)$$

$$\tilde{e}_p = (1.95 \ \pm 0.03)e \quad \text{to fit the measured } {}^{17}\text{F first-to-ground } B(E2) \quad (104)$$

The $\tilde{e}_p$ value shown here is significantly larger than the choice $\tilde{e}_p = 1.5e$ which we used in our numerical work. But Elliott and Wilsdon (9) have argued that the $^{17}$F data should be ignored in determining $\tilde{e}_p$. Their reason is that the 0.5-MeV 1/2+ level of $^{17}$F is bound by only 0.1 MeV, and this means that the relevant $\langle r^2 \rangle$ is likely to be quite different from that appropriate to other *s–d*-shell transitions for which we want to use $\mathbf{Q}^2$. Ignoring the $^{17}$F data, then, Elliott and Wilsdon determined state-independent effective charges by using $A = 17$–19 wave functions from Elliott and Flowers (105) and requiring fits to four measured $B(E2)$ values, one $B(E2)$ in each of $^{17}$O, $^{18}$O, $^{19}$O, and $^{19}$Ne. Their results, when adjusted to be appropriate for our $b^2$ at $A = 17$, are

$$\begin{aligned} \tilde{e}_n &= 0.61e \\ \tilde{e}_p &= 1.54e \end{aligned} \quad \text{to fit four measured } B(E2)\text{'s in } A = 17\text{–}19$$

These charges are close to our own standard set, $\tilde{e}_n = 0.5e$ and $\tilde{e}_p = 1.5e$. To learn whether this set remains appropriate for higher states in $A = 18$–19 and states in $A = 20$–22, we look to our own shell-model calculations in comparison with measured $E2$ data (78, 103, 104, 106–113). As we shall show in Section 5.2, the set $\{\tilde{e}_n = 0.5e, \ \tilde{e}_p = 1.5e\}$ gives rough agreement between shell-model and experimental results. To improve this agreement, while keeping the shell-model wave functions unchanged, would require a set $\{\tilde{e}_p, \tilde{e}_n\}$ which varies according to the particular $Q$ or $B(E2)$ value being calculated, and such variation would be needed even within a given nucleus. Some serious experimental uncertainties cloud the issue. (A more detailed discussion will be given in the middle of Section 5.2.2.)

For another view of effective charges, we turn to *theoretical* estimates, i.e., renormalization calculations. Enhanced charges are needed because our explicit model wave functions are restricted to $1s$ and $0d$ nucleons outside a closed 16-particle core. But of course this core is not really closed; and among the nucleons excited from it, there are protons. The $E2$ transitions involving these proton excitations account for all of the first-to-ground $B(E2)$ observed in $^{17}O$, and they account for a significant part of the first-to-ground $B(E2)$ observed in $^{17}F$. Quantitative estimates of these core-excitation effects have been made, recently with realistic effective interactions (*114–117*). There is evidence that first-order perturbation calculations are inadequate for these estimates. Calculations which sum certain classes of perturbations "to infinite order" have yielded effective charges of about $0.2e$ to $0.5e$ for $\tilde{e}_n$, and about $1.2e$ to $1.5e$ for $\tilde{e}_p$. The precise answers depend on (a) the initial and final momenta $j$ of the effectively radiating nucleon, (b) the effective interaction and single-particle energies assumed, and (c) some other aspects of the approximate calculation (e.g., the classes of perturbations included, the choice of $\hbar\omega$, and the treatment of $\langle r^2 \rangle$). As a further problem we note that, even from first-order perturbation theory, there should be some *two*-body terms in the effective operator $\mathbf{Q}^2$; higher-order treatments would give three-body terms, etc. But to our knowledge, such two-body or more-body terms have not been investigated much in renormalization calculations.[12]

*In sum*: It is not clear that (11), with state-independent values $\tilde{e}_n$ and $\tilde{e}_p$, is the best simple form for a $\mathbf{Q}^2$-operator to be used for $A = 17$–$22$ calculations within an $s$–$d$-shell vector-space. However, present knowledge does not supply us with a better form for $\mathbf{Q}^2$, nor does it supply values of $\tilde{e}_n$ and $\tilde{e}_p$ which are substantially preferable to those we have chosen. We feel that the combination $\{\tilde{e}_n = (\tilde{e}_p - e) = 0.5e$ and $\hbar\omega = 41\, A^{-1/3}\}$ is about as reasonable as any other *a priori* choice, and we find that this choice does give rough agreement with experimental $E2$ data for $A = 18$–$22$.

## 5.2. Shell-Model $E2$ Results *versus* Measured $E2$ Results

In this subsection we use Tables XVII to XX to compare shell-model $E2$ results with measured $E2$ results. Tables XVII to XVIII list quadrupole

---

[12] The size of the estimated configuration-excitation corrections, their uncertainty, and the uncertainty of experimentally determined $B(E2)$ values make it inappropriate for us to worry about some other small effects, e.g., the small contribution from the neglected spin term in $\mathbf{Q}^2$.

**TABLE XVII. Quadrupole Moments *Q*.** Here *K* is the ground-state-band quantum number, and all entries are for the lowest state of spin *J*. The results marked "Rotational" were calculated according to an adiabatic-rotational model described in Section 5.4.2.

| Nucleus | $J$ | $Q$, e fm² Expt. | Calculated $Q$,$^c$ e fm² | |
|---|---|---|---|---|
| | | | $K + {}^{17}O$ | Rotational |
| ${}^{17}O$ | 5/2 | $-2.7 \pm 0.3^a$ | $-2.6$ | |
| ${}^{18}F$ | 1 | | $-1.2$ | |
| ${}^{19}F, K = 1/2$ | 3/2 | | $-6.3$ | $-5.5$ |
| | 5/2 | $\pm 11.0 \pm 2.0^a$ | $-9.2$ | $-7.9$ |
| ${}^{19}O, K = 3/2$ | 3/2 | | 2.9 | 2.5 |
| | 5/2 | | $-0.1$ | $-0.9$ |
| ${}^{20}Ne, K = 0$ | 2 | $-24.0 \pm 3.0^b$ | $-14.3$ | $-12.8$ |
| | 4 | | $-18.2$ | $-16.3$ |
| | 6 | | $-19.6$ | $-17.9$ |
| | 8 | | $-19.8$ | $-18.8$ |
| ${}^{20}F, K = 2$ | 2 | | 7.6 | 8.4 |
| | 3 | | $-3.2$ | 0 |
| | 4 | | $-5.0$ | $-4.3$ |
| ${}^{20}O, K = 0$ | 2 | | $-4.6$ | $-4.0$ |
| | 4 | | $-1.7$ | $-5.1$ |
| ${}^{21}Ne, K = 3/2$ | 3/2 | $9.3 \pm 1.0^a$ | 10.3 | 9.4 |
| | 5/2 | | $-3.4$ | $-3.3$ |
| | 7/2 | | $-9.5$ | $-9.4$ |
| | 9/2 | | $-14.7$ | $-12.8$ |
| | 11/2 | | $-15.3$ | $-14.9$ |
| | 13/2 | | $-19.0$ | $-16.4$ |
| ${}^{21}F, K = 1/2$ | 3/2 | | $-6.5$ | $-6.3$ |
| | 5/2 | | $-10.9$ | $-8.9$ |
| | 7/2 | | $-4.4$ | $-10.4$ |
| | 9/2 | | $-13.6$ | $-11.4$ |
| | 11/2 | | $-7.2$ | $-12.0$ |
| ${}^{22}Na, K = 3$ | 3 | | 22.1 | 21.6 |

$^a$ See (*103*).
$^b$ See (*107*).
$^c$ Calculated with $\tilde{e}_p = 1.5e$, $\tilde{e}_n = 0.5e$, and $b^2 = 1.03A^{1/3}$ fm².

moments, and Tables XIX and XX list $B(E2)$ strengths. The columns headed $K+SPE$, $K+12FP$, and $MSDI$ were calculated with the shell-model Hamiltonians we have previously described. The column headed $RIP'$ was calculated with a shell-model Hamiltonian very close to $RIP$. (See footnote $b$ to Table XIX.) As we shall discuss, within ground-state bands we find rough agreement between our shell-model results and measured results. But the same cannot be said for $E2$ results involving levels outside ground-state bands.

In considering detailed experimental–theoretical comparisons, one should remember that the quality of the agreement depends on:

    (a) the goodness of our model nuclear states (Are they the "optimum" $s$–$d$-shell approximations to real nuclear states?);

    (b) the adequacy of our effective operator $\mathbf{Q}^2$; and

    (c) the accuracy of the experimental data.

Obviously, we cannot discuss explicitly all the possibilities associated with reasonable variations in these three aspects. Instead, (a) we shall concentrate on comparing experimental results to our own shell-model results, (b) we shall consider only those variations in $\mathbf{Q}^2$ obtainable by varying the state-independent values of $\tilde{e}_n$ and $\tilde{e}_p$, and (c) we shall make very few comments on the accuracy of the experimental data, except to point out some of the disagreements among results obtained by different experimenters. (Often these disagreements are too large to be consistent with the quoted experimental uncertainties.)

### 5.2.1. Quadrupole Moments: $K+{}^{17}O$ versus Experiment

Table XVII displays four measured quadrupole moments, and all our calculated $K+{}^{17}O$ results. For ${}^{17}O$, ${}^{19}F$, and ${}^{21}Ne$, the $K+{}^{17}O$ quadrupole moments agree satisfactorily with experiment; but for ${}^{20}Ne$, the $K+{}^{17}O$ result disagrees by being too small in magnitude. The calculated ${}^{20}Ne$ moment is proportional to the scalar effective charge $\tilde{e}_S = \frac{1}{2}(\tilde{e}_p + \tilde{e}_n)$, for which we used $\tilde{e}_S = 1.0e$. To force agreement by changing $\tilde{e}_S$ would entail a large increase, to $\tilde{e}_S \approx 1.7e$.

### 5.2.2. B(E2) Strengths: $K+{}^{17}O$ versus Experiment

We have separated the $B(E2)$ information into two tables. We shall start by discussing Table XIX. This table lists transitions involving only: the $s$–$d$-shell levels in $A = 17$, the lowest triad of levels in each of ${}^{18}F$ and

TABLE XVIII. Quadrupole Moments $Q$ for $^{19}$F, $^{19}$O, and $^{20}$O, from Six Different s–d-shell Models. Here $K$ is the ground-state-band quantum number, and all entries are for the lowest state of spin $J$. In the shell-model columns, stars (*) mark those values which differ substantially from the $K+^{17}O$ shell-model value (see text). The results marked "Rotational" were calculated according to an adiabatic-rotational model described in Section 5.4.2.

| Nucleus | $J$ | $Q$, e fm² Expt.[a] | Calculated $Q$,[c] e fm² | | | | | |
|---|---|---|---|---|---|---|---|---|
| | | | $K+^{17}O$ | $K+SPE$ | $K+12FP$ | $RIP$[b] | $MSDI$ | Rotational |
| $^{19}$F, $K = 1/2$ | 3/2 | | −6.3 | −6.2 | −6.2 | −6.4 | −6.0 | −5.5 |
| | 5/2 | ±11.0 ± 2.0 | −9.2 | −9.1 | −9.3 | −9.6 | −9.5 | −7.9 |
| $^{19}$O, $K = 3/2$ | 3/2 | | 2.9 | 2.7 | 0.9* | 2.5 | 2.8 | 2.5 |
| | 5/2 | | −0.1 | −0.0 | −0.6* | −0.3 | 0.2 | −0.9 |
| | 7/2 | | | −2.5 | −2.6 | −2.6 | −2.3 | −2.5 |
| | 9/2 | | | −2.2 | −1.9 | −2.0 | −1.8* | −3.4 |
| $^{20}$O, $K = 0$ | 2 | | −4.6 | −4.5 | −2.7* | −4.3 | −3.3* | −4.0 |

[a] See (103).
[b] The Hamiltonian $RIP'$ is very close to the Hamiltonian $RIP$.
[c] $Q$(e fm²) calculated with $\tilde{e}_p = 1.5e$, $\tilde{e}_n = 0.5e$, and $b^2 = 1.03A^{1/3}$ fm².

TABLE XIX. B(E2) Strengths for Selected Transitions. (For other E2 transitions, see Table XX.) Here $K$ refers to the ground-state rotational band. Unless otherwise indicated, all transitions are between the lowest states of the indicated spins. In the shell-model columns, stars (*) mark those values which differ substantially from the $K + {}^{17}O$ result (see Section 5.3). The entries marked "Rotational" are ground-state-band results, calculated according to an adiabatic–rotational model described in Section 5.4.2. In the last two columns, the ratios $B(\tilde{e}')/B(\tilde{e})$ give information on the sensitivity of the calculated $B(E2)$ strengths to the effective charges assumed. $B(\tilde{e}')$ is the strength calculated with effective charges $\tilde{e}_p = 1.2e$ and $\tilde{e}_n = 0.8e$, while $B(\tilde{e})$ is the strength calculated with $\tilde{e}_p = 1.5e$ and $\tilde{e}_n = 0.5e$. $B(E2)$ is in units of $e^2\text{fm}^4$.

| Nucleus | $J_i$ | $J_f$ | $B(E2)$ Expt.[a] | Calculated $B(E2)$[y] | | | | | | $B(\tilde{e}')/B(\tilde{e})$ | |
|---|---|---|---|---|---|---|---|---|---|---|---|
| | | | | $K+{}^{17}O$ | $K+SPE$ | $K+12FP$ | $RIP'$[b] | $MSDI$ | Rotational | $K+{}^{17}O$ | Rotational |
| ${}^{17}$F | 1/2 | 5/2 | $63.6 \pm 1.5$ | 37.7 | 37.7 | 37.7 | 37.7 | 37.7 | | 0.64 | |
| ${}^{17}$O | 1/2 | 5/2 | $6.2 \pm 0.1$ | 4.2 | 4.2 | 4.2 | 4.2 | 4.2 | | 2.56 | 2.56 |
| ${}^{18}$F | 3 | 1 | $14.7 \pm 2.2$ | 14.9 | 14.0 | 14.9 | 15.0 | 11.2 | | 1.00 | |
| | 5 | 3 | $16.7 \pm 0.8$ | 14.4 | 14.0 | 14.4 | 14.3 | 11.3 | | 1.00 | |
| ${}^{18}$O | 2 | 0 | $6.5 \pm 1.2$ | 3.0 | 2.8 | 2.1 | 2.7 | 3.0 | 2.3 | 2.56 | 2.56 |
| | 4 | 2 | $\leq 28.4$ | 2.6 | 2.4 | 1.7 | 2.3 | 2.0 | 3.3 | 2.56 | 2.56 |
| ${}^{19}$F, $K = 1/2$ | 5/2 | 1/2 | $20.8 \pm 0.7$ | 19.0 | 17.9 | 19.1 | 19.4 | 18.0 | 15.0[h] | 1.24 | 1.25[h] |
| | 3/2 | 1/2 | $27.1 \pm 9.0$ | 19.0 | 18.1 | 17.7 | 19.4 | 18.2 | 15.0[h] | 1.24 | 1.25[h] |
| | 3/2 | 5/2 | | 8.2 | 7.7 | 7.8 | 6.9 | 5.7 | 6.4[h] | 1.31 | 1.25[h] |
| | 9/2 | 5/2 | $29 \pm 5$[c] | 19.9 | 17.9 | 18.3 | 19.6 | 15.3 | 21.5[h] | 1.31 | 1.25[h] |
| | 7/2[e] | 5/2 | | 1.5 | | | | | 2.1[h] | 1.11 | 1.25[h] |
| | 7/2[e] | 3/2 | $41 \pm 21$[c] | 18.4 | | | | | 19.3[h] | 1.17 | 1.25[h] |
| | 7/2[e] | 9/2 | $67 \pm {}^{74}_{42}$ | 2.7 | | | | | 1.6[h] | | 1.25[h] |
| | 13/2 | 9/2 | $15 \pm 5$[c] | 13.7 | | | | | 23.6[h] | | 1.25[h] |
| | 11/2 | 7/2[e] | | 0.4 | | | | | 22.8[h] | | 1.25[h] |

| | $J$ | $J'$ | Exp. | | | | | | | | |
|---|---|---|---|---|---|---|---|---|---|---|---|
| ¹⁹O, $K = 3/2$ | 3/2 | 5/2 | | 8.6 | 8.3 | 7.2 | 9.4 | 7.2 | 7.8 | 2.56 | 2.56 |
| | 7/2 | 3/2 | | 2.1 | 2.0 | 1.2 | 2.0 | 2.1 | 2.2 | 2.56 | 2.56 |
| | 7/2 | 5/2 | | 3.0 | 2.6 | 3.9 | 3.1 | 3.2 | 3.3 | 2.56 | 2.56 |
| | 9/2 | 5/2 | | 2.6 | 2.5 | 2.2 | 2.4 | 2.1 | 3.3 | 2.56 | 2.56 |
| | 9/2 | 7/2 | | 2.6 | 2.5 | 2.2 | 2.1 | 2.1 | 2.1 | 2.56 | 2.56 |
| ²⁰Ne, $K = 0$ | 2 | 0 | 96 ± 14[d] / 57 ± 8[e] | 48.1 | 46.1 | 50.3 | 50.4 | 46.0 | 39.8 | 1.00 | 1.00 |
| | 4 | 2 | 49.7 ± 4.5 | 59.7 | 57.0 | 59.8 | 62.4 | 54.0 | 56.9 | 1.00 | 1.00 |
| | 6 | 4 | 89.8 ± 19.2 | 48.8 | 46.1 | 48.4 | 50.0 | 41.1 | 62.6 | 1.00 | 1.00 |
| | 8 | 6 | | 31.2 | 30.1 | 29.7 | 29.4 | 21.7* | 65.6 | 1.00 | 1.00 |
| ²⁰F, $K = 2$ | 3 | 2 | | 27.2 | 27.4 | 28.5 | 25.8 | 15.8* | 30.6 | 1.28 | 1.31 |
| | 4 | 2 | | 11.4 | 10.8 | 9.6 | 13.4 | 8.5 | 10.2 | 1.38 | 1.31 |
| | 4 | 3 | | 21.2 | 19.0 | 19.3 | 19.3 | 17.6 | 22.9 | 1.32 | 1.31 |
| | 5 | 3 | | 16.5 | 15.5 | 16.5 | 17.0 | 12.7 | 16.4 | 1.27 | 1.31 |
| | 5 | 4 | | 11.4 | 11.8 | 11.3 | 11.6 | 11.7 | 16.4 | 1.19 | 1.31 |
| ²⁰O, $K = 0$ | 2 | 0 | ≤38.7 | 4.7 | 4.2 | 3.4 | 4.8 | 4.3 | 3.9 | 2.56 | 2.56 |
| | 4 | 2 | | 1.9 | 1.2 | 1.7 | 1.5 | 0.2* | 5.6 | 2.56 | 2.56 |
| ²¹Ne, $K = 3/2$ | 5/2 | 3/2 | 63 ± 13[e] | 80.8 | | | | | 75.0 | 1.03 | 1.04 |
| | 7/2 | 3/2 | 16 ± 6[f] | 34.4 | | | | | 31.2 | 1.05 | 1.04 |
| | 7/2 | 5/2 | 24 ± 10[f] | 57.4 | | | | | 46.8 | 1.04 | 1.04 |
| | 9/2 | 5/2 | 22 ± 7[f] | 45.3 | | | | | 46.8 | 1.02 | 1.04 |
| | 9/2 | 7/2 | 21 ± 14[f] | 30.7 | | | | | 30.7 | 0.97 | 1.04 |
| | 11/2 | 7/2 | 43 ± 19[f] | 51.6 | | | | | 55.7 | 1.05 | 1.04 |
| | 11/2 | 9/2 | 13 ± 10[f] | 26.1 | | | | | 21.4 | 1.06 | 1.04 |
| ²¹F, $K = 1/2$ | 1/2 | 5/2 | 57.1 ± 2.5 | 68.6 | | | | | | 1.04 | 1.04 |

TABLE XIX (*Continued*)

| Nucleus | $J_i$ | $J_f$ | $B(E2)$ Expt.[a] | Calculated $B(E2)$[j] | | | | | | $B(\tilde{e}')/B(\tilde{e})$ | |
| --- | --- | --- | --- | --- | --- | --- | --- | --- | --- | --- | --- |
| | | | | $K+{}^{17}O$ | $K+SPE$ | $K+12FP$ | $RIP$[b] | $MSDI$ | Rotational | $K+{}^{17}O$ | Rotational |
| $^{22}$Na, $K=3$ | 4 | 3 | $94\pm15$[i] | 101.4 | | | | | 93.9 | 1.00 | 1.00 |
| | 5 | 3 | $20\pm3$[i] | 25.2 | | | | | 22.8 | 1.00 | 1.00 |
| | 5 | 4 | $59\pm8$[i] | 90.9 | | | | | 87.8 | 1.00 | 1.00 |
| | 6 | 4 | | 37.9 | | | | | 40.5 | 1.00 | 1.00 |
| | 6 | 5 | | 69.2 | | | | | 71.7 | 1.00 | 1.00 |
| $^{22}$Ne, $K=0$ | 2 | 0 | $\left\{\begin{array}{l}66\pm12^{d}\\40\pm3^{e}\end{array}\right\}$ | 54.9 | | | | | 47.9 | 1.07 | 1.07 |
| | 4 | 2 | $54.2\pm11.7$ | 72.5 | | | | | 68.4 | 1.07 | 1.07 |

[a] From (104), unless referenced otherwise. For further remarks on the experimental numbers, see Section 5.2.

[b] The Hamiltonian $RIP'$ is very close to the Hamiltonian $RIP$. For 71 transitions in $A$ = 18-20, the $B(E2)$ values were calculated from $RIP$ eigenvectors as well as from $RIP'$ eigenvectors. In 64 of these 71 cases, the answers were identical within 0.2 $e^2$fm$^4$. (In the least-square search, $RIP$ was not a completely converged result. The next iteration gave $RIP'$. The single-particle energies $\varepsilon(j)$ for $RIP'$ are within 0.02 MeV of the $RIP$ values listed in Table I. The two-body parameters $(nl + n'l')_T$ for $RIP'$ are within 6% of the $RIP$ values listed in Table II.)

[c] (106).    [d] (107).    [e] (108).    [f] (111).

[g] We imply here the 7/2 member of the ground-state band in $^{19}$F; and we identify this member with the *second* 7/2+ state in the observed spectrum (at 5.47 MeV), and with the *first* 7/2+ state in the $K+{}^{17}O$ spectrum.

[h] Although the adiabatic-rotational formula (20) is not *generally* valid for $K = \frac{1}{2}$ bands, we list rotational-model results for the $K = \frac{1}{2}$ ground-state band of $^{19}$F. The reason is that in our specialized adiabatic-rotational model, the intrinsic state for this band has $(\lambda\mu) = (60)$. This $(\lambda\mu)$ representation has $K_L = 0$ only; and this leads to a zero value for the $K$-mixing term that would otherwise invalidate formula (20) when $K = \frac{1}{2}$.

[i] These are the results of (78), adjusted in accordance with the lifetime determinations of (112). (We have adjusted the "uncertainties," too—but only roughly.)

[j] $B(E2)$ calculated with $\tilde{e}_p = 1.5e$, $\tilde{e}_n = 0.5e$, and $b^2 = 1.03A^{1/3}$ fm$^2$.

$^{18}$O, and levels within the ground-state bands of $A = 19$–22. The first-to-ground transitions[13] are listed as the first entry for each nucleus. For the first-to-ground transitions in $^{18}$F, $^{19}$F, $^{21}$Ne, and $^{22}$Na, we rate the experimental–theoretical agreement as excellent. For the first-to-ground transitions in $^{17}$F, $^{17}$O, $^{18}$O, $^{21}$F, and $^{22}$Ne, the deviations from experiment are all between 20% and 46%.

We proceed now to comment in detail on the comparison of $K+^{17}O$ and experimental $B(E2)$ values in Table XIX:

$A = 17$: For $A = 17$ the $s$–$d$-shell wave functions are Hamiltonian-independent. Thus there is only one way to bring these $s$–$d$-shell $B(E2)$ values into agreement with the quoted experimental values—change the one-body matrix elements of $\tilde{e}_k$, as we have already discussed in Section 5.1.

$A = 18$: For the first-to-ground $(3 \rightarrow 1)$ transition in $^{18}$F, Table XIX shows an experimental $B(E2)$ quoted from the compilation of Skorka *et al.* (*104*); it is their weighted average of two independently measured values, $11.7 \pm 3.1$ e$^2$fm$^4$ and $16.4 \pm 1.6$ e$^2$fm$^4$. The corresponding $K+^{17}O$ answer falls between these two measured values. For the $5 \rightarrow 3$ transition in $^{18}$F, the $K+^{17}O$ $B(E2)$ is slightly outside the quoted experimental limits; but agreement could be forced by increasing $\tilde{e}_S$ only a little, from $1.0e$ to $(1.08 \pm 0.03)e$. In $^{18}$O the $K+^{17}O$ first-to-ground $B(E2)$ is smaller than the measured value. Here agreement could be forced by increasing $\tilde{e}_n$ from $0.5e$ to $(0.75 \pm 0.15)e$.

$A = 19$: For $^{19}$F, the $K+^{17}O$ $B(E2)$'s are in satisfactory agreement with measured results for the lowest two transitions, $5/2 \rightarrow 1/2$ and $3/2 \rightarrow 1/2$. For a few of the higher transitions, there are hints that the $K+^{17}O$ $B(E2)$ values may be too small. For the $7/2 \rightarrow 9/2$ transition, the $K+^{17}O$ $B(E2)$ is only one-tenth of the lower limit on the quoted experimental result (and this is by far the worst discrepancy in Table XIX).

$A = 20$: For the $2 \rightarrow 0$ first-to-ground transition in $^{20}$Ne, Table XIX lists two conflicting experimental results. Our $K+^{17}O$ result is not far from the smaller of these two. Other experimental determinations of this $2 \rightarrow 0$ $B(E2)$ have been reported (*104, 109, 110*), but these reports do not settle the question of whether the true strength is closer to 50 or to 100 e$^2$fm$^4$. For the $4 \rightarrow 2$ transition, the $K+^{17}O$ $B(E2)$ agrees moderately well with experiment. For the $6 \rightarrow 4$ transition the $K+^{17}O$ result, 49 e$^2$fm$^4$, is outside the experimental limits on the one available measured result, $90 \pm 19$ e$^2$fm$^4$.

---

[13] By "first-to-ground transition" we mean the decay from first excited *even*-parity state to ground state.

$A = 21$: The measured $B(E2)$ of the $5/2 \rightarrow 3/2$ first-to-ground transition in $^{21}$Ne has an interesting history. The compilation of Skorka *et al.* lists $162 \pm 32$ $e^2$fm$^4$ as the best available value; but after the printing of that compilation, a much smaller value, $17 \pm 9$ $e^2$fm$^4$, was reported (*111*); and more recently yet, an intermediate value, $63 \pm 13$ $e^2$fm$^4$, has been reported (*108*). Because the experimentalists responsible for the value $17 \pm 9$ $e^2$fm$^4$ have now accepted $63 \pm 13$ $e^2$fm$^4$ as probably more accurate (*113*), we list the latter value in our Table XIX. The corresponding $K+^{17}O$ result, 80.8 $e^2$fm$^4$, is almost within the uncertainty limits of $63 \pm 13$ $e^2$fm$^4$. However, for the six other $^{21}$Ne transitions for which Table XIX lists measured $B(E2)$'s, the $K+^{17}O$ strengths are all larger than the experimentally determined values; and in four of these six cases the $K+^{17}O$ values lie outside the quoted experimental limits. For $^{21}$F, the measured first-to-ground $B(E2)$ value $57.1 \pm 2.5$ $e^2$fm$^4$ is the only result listed in the compilation of Skorka *et al.* The corresponding $K+^{17}O$ result agrees within 20%.

$^{22}$Na: For the $4 \rightarrow 3$ and $5 \rightarrow 3$ $E2$ transitions in $^{22}$Na, there is satisfactory agreement between the $K+^{17}O$ and measured $B(E2)$. For the $5 \rightarrow 4$ transition, the $K+^{17}O$ $B(E2)$ is 50% larger than the measured strength.

$^{22}$Ne: For the $2 \rightarrow 0$ $B(E2)$ in $^{22}$Ne, we have listed two conflicting experimental results. Our $K+^{17}O$ result falls between them. Recall now that there was also a conflict in experimental results for the $2 \rightarrow 0$ $B(E2)$ in $^{20}$Ne. Despite these conflicts, there is a rather unambiguous discrepancy between $K+^{17}O$ and experimental results. This discrepancy is in the *ratio* of $B(E2; 2 \rightarrow 0; {}^{20}$Ne$)$ to $B(E2; 2 \rightarrow 0; {}^{22}$Ne$)$. From (*107*) and (*108*), the sources of conflicting $B(E2)$ values for $^{20}$Ne, and also for $^{22}$Ne, we find the following *agreeing* values of the ratio $B(E2; 2 \rightarrow 0, {}^{22}$Ne$)/B(E2; 2 \rightarrow 0, {}^{20}$Ne$)$: 0.69 and 0.71. But from our $K+^{17}O$ model with $\tilde{e}_p = 1.5e$, $\tilde{e}_n = 0.5e$, and $b^2 = 1.03A^{1/3}$ fm$^2$ we get the ratio 1.14. Finally, we note that for the $4 \rightarrow 2$ transition in $^{22}$Ne, the $K+^{17}O$ $B(E2)$ is $\approx 35\%$ larger than the one available measured result, $54 \pm 12$ $e^2$fm$^4$.

In the nucleus-by-nucleus discussion just above, we have noted that our experimental–theoretical comparisons for $^{17}$O and $^{18}$O suggest an increase of $\tilde{e}_n$ to $\gtrsim 0.6e$. We shall next consider whether the Table XIX results for $A = 19$–$22$ suggest changes in $\tilde{e}_n$ or $\tilde{e}_p$. To begin this discussion, we mention some numerical work which shows that for ground-state-band transitions in $^{21}$Ne and $^{22}$Ne, our shell-model $B(E2)$'s are approximately proportional to $\tilde{e}_S{}^2$. As previously indicated, we have used $\tilde{e}_n = 0.5e$ and $\tilde{e}_p = 1.5e$ in most of our numerical calculations. But for comparison, we also calculated many of the $K+^{17}O$ $B(E2)$ values with a new set of choices:

$\tilde{e}_n = 0.8e$ and $\tilde{e}_p = 1.2e$. In this new set, $\tilde{e}_n$ and $\tilde{e}_p$ are each changed by $0.3e$, compared to their values in our "standard" set; but the scalar effective charge $\tilde{e}_S$ is left completely unchanged. The $B(E2)$ values calculated from these new charges show that many of the strengths in $^{21}$Ne and $^{22}$Ne are left practically unchanged, too. This result is demonstrated in the next-to-last column of Table XIX, where we list the ratios $K + ^{17}O$ $B(E2)$ values as calculated with the new set of effective charges, to $K + ^{17}O$ $B(E2)$ values calculated with our standard set. These ratios imply that for ground-state-band transitions in $^{21}$Ne and $^{22}$Ne, the isovector part of $\mathbf{Q}^2$ is unimportant.

Thus in Table XIX, the calculated $B(E2)$'s for $^{20}$Ne, $^{21}$Ne, $^{22}$Na, and $^{22}$Ne are all either exactly or approximately proportional to $\tilde{e}_S{}^2$. So for these nuclei, we look to our experimental–theoretical $B(E2)$ comparisons for suggestions about changing $\tilde{e}_S$. Recall that for $Q(^{20}$Ne, $2^+)$ our experimental–theoretical comparisons suggested a change to $\tilde{e}_S \approx 1.7e$, while for $Q(^{21}$Ne, $3/2^+)$ there was no call for a change from our standard choice $\tilde{e}_S \approx 1.0e$. Similarly, we find that our $B(E2)$ comparisons suggest changes in $\tilde{e}_S$ which differ from one transition to the next. On the other hand, the experimental information itself indicates that we should *not take too seriously* the criterion of matching any one measured $B(E2)$. The suggested changes in $\tilde{e}_S$ are somewhat obscured by the quoted experimental uncertainties. Furthermore, the outright conflicts between independently measured results not only present us with ambiguous matching problems, they also cast doubt on the reliability of other measured $B(E2)$ values (especially those which have not been confirmed by many independent determinations). *If* there is any trend in suggested $\tilde{e}_S$ with $A$, it is toward smaller values of $\tilde{e}_S$ with increasing $A$. One indication of this trend comes from comparison of experimental and $K + ^{17}O$ results for the ratio $B(E2; 2 \to 0, {}^{22}$Ne$)/B(E2; 2 \to 0, {}^{20}$Ne$)$. (See the discussion labeled $A = 20$, several paragraphs back.)

Next we turn to Table XX. This table shows $B(E2)$'s involving one or more of the following: $^{18}$F states outside the lowest triad, $^{18}$O states outside the lowest triad, and $A = 19$–$22$ states outside the ground-state bands. In this table an unprimed $J$-value indicates the lowest-energy even-parity state of spin $J$, a singly-primed $J$-value indicates the second-lowest even-parity state of spin $J$; etc. Square brackets enclose every measured $B(E2)$ value shown in Table XX. Each set of brackets implies that, because of $B(E2)$ *or other* information, we suspect that there is poor correlation between either the initial or final observed state, and a shell-model state employed to represent it in our $B(E2)$ calculations for that line of the table.

Most of the square-bracketed $B(E2)$'s involve observed states which we have already described in Section 3.3 as having poor or uncertain

TABLE XX. $B(E2)$ Strengths. (For other $E2$ transitions, see Table XIX.) Here a $J$-value without a prime indicates the lowest-energy state of that $A, J, T$; a $J$-value with a prime indicates the second state of that $A, J, T$; etc. For the meanings of the stars, and of the entries $B(\bar{e}')/B(\bar{e})$, see the caption to Table XVII. In the columns headed "Rotational model", the squared Clebsch–Gordan coefficient $(C)^2$, and the quantum numbers $K_i$ and $K_f$, have the same meaning as in equation (21). Many of the $K$ assignments are very speculative. $B(E2)$ is in units of $e^2\mathrm{fm}^4$.

| Nucleus | $J_i$ | $J_f$ | $B(E2)$ Expt.[a] | Calculated $B(E2)$[c] | | | | | Rotational model | | | $B(\bar{e}')/B(\bar{e})$ |
|---|---|---|---|---|---|---|---|---|---|---|---|---|
| | | | | $K+^{17}O$ | $K+SPE$ | $K+12FP$ | $RIP'^{b}$ | $MSDI$ | $(C)^2$ | $K_i$ | $K_f$ | $K+^{17}O$ |
| $^{18}$F | 2 | 1 | [ 6.2 ± 1.1] | 10.1 | 10.1 | 10.5 | 15.6* | 11.6 | | | | 1.00 |
| | 2 | 3 | [16.4 ± 5.4] | 9.1 | 8.9 | 9.4 | 10.1 | 10.7 | | | | 1.00 |
| | 1' | 1 | [ 4.2 ± 1.0] | 2.0 | 4.7* | 2.0 | 7.5* | 3.5 | | | | 1.00 |
| | 1' | 3 | | 4.2 | 5.2 | 4.1 | 2.1* | 6.4* | | | | 1.00 |
| | 1' | 2 | [ 216 ± 86] | 0.4 | 0.1 | 0.5 | 3.1* | 0.5 | | | | 1.00 |
| | 3' | 1 | | 0.2 | 0.1 | 0.2 | 0.0 | 0.0 | | | | 1.00 |
| | 3' | 3 | | 0.7 | 1.7 | 0.6 | 0.3 | 4.2* | | | | 1.00 |
| | 3' | 5 | | 0.5 | 1.3 | 0.5 | 0.6 | 5.1* | | | | 1.00 |
| | 3' | 2 | | 0.0 | 0.1 | 0.0 | 1.2 | 0.1 | | | | 1.00 |
| | 3' | 1' | | 6.2 | 6.5 | 6.3 | 6.0 | 8.3* | | | | 1.00 |
| $^{18}$O | 2' | 0 | | 0.0 | 0.1 | 1.3 | 0.2 | 0.1 | 0.2 | 2 | 0 | 2.56 |
| | 2' | 2 | | 2.1 | 3.1 | 3.9* | 3.5 | 2.7 | 0.3 | 2 | 0 | 2.56 |
| | 2' | 4 | | 0.2 | 0.5 | 1.7* | 0.4 | 0.1 | 0.01 | 2 | 0 | 2.56 |
| | 3 | 2 | | 0.0 | 0.1 | 0.3 | 0.1 | 0.0 | 0.4 | 2 | 0 | 2.56 |
| | 3 | 4 | | 1.5 | 1.6 | 1.6 | 1.8 | 2.2 | 0.1 | 2 | 0 | 2.56 |
| | 4' | 2 | | 0.4 | 0.3 | 0.1 | 0.2 | 0.8 | 0.1 | 2 | 0 | 2.56 |
| | 4' | 4 | | 0.8 | 0.8 | 0.8 | 0.8 | 0.7 | 0.4 | 2 | 0 | 2.56 |

| Nucleus | | | | | | | | | | | | |
|---|---|---|---|---|---|---|---|---|---|---|---|---|
| ¹⁹F | 0' | 2 | [22.2 ± 7.4] | 1.5 | 1.2 | 0.5 | 0.7 | 0.7 | | 0' | 0 | 2.56 |
| | 0' | 2' | | 6.4 | 7.5 | 5.5 | 6.9 | 6.2 | | 0' | 2 | 2.56 |
| | 3 | 2' | | 0.6 | 0.5 | 0.4 | 0.7 | 0.7 | 0.4 | 2 | 2 | 2.56 |
| | 4' | 2' | | 0.5 | 0.6 | 0.9 | 0.9 | 0.7 | 0.1 | 2 | 2 | 2.56 |
| | 4' | 3 | | 1.2 | 1.2 | 1.2 | 1.0 | 0.7 | 0.3 | 2 | 2 | 2.56 |
| | 7/2 | 5/2 | | 1.5 | 0.1 | 0.2 | 0.6 | 1.7 | | 1/2 | 1/2 | 1.11 |
| | (7/2)' | (7/2)' | | 0.6 | | | | | | 1/2 | 1/2 | |
| | 7/2 | 7/2 | [41 ± 21] | 18.4 | | | | | | 1/2 | 1/2 | 1.17 |
| | (7/2)' | (7/2)' | | 2.1 | 1.3* | 1.1* | 2.9* | 9.6* | | 1/2 | 1/2 | |
| | 7/2 | 9/2 | | 2.7 | | | | | | | | |
| | (7/2)' | 9/2 | [67 ± 74/42] | 1.4 | | | | | | | | |
| | 11/2 | 7/2 | | 0.4 | | | | | | | | |
| | 11/2' | (7/2)' | | 5.0 | | | | | | | | |
| ¹⁹O | (5/2)' | 1/2 | | 1.3 | 1.6 | 1.2 | 0.4 | 0.0 | 0.3 | 5/2 | 1/2 | 1.66 |
| | (5/2)' | 5/2 | | 0.2 | 0.2 | 0.4 | 0.0 | 0.3 | 0.2 | 5/2 | 1/2 | 1.45 |
| | (5/2)' | 3/2 | | 0.1 | 0.2 | 0.3 | 0.2 | 0.3 | 0.4 | 5/2 | 1/2 | 1.64 |
| | 1/2 | 3/2 | | 0.9 | 0.7 | 0.0 | 0.5 | 1.6 | | 1/2 | 3/2 | 2.56 |
| | 1/2 | 5/2 | | 2.6 | 3.0 | 1.2 | 2.2 | 2.0 | | 1/2 | 3/2 | 2.56 |
| | (5/2)' | 3/2 | | 0.4 | 0.5 | 0.1 | 0.3 | 0.3 | | 1/2 | 3/2 | 2.56 |
| | (5/2)' | 5/2 | | 0.6 | 0.5 | 1.2 | 0.6 | 1.3 | | 1/2 | 3/2 | 2.56 |
| | (3/2)' | 3/2 | | 0.6 | 1.2 | 3.3* | 1.4 | 0.5 | | 1/2 | 3/2 | 2.56 |
| | (3/2)' | 5/2 | | 0.1 | 0.0 | 1.9* | 0.1 | 0.1 | | 1/2 | 3/2 | 2.56 |
| | (3/2)' | 1/2 | | 1.8 | 1.9 | 1.7 | 3.0 | 1.6 | | 1/2 | 3/2 | 2.56 |
| | (1/2)' | 3/2 | | 0.5 | 0.4 | 0.5 | 0.2 | 0.6 | | 1/2 | 3/2 | 2.56 |
| | (1/2)' | 5/2 | | 0.3 | 0.3 | 0.5 | 0.4 | 0.6 | | | | |

**TABLE XX** (*Continued*)

| Nucleus | $J_i$ | $J_f$ | $B(E2)$ Expt.$^a$ | Calculated $B(E2)^c$ | | | | | Rotational model | | | $B(\tilde{e}')/B(e)$ |
|---|---|---|---|---|---|---|---|---|---|---|---|---|
| | | | | $K+^{17}O$ | $K+SPE$ | $K+12FP$ | $RIP^b$ | MSDI | $(C)^2$ | $K_i$ | $K_f$ | $K+^{17}O$ |
| $^{20}$Ne | 0′ | 2 | [12.3] | 9.8 | 13.7* | 1.9* | 0.0* | 10.1 | 1.0 | 0′ | 0 | 1.00 |
| | 2′ | 0 | | 0.1 | 0.1 | 0.0 | 0.0 | 0.0 | 0.2 | 0′ | 0 | 1.00 |
| | 2′ | 2 | | 2.6 | 4.0 | 1.8 | 0.6* | 4.9* | 0.3 | 0′ | 0 | 1.00 |
| | 2′ | 4 | | 0.7 | 0.9 | 0.6 | 0.1 | 0.0 | 0.5 | 0′ | 0 | 1.00 |
| | 4′ | 2 | [21.3 ± 2.8] | 2.8 | 3.5 | 3.3 | 0.2* | 3.9 | 0.3 | 0′ | 0 | 1.00 |
| | 4′ | 4 | | 0.0 | 0.0 | 0.0 | 1.3 | 4.3* | 0.3 | 0′ | 0 | 1.00 |
| | 4′ | 6 | | 8.1 | 10.3 | 5.4* | 2.4* | 0.1* | 0.5 | 0′ | 0 | 1.00 |
| | 2′ | 0′ | | 16.2 | 14.0 | 10.6* | 8.3* | 7.8* | 0.2 | 0′ | 0′ | 1.00 |
| | 4′ | 2′ | | 0.1 | 0.4 | 5.6* | 20.0* | 5.5* | 0.3 | 0′ | 0′ | 1.00 |
| | 2″ | 2 | | 0.0 | 0.1 | 0.1 | 0.1 | 0.0 | 0.3 | 2 | 0 | 1.00 |
| | 2″ | 4 | | 0.7 | 1.1 | 0.0 | 0.0 | 2.0 | 0.01 | 2 | 0 | 1.00 |
| | 2″ | 2′ | | 12.5 | 2.8* | 2.9* | 7.3* | 0.0* | 0.3 | 2 | 0′ | 1.00 |
| $^{20}$F | 1 | 2 | | 8.9 | 10.2 | 6.9 | 17.0* | 16.8* | 0.3 | 1 | 2 | 1.35 |
| | 1 | 3 | | 11.1 | 10.7 | 5.2* | 12.9 | 10.8 | 0.7 | 1 | 2 | 1.35 |
| | 2′ | 2 | | 7.5 | 8.9 | 5.4 | 9.3 | 12.7* | 0.4 | 1 | 2 | 1.34 |
| | 2′ | 3 | | 0.2 | 0.3 | 0.3 | 0.5 | 0.1 | 0 | 1 | 2 | 2.17 |
| | 2′ | 4 | | 3.6 | 3.1 | 5.5* | 2.9 | 4.8 | 0.6 | 1 | 2 | 1.55 |
| | 3′ | 2 | | 5.4 | 4.2 | 3.2* | 5.4 | 10.9* | 0.2 | 1 | 2 | 1.26 |
| | 3′ | 3 | | 3.5 | 2.4 | 1.4* | 3.7 | 0.0* | 0.3 | 1 | 2 | 1.62 |
| | 3′ | 4 | | 0.3 | 0.0 | 0.4 | 0.5 | 1.7 | 0.03 | 1 | 2 | 1.04 |
| | 3″ | 4 | | 0.5 | 1.9 | 1.7 | 0.1 | 0.4 | 0.2 | 1 | 2 | 2.56 |
| | 5 | 3′ | | 0.1 | 0.0 | 0.7 | 0.8 | 1.5 | 0.3 | 2 | 1 | 0.91 |

| Nucleus | | | | | | | | | | | |
|---|---|---|---|---|---|---|---|---|---|---|---|
| | 2' | 1 | 20.6 | 19.2 | 19.7 | 17.4 | 13.9* | | 1' | 1 | 1.26 |
| | 3' | 1 | 5.9 | 6.3 | 7.6 | 5.4 | 2.0* | | 1' | 1 | 1.37 |
| | 3'' | 1 | 4.0 | 2.8 | 2.5 | 3.5 | 2.5* | | | 1 | 1.33 |
| | 3' | 2' | 10.6 | 12.2 | 13.6 | 7.4* | 8.4 | | 1' | 1 | 1.22 |
| | 1' | 2 | 0.1 | 0.0 | 0.1 | 0.0 | 0.0 | 0.3 | 1' | 2 | 4.74 |
| | 1' | 3 | 0.8 | 0.0 | 0.1 | 0.1 | 1.5 | 0.7 | 1' | 2 | 0.15 |
| | 1' | 1 | 1.5 | 2.0 | 5.3* | 4.4* | 1.0 | | 1' | 1 | 1.69 |
| | 1' | 2' | 2.6 | 2.5 | 2.9 | 3.4 | 1.6 | | 1' | 1 | 0.83 |
| | 1' | 3' | 8.4 | 5.3* | 0.5* | 20.2* | 2.5* | | 1' | 1 | 1.15 |
| $^{20}$O | 2' | 0 | 0.6 | 0.8 | 1.9 | 0.6 | 0.4 | 0.2 | 2 | 0 | 2.56 |
| | 2' | 2 | 0.6 | 0.0 | 1.1 | 0.4 | 1.6 | 0.3 | 2 | 0 | 2.56 |
| | 4 | 2' | 1.0 | 1.4 | 0.1 | 0.9 | 2.0 | 0.01 | 0 | 2 | 2.56 |
| | 0' | 2 | 0.0 | 0.1 | 0.3 | 0.1 | 0.2 | | 0' | 0 | 2.56 |
| | 0' | 2' | 6.1 | 7.0 | 2.2* | 5.1 | 4.1* | | 0' | 2 | 2.56 |
| $^{21}$Ne | 1/2 | 5/2 | 1.5 | | | | | | 1/2 | 3/2 | 0.06 |
| | (5/2)' | | 0.1 | | | | | | 1/2 | 3/2 | |
| | (7/2)' | 3/2 | 0.2 | | | | | | 1/2 | 3/2 | |
| | (9/2)' | 5/2 | 0.5 | | | | | | 1/2 | 3/2 | |
| | (1/2)' | 5/2 | 2.7 | | | | | | (1/2)' | 3/2 | |
| $^{22}$Na | 1 | 3 | 5.5 | | | | | [0.03 ± 0.01] | | 3 | 1.00 |
| | 3' | 3 | 0.2 | | | | | | | 3 | 1.00 |
| | 3' | 1 | 66.0 | | | | | | | 3 | 1.00 |
| $^{22}$Ne | 3 | 2' | 83.8 | | | | | 0.36 | | 2 | |
| | 4' | 2' | 25.9 | | | | | 0.12 | | 2 | |
| | 4' | 3 | 56.1 | | | | | 0.27 | | 2 | |

a The experimental results for $^{19}$F are from (106); all others are from (104). The square brackets are ours; they indicate that we suspect poor correlation between either the initial or final observed state, and a shell-model state employed to represent it in our B(E2) calculations for that line of the table.

b The Hamiltonian RIP' is very close to the Hamiltonian RIP. See footnote b to Table XIX.

c B(E2) calculated with $\bar{e}_p = 1.5e$, $\bar{e}_n = 0.5e$, $b^2 = 1.03A^{1/3}$ fm$^2$.

correlation with our shell-model states. Thus, for $^{18}$F we have already explained that our comparison of measured and shell-model *spectroscopic factors* indicates that we can offer no *s–d*-shell-model partner for the lowest observed 2$^+$ state, and no *s–d*-shell-model partner for the second observed 1$^+$ state. For $^{18}$O, it is the comparison between shell-model and measured $B(E2)$ values which argues against correlation of the second observed 0$^+$ level with the second shell-model 0$^+$ level. For $^{19}$F, the two bracketed $B(E2)$ values are for transitions involving the second observed 7/2$^+$ state. Here again the poorness of experimental–theoretical correlation is deduced from $B(E2)$ comparisons. These $^{19}$F comparisons imply that the lowest two experimentally observed 7/2$^+$ states share the ground-state-band character in a way that is not reproduced by the $K+^{17}O$ model. (We shall discuss $^{19}$F more thoroughly, below.) In $^{20}$Ne, there are several observed 4$^+$ states near the excitation of the second shell-model 4$^+$ state (see the caption to Fig. 5). This multiplicity in itself raises doubts about the correlation of the second observed 4$^+$ state with the second shell-model 4$^+$ state; and then the disagreement between shell-model and measured $B(E2)$ strengths argues more definitely against this correlation. In $^{22}$Ne, the calculated excitation energies of the two lowest 1$^+$ levels do not agree with the observed energies, and this disagreement casts doubt on the correlation of experiment with theory for these 1$^+$ states.

Transitions involving the 7/2$^+$ states in $^{19}$F deserve more discussion. As we have already noted (in Section 3.3): in $^{19}$F the experimentally observed spectrum, and each of our shell-model spectra, show *two* 7/2$^+$ states in the neighborhood of excitation expected for a ground-state-band 7/2$^+$ member. In the experimental spectrum we identify the second observed 7/2$^+$ state as being the ground-state-band member; that is because it decays strongly to the 3/2$^+$ ground-state-band member, whereas the lower 7/2$^+$ level exhibits no branching to that 3/2$^+$ member. For the $K+^{17}O$ model we calculated $B(E2)$ strengths involving both the first and second 7/2$^+$ states. We identify the lowest $K+^{17}O$ 7/2$^+$ state as the ground-state-band member; that is because its calculated $B(E2)$ to the 3/2$^+$ ground-state-band member is nine times the corresponding $B(E2)$ from the next $K+^{17}O$ 7/2$^+$ state. For our other shell models we calculated $B(E2)$ values involving the lowest 7/2$^+$ state, but we did not calculate $B(E2)$ values from the second 7/2$^+$ state. However, we assume that in each model, the two lowest 7/2$^+$ states share a total 7/2$^+$ → 3/2$^+$ $B(E2)$ strength that is roughly equal to the sum of $B(E2)$ strengths calculated for the two lowest $K+^{17}O$ 7/2$^+$ states. Then from comparison with the $K+^{17}O$ results, we conclude that in the $K+SPE$, $K+12FP$, and $RIP$ shell models, the second 7/2$^+$ state has almost all of the ground-state-band character, while in the *MSDI* shell model both of the lowest 7/2⁺ states have a significant share of ground-state-band character.

In Table XIX we use labels 7/2$^g$ to denote the ground-state-band 7/2⁺ state. For these labels 7/2$^g$, we have entered the $B(E2)$ strengths associated with the second observed 7/2⁺ state and the first $K+^{17}O$ 7/2⁺ state. And since we consider that both these states deserve the label 7/2$^g$, there are no square brackets around the measured $B(E2)$ values involving 7/2$^g$. But in Table XX we use labels 7/2 and 7/2′ to denote the first and second 7/2⁺ states, respectively, these labels carrying no implication whatsoever about ground-state-band character. Hence in Table XX we have entered, straightforwardly, all of the pertinent shell-model $B(E2)$'s which we calculated. And since the observed 7/2′ state is not well correlated with the $K+^{17}O$ 7/2′ state, and not well correlated with the *MSDI* 7/2′ state, Table XX shows square brackets around the measured $B(E2)$ values involving 7/2′.

### 5.2.3. *E2 Results: K+SPE, K+12FP, RIP′, and MSDI versus Experiment*

In the foregoing subsection we compared $K+^{17}O$ $E2$ results with measured $E2$ results. Now we extend this experimental–theoretical comparison by bringing into consideration some $E2$ results calculated for $A = 18$–20 from our least-square models *K+SPE*, *K+12FP*, *RIP′*, and *MSDI*. In general, we find that $E2$ results from the unmodified realistic Hamiltonian $K+^{17}O$ agree just as closely with experimental data as do $E2$ results from the Hamiltonians *K+SPE*, *K+12FP*, *RIP′*, and *MSDI* (even though the latter four Hamiltonians were adjusted specifically to fit experimentally observed energies). Some details are mentioned in the next few paragraphs.

First we consider quadrupole moments. Table XVII includes four measured $Q$ moments in comparison with $K+^{17}O$ results. For two of these cases, we have results from our alternative Hamiltonians too. In $^{17}O$, the $J = 5/2$ quadrupole moment is the same for *all* *s–d*-shell models, provided that $Q^2$ is kept constant. In $^{19}F$, the lowest $J = 5/2$ quadrupole moment was calculated for four of our least-squares Hamiltonians, and as Table XVIII indicates, this moment was found to be insensitive to the choice among our model Hamiltonians.

Next we refer to Table XIX, which shows $B(E2)$ strengths for transitions involving only the lowest few levels in $A = 17$–18, and transitions within the ground-state bands of $A = 19$–20. For most of these transitions, the $B(E2)$ values computed from our various shell models are quite similar to each other, and the $K+^{17}O$ result generally agrees with experiment at least as well as our other model-results do. (The $^{19}$F transitions involving 7/2 states are special cases. But these transitions have already been discussed above, for our least-square Hamiltonians as well as for $K+^{17}O$.)

Finally, we refer to Table XX, which shows transitions involving levels outside the lowest triads in $A = 18$, and levels outside the ground-state bands in $A = 19$–20. Even for a single transition the experimental–theoretical comparisons in Table XX rarely favor any one of our shell models. Sometimes, for a given transition, the experimental–theoretical $B(E2)$ comparison, taken by itself, would favor one particular Hamiltonian; but often this favoring is nullified because there is other information indicating poor correlation for all models. Details are given in the next paragraph.

All of the measured $B(E2)$ values in Table XX are square-bracketed. These brackets indicate that, because of $B(E2)$ *or other* information, we suspect that either the initial or the final observed state is in poor correlation with at least one shell-model state for that line of the table. Of the nine square-bracketed results for $A = 18$–20, four are $^{18}$F results associated with correlations known to be poor for *all* our shell models, from spectroscopic-factor considerations. (Note that for $1' \rightarrow 2$ in $^{18}$F, all our shell models yield gross disagreement with the measured $B(E2)$ strengths. This disagreement confirms the generally poor experimental–theoretical correlation for one or both of the $^{18}$F levels $1'$ and 2.) In each of the five remaining square-bracketed cases in Table XX, the $E2$ results themselves indicate questionable correlation between observed and shell-model levels. In two of these five remaining cases, $0' \rightarrow 2$ in $^{18}$O, and $4' \rightarrow 2$ in $^{20}$Ne, *all* our shell models are similar in that they give gross disagreement with the quoted experimental $B(E2)$. In the other three cases, $0' \rightarrow 2$ in $^{20}$Ne, $7/2' \rightarrow 3/2$ in $^{19}$F, and $7/2' \rightarrow 3/2$ in $^{19}$F, the goodness of experimental–theoretical correlation does seem to depend on the choice among our Hamiltonians. For the $0' \rightarrow 2$ transition in $^{20}$Ne, this last-mentioned conclusion is suggested straightforwardly by the $B(E2)$ entries in Table XX, while for the $^{19}$F cases we refer to our previous discussions.

## 5.3. Comparison of $E2$ Results from Different Shell Models

Here we use Tables XVIII–XX to discuss the sensitivity of $E2$ results to the shell-model Hamiltonian. (In Section 5.2 we considered this sensitivity in connection with comparisons to measured data; but now we shall widen the discussion to include $E2$ moments and strengths for which there is no measured data.) Our conclusions in this section can be summarized briefly.

The sensitivity of the $E2$ results to the Hamiltonian parallels that of the excitation energies. Calculated results involving only the ground-state band show little sensitivity to the Hamiltonian, while results involving states outside the ground-state band tend to be more sensitive.

In Tables XVIII–XX, the $E2$ results which are sensitive to the shell-model Hamiltonian can be spotted immediately, for we have emphasized *with stars* (*) all shell-model values which differ substantially from the corresponding $K+^{17}O$ result. (The precise criteria for starring these numbers will be given below.) Table XIX lists $E2$ transitions involving only the lowest few levels in $A = 17$ and 18, and the ground-state-band levels in the heavier nuclei. For these transitions there are very few starred numbers, i.e., there is a general insensitivity to the choice among our model Hamiltonians. Table XX lists $E2$ transitions involving at least one of either the lowest few levels in $A = 18$, and levels outside the ground-state bands in heavier nuclei. Here there are many starred numbers, i.e., there is considerable sensitivity to the choice among our model Hamiltonians. In short, calculations involving only the ground-state band show little sensitivity to the Hamiltonian, while results involving states outside the ground-state band tend to be more sensitive.

Most of the entries in Table XIX are for $E2$ transitions within ground-state bands. Hence the $B(E2)$ strengths in Table XIX are generally larger than those in Table XX. However, the starred numbers in Table XX are not confined to very weak transitions. See for example the $2 \rightarrow 1$ transition in $^{18}F$, the $2' \rightarrow 0'$ and $4' \rightarrow 6$ transitions in $^{20}Ne$, and the $1 \rightarrow 2$ transition in $^{20}F$.

Some exceptions to our generalizations about sensitivity will be discussed below. But for the record, we first state our criteria for starring numbers in Tables XVIII–XX. For nuclei with at least one *s–d*-shell proton: we star those shell-model moments $Q$ whose difference from the $K+^{17}O$ result exceeds 15% or 0.5 e fm² , whichever is larger; and we star those shell-model $B(E2)$ strengths whose difference from the $K+^{17}O$ result exceeds 30% or 1.5 e²fm⁴, whichever is larger. The oxygen isotopes have $\mathbf{Q}^2$ matrix elements which are systematically smaller, because they are proportional to $\tilde{e}_n$ rather than being approximately proportional to $(\tilde{e}_n + \tilde{e}_p)/2$. Hence for oxygen isotopes, the aforementioned criteria are changed to 15% or 0.25 e fm² for $Q$ moments, and 30% or 0.4 e²fm⁴ for $B(E2)$ strengths. When the $K+^{17}O$ result is not available, we use $K+SPE$ as our standard of comparison.

In Table XIX there are only three transitions for which starred entries appear, the $8 \rightarrow 6$ transition in $^{20}Ne$, the $3 \rightarrow 2$ transition in $^{20}F$, and the

$4 \to 2$ transition in $^{20}$O. These constitute *exceptions* to the general rule that, for transitions within ground-state bands, $B(E2)$ strengths are insensitive to the choice among our model Hamiltonians. In all three of these exceptional cases, only the *MSDI* $B(E2)$ value is starred. As Table XIX shows, the *MSDI* Hamiltonian gives ground-state-band $B(E2)$ values which are generally smaller than those given by our other Hamiltonians. In the three cases of sensitivity under discussion, the generally weaker *MSDI* values drop more than 30% below the corresponding $K+^{17}O$ result.

There are a few other cases, not evident from Table XIX, which constitute additional exceptions to the general rule of insensitivity within ground-state bands. Among the $^{19}$F entries in Table XX, there is great sensitivity of $B(E2)$ strengths for the $7/2 \to 3/2$ transition. As previously discussed, this sensitivity is associated with the transfer of ground-state-band character from one $7/2^+$ state to another, according to the Hamiltonian assumed. All the other exceptions are among the $Q$-moment results, exhibited in Table XVII. Here we see sensitivity for several ground-state-band $Q$ moments in $^{19}$O and $^{20}$O, and here the $K+12FP$ model, as well as the *MSDI* model, shows some substantial deviations from the $K+^{17}O$ results. These sensitivities are somewhat surprising, in view of the apparent insensitivity of $B(E2)$ values for ground-state bands in $^{19}$O and $^{20}$O. Perhaps these quadrupole moments are more sensitive because the oxygen isotopes are less "rotational" than the nuclei with $N \approx Z$. (For hints about this less rotational behavior, see pp. 233–234 of (*12*), and the paper of Flores and Perez in (*11*).]

## 5.4. Comparison with *E*2 Results from Rotational Models

Here we compare our shell-model $E2$ results, and measured $E2$ results, with $E2$ results from models which *presuppose rotational-band structure*.

Most of our comparisons will involve the familiar strong-coupling rotational limit of the unified model (*118*). We shall refer to this strong-coupling limit as the "adiabatic-rotational" model, or simply, the "adiabatic" model. The general version of this model leads to estimates for various ratios of $Q$ moments and ratios of $B(E2)$ strengths. In order to calculate some absolute values too, we shall make use of a "specialized" adiabatic-rotational model, to be described later in this subsection.

Between our shell-model results and adiabatic-model results, there is good agreement for most $E2$ transitions and $Q$ moments within ground-state bands. Can the same be said for agreement between measured and adiabatic results? Not quite. For $A = 18$–$22$, experiments do indicate generally strong $E2$ transitions within ground-state bands. But for the measured $B(E2)$'s, there are large uncertainties attached to the "most-probable" values; these most-probable values do not bear the striking resemblance to adiabatic results that our shell-model results bear; and for a number of the pertinent transitions, no experimental values are available. For $Q$ moments the same remarks apply. In short, for ground-state-band $E2$ observables in $A = 18$–$22$, the agreement between shell model and the adiabatic model is generally *good*, but the agreement between experiment and adiabatic model is generally *rough* (just as the agreement between shell model and experiment is only *rough*).

The simplicity of the adiabatic model allows us to compute, very easily, a large set of rotational results with which to compare measured and shell-model results. Now in any adiabatic-rotational model, the wave functions depend in part on a collective angular coordinate $\Omega$, as well as on microscopic (particle) coordinates. But it has been shown analytically that certain kinds of wholly-microscopic shell models, for example $SU_3$ shell models, yield energy matrix elements and transition matrix elements quite similar to those given by the adiabatic-rotational model (*7–9, 12–14*). Thus, when one takes $SU_3$ wave functions as models for the low-lying nuclear states, one is *presupposing* a kind of rotational-band structure. Hence we classify $SU_3$ shell models as rotational; and we shall make some comparisons to $E2$ results calculated from their wave functions too. Since $SU_3$ results are usually harder to compute than adiabatic results, we shall make only a few comparisons with $SU_3$ results. These few comparisons will help us to explain some of the exceptions which we find to the general rule of similarity between our shell-model and adiabatic-model $E2$ results.

### 5.4.1. *General Features of the Adiabatic-Rotational Model*

We first write down some familiar features of the adiabatic-rotational approximation for axially symmetric deformed nuclei. According to this approximation, the rotational motion of the deformed nucleus is slow enough so that it does not perturb the "intrinsic" motion of the nucleons. Each model wave function may be written as (*118*)

$$| \alpha J K, J_z \rangle = \left[ \frac{(2J + 1)}{16\pi^2(1 + \delta_{K0})} \right]^{1/2}$$

$$\left[ D^J_{J_z, K}(\Omega)\chi_{K\alpha} + (-)^{J - J_{\mathrm{op}}} D^J_{J_z, -K}(\Omega)\chi_{-K\alpha} \right] \quad (17)$$

Here $J$ is the total angular momentum of the state, $J_z$ is the projection of $J$ on the space-fixed $z$-axis, $K$ is the projection of $J$ on the symmetry axis of the intrinsic state, and $\alpha$ specifies all other quantum numbers of the intrinsic state $\chi$. [We include $T$ and $T_z$ in $\alpha$. The exponent $J_{op}$ is an operator. For further explanation of (17), see (118).] The *intrinsic quadrupole moment* $Q'$ is defined as

$$Q'(\alpha, K) = \left(\frac{16\pi}{5}\right)^{1/2} \langle \chi_{K\,\alpha} | Q_\zeta^2 | \chi_{K\,\alpha} \rangle \tag{18}$$

where $Q_\zeta^2$ is that component of the electric quadrupole operator $\mathbf{Q}^2$ along the symmetry axis of the intrinsic state $\chi_{K\alpha}$. The quadrupole moment of a nuclear state, in the adiabatic-rotational model, is

$$Q(\alpha JK) = C_{J\,0\,J}^{J\,2\,J} C_{K\,0\,K}^{J\,2\,J} Q'(\alpha K) \equiv \frac{3K^2 - J(J+1)}{(J+1)(2J+3)} Q'(\alpha K) \tag{19}$$

and the $B(E2)$ between two states within the same rotational band is

$$B(E2)_{i \to f} = [C_{K\,0\,K'}^{J_i\,2\,J'}]^2 \frac{5}{16\pi} (Q')^2 \qquad \text{if } K = 0 \text{ or } K > 1 \tag{20}$$

Equation (20) is a special case of the result for transitions between any two bands,

$$B(E2)_{i \to f} = [C_{K_i\,K_f - K_i\,K_f'}^{J_i\,2\,J_f'}]^2 g(\alpha_i K_i, \alpha_f K_f) \qquad \begin{aligned} &\text{if } K_i = 0 \text{ or } K_f = 0 \\ &\text{or } (K_i + K_f) > 2 \end{aligned} \tag{21}$$

Here the function $g(\alpha_i K_i, \alpha_f K_f)$ depends on the initial and final bands, but not on the $J_i$ and $J_f$ within these bands. This function $g$ is symmetric under interchange of its arguments $\alpha_i K_i$ and $\alpha_f K_f$.

From the expressions (19)–(21), one can determine relative $Q$ values for states within the same rotational band, relative $B(E2)$ values for transitions within the same band, and relative $B(E2)$ values for transitions within a set connecting two different bands. But, given a way to estimate the intrinsic quadrupole moment $Q'$, one could use (19)–(20) to calculate absolute $Q$ and $B(E2)$ values within a band.

### 5.4.2. A Specialized Adiabatic-Rotational Model

Here we describe the set of assumptions which we have used to estimate $Q'$ for ground-state bands. These assumptions were suggested to us by K. H. Bhatt.

The basic premises are:

(a) that the intrinsic wave function $\chi_{K\alpha}$ is a determinant describing the occupation of $A - 16$ definite single-nucleon $1s$–$0d$ states;

(b) that the occupied states are Hartree–Fock states (the lowest-energy Hartree–Fock orbit being filled first, the second-lowest-energy orbit being filled next); and

(c) that each of these single-nucleon states has expectation value $\langle r^2 Y_0^2 \rangle$ equal to an *eigenvalue* of $r^2 Y_0^2$ in the $s$–$d$ shell.

For $^{20}$Ne and $^{21}$Ne, it has been found that Hartree–Fock calculations with $A - 16$ active $s$–$d$-shell particles, and with "reasonable" nucleon–nucleon forces, yield axially symmetric solutions in which states of the lowest few orbits satisfy (c) closely (*12, 119, 120*). Assumption (c) involves the single-nucleon *mass* quadrupole operator. To discuss the eigenvalues and expectation values of this operator, we consider a dimensionless quantity $\varepsilon$ defined so that, for a single-nucleon state,

$$(16\pi/5)^{1/2}\langle r^2 Y_0^2 \rangle = \varepsilon b^2 \tag{22}$$

Then in the vector space of single-nucleon $1s$–$0d$ harmonic-oscillator functions, the eigenvalues of $r^2 Y_0^2$ correspond (*8*) to $\varepsilon = 4, 1, 1, -2, -2, -2$. Each of these six eigenvalues is four-fold degenerate, because single-nucleon spin and isospin projections do not affect $\langle r^2 Y_0^2 \rangle$. In the $A \approx 20$ Hartree–Fock solutions, the lowest four active single-nucleon states have $\varepsilon \approx 4$, with $j_\zeta \equiv \pm k = \pm 1/2$ and $t_0 = \pm 1/2$. The next-lowest states have $\varepsilon \approx 1$ with $j_\zeta = \pm 3/2$ and $t_0 = \pm 1/2$. Thus for ground-state bands in $A = 18$–22, it is reasonable to form an $s$–$d$-shell $\chi$ in the following way. First, place as many of the active nucleons as possible, i.e., as many as are allowed by the Pauli principle, in the $\varepsilon = 4$ orbit. Then place all remaining nucleons in the $\varepsilon = 1$, $k = 3/2$ orbit (if that is allowed by the Pauli principle). To calculate $Q'$ for such an intrinsic state $\chi$, we assume effective charges $\tilde{e}_n = 0.5e$ and $\tilde{e}_p = 1.5e$. Then the effective electric quadrupole moment of each active nucleon $k$ is $\tilde{e}_k \varepsilon_k b^2$; and the total intrinsic quadrupole moment $Q'$ is a sum over the contributions from $A - 16$ active nucleons:

$$Q' = \sum_{k=1}^{A-16} \tilde{e}_k \varepsilon_k b^2$$

Here is an example. For the ground-state band in $^{21}$Ne, we choose $\chi$ to be a determinant describing two protons in the $\varepsilon = 4$ orbit, two neutrons in the $\varepsilon = 4$ orbit, and one neutron in the $\varepsilon = 1$, $k = 3/2$ orbit. Then the intrinsic electric quadrupole moment is estimated to be

$Q'(^{21}\text{Ne, ground-state band})$
$$= [(2 \times 1.5e \times 4b^2) + (2 \times 0.5e \times 4b^2) + (1 \times 0.5e \times b^2)]$$
$$= 16.5eb^2$$

Similarly, in $^{19}$F the ground-state band is estimated to have $Q' = 10eb^2$, and in $^{22}$Ne the ground-state band is estimated to have $Q' = 17eb^2$.

When $Q'$ is estimated as just described, it has the maximum value allowed by the Pauli principle for an intrinsic state in which $A - 16$ active nucleons are all restricted to the $1s$–$0d$ shell. Thus, we have described a specialized adiabatic-rotational model which *maximizes* ground-state-band $Q$ moments and $B(E2)$ strengths, subject to the condition that $\chi$ is a $1s$–$0d$-shell function.

### 5.4.3. *Comparisons with Rotational-Model E2 Results in Ground-State Bands*

Tables XVII–XIX show some ground-state-band $Q$ moments and $B(E2)$ strengths calculated according to the above-described specialized adiabatic-rotational model. In order to apply this model, we have assumed the quantum numbers $K$ listed in the tables. (In some cases, especially outside ground-state bands, the listed $K$ values may be poor assignments for the observed or shell-model states on the same line of the table.)

We start with a brief comparison of experimental and adiabatic $E2$ results. Tables XVII–XIX illustrate our earlier remarks concerning the scanty, uncertain evidence for deviations between adiabatic and experimental results. Tables XVII–XIX show that there are no cases in $A = 18$–$22$ where measured values exist for a series of $Q$ moments within the same ground-state band, and there are rather few cases where measured values exist for a series of $B(E2)$ values within the same ground-state band. Where such $B(E2)$ series *have* been measured, the ratios of "most-probable" $B(E2)$ values, as determined from experiment, often deviate considerably from adiabatic-rotational ratios. However, the quoted experimental uncertainties are usually large, so large that they admit the possibility of consistency between experimental and adiabatic ratios. In some cases the disagreements exceed the quoted experimental uncertainties; but, because we know that experimental uncertainties are often underestimated, we do not know how seriously one should take this experimental evidence for deviations from adiabatic-rotational behavior. There is one kind of deviation which has excited particular interest lately. It involves the relation between $B(E2;$

$2^+ \to 0^+$) and $Q(2^+)$ within a $K = 0$ ground-state band. For such bands, consider the two quantities $-7Q(2^+)/2$ and $[16\pi B(E2; 2^+ \to 0^+)]^{1/2}$. According to the adiabatic model, both these quantities are equal to the intrinsic quadrupole moment $Q'$, and therefore the adiabatic model predicts their ratio as unity. Our $K+^{17}O$ shell model yields a ratio 1.02 for $^{20}$Ne, but the experimental data of (*107*) yields a ratio 1.22 $\pm$ 0.15 for $^{20}$Ne, and 1.28 $\pm$ 0.20 for $^{22}$Ne.

In the remainder of this section, we shall compare shell-model and adiabatic $E2$ results within ground-state bands. Tables XVII–XIX show that in general, there is very striking similarity between ground-state-band $E2$ results calculated from our shell-model wave functions, and ground-state-band $E2$ results calculated from our specialized adiabatic-rotational model. Notice that in several cases where our $K+^{17}O$ shell-model results disagree with measured results, these $K+^{17}O$ do agree approximately with the adiabatic-rotational results. Examples are the quadrupole moment of the first-excited state in $^{20}$Ne (Table XVII) and the $B(E2)$ strengths in $^{21}$Ne (Table XIX).

Most of our discussion here will center on *exceptions* to the general rule of striking similarity between adiabatic-model and $K+^{17}O$ shell-model results. One exception appears among the $B(E2)$'s for $^{20}$Ne. For this nucleus, the rotational model agrees with measured results in giving a ratio $B(E2, 6 \to 4)/B(E2, 2 \to 0)$ that is distinctly greater than unity; but all our shell models give a ratio $\approx 1$. Other notable deviations between our $K+^{17}O$ and adiabatic-rotational results occur for: the $13/2 \to 9/2$ and $11/2 \to 7/2$ $B(E2)$ in $^{19}$F; the $5 \to 4$ $B(E2)$ in $^{20}$F; the $4 \to 2$ $B(E2)$ in $^{20}$O; the $Q$ for $J = 9/2$ in $^{19}$O; the $Q$ for $J = 3$ in $^{20}$F; the $Q$ for $J = 4$ in $^{20}$O; and the $Q$ values for $J = 7/2$ and $J = 11/2$ in $^{21}$F. In the discussion to follow, we shall speculate on reasons for the pattern of agreement and disagreement seen in our comparison between shell-model and rotational-model results.

When we use the term "rotational band," we imply that states within the same band, while differing in their total angular momentum $J$, are all associated in some systematic way with a common intrinsic state $\chi$. The adiabatic-rotational model, in which wave functions depend explicitly on a collective angular coordinate $\Omega$, may be viewed as an approximation to a "*J*-projection" model. This is a *microscopic* model in which states belonging to the same band have multiparticle shell-model wave functions all formed by projecting good-*J* parts out of the same axially symmetric deformed intrinsic state (*7–9, 12–14*).[14] We presume that, at least for some

---

[14] Here, and throughout our article, the word *projecting* implies *projecting and normalizing*.

nuclei, our $K+^{17}O$ ground-state-band wave functions could be approximately reproduced by projecting the good-$J$ parts out of axially symmetric intrinsic states $\chi'(K+^{17}O, A, J)$; and similarly for our other shell models. Thus, when we find that one of our shell-model $E2$ results differs substantially from our specialized-adiabatic result, then we presume the reason to be that

(a) the intrinsic state $\chi$ in our specialized adiabatic model differs substantially from the intrinsic state $\chi'$ that is operative in our shell model; or

(b) although the $\chi$ used in our specialized adiabatic model is a good choice, the adiabatic-model $E2$ formulas give a poor approximation to the result that would be obtained by using model states "$J$-projected" from $\chi$; or

(c) the relevant shell-model states are not related to each other in such a way that they could be well approximated by $J$-projections from one common axially symmetric intrinsic state.

We shall next explain why we suspect that, within ground-state bands, point (b) plays an important role in producing exceptions to the general rule of similarity between our shell-model and adiabatic-rotational $E2$ results.

If *only* point (a) were responsible for the deviations between our shell-model and specialized-adiabatic results, then there would be differences in absolute values, but the two models would give similar ratios of $Q$ moments within a band, similar ratios of $B(E2)$ strengths for transitions within a band, etc. But we find that within ground-state bands, the few prominent deviations between our shell-model and adiabatic results involve disagreements in relative as well as absolute values. For example, in $^{20}$Ne the adiabatic model gives monotonically increasing $B(E2)$ strengths as one proceeds upward through the $2 \to 0$, $4 \to 2$, $6 \to 4$, and $8 \to 6$ transitions; but our shell-model $B(E2)$ strengths increase from $2 \to 0$ to $4 \to 2$, and then turn around to decrease monotonically. Also, remember that we chose our specialized adiabatic-rotational model to yield the maximum $Q$ and $B(E2)$ values possible within the framework of an adiabatic model with $1s$–$0d$-shell intrinsic states. However, our $s$–$d$-shell model answers very often exceed these adiabatic maxima. Thus, we see that an $s$–$d$ adiabatic model is often inadequate for estimating the maximum $Q$ and $B(E2)$ allowed by an $s$–$d$ shell model; and more generally, we conclude that at least *some* of the deviation between our shell-model and adiabatic-rotational results stems from point (b)—the difference between adiabatic-rotational results and *J*-projected shell-model results, when both are obtained from the same intrinsic state.

This "inadequacy" of the adiabatic-rotational model can be easily demonstrated for $B(E2)$ transitions in $^{20}$Ne. For this nucleus, the intrinsic state which we use in our specialized adiabatic-rotational model is identical to the *s–d*-shell $SU_3$ state of maximum weight (*8*); the good-*J* projections from this intrinsic state are simply the $J = 0, 2, 4, 6, 8$ wave functions of the $(\lambda\mu) = (80)$ representation; and we can easily produce the $Q$ moments and $B(E2)$ values for these (80) states. Thus for $^{20}$Ne, we can quantitatively compare a set of adiabatic and exact-*J*-projection $E2$ results obtained from the same intrinsic state—an $SU_3$ intrinsic state. In Table XXI we list for comparison: the $^{20}$Ne $B(E2)$'s calculated from the $SU_3$ adiabatic-rotational model; the $B(E2)$'s calculated from the $SU_3$ *J*-projected model; and the $B(E2)$'s calculated from our $K+^{17}O$ shell model. Since the $SU_3$ adiabatic strengths deviate from the $SU_3$ *J*-projected strengths, we conclude that the general adiabatic approximation itself, and not just our choice of $\chi$, causes some of our specialized-adiabatic $B(E2)$ answers to disagree with our $K+^{17}O$ $B(E2)$ answers.

Table XXI shows that within the ground-state band of $^{20}$Ne the $K+^{17}O$ $B(E2)$'s, while not at all proportional to the specialized-adiabatic $B(E2)$'s, *are* quite closely proportional to the $SU_3$ *J*-projected $B(E2)$'s. The $K+^{17}O$

**TABLE XXI.** $B(E2)$ **Strengths Involving the Lowest** $0^+$, $2^+$, $4^+$, $6^+$, $8^+$ **States in** $^{20}$Ne. All the results shown here were calculated from *s–d*-shell models with $\tilde{e}_p = 1.5e$, $\tilde{e}_n = 0.5e$, and $b^2 = 1.03A^{1/3}$ fm$^2$. The listed $B(E2)$'s are in units of $e^2$fm$^4$.

| $J_i \rightarrow J_f$ | $B(E2)$ from $(\lambda\mu) = (80)$ rotational models | | $B(E2)$ from $K+^{17}O$ |
|---|---|---|---|
| | Adiabatic[a] | *J*-projected[b,c] | |
| $2 \rightarrow 0$ | 39.8 | 54.7 | 48.1 |
| $4 \rightarrow 2$ | 56.9 | 69.3 | 59.7 |
| $6 \rightarrow 4$ | 62.6 | 58.7 | 48.8 |
| $8 \rightarrow 6$ | 65.6 | 34.8 | 31.2 |

[a] This model is described in Section 5.4.2.

[b] If one takes $SU_3$ (*80*) *J*-projections as models for the $J = 0, 2, 4, 6, 8$ states in $^{20}$Ne, then one is assuming a kind of "band" structure for $^{20}$Ne. Hence we classify the *J*-projected $SU_3$ shell model as a *rotational* model. For more extensive discussion of the rotational features of $SU_3$ states, see (*7–9*).

[c] The value listed for $B(E2; 2 \rightarrow 0)$ was calculated from the expression $b^4\lambda(\lambda + 3)/4\pi$. The ratios of the other $B(E2)$'s to $B(E2; 2 \rightarrow 0)$ were taken from Table 5.IIa of (*8*).

answers are $(86 \pm 3)\%$ of the $SU_3$ $J$-projected answers. (The $SU_3$ $Q$ *moments* for $^{20}$Ne will be discussed shortly.)

We shall next mention a few $SU_3$ results which indicate that in some cases the differences between adiabatic and $J$-projected $E2$ matrix elements lead to *fortuitous agreement* between our adiabatic-rotational and $K+^{17}O$ $E2$ matrix elements. The $SU_3$ results to be mentioned are all based on formulas and arguments in (7, 8, 121).

For the $2 \to 0$ transition in $^{20}$Ne, the $J$-projected $SU_3$ (80) model gives the maximum $B(E2)$ possible within the framework of any $s$–$d$ *shell* model, and our $SU_3$ (80) adiabatic-rotational model gives the maximum $B(E2)$ possible within the framework of any $s$–$d$ *adiabatic-rotational* model. We shall refer to the (80) *shell-model* maximum simply as "the $SU_3$ maximum." For the $^{20}$Ne $2 \to 0$ $B(E2)$, it can be shown that the $SU_3$ maximum is exactly $[(\lambda + 3)/\lambda] = 11/8$ times the (80) adiabatic-rotational answer. As Table XXI shows, our $K+^{17}O$ $2 \to 0$ $B(E2)$ falls short of the $SU_3$ maximum. This is because the $K+^{17}O$ $^{20}$Ne states mix in representations other than (80). In contrast, our specialized-adiabatic $B(E2)$ falls short of the $SU_3$ maximum *not* because our specialized-adiabatic $\chi$ mixes in other $(\lambda\mu)$ representations [indeed, $\chi$ is pure (80)], but because of a general inadequacy of the adiabatic approximation. Thus, we know that for this $2 \to 0$ $^{20}$Ne transition, the rough agreement between $K+^{17}O$ and specialized-adiabatic $B(E2)$ is partly fortuitous.[15]

For transitions between higher levels in the $^{20}$Ne ground-state band, we see from Table XXI that the ratio of (80)-shell-model $B(E2)$ to (80)-adiabatic-model $B(E2)$ does not stay near the ratio $11/8$ that holds exactly for the $2 \to 0$ transition. But for electric quadrupole moments $Q$, it can be shown that for *every* level in the $^{20}$Ne ground-state band, the ratio of (80)-shell-model $Q$ to (80)-adiabatic-model $Q$ is exactly $[(2\lambda + 3)/2\lambda] = 19/16$. Now suppose that, for the purpose of computing $E2$ matrix elements, our $K+^{17}O$ $^{20}$Ne ground-state-band wave functions could be well approximated by (80) shell-model wave functions. Then within the ground-state band of $^{20}$Ne, the $K+^{17}O$ $Q$ values, *unlike* the $K+^{17}O$ $B(E2)$ strengths, would have approximately the same ratios to each other as are predicted by the adiabatic-rotational model. And this result is just what Table XVII shows; i.e., the relative values of $K+^{17}O$ $Q$ moments are similar to the relative values of

---

[15] A similar comment holds for the lowest $2 \to 0$ transition in $^{18}$O. There an $SU_3$ (40) model gives the maximum $B(E2)$ possible in an $s$–$d$-shell model, and this maximum is exactly $[(\lambda + 3)/\lambda] = 7/4$ times the specialized-adiabatic answer shown in Table XIX.

adiabatic-rotational $Q$ moments, even though the $K+^{17}O$ $B(E2)$ ratios are quite different from the adiabatic-rotational $B(E2)$ ratios. Also, just as we found for the $2 \to 0$ $B(E2)$ values, the $K+^{17}O$ and specialized-adiabatic $Q$ moments in $^{20}Ne$ all fall short of the $SU_3$ maxima, but the reason is different, for the two different models, $K+^{17}O$ and adiabatic.

We have earlier mentioned that, besides the $^{20}Ne$ $B(E2)$ strengths, there are some other ground-state-band cases showing notable deviations between our adiabatic-rotational and $K+^{17}O$ $E2$ results. Are the other deviations *also* to be associated with departures of adiabatic answers from the corresponding $J$-projected shell-model answers? And why are the deviations not more prevalent?

We shall discuss these matters qualitatively, in connection with the following consideration. Since the adiabatic-rotational model would provide answers for arbitrarily high nuclear spins (beyond the upper limit for which an *s–d shell* model allows states), it seems reasonable to expect that adiabatic-rotational and shell-model $E2$ results may be in rough agreement for low spins, yet diverge for spins near the upper limit for an *s–d* shell model. A more sophisticated analysis of the low-spin versus high-spin situation is given in (*12*) [see especially pp. 235–241]. There it is shown why adiabatic-model matrix elements are expected to deviate from exact-projection matrix elements when the initial and final states involve high spins $J$. [But for *some* low-spin cases, there may be large deviations too, depending on the behavior of the function $d^J_{KK'}$, and the behavior of "the overlap function" (*12*).] Here are some of the maximum spins $J$ allowed by an *s–d*-shell model: 4 for $^{18}O$; 13/2 for $^{19}F$; 11/2 for $^{19}O$; 8 for $^{20}Ne$; 8 for $^{20}F$; 6 for $^{20}O$; 19/2 for $^{21}Ne$; 17/2 for $^{21}F$; 11 for $^{22}Na$; and 10 for $^{22}Ne$.[16] Our Tables XVII–XIX include only a few rotational-model entries involving states which have spins so high that they come within 2 units of the *s–d*-shell maxima. These few high-spin cases are: the $^{20}Ne$ $J = 6$ and $J = 8$ entries already discussed; the $13/2 \to 9/2$, $11/2 \to 7/2$, and $11/2 \to 13/2$ $B(E2)$ in $^{19}F$; the $4 \to 2$ $B(E2)$ in $^{18}O$; the $J = 9/2$ quadrupole moment for $^{19}O$ (see Table XVIII); the $J = 4$ quadrupole moment for $^{20}O$ (see Table XVII); and the $4 \to 2$ $B(E2)$ for $^{20}O$ (see Table XIX). We see that most of the exceptional deviations between our ground-state-band shell-model results and our adiabatic results occur for these particular "high-spin" cases.

---

[16] These maxima are the same whether one considers all possible *s–d*-shell states [see (*6*)], or only the representation $(\lambda\mu)$ and spin $S$ expected to be most important for low-lying states [see (*8, 9*)].

Among the cases of exceptional deviations listed earlier, there are three cases outside the aforementioned "high-spin" group. These three are: the $5 \rightarrow 4$ $B(E2)$ in $^{20}$F, and the $J = 7/2$ and $J = 11/2$ quadrupole moments of $^{21}$F. Perhaps the deviations in these three instances are due in part to some behavior of $d_{KK'}^J$, or some behavior of "the overlap function" (12), which causes the adiabatic-rotational model to give matrix elements quite different from those of the corresponding $J$-projected model. However, we have reason to suspect that, for the relevant $^{20}$F and $^{21}$F states, our shell-model wave functions are poorly described by $J$-projections from a common axially symmetric intrinsic state $\chi$.

In $^{20}$F, there is significant competition between the $K = 2$ $(S = 1)$ and $K = 1$ $(S = 0)$ possibilities for a $Q'$-maximizing state $\chi$ (122). It may well be that this competition is $J$-dependent in a way that leads to significant departures from a common-intrinsic-state pattern.

In the case of $^{21}$F, the papers of Johnstone (123) and of Bhatt, Ball, and Parikh (124) lead us to suspect that for the lowest 7/2 and lowest 11/2 states, our shell-model wave functions are poorly described by $J$-projections out of the (prolate) intrinsic state which is appropriate for other $J$-values in the ground-state band. Neither of these papers, (123) or (124), involves $s$–$d$-shell nuclei, but both papers tabulate numbers which show explicitly how shell-model Hamiltonian eigenfunctions *overlap* with wave functions $J$-projected from prolate intrinsic states $\chi$. Also, both papers report odd-$A$ cases in which, for $K = 1/2$ prolate intrinsic states, the goodness of overlap alternates with $J$ in the following way. The overlap is large for $J = 1/2$, $3/2, 5/2, \ldots, j' - 1, j', j' + 2, j' + 4$, etc.; but the overlap is small for $J = j' + 1, j' + 3, j' + 5$, etc. Here $j'$ has two coincidental features: it is the single-particle $j$-value which is *dominant* for the odd neutron or odd proton, and it is the largest $j$-values allowed in the shell-model calculation for that odd nucleon. [The influence of $J$ and $j'$ on the "goodness of overlap" can be understood, qualitatively, from an argument given in (125) and from the formulas in (124) and (125).] In our $s$–$d$-shell-model calculations, $j'$ is 5/2. Hence we suspect that, in the $K = 1/2$ ground-state band of $^{21}$F, our shell-model wave functions for $J = 1/2, 3/2, 5/2, 9/2$, and 13/2 have large overlaps with wave functions $J$-projected from a common prolate intrinsic state. But for $J = 7/2$ and $J = 11/2$, we expect small overlaps with wave functions $J$-projected from that prolate intrinsic state.

Before proceeding to $E2$ results involving states outside the ground-state band, we shall comment on how our specialized adiabatic-rotational model helps us to predict the dependence of shell-model $E2$ results on the choice of effective charges $\tilde{e}_n$ and $\tilde{e}_p$.

### 5.4.4. *The Rotational Model and the Dependence of E2 Results on Effective Charges*

Our specialized adiabatic-rotational model provides a convenient way of estimating how some shell-model $E2$ results depend on $\tilde{e}_n$ and $\tilde{e}_p$. According to the adiabatic model, quadrupole moments are proportional to the intrinsic moment $Q'$, and $B(E2)$ strengths within a band are proportional to $(Q')^2$. But for ground-state bands of light $s$–$d$-shell nuclei, our specialized adiabatic-rotational model gives $Q'$ as a linear combination of $\tilde{e}_n$ and $\tilde{e}_p$ (see Section 5.4.2). Thus there is a very simple dependence of specialized-adiabatic $E2$ observables on $\tilde{e}_n$ and $\tilde{e}_p$. We find that the same simple dependence applies very closely to *shell-model E2* results, at least for the $K+^{17}O$ Hamiltonian.

Some evidence for this conclusion is shown in the last two columns of Table XIX. These columns list the $K+^{17}O$ and specialized-adiabatic answers for the ratio of $B(\tilde{e}')/B(\tilde{e})$, where $B(\tilde{e}')$ is the strength calculated with $\tilde{e}_n = 0.8e$ and $\tilde{e}_p = 1.2e$, and $B(\tilde{e})$ is the strength calculated with $\tilde{e}_n = 0.5e$ and $\tilde{e}_p = 1.5e$. [For oxygen isotopes and $T = 0$ nuclei, the agreement of these two columns is trivial, because the form of our $\mathbf{Q}^2$ guarantees that for oxygen nuclei all $B(E2)$ are exactly proportional to $\tilde{e}_n{}^2$, and for $T = 0$ nuclei all $B(E2)$ are proportional to $\tilde{e}_S{}^2 = (\tilde{e}_n + \tilde{e}_p)^2/4$. But for $^{20}F$ and $^{21}Ne$ the agreement is nontrivial.] In our calculations of quadrupole moments, we have tried the $\tilde{e}'$ effective charges for two states of $^{19}F$, $J = 3/2$ and $J = 5/2$. The $K+^{17}O$ model gives $Q(\tilde{e}')/Q(\tilde{e})$ equal to 1.12 for the $J = 3/2$ state, and 1.10 for the $5/2$ state. The rotational model predicts $Q(\tilde{e}')/Q(\tilde{e}) = 1.12$ for both states.

### 5.4.5. *Shell-Model versus Rotational-Model E2 Results Outside Ground-State Bands*

Table XX shows $B(E2)$ results for transitions involving levels outside ground-state bands. Some of the listed $K_i$ and $K_f$ assignments here are very speculative (i.e., some of them may be quite inappropriate for the observed states and shell-model states on that line of the table). But in any case, these listed $K$ values are the ones we have used in applying equation (21) so as to get relative values of adiabatic-rotational-model $E2$ strengths, for comparison with relative values of shell-model $E2$ strengths.

The large right-hand brackets in Table XX show sets of transitions connecting the same pair of bands, $\alpha_i K_i \rightleftarrows \alpha_f K_f$. Within each of the right-bracketed sets for which $K_i = 0$ or $K_f = 0$ or $|K_i + K_f| > 2$, the adiabatic-rotational model predicts $B(E2)$ strengths proportional to the squared

Clebsch–Gordan coefficients of (21). The numerical values of these squared Clebsch–Gordan coefficients are listed in Table XX, and their ratios may be compared with ratios of the shell-model $B(E2)$ values listed in other columns of the table.

For the $E2$ transitions in Table XX, there is generally poor agreement between the rotational-model predictions and our shell-model results. In many cases the fluctuation of $B(E2)$ strength for the same transition, but from one shell-model to the next, exceeds the variation from one transition to the next as predicted by the adiabatic-rotational model. Furthermore, there is no one choice, among our shell models, which gives agreement with the adiabatic-rotational predictions. Indeed, we see the following general kind of disagreement between results from the rotational model and any particular shell model: within a right-bracketed set, the shell model gives larger variations in strength, from one transition to another, than the rotational model does.

We see two faint glimmers of qualitative agreement within the $K_i = 1$, $K_i = 2$ right-bracketed set for $^{20}$F. Table XX shows that in this set there are two transitions for which the rotational model predicts forbidden or almost-forbidden $E2$ decay. For these two $^{20}$F transitions, $2' \rightarrow 3$ and $3' \rightarrow 4$, our shell models generally do give unusually small strengths. (But for a counterexample, see the $4 \rightarrow 2'$ transition in $^{20}$O.)

Table XX includes one striking example of good agreement between shell-model and rotational-model results. This agreement is in $^{22}$Ne, for the relative $E2$ strengths interconnecting the three lowest states within the first-excited ($K = 2$) band. Within this band, the $K+^{17}O$ model gives $B(E2)$ strengths which, in units of $e^2 fm^4$, are all close to 20 times the listed ($C^2$) values. But the agreement is *only* in relative values. We know that a simple adiabatic-rotational model could not reproduce the magnitude of the shell-model $B(E2)$ strengths within this presumed $K = 2$ band, for even if we assume that the $Q'$ of this band is the maximum allowed by an $s$–$d$-shell adiabatic model (i.e., even if we assume that $Q'$ is the same as our $Q'$ for the $K = 0$ ground-state band in $^{22}$Ne), the adiabatic-model $B(E2)$ values within this $K = 2$ band would be only about half of the $K+^{17}O$ $B(E2)$ values.

In short, although we find generally good agreement between shell-model and rotational-model $E2$ results within ground-state bands, we find generally poor agreement for $E2$ transitions involving levels outside ground-state bands. Recall that there is a parallel situation when it comes to the sensitivity of our shell-model $E2$ results to the choice among our alternative model Hamiltonians. We found generally weak sensitivity for shell-model

*E*2 results within ground-state bands, but we found rather strong sensitivity for shell-model *E*2 results involving levels outside ground-state bands.

There is a plausible explanation for both these parallel situations. This plausible explanation is connected with the dominance of $Q'$-maximizing $(\lambda\mu)$ components in our ground-state-band shell-model states. Such a dominance is probably common to all our alternative shell models, but the alternative Hamiltonians differ in the admixtures which they introduce into these predominantly pure $(\lambda\mu)$ bands. Now the *mass* quadrupole operator cannot connect states belonging to different $(\lambda\mu)$ representations (*7, 8*). Hence, to the extent that the effective electric quadrupole operator has matrix elements like those of the mass quadrupole operator, small admixtures to a predominantly pure $(\lambda\mu)$ band will have no first-order effects on $B(E2)$ strengths within the band. In contrast, there *will* be first-order effects on $B(E2)$ values for transitions between bands which differ in their predominant $(\lambda\mu)$ character. These considerations explain why we find that shell-model $B(E2)$ strengths within ground-state bands are rather insensitive to the choice among our model Hamiltonians, while *inter*band $B(E2)$ strengths tend to be sensitive to the choice among our model Hamiltonians. Also, these considerations explain why interband *E*2 strengths often show substantial Hamiltonian-dependent deviations from the relative $B(E2)$ predictions of the adiabatic-rotational model [since this model presumes that the nuclear levels fall neatly into sets of rotational bands, every state within the same band involving the same mixture of $(\lambda\mu)$ representations].[17,18]

The foregoing $(\lambda\mu)$ arguments suggest that the model-sensitivity which we find for interband *E*2 transitions will extend to other kinds of interband transitions, and indeed, to any observable for which the pertinent operator does not particularly favor connections between nuclear states of the same dominant $(\lambda\mu)$. Perhaps this consideration explains in part the model-

---

[17] Sometimes *J*, *T*, and the "leading representation" $(\lambda\mu)$ do not uniquely determine a state. Then our simple arguments above need some amendment, and the conclusions tend to become weaker.

[18] The $(\lambda\mu)$ explanation offered in our text *supplements* a more general and obvious explanation for the difference in Hamiltonian-sensitivity between ground-state-band $B(E2)$'s and intraband $B(E2)$'s. (However, we believe that this more general explanation is, by itself, inadequate to account for the cases we discuss.) Here is the more general explanation. Levels close to other levels of the same *J* are likely to have wave functions which are "sensitive to the details of the model Hamiltonian." But levels outside ground-state bands lie in a higher and *denser* region of the energy spectra, compared to ground-state-band levels. Hence when a transition involves at least one level outside the ground-state band, its strength is likely to be more Hamiltonian-sensitive than the strengths for transitions between two ground-state-band levels.

sensitivity which we find for spectroscopic factors $S$, even for transfers connecting one ground-state band to another. (See Table XV and Section 3.4). This consideration will be relevant again in Section 6, when we discuss the model-sensitivity of $B(M1)$ strengths.

## 6. MAGNETIC MOMENTS AND $M1$ TRANSITIONS

We have used the shell-model wave functions associated with our several model Hamiltonians to compute magnetic dipole moments and $M1$ transition strengths. In this section we shall discuss topics similar to those in our preceding $E2$ section; but for several reasons, we consider this $M1$ discussion to be less satisfying than the $E2$ discussion. *First*, there is much less experimental information on $B(M1)$'s than on $B(E2)$'s. *Second*, the $B(M1)$'s are much more sensitive to the choice among our model Hamiltonians (but we see no clear pattern to the sensitivity). *Third*, for $B(M1)$'s there is very little resemblance between shell-model and rotational-model results.

At first sight the calculated $B(M1)$'s seem to vary erratically in size, from one transition to another. However, because of simplicities in the $M1$-operator, we are able to offer some qualitative explanations for the variations of strengths seen among these calculated $B(M1)$'s.

### 6.1. The Effective $M1$ Operator

In calculating magnetic dipole moments and transition strengths, we use the operator

$$\mathbf{M}^1 = \sum_{k=1}^{A} \frac{1}{2} \{ \mathbf{l}(k) + (g_p + g_n)\mathbf{s}(k) \} + \sum_{k=1}^{A} t_z(k) \{ \mathbf{l}(k) + (g_p - g_n)\mathbf{s}(k) \} \quad (23)$$

Here $\mathbf{l}(k)$ and $\mathbf{s}(k)$ are the orbital and spin angular momenta of nucleon $k$; the units are nuclear magnetons $\mu_N$, and we use the convention $t_z = +1/2$ for a proton and $t_z = -1/2$ for a neutron. In writing (23) we have assumed the free-nucleon orbital gyromagnetic ratios (unity for each proton, zero for each neutron), and in our computations we also assume the free-nucleon spin gyromagnetic ratios,

$$g_p = 5.58, \qquad g_n = -3.82 \quad (24)$$

Thus, we assume that the effective $M1$ operator is simply the *unperturbed free-nucleon* $M1$ operator. The magnetic dipole moment $\mu$ of a many-particle shell-model state $\psi^{JT}(\alpha)$ is calculated from

$$\mu = \langle \alpha\, J\, T \quad J_z = J \quad T_z \mid M_0^1 \mid \alpha\, J\, T \quad J_z = J \quad T_z \rangle \tag{25}$$

Here the subscript zero on $M_0^1$ means that one takes the $z$-component of the vector operators $\mathbf{l}$ and $\mathbf{s}$ in (23). The strength of an $M1$ transition from $\psi^{J_i T_i}(\alpha_i)$ to $\psi^{J_f T_f}(\alpha_f)$ is described by

$$B(M1)_{i\to f} = (2J_i + 1)^{-1}\langle \alpha_f\, J_f\, T_f\, T_{z_f} \,\|\, \mathbf{M}^1 \,\|\, \alpha_i\, J_i\, T_i\, T_{z_i}\rangle^2 \tag{26}$$

where the double-barred matrix element (5) is reduced with respect to $J$ but not $T$.

In using (23) and (24) as an effective $M1$ operator we are following the usual procedure of ignoring several complicating effects. We shall review these briefly, mentioning some theoretical estimates of the associated corrections. First, we mention the complications arising from meson exchange currents (etc.), and from momentum dependence of the bare nucleon–nucleon force. These are *related* sources, whose effect on magnetic moments may be several tenths of a nuclear magneton, for nuclei in the mass range of our interest (*125*). Then there are "configuration excitations", i.e., low-energy excitations excluded from the *s–d*-shell vector space in which our $\psi^{JT}(\alpha)$ are calculated. Estimates have been made for configuration-excitation corrections to the magnetic moments of $^{17}O$ and $^{17}F$, and again, the perturbations may amount to several tenths of a nuclear magneton (*126–128*). These configuration-excitation corrections are based on second-order perturbation calculations, rather than first-order. This is because the free-nucleon $M1$ operator, (23)–(24), cannot connect single-nucleon states which differ in principal quantum number. Now, a calculation of the second-order-perturbed matrix elements in each one-body state is equivalent to a calculation of the one-body part of the second-order-perturbed effective operator. But to second order, the configuration-perturbed magnetic-dipole operator includes two-body and three-body parts, as well as the one-body part that is operative in $A = 17$. These two- and three-body parts have not been calculated, to our knowledge. Finally, we note that for the one-body part of $\mathbf{M}^1$, there are some theoretical arguments for keeping the isoscalar term, the first sum in (23), almost unperturbed (*127–130*).

There is also some empirical information about the usefulness of the unperturbed $M1$ operator (23)–(24). For $p$-shell nuclei ($A = 5$–15), magnetic moments and $B(M1)$ strengths have been calculated with the operator

(23)–(24) for a variety of $(0s)^4(0p)^{4-4}$ models. The results have generally been in good agreement with measured magnetic moments (within $\approx 0.1\mu_N$), and in fair agreement with measured $B(M1)$ strengths (within $\approx 0.4\mu_N{}^2$) (*131, 132*). For the $s$–$d$ shell there are some empirical arguments (*133*), as well as theoretical arguments, for keeping the isoscalar term in $\mathbf{M}^1$ almost unperturbed. For even-parity states in $^{31}$P (where $M1$ moments and transition strengths derive most of their strength from $s_{1/2} \rightarrow s_{1/2}$ contributions), there are suggestions that the agreement between shell-model results and experimental data could be improved if the isovector term of $\mathbf{M}^1$ [i.e., the second sum in (23)] were multiplied by $\sim 0.6$ (*134*).[19] In the light $sd$-shell region itself, the measured ground-state magnetic moments of $^{17}$F and $^{17}$O are reproduced within 2% by the assumptions (23)–(24). For $A = 18$–22 our comparisons with experimental data do not recommend any clearly favored alternatives to (23)–(24). (These $A = 18$–22 comparisons will be discussed in Sections 6.2 and 6.4.)

## 6.2. Magnetic Moments: Shell Model *versus* Experiment

Table XXII shows measured values of magnetic moments $\mu$ for the nuclei $A = 17$–22, and lists the corresponding shell-model values calculated via (23)–(25) with $K + {}^{17}O$ wave functions. The agreement of theory with experiment is generally quite good. We shall next comment on the implications of this agreement for our $\mathbf{M}^1$-operator (23)–(24), and for our $K + {}^{17}O$ wave functions.

In Table XXII the $A = 17$ states and $A = 18$ states have shell-model wave functions which are completely determined by our restriction to an $s$–$d$-shell basis. Hence, comparisons between their observed and calculated magnetic moments constitute clear-cut tests of (23)–(24) as an effective $M1$-operator for $s$–$d$-shell models. As Table XXII shows, all three of these "clear-cut" tests are passed very well; but of course they are only consistency checks—they do not prove that the operator (23)–(24) is correct in all its features. The agreements with measured $^{17}$F and $^{17}$O magnetic moments check the following parts of $\mathbf{M}^1$: the one-body $d_{5/2}$–$d_{5/2}$ part of the isoscalar term, and the one-body $d_{5/2}$–$d_{5/2}$ part of the isovector term. The measured moment of the $J = 5$, $T = 0$ state in $^{18}$F equals the sum of the measured $^{17}$F and $^{17}$O $J = 5/2$ moments. This experimental result is a check on the following feature of $\mathbf{M}^1$: that the isoscalar term should have no two-

---

[19] These suggestions are based on calculations with a truncated $s$–$d$-shell basis.

**TABLE XXII. Magnetic Moments $\mu$—Comparison between Measured Results and $K+{}^{17}O$ Shell-Model Results.** All entries refer to the lowest-energy state of the indicated spin. Experimental uncertainties are less than 0.01 $\mu_N$, except where indicated otherwise.

| Nucleus | $J$ | Magnetic moment $\mu$, in units of nuclear magnetons ($\mu_N$) | |
| | | Expt.[a] | $K+{}^{17}O$ |
|---|---|---|---|
| ${}^{17}F$ | 5/2 | 4.72 | 4.79 |
| ${}^{17}O$ | 5/2 | −1.89 | −1.91 |
| ${}^{18}F$ | 5 | 2.86 ± 0.03[b] | 2.88 |
| ${}^{19}F$ | 1/2 | 2.63 | 2.87 |
| | 5/2 | $\begin{cases} 3.59 \pm 0.02 \\ 3.69 \pm 0.04 \end{cases}$ | 3.55 |
| ${}^{20}F$ | 2 | 2.09 | 1.92 |
| ${}^{21}Ne$ | 3/2 | −0.66 | −0.90 |
| ${}^{22}Na$ | 3 | 1.75 | 1.82 |

[a] Except for ${}^{18}F$, the values listed are from (103).
[b] This value is the mean of three recent measurements. It is given in (133).

body part connecting $(d_{5/2})^2 J = 5$ states. But even if we take these checks very seriously, the "true" effective $\mathbf{M}^1$ operator could still differ from (23)–(24) in several ways, e.g., by deviating from the coefficients of l and s shown in (23), or by including $j$-dependent one-body operators which cannot be represented by linear combinations of l and s, or by including some non-negligible two-body operators.

Although the aforementioned $A = 17$ and 18 checks are rather special, they do discourage any simple wholesale changes in (23)–(24). In particular, the ${}^{17}F$ and ${}^{17}O$ checks discourage any overall rescaling of either the isovector or the isoscalar term in $\mathbf{M}^1$.

All the other shell-model $\mu$-values in Table XXII depend on the $K+{}^{17}O$ Hamiltonian, as well as on the $\mathbf{M}^1$-operator (23)–(24). Thus, "errors" from these two sources are intertwined. However, we can at least consider which features of $\mathbf{M}^1$, and which features of the wave functions, affect calculated values of $\mu$. For example, each of the ${}^{19}F$ moments shown in Table XXII has important contributions from both the isoscalar and isovector parts of $\mathbf{M}^1$. This can be seen by rewriting (23)–(24) as

$$\mathbf{M}^1 = (0.50\mathbf{L} + 0.88\mathbf{S}) + \sum_{k=1}^{A} t_z(k)\{\mathbf{l}(k) + 9.40\mathbf{s}(k)\}$$

$$= (0.50\mathbf{J} + 0.38\mathbf{S}) + \sum_{k=1}^{A} t_z(k)\{\mathbf{l}(k) + 9.40\mathbf{s}(k)\} \qquad (27)$$

Then, since $^{19}$F has $|S_z| \leq 3/2$, we see that the isoscalar contribution to our calculated $\mu$ cannot exceed $0.82\mu_N$ for $J = 1/2$, and cannot exceed $1.82\mu_N$ for $J = 5/2$. Hence the isovector contributions to these calculated moments must be at least $2.05\mu_N$ and $2.73\mu_N$, respectively. These results indicate that any major quenching of the isovector part in (27) would ruin the experimental–theoretical agreement shown in Table XXII for these $^{19}$F states.

In Table XXII, the worst percentage disagreement between experiment and theory is for the $\mu$-value of $^{21}$Ne. Here the relatively small magnitude of $\mu$ hints that cancellations among the $L_z$, $S_z$, and isovector contributions make this calculated moment particularly sensitive to the numerical coefficients in (27), and particularly sensitive to characteristics of the theoretical wave function.

In some cases a comparison of theoretical and observed magnetic moments yields very little information about the detailed wave function of a nuclear state. The $J = 3$, $T = 0$ state of $^{22}$Na illustrates this. Here $\mu$ involves only the isoscalar contribution, which has as its theoretical value $1.5 + 0.38\langle S_z \rangle$. A pure $(d_{5/2})^6$ $J_z = J = 3$ state[20] would have $S_z = (J_z/5) = 0.60$; it would therefore have a calculated $\mu$ of $1.73\mu_N$. This is very close to the $K+^{17}O$ value listed in Table XXII, and it is also close to the measured value. But of course many different $J = 3$ $s$-$d$-shell states could have the same $\langle S_z \rangle$, and therefore the same calculated $\mu$. In fact, our $K+^{17}O$ wave function for the $^{22}$Na ground state is only 15% $(d_{5/2})^6$.

## 6.3. Magnetic Moments: Variation Among Shell-Model Results from Different Hamiltonians

Table XXIII allows some comparison among theoretical magnetic dipole moments, all calculated from the same effective operator (23)–(24) but from a variety of $s$-$d$-shell wave functions. Columns 4–6 show how the $A = 18$–20 results vary when calculated with shell-model eigenfunctions of

---

[20] For a $j^n$ state, the formal expressions for the expectations $\langle S_z \rangle$ and $\langle J_z \rangle$ are identical except that the $\langle S_z \rangle$ expression has a factor $\langle j \| \mathbf{s}^1 \| j \rangle$ where the $\langle J_z \rangle$ expression has $\langle j \| \mathbf{j}^1 \| j \rangle$. But when $j = l + s$, we have $\langle j \| \mathbf{s}^1 \| j \rangle = (S/j) \langle j \| \mathbf{j}^1 \| j \rangle$.

TABLE XXIII. Magnetic Moments—Comparison among Results from Various *s–d*-Shell Models. All entries are for the lowest-energy state of the indicated spin.

| Nucleus | J | | Magnetic moment in units of nuclear magnetons | | | | | |
|---|---|---|---|---|---|---|---|---|
| | | Expt.[a] | Shell models | | | Rotational models | | |
| | | | $K+{}^{17}O$ | RIP | MSDI | Adiabatic (Hartree–Fock $\chi$)[g] | J-projected (Hartree–Fock $\chi$)[g] | J-projected ($SU_3$) |
| $^{18}$F | 1 | | 0.84 | 0.84 | 0.75 | | | |
| | 2 | | 1.12 | 1.10 | 1.08 | | | |
| | 3 | | 1.87 | 1.87 | 1.83 | | | |
| $^{19}$F | 1/2 | 2.63 | 2.87 | 2.86 | 2.89 | 2.82[b,c] | 2.89[d] | 2.79[e,f] |
| | 3/2 | | −1.41 | −1.59 | −1.66 | | −1.58[d] | |
| | 5/2 | $\left\{\begin{array}{l}3.59 \pm 0.02\\ 3.69 \pm 0.04\end{array}\right\}$ | 3.55 | 3.56 | 3.84 | 3.86[b] | 3.78[d] | 3.46[f] |
| $^{19}$O | 3/2 | | −0.86 | −0.78 | −0.93 | | | −0.82[e] |
| | 5/2 | | −1.45 | −1.41 | −1.71 | | | |
| $^{20}$F | 2 | 2.09 | 1.92 | 1.80 | −2.22 | | 1.54[d] | 0.86[f] |
| $^{21}$Ne | 3/2 | −0.66 | −0.90 | | | −0.58[c] | −0.57[d] | −0.88[e] |
| $^{22}$Na | 3 | 1.75 | 1.82 | | | | 1.75[d] | 1.78[e] |

[a] From (103). Experimental uncertainties are less than 0.01 $\mu_N$, except where indicated otherwise.   [b] (137).   [c] (12).   [d] (14).   [e] (122).   [f] (9).

[g] The intrinsic $\chi$ functions used in calculating the adiabatic-rotational results referenced b and/or c are very close to those used to calculate the J-projected results referenced d (12).

the Hamiltonians $K+{}^{17}O$, $RIP$, and $MSDI$. For the same state, these three models yield $\mu$ values which agree within $0.4\mu_N$. (The other columns of Table XXIII, marked "Rotational models," will be discussed later.)

## 6.4. Magnetic Dipole Transitions

Table XXIV shows $B(M1)$ strengths calculated from eigenvectors of our shell-model Hamiltonians $K+{}^{17}O$, $RIP$, and $MSDI$. There are rather few experimentally determined $B(M1)$ values in the light $s$–$d$-shell region, and we have included in Table XXIV most of the relevant ones listed in the compilation of Skorka et al. (104), plus a few recent determinations for ${}^{21}$Ne and ${}^{22}$Na.

Comparisons of our calculated $B(M1)$'s with the few measured values in Table XXIV show that the calculated $B(M1)$'s tend to exceed the observed values by a factor of $\gtrsim 2$. But, as also shown by the table, the measured $B(M1)$'s are rather uncertain, and the pertinent calculated $B(M1)$'s are sensitive to the choice among our model Hamiltonians.

We shall return to the question of sensitivity at the end of this section, and in Section 6.5. But first, we offer some qualitative explanations for the pattern of weakness and strength seen among the shell-model entries in Table XXIV.

For the transitions we have investigated, there is a preponderance of weak $B(M1)$ values. Many of the extremely weak transitions are $(T = 0)$ $\rightarrow (T = 0)$ transitions. In these cases Morpurgo's rule (135) applies: the isovector contribution vanishes, and then, since the $J$ part of the isoscalar term in (27) cannot connect two orthogonal states, the effective $\mathbf{M}^1$ operator reduces to just 0.38S, a small term. Since these $(T = 0) \rightarrow (T = 0)$ $B(M1)$'s are so small, we have used an exponential form to list them in Table XXIV. For the other entries in the table, it is fair to conclude that any $B(M1)$ $\gtrsim 0.5\mu_N{}^2$ is dominated by the isovector part of $\mathbf{M}^1$.

In ${}^{18}$O, the weakness of the $2' \rightarrow 2$ and $3 \rightarrow 2$ $M1$-transitions could be explained qualitatively by assuming that the pertinent $s$–$d$-shell wave functions are dominated by those configurations which lie lowest in energy when only the one-body part of the Hamiltonian is considered. That is, assume $(d_{5/2})^2$ for $J = 2$, $d_{5/2}s_{1/2}$ for $J = 2'$, and $d_{5/2}s_{1/2}$ for $J = 3$. With pure configurations such as these, the $2' \rightarrow 2$ and $3 \rightarrow 2$ $B(M1)$ would vanish by angular-momentum selection rules. Along the same lines, the configurations $(d_{5/2})^2$ for $J = 4$, and $d_{5/2}d_{3/2}$ for $J = 4'$, would give a nonzero $M1$ matrix element for $4' \rightarrow 4$, and so would explain the much greater strength of the $4' \rightarrow 4$ transition compared with $2' \rightarrow 2$ and $3 \rightarrow 2$. These

TABLE XXIV. B(M1) Strengths. A J-value without a prime indicates the lowest-energy state of that A, J, T; a J-value with one prime indicates the second state of that A, J, T; etc. An entry nc means that the shell-model value was not calculated. In the last three columns, the squared Clebsch–Gordan coefficient $(C)^2$, and the rotational-model quantum numbers $K_i$ and $K_f$, have the same meaning as in equation (28). Some of the K assignments are very speculative. We list $(C)^2$ values for those groups of transitions within which the simple rotational-model formula (28) may be applicable for predicting relative B(M1) values. (Its prediction is that, within each group, the B(M1) values will be proportional to the $(C)^2$ values.)

| Nucleus | Initial and final states | | | | B(M1) in units of $(\mu_N)^2$ | | | | Rotational-model numbers | | |
|---|---|---|---|---|---|---|---|---|---|---|---|
| | $T_i$ | $T_f$ | $J_i$ | $J_f$ | Expt. | $K+^{17}O$ | RIP | MSDI | $(C)^2$ | $K_i$ | $K_f$ |
| $^{18}$F | 0 | 0 | 1′ | 1 | | $<0.01 \times 10^{-2}$ | $0.49 \times 10^{-2}$ | $1.08 \times 10^{-2}$ | | | |
| | 0 | 0 | 2 | 1 | | $0.09 \times 10^{-2}$ | $0.01 \times 10^{-2}$ | $0.02 \times 10^{-2}$ | | | |
| | 0 | 0 | 2 | 3 | | $2.85 \times 10^{-2}$ | $2.39 \times 10^{-2}$ | $1.26 \times 10^{-2}$ | | | |
| | 0 | 0 | 3′ | 3 | | $0.01 \times 10^{-2}$ | $0.08 \times 10^{-2}$ | $0.14 \times 10^{-2}$ | | | |
| | 0 | 0 | 1′ | 2 | | $0.19 \times 10^{-2}$ | $1.09 \times 10^{-2}$ | $0.22 \times 10^{-2}$ | | | |
| | 1 | 0 | 0 | 1 | | 17.2 | 14.0 | 18.5 | | | |
| | 0 | 1 | 1′ | 0 | | 0.37 | 0.79 | 0.04 | | | |
| | 0 | 1 | 1″ | 0 | | 0.32 | 0.61 | 0.15 | | | |
| $^{18}$O | 1 | 1 | 2′ | 2 | | 0.09 | 0.20 | 0.09 | 0 | 2 | 0 |
| | 1 | 1 | 3 | 2 | | 0.11 | 0.08 | 0.13 | 0 | 2 | 0 |
| | 1 | 1 | 4′ | 4 | | 1.09 | 1.08 | 0.91 | 0 | 2 | 0 |
| | 1 | 1 | 3 | 2′ | | 0.36 | 0.47 | 0.48 | | 2 | 2 |
| $^{19}$F | 1/2 | 1/2 | 3/2 | 1/2 | | 0.03 | 0.03 | 0.21 | | 1/2 | 1/2 |
| | 1/2 | 1/2 | 3/2 | 5/2 | | 3.97 | 3.56 | 3.58 | | 1/2 | 1/2 |
| | 1/2 | 1/2 | (5/2)′ | 3/2 | | 0.01 | 0.14 | 0.05 | 0 | 5/2 | 1/2 |
| | 1/2 | 1/2 | (5/2)′ | 5/2 | | 0.03 | 0.11 | 0.21 | 0 | 5/2 | 1/2 |

**TABLE XXIV** (*Continued*)

| Nucleus | Initial and final states | | | | | $B(M1)$ in units of $(\mu_N)^2$ | | | | Rotational-model numbers | | |
|---|---|---|---|---|---|---|---|---|---|---|---|---|
| | $T_i$ | $T_f$ | $J_i$ | $J_f$ | Expt. | $K+^{17}O$ | RIP | MSDI | $(C)^2$ | $K_i$ | $K_f$ |
| $^{19}O$ | 3/2 | 3/2 | 3/2 | 5/2 | $0.032 \pm 0.001$[a] | <0.01 | 0.05 | 0.02 | 0.40 | 3/2 | 3/2 |
| | 3/2 | 3/2 | 7/2 | 5/2 | | 0.03 | 0.07 | 0.03 | 0.36 | 3/2 | 3/2 |
| | 3/2 | 3/2 | 9/2 | 7/2 | | 0.10 | 0.05 | 0.04 | 0.40 | 3/2 | 3/2 |
| | 3/2 | 3/2 | 1/2 | 3/2 | | 0.07 | 0.13 | 0.10 | 1.00 | 1/2 | 3/2 |
| | 3/2 | 3/2 | (3/2)' | 3/2 | | <0.01 | 0.17 | 0.03 | 0.40 | 1/2 | 3/2 |
| | 3/2 | 3/2 | (5/2)' | 3/2 | | 0.17 | 0.18 | 0.25 | 0.07 | 1/2 | 3/2 |
| | 3/2 | 3/2 | (3/2)' | 5/2 | | 0.17 | 0.53 | 0.06 | 0.60 | 1/2 | 3/2 |
| | 3/2 | 3/2 | (5/2)' | 5/2 | | <0.01 | 0.01 | <0.01 | 0.46 | 1/2 | 3/2 |
| | 3/2 | 3/2 | 7/2 | 5/2 | | 0.02 | 0.03 | 0.03 | 0.36 | 3/2 | 1/2 |
| | 3/2 | 3/2 | (3/2)' | (5/2)' | | 0.31 | 0.09 | 0.49 | | 1/2 | 1/2 |
| $^{20}Ne$ | 0 | 0 | 2' | 2 | | $0.03 \times 10^{-3}$ | $0.47 \times 10^{-3}$ | $0.49 \times 10^{-3}$ | 0 | 0' | 0 |
| | 0 | 0 | 4' | 4 | | $0.02 \times 10^{-3}$ | $2.69 \times 10^{-3}$ | $4.78 \times 10^{-3}$ | 0 | 0' | 0 |
| $^{20}F$ | 1 | 1 | 3 | 2 | | 0.06 | 0.29 | 0.09 | 0.24 | 2 | 2 |
| | 1 | 1 | 4 | 3 | | 0.10 | 0.15 | 0.24 | 0.33 | 2 | 2 |
| | 1 | 1 | 5 | 4 | | 0.43 | 0.80 | 1.23 | 0.38 | 2 | 2 |
| | 1 | 1 | 1 | 2 | | 5.06 | 4.96 | 7.08 | 1.00 | 1 | 2 |
| | 1 | 1 | 2' | 2 | | 0.09 | 0.01 | 0.17 | 0.33 | 1 | 2 |
| | 1 | 1 | 3' | 2 | | 0.05 | 0.11 | 0.05 | 0.05 | 1 | 2 |
| | 1 | 1 | 2' | 3 | | 4.30 | 4.46 | 4.16 | 0.67 | 1 | 2 |
| | 1 | 1 | 3' | 3 | | <0.01 | 0.04 | 0.08 | 0.42 | 1 | 2 |
| | 1 | 1 | 3' | 4 | | 2.60 | 0.94 | 3.34 | 0.54 | 1 | 2 |

| | T | $J_i$ | $J_f$ | Exp | | | | | $J'$ | $J$ |
|---|---|---|---|---|---|---|---|---|---|---|
| | 1 | 2′ | 1 | | 0.09 | 0.03 | 0.14 | 0.30 | 1 | 1 |
| | 1 | 3′ | 1 | | 0.03 | 0.03 | <0.01 | 0.38 | 1 | 1 |
| | 1 | 1′ | 2 | | 0.46 | 1.69 | 0.15 | | 1′ | 2 |
| | 1 | 1′ | 1 | | nc | 0.26 | 0.09 | | 1′ | 1 |
| | 1 | 0 | 1 | | 0.01 | 0.03 | <0.01 | | 1′ | 1 |
| | 2 | 2′ | 1 | | 1.45 | nc | nc | | | |
| $^{20}$O | 2 | 2′ | 2 | | 0.03 | 0.01 | 0.07 | | | 0 |
| $^{21}$Ne | 1/2 | 5/2 | 3/2 | 0.06 ± 0.02[b] | 0.17 | nc | nc | 0.27 | 3/2 | 3/2 |
| | 1/2 | 7/2 | 5/2 | 0.13 ± 0.04[b] | 0.28 | nc | nc | 0.36 | 3/2 | 3/2 |
| | 1/2 | 9/2 | 7/2 | 0.22 ± 0.06[b] | 0.35 | nc | nc | 0.40 | 3/2 | 3/2 |
| | 1/2 | 11/2 | 9/2 | 0.14 ± 0.05[b] | 0.35 | nc | nc | 0.42 | 3/2 | 3/2 |
| | 1/2 | 1/2 | 3/2 | | 0.72 | nc | nc | | | 3/2 |
| | 1/2 | (3/2)′ | 3/2 | | 0.41 | nc | nc | | | 3/2 |
| | 1/2 | (1/2)′ | 3/2 | | 0.13 | nc | nc | | | 3/2 |
| $^{22}$Na | 0 | 4 | 3 | $(0.37 \pm .08) \times 10^{-3}$[c] | $0.91 \times 10^{-3}$ | $1.30 \times 10^{-3}$ | nc | 0.19 | 3 | 3 |
| | 0 | 5 | 4 | $(0.50 \pm .08) \times 10^{-3}$[c] | $1.60 \times 10^{-3}$ | $2.50 \times 10^{-3}$ | nc | 0.29 | 3 | 3 |
| | 0 | 4′ | 3 | | $<0.01 \times 10^{-3}$ | $0.04 \times 10^{-3}$ | nc | | | 3 |
| | 0 | 5′ | 4 | | $0.05 \times 10^{-3}$ | $0.05 \times 10^{-3}$ | nc | | | 3 |

[a] From (104).
[b] From (111).
[c] These are the results of (78), adjusted in accordance with the lifetime determinations of (112). (We have adjusted the "uncertainties," too, but only roughly.)

simplifying assumptions about our $^{18}O$ shell models do in fact hold approximately, in most cases. An exception is that, in the $K+^{17}O$ model, the states 2 and 2′ both have big $(d_{5/2})^2$ and big $d_{5/2}s_{1/2}$ components. The $K+^{17}O$ 2′ → 2 $B(M1)$ is weak because (a) its $(d_{5/2})^2 → (d_{5/2})^2$ and $d_{5/2}s_{1/2} → d_{5/2}s_{1/2}$ contributions each have rather small magnitude, (b) these small contributions are of opposite sign, and (c) the contributions from other configurations almost cancel among themselves.

For $^{19}F$, the qualitative behavior of the first two entries in Table XXII can be understood in terms of an $L$ selection rule. We have already pointed out that only the isovector part of $\mathbf{M^1}$ can be expected to produce strong transitions. In this isovector part, the $t_z(k)s(k)$ term (which is usually the dominant term) cannot connect different $L$-values. Using the wave functions of a previous paper (105), Elliott (7) has shown that in the ground-state band of $^{19}F$, the $J = 1/2$ member state is dominated by components having $L = 0$, while both the $J = 3/2$ and $J = 5/2$ members are dominated by $L = 2$. These $L$ characteristics would explain why, in the ground-state band of $^{19}F$, we get a weak $B(M1)$ for $3/2 → 1/2$ but a strong $B(M1)$ for $3/2 → 5/2$.

The weakness of $B(M1)$'s for the $(5/2)′ → 3/2$ and $(5/2)′ → 5/2$ transitions in $^{19}F$ may be explained, qualitatively, in terms of a rotational-model selection rule forbidding $M1$ transitions with $\Delta K > 1$. (Further rotational-model considerations will be discussed in Section 6.5.)

In $^{19}O$ and $^{20}O$, the general weakness of the tabulated $B(M1)$ values can be explained by arguments combining two rules: (a) the angular-momentum selection rule that forbids $M1$ transitions connecting $(d_{5/2})^n$ components with $(d_{5/2})^{n-1}s_{1/2}$ components, and (b) the rule that $M1$ transitions are forbidden between two orthogonal $j^n$ states of identical particles (136). [For the particular $^{18}O$ transitions listed in Table XXIV, rule (b) is inoperative. That is because at least one of the states, final or initial, is always dominated by configurations other than $(d_{5/2})^2$.] For some of the low-lying $^{19}O$ states, $j^n = (d_{5/2})^3$ components are dominant; then rule (b) applies. But for $J = 1/2$ and $J = 7/2$, the antisymmetry requirement allows $(d_{5/2})^2s_{1/2}$ but disallows $(d_{5/2})^3$; then rule (a) applies. Thus we see that the dominance of $(d_{5/2})^n$ and $(d_{5/2})^n s_{1/2}$ configurations can account for the generally small $B(M1)$ strengths calculated for transitions among the low-lying states of $^{19}O$ and $^{20}O$. As the excitation increases, one expects other configurations to increase in importance; and indeed, Table XXIV shows some larger-than-usual $B(M1)$ values for transitions between higher-lying states in $^{19}O$.

Among the $^{20}F$ transitions in Table XXIV, there are three with especially strong $B(M1)$: 1 → 2, 2′ → 3, and 3′ → 4. There is a plausible explanation for these in terms of rotational-band considerations. We conjecture that the $J = 2, 3, 4$ states in $^{20}$F belong to a band having $K = 2$ (with $K_L = 1$ and $K_S = 1$); and we conjecture that the $J = 1, 2', 3'$ states belong to a band having $K = 1$ (with $K_L = 1$ and $K_S = 0$) (*122*). Then the prominent $1 \rightarrow 2$ transition would be a transition between the two band heads, $2' \rightarrow 3$ would be a transition between the second states of each ·band, and $3' \rightarrow 4$ would be a transition between the third states of each band. Let us further presume that the predominant $L$-value is 1 for the lowest member in each of these two bands, 2 for the second members, etc. Then by invoking the $\Delta L = 0$ selection rule which holds for the last term in the $\mathbf{M}^1$-operator (27), we can understand the greater strength of these $1 \rightarrow 2, 2' \rightarrow 3$, and $3' \rightarrow 4$ $B(M1)$'s compared with other $B(M1)$'s in $^{20}$F.

This ends our offering of qualitative explanations for the pattern of weakness and strengths seen among the shell-model $B(M1)$'s in Table XXIV. We stress these facts: (a) many of these explanations are very rough, (b) in most cases we have not analyzed our shell-model wave functions sufficiently to check whether these explanations apply to our particular models, and (c) where we *have* checked, we have sometimes found that there are substantial contributions from terms besides those considered in the simple explanation, but these additional terms almost cancel each other.

We return now to discuss the sensitivity of shell-model $B(M1)$ strengths to the choice among our model Hamiltonians. We have already noted that we find shell-model $B(M1)$ strengths to be sensitive in this way. Thus at the close of Section 3.4.5, we remarked that the $(\lambda\mu)$ selection rule which accounts for the insensitivity of ground-state-band $E2$ strengths does not apply to $M1$ strengths. And early in this section we indicated that the fluctuation of shell-model $B(M1)$ values, from one shell model to the next, inhibits us from making any firm general statements about the agreement between shell-model and measured $M1$ strengths. At this point we simply call attention to some of the striking cases of model-sensitivity shown in Table XXIV. One example is the next-to-last transition listed for $^{18}$F: $(JT = 1'0) \rightarrow (JT = 01)$. Its $B(M1)$ is 0.37, 0.79, and $0.04\mu_N{}^2$, for the $K+^{17}O$, $RIP$, and $MSDI$ Hamiltonians respectively. Other examples are the $(3/2)' \rightarrow 5/2$ transition in $^{19}$O, with $B(M1)$ of 0.17, 0.53, and $0.06\mu_N{}^2$, the $3' \rightarrow 4$ transition in $^{20}$F, with $B(M1)$ of 2.60, 0.94, and $3.34\mu_N{}^2$, and the $1' \rightarrow 2$ transition in $^{20}$F, with $B(M1)$ of 0.46, 1.69, and $0.15\mu_N{}^2$. Even the generally strong transitions often show considerable variation in $B(M1)$, from one Hamiltonian to another. However, for a given mass value the *strongest* $M1$ transition does retain its identity, independent of the choice among our model Hamiltonians.

## 6.5. Comparison to *M*1 Results from Rotational Models

Here we compare relative $B(M1)$ values as calculated from our shell models, with relative $B(M1)$ values as calculated from the adiabatic-rotational model. Also, we compare shell-model magnetic moments $\mu$ with those calculated from two kinds of rotational-model wave functions: adiabatic-rotational wave functions, and microscopic wave functions obtained by $J$-projection out of deformed intrinsic states.

We consider first the adiabatic-rotational model for axially symmetric nuclei, with wave functions $\mid \alpha\,J\,K,\,J_z\rangle$ as in (17). The strength of an $M1$ transition from an initial unpolarized "pure-band" state $\mid \alpha_i\,J_i\,K_i\rangle$, to a final unpolarized pure-band state $\mid \alpha_f\,J_f\,K_f\rangle$, is given by (*118*)

$$B(M1)_{i \to f} = f(\alpha_i K_i,\,\alpha_f K_f)[C^{J_f}_{K_i\;1_{K_i-K_f},\;K_f}]^2 \quad \text{if } K_i \neq 1/2 \text{ or } K_f \neq 1/2 \quad (28)$$

Here the factor $f(\alpha_i K_i,\,\alpha_f K_f)$ depends on the initial and final bands, but not on the $J_i$ and $J_f$ within these bands. The function $f$ is symmetric under exchange of its arguments $\alpha_i K_i$ and $\alpha_f K_f$.

In order to apply (28), we assign the quantum numbers $K_i$ and $K_f$ shown in Table XXIV. These $K$ assignments are based simply on the sequences of spin values in the energy spectra, and on the kind of bands expected according to a Nilsson model in which the lowest single-particle orbit has $k = 1/2$, followed by $k = 3/2$ and then $k = 5/2$. We have listed $K$ assignments even in cases where we suspect that the pure-rotational-band picture is a poor one (e.g., for the oxygen isotopes, and quite generally, for levels outside ground-state bands and levels close to other levels of the same $J$).

The large right-hand brackets in Table XXIV mark sets of transitions connecting the same pair of bands, $\alpha_i K_i \rightleftarrows \alpha_f K_f$. Within each bracketed set for which $K_i \neq 1/2$ or $K_f \neq 1/2$, the adiabatic-rotational model would give $B(M1)$ values proportional to the squared Clebsch–Gordon coefficients of (28). The numerical values of these squared coefficients are listed in Table XXIV. Clearly, their ratios do not go very far toward explaining the ratios of $B(M1)$ values calculated in our various shell models. Some specific examples are pointed out in the following paragraphs.

We focus attention first on ground-state bands, where our $K$ assignments are likely to be more trustworthy. Here are two examples which show rough agreement between the ratios of shell-model $B(M1)$ values, and the ratios of squared Clebsch–Gordan coefficients: the $K = 3/2$ ground-state-band results for $^{21}$Ne, and the $K = 3$ ground-state-band results for

$^{22}$Na. Two examples of poor agreement are the $K = 3/2$ ground-state-band results for $^{19}$O, and the $K = 2$ ground-state-band results for $^{20}$F. In these latter two examples we see the same kinds of features as noted previously for interband $E2$ transitions. That is, the percentage variation in $B(M1)$ strength, for the same transition but from one shell model to the next, often exceeds the variation from one transition to the next as predicted by the rotational model. Also, for a given effective Hamiltonian, the shell-model $B(M1)$ values often show more variation in strength, from one transition to the next, than the rotational model does.

As to intraband transitions, for these too Table XXIV shows generally poor agreement between shell-model and rotational-model $B(M1)$ ratios. For example, look at the right-bracketed set of six $^{19}$O transitions assigned as connecting the lowest $K = 1/2$ band to the lowest $K = 3/2$ band. Within this right-bracketed set, the squared Clebsch–Gordan coefficients seem to bear no relation at all to the $K+^{17}$O $B(M1)$'s. For $^{19}$F, we do see a glimmer of agreement; it is in the weakness of our shell-model strengths for $M1$ transitions connecting the presumed $K = 5/2$ and presumed $K = 1/2$ bands. The rotational model says that these transitions are $K$-forbidden. However, the rotational model also predicts $K$-forbiddenness for the $^{18}$O transitions which we have assigned as $(K_i = 2)$ $\rightarrow (K_f = 0)$, yet our shell models give a strong $B(M1)$ for the $4' \rightarrow 4$ transition in this group.

Surely some of these disagreements should be attributed to the poorness of the "pure-$K$ band" assumptions which we have made in applying the adiabatic-rotational model. As we have noted earlier, small impurities in predominantly pure $(\lambda\mu)$ states will have first-order effects on the $B(M1)$ strengths between such states. Another source of disagreement may be the "inadequacy" of the adiabatic-rotational model as an approximation to the microscopic $J$-projected rotational model. (See Section 5.4.3.) We have found almost no published information which allows comparison of adiabatic-rotational $B(M1)$ strengths with exact $J$-projection $B(M1)$ strengths when both models incorporate the same intrinsic state. Hence, we do not know whether the discrepancies between such strengths would be significant in accounting for the differences that we find between adiabatic-rotational $B(M1)$ results and our own $s–d$-shell-model $B(M1)$ results.[21]

---

[21] One may also ask: Are there *other* rotational models which we should consider?— models which might lead to results better suited to comparison with our shell-model results? In this connection we note that there is no clear relation between the $L$ selection rule arguments which we used in Section 4.4, and the $J$-dependent Clebsch–Gordan coefficients of the adiabatic-rotational formula (28). Nor does the exact-$J$-

In the case of magnetic moments, there *is* published information which allows comparison between results calculated from adiabatic-rotational wave functions, and results calculated from *J*-projected wave functions, when both models involve similar intrinsic states (*12, 14, 133*). Table XXIII includes, under the heading "Rotational models," some magnetic moments calculated from three kinds of "pure-band" wave functions:

(i)   adiabatic-rotational wave functions, with intrinsic states $\chi$ which are the lowest energy axially symmetric Hartree–Fock solutions for a Gaussian–Rosenfeld nucleon–nucleon potential;

(ii)  microscopic wave functions obtained by *J*-projection from Hartree–Fock intrinsic states, the latter very similar (*12*) to those used in (i); and

(iii) microscopic wave functions obtained by *J*-projection from the $SU_3$ state of maximum weight.

There are three states in Table XXII for which magnetic moments of types (i) and (ii) are available. Judging from these, one would conclude that for ground-state magnetic moments in light *s*–*d*-shell nuclei, the adiabatic model is a good approximation to the *J*-projected model when both models incorporate the same intrinsic state.

The magnetic moments calculated from the $SU_3$ shell-model wave functions of item (iii) are associated with the same general model as the $SU_3$-shell-model *Q* values and $B(E2)$ strengths which we discussed earlier (see Section 5). In two cases, $J = 5/2$ for $^{19}$O and $J = 3/2$ for $^{21}$Ne, there seems to be a significant difference between the $\mu$ calculated from this $SU_3$ shell model, and the $\mu$ calculated from our shell-model Hamiltonian eigenvectors. In the four other cases where Table XXII allows comparison, there is good agreement between the $SU_3$ results and our shell-model results.

## 7. SUMMARY AND FINAL REMARKS

Using the full basis of Pauli-allowed $(0s)^4(0p)^{12}(1s, 0d)^{4-16}$ states, we have performed conventional shell-model calculations for $A = 18$–22. Seven

---

projection scheme have any clear relation to our *L*-selection-rule arguments. However, there is an alternative projection scheme to be considered, the "*L*-projection" scheme. In this latter scheme, states of good *L* are projected from orbital intrinsic states having quantized $K_L$; then a resulting good-*L* state is coupled with a charge-spin function of quantized *S* so as to produce a good-*J* state. [For a more detailed description of the *L* projection scheme, see (*8, 9*)].

alternative $(1 + 2)$-body Hamiltonians were used. Two of these were the "realistic" Hamiltonians proposed by Kuo and Brown (*3*) and Kuo (*4*) for $A = 18$. Three other Hamiltonians were obtained by allowing various adjustments in these Kuo (*3*) and Kuo-Brown (*4*) Hamiltonians, the adjustable parameters being fixed by requiring least-squares fits to measured energies in $A = 17$–22. The remaining two Hamiltonians were also obtained via least-square fits, but these Hamiltonians were not based on "realistic" forces. They each involved a phenomenological central force (a different form, for the two different Hamiltonians), and in each case the one-body parts of the Hamiltonian, and the strengths of selected two-body parts, were determined by least-square search. Besides energies, we calculated the following additional kinds of shell-model "observables": spectroscopic factors $S$ for single-nucleon transfer, $E2$ moments and transition strengths, and $M1$ moments and transition strengths. The electromagnetic moments and strengths were calculated with rather simple, standard effective operators (see below). The shell-model observables calculated from our alternative Hamiltonians were compared with each other, with experimental results, and with results obtained from various rotational models.

In summarizing the results we find it convenient to distinguish between two groups of calculated observables. We start with GROUP 1. This group includes *static properties* and *transition strengths* involving only these states: the three $A = 17$ states, the lowest triad of states in each of $^{18}$F and $^{18}$O, and all those states in $A = 19$–22 which seem to be members of ground-state rotational bands.

The measured energies of these GROUP-1 states were included as fitting criteria in the searches which determined our five "least-square" Hamiltonians. These measured excitation energies are fairly well reproduced by all seven of our Hamiltonians; and so it follows that these excitation energies are rather insensitive to the choice among our Hamiltonians. In particular, the qualitative deviations of the observed energies from $J(J + 1)$ behavior are generally reproduced by all seven of our Hamiltonians.

Spectroscopic factors $S$ for these GROUP 1 states are more sensitive to the choice among our model Hamiltonians. We do *not* consistently find that if a given Hamiltonian yields the best agreement with the measured energy of an observed state, the same Hamiltonian will yield similarly superior agreement for $S$-factors involving that state. Generally speaking, the "realistic" Kuo Hamiltonian (*4*) reproduces experimentally determined $S$-factors at least as well as our other Hamiltonians do. (In considering these theoretical-*vs*-experimental comparisons, we bear in mind that experimentally determined $S$-factors are uncertain by $\gtrsim 30\%$.)

Shell-model $B(E2)$'s and $Q$ moments were calculated by assuming that the effective $E2$ operator has the usual one-body form, with total effective charges $\tilde{e}_p = 1.5e$ and $\tilde{e}_n = 0.5e$. Shell-model $B(M1)$ and $\mu$ values were calculated with the free-nucleon $M1$ operator. Our comparisons with experimental $E2$ and $M1$ data do not indicate any clearly superior alternatives to these simple choices of effective electromagnetic operators. For GROUP-1 $E2$ and $M1$ observables, the agreement between our model results and experiment is fair. (As in the case of spectroscopic factors, there are inconsistencies in the experimental determinations, and these inconsistencies cast doubt on the reliability of *all* the experimental determinations. This doubt complicates our assessment of theoretical-*vs*-experimental agreement.)

How sensitive are the calculated $E2$ and $M1$ observables to the effective Hamiltonian assumed? In GROUP 1, the computed $Q$ moments, $B(E2)$ strengths, and $\mu$ moments, like the excitation energies, are generally insensitive to the choice among our seven Hamiltonians. In contrast, the $B(M1)$ strengths are generally sensitive to the choice among our Hamiltonians. Since there are rather few experimental determinations of $B(M1)$ for $A = 18$–22, we have not formed any conclusion about which Hamiltonian yields the best agreement between shell-model and measured $B(M1)$ strengths.

In general then, our shell models yield fair-to-good agreement with measured results for the observables in GROUP 1: observables involving only the $A = 17$ levels, the lowest triad of levels in each of $^{18}$F and $^{18}$O, and ground-state-band levels in $A = 19$–22.

The above pleasant conclusion does not apply to GROUP-2 observables. In GROUP 2 we include all observables involving any states other than those mentioned in our definition of GROUP 1. For example, GROUP 2 includes any transition strength connecting two levels which are both outside the ground-state band; also, GROUP 2 includes any transition strength connecting a ground-state-band level with a level outside the ground-state band. We find that our models do yield some states whose calculated energies and spectroscopic factors agree fairly well with measured GROUP-2 data. Hence we can propose very plausible shell-model partners for some of the observed nuclear states outside ground-state bands. However, in these GROUP-2 experimental–theoretical partnerships, the agreement of shell-model excitation energies and $S$-factors is generally poorer than what is found for ground-state-band levels. For some of the low-lying observed states outside ground-state bands, we do not find plausible shell-model partners, while for some of the low-lying shell-model states outside ground-state bands, we do not find plausible observed partners. Indeed, since the GROUP-2 excitation energies and spectroscopic factors tend to be rather sensitive to the choice among our model Hamiltonians, the existence of a plausible partnership sometimes depends on the model Hamiltonian being considered. Suppose we were to base our preference in Hamiltonians on their success in reproducing measured energies and S-factors outside ground-state bands. Then we would say that the Kuo and Kuo–Brown "realistic" Hamiltonians, and our three adjusted versions of these realistic Hamiltonians, are somewhat superior to our two central-force Hamiltonians. As to GROUP-2 electromagnetic observables, they are generally sensitive to the choice among our model Hamiltonians, and there are almost no cases where we can claim agreement between shell-model and measured results.

Besides comparing with experiment, we have compared our shell-model $E2$ and $M1$ results with results calculated from a variety of rotational models. By "rotational model" we mean any model inspired by the idea that if a group of nuclear levels exhibits excitation energies roughly proportional to $J(J + 1)$, this behavior implies a series of nuclear states all similar to each other in "intrinsic" motion but differing in their collective-rotational energy.

The simplicity of the *strong-coupling* rotational model allows us to compute, quite easily, a large set of rotational-model results with which to compare our shell-model results. For levels within ground-state bands, we made some special assumptions about the intrinsic state $\chi$ in strong-coupling wave functions. These assumptions allowed us to compute absolute (not just relative) $Q$ moments and $B(E2)$ strengths within ground-state bands. We find many striking similarities between the ground-state-band $E2$ results calculated from this specialized strong-coupling rotational model, and the results calculated from our shell models. However, for $E2$ observables involving levels *outside* ground-state bands, and for all $M1$ observables, there is rather little resemblance between strong-coupling rotational-model predictions and our shell-model results.

We have suggested several reasons for the pattern of agreement and disagreement seen in our comparison of shell-model results and strong-coupling-rotational results. Some of these suggestions involve the importance of contributions from the "leading $SU_3$ representation" $(\lambda\mu)$ in our shell-model wave functions. Other suggestions bring into consideration the partly macroscopic character of strong-coupling wave functions, as contrasted to the fully microscopic character of our shell-model Hamiltonian eigenfunctions. To explore the role of these macroscopic–microscopic differences, we have also made some comparisons with results from fully microscopic rotational models, e.g., $SU_3$.

Our section on nuclear binding energies is very short—short because the shell-model results inspired us with very few ideas we thought worth communicating. Each of our seven Hamiltonians yields a set of nuclear binding energies with respect to $^{16}O$. The differences among these sets, and also the differences between theory and experiment, could be very much diminished by adding small state-independent shifts to the diagonal one-body and two-body matrix elements of our model Hamiltonians. Such shifts would not affect the computed excitation spectra for fixed $A$.

Our $(0s)^4(0p)^{12}(1s, 0d)^{A-16}$ models can describe only *even*-parity states. But the experimentally observed spectra for $A = 18$–$22$ include (a) some low-lying odd-parity levels, and (b) some low-lying even-parity levels which apparently defy description in terms of a $(0s)^4(0p)^{12}(1s, 0d)^{A-16}$ basis. Recently many of these states (a) and (b) have been described quite successfully by shell models which use a basis of all $(0s)^4(0p_{3/2})^8(0p_{1/2}, 0d_{5/2}, 1s_{1/2})^{A-12}$ states (*88–89, 138–140*). Furthermore, these same models account for the principal properties of nearly all the observed even-parity states which our $(0s)^4(0p)^{12}(1s, 0d)^{A-16}$ models describe. Note that the $(0s)^4(0p_{3/2})^8$ $(0p_{1/2}, 0d_{5/2}, 1s_{1/2})^{A-12}$ basis excludes $0d_{3/2}$ nucleons but allows $0p_{1/2}$ holes. In such "open-core" calculations there is the possibility that low-lying states may be contaminated by spurious center-of-mass motion (*141*). In the open-core studies which we cite (*88–89, 138–140*) this center-of-mass problem has not yet been formally investigated. However, the gratifying success of these models in reproducing observed data hints that spurious center-of-mass motion does not seriously affect most of the nuclear energies, moments, and transition strengths being calculated. In the $A = 18$–$22$ region there are of course some observed properties, and a few low-lying states, which defy description in terms of any basis without $d_{3/2}$ nucleons. Hence studies in the two vector spaces, $(0s)^4(0p)^{12}(1s, 0d)^{A-16}$ and $(0s)^4(0p_{3/2})^8$-$(0p_{1/2}, 1s_{1/2}, 0d_{5/2})^{A-12}$, complement each other.

As the present study indicates, there are a *variety* of $(1 + 2)$-body Hamiltonians, acting within the full $(0s)^4(0p)^{12}(1s, 0d)^{A-16}$ space, which are almost equally successful in reproducing the observed properties of even-parity $A = 18$–$22$ states. Among the successful alternatives, "realistic" Hamiltonians have obvious philosophical advantages. However, present-day realistic-effective-interaction techniques are not good enough to make us confident about any one Hamiltonian that is produced by these techniques. We do not know whether the next few years of realistic-interaction theory and the next few years of experimental measurements will be of significant help either for finding better Hamiltonians for $(0s)^4(0p)^{12}(1s,$ $0d)^{A-16}$ calculations, or for discriminating among Hamiltonians which are now equally satisfactory in this space. Indeed, we do not know how well one should *expect* to reproduce the observed properties of $A = 18\text{–}22$, using a $(0s)^4(0p)^{12}(1s, 0d)^{A-16}$ vector space with a $(1 + 2)$-body Hamiltonian operator and simple one-body transition and moment operators.

The present paper is a lengthy illustration of the fact that relatively simple models can reproduce many observed features of truly complicated states. In brief, the reasons for this fact are thought to be the following:

(a) all observed features are related to matrix elements

$$\langle \psi_{\text{true}} \mid \mathscr{O}_{\text{true}} \mid \psi'_{\text{true}} \rangle;$$

(b) it is possible to define transformations which map the true nuclear wave functions onto model wave functions, and related transformations which map true operators onto effective operators, such that

$$\langle \psi_{\text{model}} \mid \mathscr{O}_{\text{eff}} \mid \psi'_{\text{model}} \rangle = \langle \psi_{\text{true}} \mid \mathscr{O}_{\text{true}} \mid \psi'_{\text{true}} \rangle \qquad (29)$$

(c) there *exist* some tractable shell-model vector spaces, and some interesting classes of observables, which allow practical applications of (29).

In (c), a "practical application" is one in which $H_{\text{eff}}$ is a simple $(1 + 2)$-body operator, and in which the pertinent effective transition and moment operators are simple. In any practical application, (29) becomes, of course, an *approximate* equality.

The study and elaboration of points (a)–(c) has been variously termed "the theory of nuclear models," "foundations of the nuclear shell model," "the reaction matrix in nuclear shell theory," and "effective-operator theory." (Since the Hamiltonian is an operator, we let "effective-operator theory" *include* "effective-interaction theory.") Many effective-operator studies have suggested mathematical techniques for constructing effective operators, and numerous studies have utilized these techniques to produce specific operators suitable for use in particular shell-model vector-spaces (*3–4, 22, 85–86, 115–117, 126–128, 143*). The general and specific problems in this field are far from being solved in detail. Still, in any shell-model study such as ours, the procedures are to some extent suggested, justified, and illuminated by effective-operator theory. The calculated results should be interpreted accordingly. It is in recognition of the importance and relevance of effective-operator concepts that we have chosen to end our article by drawing attention to them.

## ACKNOWLEDGMENTS

We thank P. W. M. Glaudemans for participating in the initial stages of work with the *MSDI* Hamiltonian form. The least-square searches described in this paper represent the first large-scale use of a computer code written by Glaudemans and one of us (B.H.W.). We are grateful to K. H. Bhatt for suggesting the "specialized adiabatic-rotational model" (see Section 5.4.2), and for many educational discussions about rotational models. We thank the following physicists for helpful conversations, preprints, and other unpublished information: J. H. Aitken, R. M. Diamond, H. T. Fortune, A. J. Howard, T. T. S. Kuo, L. M. Polsky, J. G. Pronko, F. W. Prosser, D. J. Pullen, P. A. Quin, C. Rolfs, and E. K. Warburton. We are grateful to Althea Tate for typing the many drafts and the final manuscript of this article.

## APPENDIX A.  SHELL-MODEL BASIS AND METHODS

### A.1.  Coupling Scheme

The shell-model basis for our calculations forms a complete set of states for the configurations

$$[(0s_{1/2})^4(0p_{3/2})^8(0p_{1/2})^4]\,(0d_{5/2})^{n_1}(1s_{1/2})^{n_2}(0d_{3/2})^{n_3} \qquad (A1)$$

where $n_1 + n_2 + n_3 = A - 16$. The shells within square brackets in (A1) form a closed core, the same for all basis states. For many purposes this common closed core can be ignored, and we shall ignore it is specifying our basis states. Each basis state is then specified by the ordered coupling scheme for its three active shells,

$$\psi_{\text{basis}}^{JT} \sim [(\psi_1^{\gamma_1} \times \psi_2^{\gamma_2})^{J_{12}T_{12}} \times \psi_3^{\gamma_3}]^{JT} \qquad (A2)$$

and by the quantum numbers within each active shell, e.g.,

$$\psi_1^{\gamma_1} \equiv \psi^{J_1 T_1}[(0d_{5/2})^{n_1}s_1 t_1 x_1] \qquad (A3)$$

In (A2) and (A3) the subscript 1 refers to the $0d_{5/2}$ shell, subscript 2 to the $1s_{1/2}$ shell, and subscript 3 to the $0d_{3/2}$ shell. The superscript $\gamma_i$ denotes the angular momentum and also the isospin contributed by nucleons in shell $i$. The symbol $\times$ implies coupling via the usual Clebsch–Gordan coefficients.

In the "single-shell" functions (A3), $s$ denotes seniority, $t$ denotes reduced isospin, and $x$ is a quantum number used to distinguish among $j^n$-states which are identical in all the other quantum numbers explicitly listed.

## A.2. Phases

Although the formulas programmed into our shell-model codes are most easily derived with a second-quantized formalism (5), we shall state our phase conventions in terms of a corresponding first-quantized description. We shall include information for the general case of $n$ particles (even though for this paper only the one- and two-body phases are strictly relevant, because we exhibit two-body matrix elements but not many-body wave functions).

There are three kinds of phase conventions to discuss: those assumed for the single-particle states, those used in coupling single-particle states to form single-shell states, and those used in coupling single-shell states to form multishell states. The phases of the single-particle states $\varphi$ are given by an $l–s$ (rather than $s–l$) coupling scheme:

$$\varphi_{m_j}^{lsj} = (\boldsymbol{\theta}^l \times \boldsymbol{\chi}^s)_m^{\ j} = \sum_{m_l m_s} C_{m_l\ m_s\ m_j}^{l\ \ 1/2\ \ j}\ \theta_{m_l}^l\ \chi_{m_s}^{1/2} \tag{A4}$$

where

$$\boldsymbol{\theta}^l = R_{nl}(r)\ \mathbf{Y}^l(\Omega) \tag{A5}$$

with $R_{nl}(r)$ positive at small values of $r$. Here $\mathbf{Y}^l(\Omega)$ is the usual (101) spherical harmonic tensor operator of rank $l$. All single-shell functions such as (A3) are defined with phases specified by available fractional-parentage coefficients. These fractional-parentage coefficients were calculated by a method due to J. N. Ginocchio and J. B. French (142), with a computer program written mainly by Joan Rayburn of Oak Ridge National Laboratory. All multishell states (A2) are coupled and antisymmetrized so that there is a _positive_ coefficient for the term having particle numbers assigned as in

$$\{[\boldsymbol{\psi}_1^{\gamma_1}(1, \ldots, n_1) \times \boldsymbol{\psi}_2^{\gamma_2}(n_1 + 1, \ldots, n_1 + n_2)]^{J_{12}T_{12}} \times \boldsymbol{\psi}^{\gamma_3}(n_1 + n_2 + 1, \ldots,$$
$$n_1 + n_2 + n_3)\}^{JT} \tag{A6}$$

These single-shell and multishell conventions are consistent with a scheme in which all antisymmetrized two-nucleon wave functions are given by

$$|abJT\rangle = 2^{-1/2}(1 + \delta_{ab})^{-1/2}\{[\boldsymbol{\varphi}^a(1) \times \boldsymbol{\varphi}^b(b)]^{JT} - [\boldsymbol{\varphi}^a(2) \times \boldsymbol{\varphi}^b(1)]^{JT}\} \tag{A7}$$

Here $a$ implies the single-particle ranks $l_a$ and $t = 1/2$ and $j_a$, but not their $z$-projections.

The phases of the Kuo (4) and Kuo–Brown (3) interactions agree with our conventions (A1)–(A7); and our least-squares searches are arranged so that their Hamiltonian solutions are expressed in terms of a set of two-body matrix elements appropriate to the phases we have just described. Of course we must choose these same phase conventions, again, when we specify the single-particle matrix elements that define any given one-body operator (such as the magnetic dipole operator) whose many-body matrix elements are to be evaluated between Hamiltonian eigenstates.

## A.3. Computer Codes

All of the calculations reported here were run on the IBM 360/75.

The evaluation of Hamiltonian matrix elements between our many-body basis states, and the evaluation of moment and transition matrix elements between many-body Hamiltonian eigenstates, were handled by the Oak Ridge–Rochester shell-model codes. These codes and their underlying formalism have been described in detail in a recent publication (5). The basis-state representation described by (A1)–(A3) and (A6) was chosen because the Oak Ridge–Rochester shell-model codes are programmed for this kind of representation. These codes are based on a second-quantized formalism which takes advantage of the rotational symmetries of the basis states, and the rotational symmetries of the Hamiltonian operator which is to be diagonalized. In our computer calculations, the essential practical limitation is the size of the Hamiltonian matrix (for fixed $A, J, T$). The largest matrix in the present $s$-$d$-shell study was $537 \times 537$. It took approximately 26 minutes to construct this matrix on the IBM 360/75, and then another 33 minutes to obtain its lowest four eigenvalues and the associated eigenvectors. For $A = 23$, the dimensionalities of many of the full $s$-$d$-shell matrices exceed $1000 \times 1000$ (6). The treatment of such $1000 \times 1000$ matrices is impractical with presently available computers and codes.

In a shell-model Hamiltonian matrix for fixed $A, J, T$, each many-body element is expressible as a linear combination of the one-body and two-body matrix elements. The coefficients in this linear combination are "geometric"; they involve combinations of Racah coefficients, and they depend on the many-body basis states but are independent of the $(1 + 2)$-body Hamiltonian operator. In the course of computing an $A, J, T$ matrix for a definite Hamiltonian operator, the Oak Ridge–Rochester programs can write a magnetic tape which stores the numerical values of these geometric coefficients for each element in the $A, J, T$ matrix. By saving and remounting this tape, we are able to quickly construct the numerical $A, J, T$ matrix for a new Hamiltonian operator. For our present calculation we save a tape which stores the geometric coefficients for *many* $A = 17$–22 matrices.

This tape of geometric coefficients serves as input to the computer program which finds our "least-square" Hamiltonians. This least-square search program was written by P. W. M. Glaudemans and one of us (B. H. W.). Since a description of this program has not been published, we shall outline its main features below.

Starting with an initial Hamiltonian $H^{(0)}$, the search program produces by iteration a sequence of Hamiltonians $H^{(1)}$, $H^{(2)}$, $H^{(3)}$, etc. The aim is to converge toward a Hamiltonian whose eigenvalues minimize the mean square deviation,

$$D \equiv (\text{rms})^2 = \frac{1}{L} \sum_{\lambda=1}^{L} [E_\lambda^{\text{expt}} - E_\lambda^{\text{calc}}]^2 \tag{A8}$$

between a selected set of $L$ experimentally observed energies and their shell-model counterparts. Throughout the search, these shell-model counterparts are to be computed in the same vector space as was used to generate the aforementioned geometric-coefficient tape, which is mounted as input. The $L$ observed levels are specified by card input.

The initial Hamiltonian $H^{(0)}$ can be any $(1 + 2)$-body operator; it is specified (on card input) by its pertinent one-body and two-body matrix elements. But all subsequent Hamiltonians $H^{(1)}$, $H^{(2)}$, $H^{(3)}$, etc. are constrained to a *particular variational form*—a form that is specified by yet another set of input data. The present code is designed to handle problems in which these subsequent Hamiltonians have the general form

$$H = H_0 + \sum_{p=1}^{P} h_p H_p$$

Here each $h_p$ is an adjustable Hamiltonian parameter, and each $H_p$ is a $(1 + 2)$-body operator that is specified by program input and held fixed throughout the iterative search. Thus, for $H^{(i)}$—the $i$th Hamiltonian in the iterative sequence $H^{(1)}$, $H^{(2)}$, $H^{(3)}$, etc.—we have

$$H^{(i)} = H_0 + \sum_{p=1}^{P} h_p^{(i)} H_p \qquad \text{when } i > 0 \tag{A9}$$

The fixed Hamiltonian operators $H_p$ of (A9) are specified, by card input, in terms of their one-body and two-body matrix elements. The many-body matrix elements of $H^{(0)}$, $H^{(i>0)}$, and each $H_p$ can be very rapidly evaluated with the help of the geometric coefficients.

To minimize $D$ of (A8), the program searches for a set of parameters $h_1, \ldots, h_p$ such that each $\partial D/\partial h_p$ will be zero. Suppose we have already found $H^{(i)}$. For the set of parameters which would define $H^{(i+1)}$, the first derivatives of $D$ would be

$$\left(\frac{\partial D}{\partial h_p}\right)^{(i+1)} = \frac{1}{L} \sum_{\lambda=1}^{L} 2[E_\lambda^{\text{expt}} - E_\lambda^{\text{calc}(i+1)}] \left(\frac{\partial E_\lambda^{\text{calc}}}{\partial h_p}\right)^{(i+1)} \qquad (A10)$$

To find $H^{(i+1)}$, the search program constructs and solves a system of $P$ equations. Each equation sets to zero an *estimate* of one of the first derivatives (A10), expressed in terms of the unknown parameters $h_1^{(i+1)}, h_2^{(i+1)}, \ldots, h_P^{(i+1)}$.

The course of a single iteration $i \to (i+1)$ proceeds as follows. First, the program uses the one- and two-body matrix elements defining $H^{(i)}$, together with the geometric coefficients, to construct the many-body $A, J, T$ matrices of $H^{(i)}$. Second, the program partially diagonalizes these matrices, so as to get the pertinent $L$ eigenvalues $E_\lambda^{\text{calc}(i)}$ and their eigenvectors $\psi_\lambda^{(i)}$. Third, the program estimates each derivative $(\partial D/\partial h_p)^{(i+1)}$ by using the following approximations[22]:

$$E_\lambda^{\text{calc}(i+1)} \equiv \langle \psi_\lambda^{(i+1)} | H^{(+1)} | \psi_\lambda^{(i+1)} \rangle$$

$$\approx \langle \psi_\lambda^{(i)} | H^{(i+1)} | \psi_\lambda^{(i)} \rangle \equiv H_0 + \sum_{p=1}^{P} h_p^{(i+1)} \langle \psi_\lambda^{(i)} | H_p | \psi_\lambda^{(i)} \rangle \qquad (A11)$$

$$\left(\frac{\partial E_\lambda^{\text{calc}}}{\partial h_p}\right)^{(i+1)} \equiv \left\langle \psi_\lambda^{(i+1)} \left| \frac{\partial H}{\partial h_p} \right| \psi_\lambda^{(i+1)} \right\rangle$$

$$\approx \left\langle \psi_\lambda^{(i)} \left| \frac{\partial H}{\partial h_p} \right| \psi_\lambda^{(i)} \right\rangle \equiv \langle \psi_\lambda^{(i)} | H_p | \psi_\lambda^{(i)} \rangle \qquad (A12)$$

The program calculates the many-body matrix elements $\langle \psi_\lambda^{(i)} | H_p | \psi_\lambda^{(i)} \rangle$, on the right of (A11) and (A12), again with the help of the geometric coefficients. When the approximations (A11) and (A12) are inserted in (A10), and when the resulting estimates of $(\partial D/\partial h_p)^{(i+1)}$ are set equal to zero, we get a set of $P$ linear equations involving the unknowns $h_1^{(i+1)}, h_2^{(i+1)}, \ldots, h_P^{(i+1)}$. The search program solves these linear equations, to get the new set of parameters $h_p^{(i+1)}$. These new parameters define $H^{(i+1)}$, and one is then ready for the next iteration.

---

[22] We have written (A11) and (A12) for the case in which every $E_\lambda^{\text{calc}}$ is an eigenvalue, rather than a difference between two eigenvalues. (Generalizations are easy.)

# REFERENCES

*1.* Preliminary accounts of the present work are given by E. C. Halbert, J. B. McGrory, and B. H. Wildenthal, *Phys. Rev. Letters*, **20**:112 (1968); and E. C. Halbert, in *Third Symposium on the Structure of Low-Medium Mass Nuclei* (J. P. Davidson, ed.), University Press of Kansas, Lawrence and London (1968).

*2.* T. Hamada and I. D. Johnston, *Nucl. Phys.*, **34**:382 (1962).

*3.* T. T. S. Kuo and G. E. Brown, *Nucl. Phys.*, **85**:40 (1966).

*4.* T. T. S. Kuo, *Nucl. Phys.*, **A103**:71 (1967); and private communication.

*5.* J. B. French, E. C. Halbert, J. B. McGrory, and S. S. M. Wong, in *Advances in Nuclear Physics*, Vol. 3 (M. Baranger and E. Vogt, eds.), Plenum Press, New York (1969).

*6.* T. Sebe and M. Harvey, Atomic Energy of Canada Ltd., Chalk River, Report No. 3007 (1968).

*7.* J. P. Elliott, *Proc. Roy. Soc. (London)*, **A245**:128 and 562 (1958); and J. P. Elliott and M. Harvey, *Proc. Roy. Soc. (London)*, **A272**:557 (1963).

*8.* M. Harvey, in *Advances in Nuclear Physics*, Vol. 1 (M. Baranger and E. Vogt, eds.), Plenum Press, New York (1968).

*9.* J. P. Elliott and C. E. Wilsdon, *Proc. Roy. Soc. (London)*, **A302**:509 (1968).

*10.* Y. Akiyama, A. Arima, and T. Sebe, *Nucl. Phys.*, **A138**:273 (1969).

*11.* M. Moshinsky, in *Physics of Many-Particle Systems* (E. Meeron, ed.) Gordon and Breach, New York (1966). See also, for example: J. Flores and R. Perez, *Phys. Letters*, **26B**:55 (1967); and M. Berrondo and J. Pineda, *Rev. Mex. Fis.*, **16**:17 (1967).

*12.* G. Ripka, in *Advances in Nuclear Physics*, Vol. 1 (M. Baranger and E. Vogt, eds.), Plenum Press, New York (1968).

*13.* M. R. Gunye and C. S. Warke, *Phys. Rev.*, **156**:1087 (1967); and *Phys. Rev.*, **164**:1264 (1967).

*14.* M. R. Gunye and C. S. Warke, *Phys. Rev.* **159**:885 (1967).

*15.* I. P. Johnstone and H. G. Benson, *Nucl. Phys.*, **A134**:68 (1969).

*16.* M. H. Macfarlane (private communication).

*17.* A. Arima, S. Cohen, R. D. Lawson, and M. H. Macfarlane, *Nucl. Phys.*, **A108**:94 (1968).

*18.* B. H. Wildenthal, J. B. McGrory, E. C. Halbert, and P. W. M. Glaudemans, *Phys. Letters*, **26B**:692 (1968).

*19.* C. Maples, G. W. Goth, and J. Cerny, *Nucl. Data*, **A2**:429 (1966).

*20.* R. Arvieu and S. A. Moszkowski, *Phys. Rev.*, **145**:830 (1966).

*21.* T. T. S. Kuo, private communication; to be published in *Fourth Symposium on the Structure of Low-Medium Mass Nuclei* (J. P. Davidson, ed.), University Press of Kansas.

*22.* B. R. Barrett and M. W. Kirson, *Nucl. Phys.*, **A148**:145 (1970); L. Zamick, *Phys. Rev. Letters*, **23**:1406 (1969).

*23.* S. Cohen, R. D. Lawson, and S. P. Pandya, *Nucl. Phys.*, **A114**:541 (1968).

*24.* S. Cohen, E. C. Halbert, and S. P. Pandya, *Nucl. Phys.*, **A114**:353 (1968).

*25.* P. W. M. Glaudemans, P. J. Brussaard, and B. H. Wildenthal, *Nucl. Phys.*, **A102**:593 (1967); and an addendum explaining the phase conventions (P. W. M. Glaudemans, private communication).

26. B. H. Wildenthal, J. B. McGrory, E. C. Halbert, and P. W. M. Glaudemans, *Phys Letters*, **27B**:611 (1968).
27. L. M. Polsky, C. H. Holbrow, and R. Middleton, *Phys. Rev.*, **186**:966 (1969).
28. J. L. Wiza, R. Middleton, and P. V. Hewka, *Phys. Rev.* **141**:975 (1966).
29. B. Zeidman and T. H. Braid, *Phys. Letters*, **16**:139 (1965).
30. R. W. Ollerhead, J. S. Lopes, A. R. Poletti, M. F. Thomas, and E. K. Warburton, *Nucl. Phys.*, **66**:161 (1965).
31. D. Rendic, B. Antolkovic, G. Paic, M. Turk, and P. Tomas, *Nucl. Phys.*, **A117**:113 (1968).
32. M. F. Thomas, J. S. Lopes, R. W. Ollerhead, A. R. Poletti, and E. K. Warburton, *Nucl. Phys.*, **78**:298 (1966).
33. J. W. Olness and D. H. Wilkinson, *Phys. Rev.* **141**:966 (1966).
34. D. D. Tolbert, P. M. Cockburn, and F. W. Prosser, Jr., *Phys. Rev., Letters* **21**:1535 (1968).
35. D. D. Tolbert, Ph.D. Thesis, University of Kansas, 1968.
36. F. K. P. Jackson, P. G. Lawson, K. Bharuth-Ram, N. G. Chapman, and K. W. Allen, *Bull. Am. Phys. Soc.*, **13**:1370 (1968).
37. J. H. Aitken, K. W. Allen, R. E. Azuma, A. E. Litherland, and D. W. O. Rogers, *Phys. Letters* **28B**:653 (1969).
38. R. G. Hirko, R. A. Lindgren, A. J. Howard, W. Scholz, and D. A. Bromley, in "Proceedings of the International Conference on Nuclear Structure, Tokyo, 1967" (J. Sanada, ed.), *Suppl. J. Phys. Soc. Japan*, **24**:653 (1968).
39. J. H. Aitken, R. E. Azuma, and A. E. Litherland, *Physics in Canada*, **24**:40 (1968).
40. J. H. Aitken, private communication.
41. M. G. Silbert and N. Jarmie, *Phys. Rev.*, **123**:221 (1961).
42. H. Smotrich, K. W. Jones, L. C. McDermott, and R. E. Benenson, *Phys. Rev.*, **122**:232 (1961). The $^{19}$F excitation energies listed in this 1961 paper should be increased by 18 keV; this change is indicated by an improved value of the $^{15}$N–$^{19}$F mass difference.
43. J. H. Aitken, R. E. Azuma, A. E. Litherland, A. W. Charlesworth, D. W. O. Rogers, and J. J. Simpson, private communication.
44. D. W. O. Rogers, J. H. Aitken, R. E. Azuma, and A. E. Litherland, *Bull. Am. Phys. Soc.*, **14**:123 (1969).
45. C. O. Lennon, P. R. Alderson, J. L. Durell, L. L. Green, and I. M. Naqib, *Phys. Letters*, **28B**:253 (1968).
46. H. G. Benson and B. H. Flowers, *Nucl. Phys.*, **A126**:305 (1969).
47. K. W. Allen and P. G. Lawson, reported in Ref. 46.
48. M. R. Wormald and I. F. Wright, in *Contributions to the International Conference on Properties of Nuclear States, Montreal, 1969*, University of Montreal Press, Montreal (1969), p. 111.
49. J. L. Wiza and R. Middleton, *Phys. Rev.*, **143**:676 (1966).
50. J. C. Armstrong and K. S. Quisenberry, *Phys. Rev.*, **122**:150 (1961).
51. K. Yagi, Y. Nakajima, K. Katori, Y. Awaya, and M. Fujioka, *Vucl. Phys.*, **41**:584 (1963).
52. F. A. El Bedewi, M. A. Fawzi, and N. S. Rizk, in *Comptes Rendes du Congres Intern. de Physique Nucléaire*, Vol. II (P. Gugenburger, ed.) Editions du Centre National de la Recherche Scientifique, Paris (1964), p. 432.

*53.* R. Moreh and A. A. Jaffe, *Proc. Phys. Soc.* (*London*), **84**:330 (1964).

*54.* G. Wickenberg, S. Hjorth, N. G. E. Johansson, and B. Sjogren, *Arkiv Fysik*, **25**:191 (1963).

*55.* R. Siemssen, L. L. Lee, and D. Cline, *Phys. Rev.*, **140B**:1258 (1965).

*56.* J. A. Kuehner and R. W. Ollerhead, *Phys. Letters*, **20**:301 (1966).

*57.* J. D. Pearson, E. Almqvist, and J. A. Kuehner, *Can. J. Phys.*, **42**:477 (1964).

*58.* A. E. Litherland, in *Nuclear Structure and Electromagnetic Interactions* (N. MacDonald, ed.) Plenum Press, New York (1965).

*59.* T. Lauritsen and F. Ajzenberg-Selove, *Energy Levels of Light Nuclei*, National Academy of Sciences, National Research Council, May, 1962.

*60.* R. C. Bearse, H. T. Fortune, G. C. Morrison, J. L. Yntema, and B. H. Wildenthal, to be published.

*61.* P. A. Quin, "Status Report on $^{20}$F—August, 1969" (unpublished); and P. A. Quin, private communication.

*62.* P. A. Quin, Ph.D. Thesis, University of Notre Dame, 1968; P. A. Quin, G. A. Bissinger, and P. R. Chagnon, to be published.

*63.* P. A. Quin, A. A. Rollefson, G. A. Bissinger, C. P. Browne, and P. R. Chagnon, *Phys. Rev.*, **157**:991 (1967); G. A. Bissinger, R. M. Mueller, P. A. Quin, and P. R. Chagnon, *Nucl. Phys.*, **A90**:1 (1967).

*64.* P. Spilling, H. Gruppelaar, H. F. de Vries, and A. M. J. Spits, *Nucl. Phys.*, **A113**:395 (1968). The 1.885-MeV gamma ray which they attributed to the 4.082 → 2.195 MeV transition could be associated, instead, with the 5.413 → 3.526 MeV transition.

*65.* R. L. Hershberger, M. J. Wozniak, Jr., and D. J. Donahue, *Phys. Rev.*, **186**:1167 (1969).

*66.* R. Hardell and A. Hasselgren, *Nucl. Phys.*, **A123**:215 (1969).

*67.* F. A. El Bedewi, *Proc. Phys. Soc.* **A69**:221 (1956).

*68.* R. Moreh, *Nucl. Phys.* **70**:293 (1965).

*69.* R. Middleton, in *Proc. Conf. Direct Interactions and Nuclear Reaction Mechanisms* (E. Clementel and C. Villi, eds.) Gordon and Breach, New York and London (1963), p. 435.

*70.* C. Rolfs, private communication. Recent work on $^{21}$Ne shows that a previous experimental determination of $J = 3/2$ for the 2.79-MeV state was incorrect; the new assignment is $1/2^-$. Also, a previous assignment of $J^\pi = 5/2^+(3/2^+)$ to the 3.74-MeV level is changed to a definite $J^\pi = 5/2^+$, and the state at 6.45-MeV is now assigned $13/2^{(+)}(9/2)$.

*71.* D. J. Pullen, A. Sperduto, and E. Kashy, *Bull. Am. Phys. Soc.*, **10**:38 (1965); and D. J. Pullen (private communication).

*72.* A. J. Howard, J. G. Pronko, and C. A. Whitten, Jr., *Phys. Rev.*, **184**:1094 (1969); and private communication.

*73.* M. B. Burbank, G. G. Frank, N. E. Davison, G. C. Neilson, S. S. M. Wong, and W. J. McDonald, *Nucl. Phys.*, **A119**:194 (1968). We believe that the shell-model *S*-values which they list should be multiplied by a factor of two.

*74.* F. X. Haas, C. H. Johnson, and J. K. Bair, Physics Division Annual Progress Report, Oak Ridge National Laboratory (1968), p. 60.

*75.* H. F. Lutz, J. J. Wesolowski, L. F. Hansen, and S. F. Eccles, *Nucl. Phys.*, **A95**:591 (1967).

*76.* P. Horvat, *Nucl. Phys.*, **52**:410 (1964). In his section headed "Concluding Remarks," Horvat assigns $J^\pi = 5/2^+$ to the ground state of $^{21}$F.

77. D. C. Hensley, *Phys. Letters*, **27B**:644 (1968).

78. E. K. Warburton, A. R. Poletti, and J. W. Olness, *Phys. Rev.*, **168**:1232 (1968).

79. P. Paul, J. W. Olness, and E. K. Warburton, *Phys. Rev.*, **173**:1063 (1968).

80. W. Kutschera, D. Pelte, and G. Schrieder, *Nucl. Phys.* **A111**:529 (1968).

81. B. H. Wildenthal and E. Newman, *Phys. Rev.* **175**:1431 (1968).

82. A. J. Howard, J. G. Pronko, and R. G. Hirko, *Nucl. Phys.*, **A150**:609 (1970).

83. W. Scholz, P. Neogy, K. Bethge, and R. Middleton, *Phys. Rev. Letters*, **22**:949 (1969).

84. F. Everling, L. A. Konig, J. H. E. Mattauch, and A. H. Wapstra, *Nucl. Phys.*, **18**: 529 (1960). The experimental energies in this 1960 paper, when corrected for Coulomb contributions as described in our Section 4, provided the empirical nuclear binding-energy differences which we used in our least-square searches and in the first columns of Figs. 2–9.

85. R. J. Eden and N. C. Francis, *Phys. Rev.*, **97**:1366 (1955); H. A. Bethe, *Phys. Rev.*, **103**:1353 (1956); and M. H. Macfarlane, in *Nuclear Structure and Nuclear Reactions* (M. Jean and R. A. Ricci, eds.) Academic Press (1969).

86. For reviews and further references, see the 1968 Les Houches summer school lectures of S. A. Moszkowski (p. 1), M. Baranger (p. 151), and C. W. Wong (p. 205), all published in *Nuclear Physics* (C. de Witt and V. Gillet, eds.), Gordon and Breach Science Publishers, New York (1969); and see J. P. Elliott, in *Proceedings of the International Conference on Properties of Nuclear States, Montreal*, 1969, University of Montreal Press (1969).

87. S. Cohen, R. D. Lawson, and J. M. Soper, *Phys. Letters*, **21**:306 (1966); and R. D. Lawson and J. M. Soper, in *Proceedings of the International Nuclear Physics Conference, Gatlinburg, Tennessee*, 1966 (R. L. Becker, C. D. Goodman, P. H. Stelson, and A. Zucker, eds.) Academic Press, New York and London (1967).

88. A. P. Zuker, *Phys. Letters*, **23**:983 (1969).

89. A. P. Zuker, B. Buck, and J. B. McGrory, *Phys. Rev. Letters*, **21**:39 (1968); and *Bull. Am. Phys. Soc.*, **14**:35 (1969).

90. M. H. Macfarlane and J. B. French, *Rev. Mod. Phys.*, **32**:567 (1960).

91. M. H. Macfarlane, in *Proceedings of the International Conference on Properties of Nuclear States, Montreal*, 1969, University of Montreal Press (1969).

92. G. R. Satchler, in *Lectures in Theoretical Physics, Vol. VIII-C—Nuclear Structure Physics* (P. D. Kunz, D. A. Lind, and W. E. Brittin, eds.), University of Colorado Press, Boulder (1966); and N. Austern, *Direct Nuclear Reaction Theories*, Wiley-Interscience, New York (1970).

93. G. R. Satchler, *Ann. Phys. (N.Y.)*, **3**:275 (1958).

94. A. Arima, H. Horiochi, and T. Sebe, *Phys. Letters*, **24B**:129 (1967).

95. T. Engeland and P. J. Ellis, *Phys. Letters*, **25B**:57 (1967).

96. H. G. Benson and B. H. Flowers, *Nucl. Phys.*, **A126**:332 (1969).

97. R. Middleton, L. M. Polsky, C. H. Holbrow, and K. Bethge, *Phys. Rev. Letters*, **21**:1398 (1968).

98. P. Federman and I. Talmi, *Phys. Letters* **15**:165 (1965).

99. The general procedure is the same as that used by P. W. M. Glaudemans, G. Wiechers, and P. J. Brussaard, *Nucl. Phys.*, **56**:529 (1964).

100. A. de-Shalit and Igal Talmi, *Nuclear Shell Theory*, Academic Press, New York and London (1963), p. 345.

101. E. U. Condon and G. H. Shortley, *Theory of Atomic Spectra*, Cambridge University Press, London (1935).

*102.* S. A. Moszkowski, *Handbuch der Physik*, Vol. 39 (S. Flügge, ed.), Springer-Verlag, Berlin, Germany (1957), p. 411.

*103.* V. S. Shirley, in *Hyperfine Structure and Nuclear Radiations* (E. Matthias and D. A. Shirley, eds.), North-Holland Publishing Company, Amsterdam (1968), p. 985.

*104.* S. J. Skorka, J. Hertel, and T. W. Retz-Schmidt, *Nucl. Data*, 2:347 (1967).

*105.* J. P. Elliott and B. H. Flowers, *Proc. Roy. Soc.*, A229:536 (1966).

*106.* F. W. Prosser, Jr., private communication (based on the thesis work of Ref. 35).

*107.* K. Nakai, F. S. Stephens, and R. M. Diamond, *Nucl. Phys.*, A150:114 (1970).

*108.* D. Schwalm and B. Povh, in *Contributions to the International Conference on Properties of Nuclear States, Montreal, 1969*, University of Montreal Press (1969), p. 15 of Addendum.

*109.* H. Grawe and K. P. Lieb, *Nucl. Phys.*, A127:13 (1969).

*110.* J. H. Anderson and R. C. Ritter, *Nucl. Phys.*, A128:305 (1969).

*111.* J. G. Pronko, C. Rolfs, and H. J. Maier, *Phys. Rev.*, 186:1174 (1969).

*112.* K. W. Jones, A. Z. Schwarzschild, E. K. Warburton, and D. B. Fossan, *Phys. Rev.*, 178:1773 (1969).

*113.* J. G. Pronko, R. A. Lindgren, and D. A. Bromley, *Nucl. Phys.*, A140:465 (1970), (see their Table 5); and C. Rolfs, private communication.

*114.* S. Siegel and L. Zamick, *Phys. Letters*, 28B:450 (1969).

*115.* M. Harvey, "Effective Operators in the Nuclear Shell Model", Chalk River Report AECL-3366 (1969).

*116.* S. Siegel and L. Zamick, *Nucl. Phys.*, A145:89 (1970).

*117.* H. A. Mavromatis and B. Singh, *Nucl. Phys.*, A139:451 (1969).

*118.* J. P. Elliott, "Collective Motion in Nuclei," notes compiled by M. H. Macfarlane from lectures delivered at the Dept. of Physics, University of Rochester, 1958, [AT(30-1)-875].

*119.* A. P. Stamp, *Nucl. Phys.*, A105:627 (1967); and J. C. Parikh, *Phys. Letters*, 25B:181 (1967).

*120.* K. H. Bhatt, private communication.

*121.* L. S. Hsu, *Phys. Letters*, 25B:588 (1967). We have used the following result from this letter: When all active nucleons belong to one major shell, and when $T = 0$ or all active nucleons are identical, the maximum possible $B(E2; 2^+ \rightarrow 0^+)$ is proportional to the maximum value of $[\lambda^2 + 3\lambda + \mu^2 + 3\mu + \lambda\mu - (3/4) L(L + 1)]$.

*122.* C. E. Wilsdon, Ph.D. Thesis 1965, Sussex (as quoted in Ref. 8).

*123.* I. P. Johnstone, *Nucl. Phys.*, 110:429 (1968).

*124.* K. H. Bhatt, J. B. Ball, and J. C. Parikh, *Phys. Rev.*, 178:1632 (1969).

*125.* R. J. Blin-Stoyle, in *Selected Topics in Nuclear Spectroscopy* (B. J. Verhaar, ed.), North-Holland Publishing Company, Amsterdam and John Wiley & Sons, New York (1964), p. 226; A. Bohr and B. R. Mottelson, *Nuclear Structure*, W. A. Benjamin, New York and Amsterdam (1969), Vol. 1, pp. 336–340 and 390–394; and M. Chemtob, *Nucl. Phys.*, A123:449 (1969).

*126.* H. A. Mavromatis, L. Zamick, and G. E. Brown, *Nucl. Phys.*, 80:545 (1966).

*127.* H. A. Mavromatis and L. Zamick, *Phys. Letters*, 20:171 (1966).

*128.* L. Zamick, "Lectures on Effective Interactions, Effective Operators and Closed Shell Nuclei," University of California report UCRL-18445, August 1, 1968.

*129.* M. Ichumira and K. Yazaki, *Nucl. Phys.*, 63:401 (1965).

*130.* R. G. Sachs, *Phys. Rev.*, 69:611 (1946); and *Nuclear Theory*, Addison-Wesley Publishing Company, Cambridge, Massachusetts (1953), p. 246.

*131.* S. Cohen and D. Kurath, *Nucl. Phys.*, **73**:1 (1965).

*132.* E. C. Halbert, Y. E. Kim, J. B. McGrory, and T. T. S. Kuo, *Procs. International Nuclear Physics Conference, Gatlinburg, Tennessee, 1966* (R. L. Becker, C. D. Goodman, P. H. Stelson, and A. Zucker, eds.), Academic Press, New York and London (1967), p. 531.

*133.* J. Bleck, D. W. Haag, W. Leitz, R. Michaelsen, W. Ribbe, and F. Sichelschmidt, *Nucl. Phys.*, **A123**:65 (1969).

*134.* P. W. M. Glaudemans, A. E. L. Dieperink, R. J. Keddy, and P. M. Endt, *Phys. Letters*, **28B**:645 (1969).

*135.* G. Morpurgo, *Phys. Rev.*, **110**:721 (1958).

*136.* See p. 409 of Ref. 100.

*137.* L. Zamick and G. Ripka, *Phys. Letters*, **23**:347 (1966).

*138.* A. P. Zuker, B. Buck, and J. B. McGrory, in *Contributions to the International Conference on Properties of Nuclear States*, University of Montreal Press (1969), p. 193.

*139.* J. B. McGrory, *Phys. Letters*, **31B**:339 (1970).

*140.* J. B. McGrory and B. H. Wildenthal, *Bull. Am. Phys. Soc.*, **15**:543 (1970).

*141.* J. P. Elliott and T. H. R. Skyrme, *Proc. Roy. Soc.* **A232**:561 (1955); and E. U. Baranger and C. W. Lee, *Nucl. Phys.*, **22**:157 (1961).

*142.* J. B. French, in *Many-Body Description of Nuclear Structure and Reactions* (C. Bloch, ed.), Academic Press, New York and London, (1966).

*143.* M. B. Johnson and M. Baranger, *Ann. Phys.* (*N.Y.*), to be published.

# INDEX